Betrand Wong

Logic and Deduction

Also of interest

Mathematical Logic.
An Introduction
Daniel Cunningham, 2023
ISBN 978-3-11-078201-1, e-ISBN (PDF) 978-3-11-078207-3

Sets, Functions, Measures.
Volume 1 Fundamentals of Set and Number Theory
Valeriy K. Zakharov, Timofey V. Rodionov, 2018
ISBN 978-3-11-055008-5, e-ISBN (PDF) 978-3-11-055094-8
Volume 2 Fundamentals of Functions and Measure Theory
Valeriy K. Zakharov, Timofey V. Rodionov, Alexander V. Mikhalev, 2018
ISBN 978-3-11-055009-2, e-ISBN (PDF) 978-3-11-055096-2

Philosophy of Mathematics
Thomas Bedürftig, Roman Murawski, 2018
ISBN 978-3-11-046830-4, e-ISBN (PDF) 978-3-11-046833-5

Data Science.
Time Complexity, Inferential Uncertainty, and Spacekime Analytics
Ivo D. Dinov, Milen Velchev Velev, 2021
ISBN 978-3-11-069780-3, e-ISBN (PDF) 978-3-11-069782-7

De Gruyter Series in Logic and Its Applications
Edited by Denis R. Hirschfeldt, Amador Martin-Pizarro, Ieke Moerdijk, Itay Neeman, Anand Pillay
ISSN 1438-1893

Bertrand Wong

Logic and Deduction

Applications to Theoretical Physics, and Number Theory

DE GRUYTER

Mathematical Subject Codes MCS. 03 Bxx, MCS. 03-02

Author
Dr. Bertrand Wong
Department of Science and Technology
Eurotech
Singapore

ISBN 978-3-11-914531-2
e-ISBN (PDF) 978-3-11-221436-7
e-ISBN (EPUB) 978-3-11-221463-3

Library of Congress Control Number: 2025936416

Bibliographic information published by the Deutsche Nationalbibliothek
The Deutsche Nationalbibliothek lists this publication in the Deutsche Nationalbibliografie; detailed bibliographic data are available on the Internet at http://dnb.dnb.de.

Cover image: Just_Super/iStock/Getty Images Plus
Typesetting: Integra Software Services Pvt. Ltd.
Printing and binding: CPI books GmbH, Leck

www.degruyter.com
Questions about General Product Safety Regulation:
productsafety@degruyterbrill.com

Preface

The contents of this book are the result of many years of research and thought about a very important subject, namely, logic.

The ideas in the book represent some of the most important and deepest thoughts of the author. The author takes a daring, no-holds-barred examination of a subject that has both fascinated and baffled him. Many people, especially intellectuals such as philosophers and mathematicians, think that logic is everything. After reading this book they would probably rethink their stand on logic. Though logic and logical reasoning as such are very important in life, we have to be careful about its use if we were not to be entrapped in paradoxes.

The ideas in the book are expected to make the reader realize things he has not known before. It should be a mind-expanding adventure for him. Though the author appears to be disparaging logic at times he is merely showing the reader that the utilization of logic and logical reasoning, though very important in their own right, should be so done with discretion and an open and flexible mind-set. Do not be carried away by logic and keep arguing against or belittling the ideas of others for no matter how silly their ideas might appear to one there could be some merit in them that one might not be aware of.

On the other hand, the power of logical reasoning should not be underestimated. The sheer confident application of logical reasoning has in fact enabled the author to provide the possible solutions to three famous unsolved mathematical puzzlers, the twin prime conjecture, the Goldbach conjecture, and the Navier-Stokes equation and turbulence, which have challenged and humiliated the efforts of some of the best brains in mathematics for many years. The author has included the possible solutions to these three famous problems for the great benefit of the intellectually curious and challenged reader.

The most important thing is that the book offers some tip on how to improve one's powers of logical reasoning, besides being an example itself of logical reasoning to the reader.

It is on the whole a very serious and ambitious attempt at solving the problem of the inconsistency of logic which could be regarded as the cause of practically all the conflicts in society.

There are also plenty of documents and articles in the appendices which contain very important thought-provoking ideas, which should not be neglected.

The range of subjects covered in the book, as the reader should be able to see, is ambitious and wide, ranging from social sciences, hard sciences, to mathematics, and so on. There are much original and thought-provoking ideas in the book. The author has incorporated difficult, abstract ideas from, e.g., quantum theory, relativity, and mathematics, to have the reader exposed and accustomed to difficult, abstract concepts, so that their confidence in grappling with abstruse matters may increase – the author believes that if the reader could master such abstruse concepts as have been

https://doi.org/10.1515/9783112214367-202

presented in the book, they are more likely to be able to master other abstract concepts. There are also plenty of important practical ideas in the book, e.g., on how logic should be dealt with, and the proposed solutions to our economic problems in the appendices. The author has made a very serious attempt to cover practically everything important "under the sun", not excluding the supernatural or spiritual.

Many happy thoughts to all of you readers and hoping that you could all help to do something about the apparently unsatisfactory state of human affairs.

Bertrand Wong, Ph.D.

Contents

Part Two

Author's Message

As per the Preface and Contents List of the book, the book represents a unique contribution and appeal to the society at large. It touches on the very important practical aspects of logic, especially how logic could be properly and wisely used to avoid or prevent conflicts in inter-human relations, in the social setting, and in the work environment, e.g., in scientific or professional work where success depends much on it.

The author believes that it would be very difficult to find a comparable book of such a practical nature on logic. He has been a keen student and observer of logic and logical reasoning for a long time. He has encountered much problems and frustration with the use of logic in relation to his interaction and dealings with his fellow-beings. The ideas in the book have been culled from his own personal experiences, own research and study, and considerable amount of time devoted to thinking about logic and logical reasoning.

The author has studied the more popular works on logic, e.g., those of the famous logician, Kurt Gödel. The book contains his heart-felt original thoughts and proposals for logic, culled from decades of thinking, analysis, and direct experience with the practical applications of logic. It is the result of the author's keen desire to see that as many people in this world as possible would be able to think and reason objectively, clearly, and effectively in order to achieve social harmony and peace and avoid social rifts and conflicts. The book is a logic made easy, practical, and useful kind of book, with the author sharing the fruits of his experiences with his own intellectual awakening and expansion with the reader. It is really an impassioned "how to" book for everyone who wishes to embark on the journey toward making the world a much better place to live in through the right and proper use of logic in their everyday life, instead of misusing logic. The book is in fact aimed at opening and widening the minds of readers, stimulating deeper thoughts in them, and spurring them to do the right and proper things with logic.

In his effort at producing a highly effective and readable book, the author has attempted his best to write as clearly, jargon-free, and easily understood as possible. Highly abstract ideas, e.g., Godel's theorems, quantum theory, calculus, and relativity, have been reduced to the simplest layman terms possible, for easier comprehension and assimilation. One of the strong selling points of the book would be the author's possible solutions to three very difficult, unsolved mathematical problems, all of which had already been published in international journals. Included are also two very important published articles on the author's proposed solutions to our economic problems, which should appeal to everyone. All these articles are presented as examples of mathematical and economical reasoning.

There are much human interest and challenging ideas in the book, e.g., some of the methods of logical deduction used by the famous fictitious crime buster, Sherlock Holmes, and the wily, crafty premier of the Three Kingdoms period, Chuko Liang, the interesting political ideas of the famous statesman, Machiavelli, ethical concepts, eco-

https://doi.org/10.1515/9783112214367-204

nomic concepts, and a host of other subjects which have a direct relationship with our everyday life.

The author has purposefully included the normally very difficult and abstract subjects such as quantum theory, relativity, and calculus, but presented in the book in a simplified, made easy, format, instead of shying away from them, as could normally be the case. This is for the purpose of encouraging the "normally frightful of such difficult subjects" readers to embrace such difficult subjects, comprehend them, and be able to reason with them. It is to instill in the reader the confidence of understanding and mastering such difficult subjects and being able to carry out logical reasoning with them. There are a number of tips on how sound logical reasoning could be achieved, and on how the problems of impasses in logical reasoning could be overcome.

Furthermore, to add some humor and fun, instead of putting the reader to sleep, there are some witticisms and interesting quotes. The book has been written in an engaging, conversational style as far as possible so that the reader may feel that the author is talking directly to him one-on-one.

The book could be regarded as the author's most important, his magnum opus, his life's work. It incorporates the author's practical philosophy on logic and logical reasoning in relation to the society at large. It represents the author's heart-felt desire to share this very important practical philosophy with as much people as possible out there in the society at large.

The book is a treatise on logic and logical reasoning. It provides the answers to such important questions as: What is logic? What is the proper way of logical reasoning? How could harmony and peace in society be achieved through the proper utilization of logic?

Part One

1 Viewpoints on Logic

Logic has to be convincing.

The Scientist's View of Logic

Logic has to have rigor.

The Mathematician's View of Logic

Logic is common sense.

The Ordinary Person's View of Logic

What actually is logic? We have a brain. We think. As long as we think, there should be logic. To see logic is to be aware of the implication or implications of a truth or some truths or a fact or some facts (or even an untruth or some untruths or a lie or some lies) to be aware of the link or links of the truth(s) or fact(s) with another truth or other truths. For example, truth A implies truth B; truths A and B imply truth C or (perhaps) truths C and D; truth A implies truths B, C, and D (or more) or truth A implies truth B, truth B implies truth C, truth C implies truth D, the implication (or link) may continue further from truth D, and so on. The truth(s) or fact(s) might be presented in any form, e.g., in the English language, even in the sign language or in the mathematical language or other forms of symbolism and so might the logical deductions or implications. Logic is thus implication, implication, and more implications. From a single truth or fact, we might arrive at more truths or facts (i.e., implications). That is how human knowledge grows. Also, from a number of truths or facts or propositions or statements (which might not be truths but downright falsehoods) implications (which might themselves be falsehoods if they were the results of falsehoods) might be realized. Every person should understand what the statements produced by other people and the "statements" produced by nature as well imply. Logic is really implication. It would be great if our minds were more often than not alert to all such implications. This would indeed be a mind steeped in logic. But there is a serious problem, viz. what is considered as an implication (or link) by one person might not be considered so by another person. This is the greatest problem of logic. How should this problem be dealt with? This book attempts to settle this problem. Reasoning could either proceed from causes to effects (or results) or vice versa, i.e., proceed from effects to causes. The longer the chain of reasoning is, the more the premises there are and/or the more complex the premises are, the more difficult the reasoning would be; in such instances, the reasoner has to hold many points or data in his mind at the same time while performing the act of reasoning; therefore, a good memory goes hand-in-hand with good reasoning power. However, one should

https://doi.org/10.1515/9783112214367-001

only discuss logic with a person who is really capable of logical reasoning. But the problem is that it is difficult to judge whether a person is really capable of logical reasoning. This problem is compounded by the possibility that the person is really incapable of logical reasoning regarding his very own “logic” as the correct or valid one while dismissing the true logician’s logic as “nonsense.” This means that logic serves no useful purpose at all if it fails to persuade or influence another person. Logic should thus not be used or applied blindly but with discretion. Logic is only justified by the end or the result. Disputes and conflicts are manifestations of the failure or weakness of logic. Winning an argument or dispute and making the other party feel that he is wrong or foolish may be an achievement for a person well-versed in logic, but if the out-argued or outwitted person takes it badly or feels hurt as a result he may hold a grudge against his victor and may even seek revenge. To avoid such an undesirable consequence everyone should use or apply logic wisely if at all so that there would be mutual respect and harmony; logic should never be used to make another person look like a fool but should only be used to make the other party accept it willingly and happily – if this happens logic would have played its rightful role, viz., bring about better changes and more happiness. Unfortunately, many egoistic people use logic to brow-beat others to make themselves feel good while the latter feel bad, resulting in unhappiness and disharmony. Logic could simply be summed up as follows – it is reason, and there are many reasons governing human conduct; and reasons explain things and give understanding. One interesting poser: What is logic for really? Would logic be of great importance if there were only one lone surviving logical human being in the whole world after the rest of the human race had perished?

The Author’s View of Logic

Man is so intelligent that he could make his stupidity look like intelligence.
Anonymous

2 Introduction

Logic is the art of reasoning correctly, the study of deductive argument, or correctness of reasoning. The prime concept of logic is that of a valid argument or statement where, if the premises are true, then the conclusion must also be true.

We use logic everyday whether we realize it or not. We may use it correctly or misuse it. To survive we all have to use it. The ability to use logic well, reason well, understand, and achieve results is apparently the most important human ability. Education is supposed to train us well in this ability besides equipping us with the required knowledge and inculcating the correct values in us. Despite all this, logic is often not correctly or properly used. Why is this so?

Logic can be complex and abstruse. It can be difficult. For instance, many have difficulty mastering the logic of mathematics. We may not like hard thinking, hard logical reasoning, but we often find ourselves in a predicament where we have not much choice but to carry it out and make the best out of it, e.g., when we find ourselves in a difficult calculus class or when we are faced with an intractable engineering problem.

Relatively few are the ones who excel in it. Many people are just able to get by utilizing simple common sense or simple logic. To be able to excel in it could earn a person respect, admiration, success, and satisfaction.

Logic may be difficult but it can be fascinating and challenging. Since we all live it and breathe it we may as well accept it with open arms and make the best out of it. We may as well try to go deeper into this mysterious character and try to understand, master it as best as we can so that we may become the better for it.

Often, one may arrive at a conclusion after a chain of reasoning, which to one is evidently correct and valid, only to face objection from another party, who proclaims that our reasoning and conclusion are not valid. If the objector offers some reason or reasons for the lack of validity, e.g., some contradiction is found within the reasoning, some counter-example is found, some data or fact is incorrect, or some other data or fact has been overlooked, then the objection may be justified, but often the objector could not offer any such reasons for the objection, implying that the lack of objectivity is but the result of an intuitive feeling on the objector's part, an opinion, and this is the problem as without being able to provide any true reasons for his objection the objector would normally be unable to convince the other party that his reasoning and conclusion are incorrect (though if the objector were considered or perceived to be an expert in the field the other party might feel some doubt or nervousness about his

https://doi.org/10.1515/9783112214367-002

reasoning and conclusion). This is a serious problem with logic. Mere opinions should never be confused with logic though apparently this quite often happens.

More will follow hereafter.

Even inanimate objects may have their own logic which we humans fail to notice or understand.

Anonymous

3 Mind and Logic

Mind and logic are two inseparable things. In order that there is logic there has to be mind. Logic is the relevant relationship between objects or statements or premises. Logic is the result of intuition or “feeling of truth” or “awareness of truth.”

Many things are logical because our intuition perceives them to be logical. If, however, people question the logic of these things, which, to us, are logical, we would start doubting their logic and would think about these things.

In order that logic exists, there must be perception or awareness or experience. Only when there are facts or truths, or assumptions, such as premises or statements, would there be logic.

https://doi.org/10.1515/9783112214367-003

4 More on Logic

A mathematician plays with logic through the use of symbols, while a philosopher plays with logic through the medium of words.

In order that logic can function, there should be facts, which are knowledge derived from experience or empirical knowledge. These facts form the premises or statements upon which a new premise or statement can be formed. This act of deriving a new premise or statement from given premises or statements is the act of reasoning or the act of logical thinking. There are systems of logic developed by philosophers such as Aristotle, hence, Aristotelian logic. These systems set up the types of premises which can be derived through logical thinking. They also expose the importance of language or symbols in logical thinking. Ludwig Wittgenstein, the Danish philosopher, had in his Tractacus, acclaimed for language and symbols great importance in logic, so had Russell, the British philosopher.

Logic is based on intuition, the sixth sense. What are logical, from an objective point of view, should be universally true, i.e., other people with enough of common sense should agree that such and such a thing is logical in order to acclaim to that thing its logic or sense. Anything that cannot be universally agreed upon should hence be illogical or nonsensical. This brings about a problem. If, e.g., you are one of those original thinkers who believe that something is logical, but your fellowmen fail to agree with you because they are too stupid, would you hesitate to claim that it is logical? It is understandable that you would start doubting its logic. But if you could find another human being who could agree with you, you most probably would not doubt its logic. This statement of the author is based on intuition like the logic he is discussing; it is subject to the benefit of the reader's doubt. The author is here trying to philosophize, to be logical, using his intuition, his sixth sense, as a result. It can be said that nothing is absolutely certain; nobody should be so certain that he is always right, that he is always logical, if to be right is to be logical. Here is a witticism: It is certainly certain that nothing is absolutely certain. (This statement appears self-contradictory.)

Look at the five senses we possess, viz., sight, hearing, taste, smell, and touch. The various stimuli that give rise to each of these sensations produce in the body impulses which travel to the brain, producing impressions there, and it is these impressions which are the senses so derived. But can we always trust our senses, our body or our brain upon which impulses act? This is the question that is difficult to answer. If we can always trust our senses, why? If not, why? Certainly trust, and distrust, are based on intuition. What is our intuition based on? How can we be certain that our intuition does not fail us? How can we be certain that our logic, which is based on our intuition, is reliable?

https://doi.org/10.1515/9783112214367-004

5 Logic and Reality in Life

What is logic? And, what is reality? To some, these two are one and the same; what is logical is real, i.e., logic is real. But, what is real need not be logical, e.g., the ideas of foolish, illogical people are real and could have far-reaching, harmful effects on society despite their lack of logic.

To know what the two terms, logic and reality, imply, it would be appropriate to state what they mean. Logic could be regarded as the "beauty" of ideas or concepts. Logical ideas or concepts give some kind of "pleasure" to the mind which conceives them, a pleasure which is somewhat akin to the pleasure that one gets from viewing a beautiful piece of art, for example. This kind of "pleasure" is apparently universal; in other words, the minds which could understand the ideas or concepts would experience this "feeling." One may wonder: What is the importance of logic? It is hardly worth the while for logical ideas or concepts to be able to invoke in intelligent minds just the feeling: Aha! These ideas sound great! As we know, logic has utilitarian value. It is an aid to practically every aspect of our life from the not so serious side like play to the more serious aspect such as work. The successful exercise of logical reasoning would result in achievements which make a person happy. With the proper use of logic or logical reasoning, one could arrive at more ideas or conclusions, with which one could make the most appropriate decisions or take the most appropriate actions which could bring the desired result. In other words, logic is the means to an end. Reality refers to the actual state of things, and in the case here, the actual state of logic. However, true logic, which could be regarded as valid common sense, may not exist, may not be real, sometimes, but may only be an illusion, a wrong impression.

Would the most logical person be the person most capable of achieving success in any field he chooses? In reality, this may not be so. In a field of activity involving straight, objective, logical reasoning such as philosophy, mathematics, or science, efficiency in logical reasoning could carry a person far. But, in a field of activity especially involved in dealing with other people, who may not be that logical, e.g., in business and politics, the employment of logical reasoning may not prove so fruitful. In such activities, which have a social element, wherein the ability to play with the moods and emotions of the other party or parties concerned is relatively important, the self-cultivation of certain personal characteristics such as stylishness of dress, linguistic style, and the projection of an impressive personal image may play a far more important role than an impressive intellect. This is a reality of life and it is no surprise that many a person with a brilliant intellect fails in the "social" sphere due to a lack of such personal advantages.

Some intellectuals seem to think that really logical reasoning could be employed to influence or persuade anyone. This is a misconception. Consider this. In a society of fools would the person of genius be looked upon with esteem for his genius? It would not be surprising here that the greatest fool would be held with the highest esteem by

https://doi.org/10.1515/9783112214367-005

his fellow-fools. It could hence be concluded that the person with the most logical intellect should avoid dealing with people lacking with logic as much as possible – he should seek out like-minded people to do business with, if his logical reasoning were to be appreciated, accepted, and utilized for a fruitful end. Certainly, one should not attempt to teach another person higher calculus if the latter could not even master elementary algebra! In certain situations, where the party or parties concerned lack the brain power or the power of reasoning, it would be "illogical", pointless, to rely on the power of logic. This is especially so when the emotions of the parties concerned run high, for emotion is the antithesis of logical thinking.

How could a person with the most logical of minds be able to co-exist amicably with his rather average fellowmen? How could brilliance go hand-in-hand with the mediocre and the "not that logical"? It would not be much of a contradiction to state that, in this instance, it is stupid to be brilliant, it would be brilliant to be stupid, to act stupid. The brilliant guy, if he were practical, shrewd, and adaptable, should be able to "hide" his brilliance and "act" ordinary or even the fool if the situation warranted it. However, not all brilliant people possess the qualities of practical-ness, shrewdness, and adaptableness – in this instance, avoidance of the mediocrities and the inferior intellects may be the best solution since the brilliant person is unable to blend with them and gain acceptance by them. To be practical, shrewd and adaptable requires some degree of low animal cunning, pettiness of mind, and shamelessness (or possession of a "thick skin") – bluff and showy-ness, something which a person of honor would generally not care for, are normally required.

The definition and interpretation of logic could be fuzzy and vague. Some may find it difficult to describe logic, e.g., someone may describe it as a sensation that something seems right, seems to make sense. This will be a subjective feeling, and logic, if it is just a "subjective feeling", seems more like an "opinion" than a statement of truth. Some may regard logic as "common sense", a kind of "intuitive feeling" which will be felt by all rational people when encountering certain statements, premises, or ideas. A better definition of logic, which will make logic appear less controversial, is, probably, a set of ideas or reasoning which enables a person to tell or forecast the outcome of a certain set of events. For example, we notice that Mr. A is angry and unhappy. We know that Mr. B is personally disliked by Mr. A. If Mr. B now approaches Mr. A for a small loan of, say, $100.00, will he be able to obtain the loan? A logical person will probably conclude that Mr. B will not obtain the loan. His logic will be practically confirmed to be correct if Mr. B were seen trying to get the loan and was turned down. If, on the other hand, Mr. B managed to secure the loan, then the logic of the person's conclusion will be taken to be incorrect. This kind of "experimental" proof, or proof by results, will be the least controversial proof of the soundness or validity of a logic and the most practical. A further example should demonstrate this point to perfection. Logic concludes that if Rod A is longer than Rod B and Rod B is longer than Rod C, then Rod A will be longer than Rod C. How do we prove the absolute soundness of this logic, without causing any controversies? Well, by very simply conducting an experiment, by bringing together three rods of different lengths, naming the Rods A, B, and C according to the given premises and finally bringing together Rods A and C, placing them side by side and comparing their respective lengths. Unless one is blind or has no eyesight, one could hence see with one's very own eyes that Rod A will be indeed longer than Rod C, and there will not be any doubt or controversy about this inevitable fact. The result of this "experiment" will indubitably prove that the conclusion, Rod A is longer than Rod C, is logical and correct.

The problem will only arise when we apply logic to the more abstract concepts which have little to do with practicability or reality, especially concepts concerning objects which are intangible, objects which cannot be "experimented" with. Consider the following example. Book A is more difficult to read than Book B. Therefore, more people will prefer Book B to Book A. It is very difficult to prove or confirm the soundness of this logical reasoning, as levels of reading ability and personal likes and dislikes are very difficult to measure or gauge (unlike the above-mentioned measurable rods of varying lengths), the former will depend on one's command of the language, education, and intelligence while the latter will depend on one's culture, family, and social background.

It seems that logic has often been overrated. As mentioned earlier, the "sense" of logic is a kind of "intuitive feeling" that something seems "right" or "correct." Often this "sense" of logic could be validated by the "connectedness" of real, tangible objects. Logic involving intangible objects, objects which could not be seen or observed, could be very abstruse and difficult to validate. It is probably correct to say that it is better to arrive at the truth through the utilization of pure intuition rather than logic (which is dependent on or a product of the intuition). If we could find a shorter, more efficient way of arriving at the truths, we should do so.

The exercise of logical reasoning in human affairs appears to be a rather arbitrary one. What is logical to one person might not be regarded so by another person, though there is much consensus amongst people in many instances in logical thinking; in other words, many people could be found to share the same "logical" thoughts regarding some particular matters, but the consensus might not be there where "controversial" matters are concerned.

Logical reasoning could be fallacious. For example, in science and economics, the experts study phenomena and observe patterns, and on the basis of these patterns, they forecast or predict future events, e.g., the arrival of comets and the timing of share price increase or decrease. The logical mind here makes the assumption that all natural/economic phenomena have a pattern, and once the pattern is noted, the timing of the recurrence of these same phenomena could be predicted. What is the logical basis of this assumption? The second assumption is that this pattern does not change with time. Here again, what is the logical basis of this assumption? However, one should not discount the possibility that this said pattern changes with time. So, holding the assumption that an observed pattern remains the same in the future might not be logical. Taking an analogy here to illustrate this point, should we assume that an apparently kind person who would not even kill an ant is incapable of committing a murder? There had been instances whereby apparent paragons of virtue and high morals committed serious crimes, and all those who knew him were greatly surprised or shocked. A logical thought based on a fallacious assumption should not be regarded as logical, though for practical purposes it might be allowed (we should only allow ourselves the luxury of such a "logical" thought by carefully taking note that the assumption upon which it is based is only tentative).

Logic could be regarded as the system of rules for thinking. Some of these rules could be highly specialized and arcane while others could be simple. We humans have created rules and procedures for practically everything under the sun, e.g., rules for behaving in our society, which we call etiquette, rules for sports and games, e.g., rules for soccer and chess, rules or laws for preventing criminal behavior and disciplining and punishing criminals, rules for mathematical computations, and rules for driving . Of course, the rules for thinking, viz., logic, are apparently also one of the many creations of the human intellect, though many might believe that logic is the gift of the Divine. Logic could also be regarded as the guide for human action and should be based objectively on facts if it were to serve any useful purpose. The problem appears to be not so much in having a logical mind, but, rather in having a truly objective mind, for how good is a logical thought if it is based on fallacious or subjective assumptions? Such a "logical" thought would be the "incorrect" thought, which would not bring about the desired outcome or result (except perhaps by fluke). As mentioned earlier, a logical thought should lead to a desired result or a predicted outcome. If it does not do so, then by this criterion, the logic of the thought is dubious.

There seems to be no such thing as pure truths or pure logic. What are regarded as true or logical appear relatively so only. Truths and logic are the result of the perception of the mind. To be able to prove the existence of truths or logic we should first be able to prove the existence of the mind, which is the only object capable of perceiving the truth or the logical. If we cannot prove the existence of the mind, how could we say that we can prove the existence of truths or logic?

To prove the existence of matter, any tangible object, the simplest way is to touch it, hold it, or play around with it. Can we do likewise with the mind, which is an intangible object? Definitely not. Moreover, the mind has two aspects, viz., the conscious and the unconscious aspects. It is believed that dreams and psychological problems originate from the unconscious aspect of the mind. The mind is capable of imagination as well. How could we be certain that a perceived truth or logic is not a dream or a figment of the imagination, whereby it is not real? We should therefore exercise great caution when we deal with logic and truths. However, subjectivity and bias seem rather common, while the really objective person seems to be a rare breed. As logic is apparently relative and variable, an objective outlook is highly important when dealing with it so that there would be effective results. A truly objective mind should be detached from the ego and be flexible in adopting new, logical ideas in view of the new facts and discarding old ideas (once held to be valid or true) when their validity is disproved by the new facts.

There seem to be much conflicts and controversies in human affairs, which is perhaps a manifestation of the limitations of the human mind which include constraints such as prejudices, egoistical tendencies, and overactive imagination. It could be observed that what could not be explained by factual evidences are often attributed to the Divine or God, which is not really a logical act. A person could creatively use his imagination to come up with reasons to support a postulation or stand, reasons which may or may not be valid, reasons whose validity may not be confirmable by hard or physical evidences. Though imagination could work wonders for man, it could also work against him – it is a "double-edged sword". In the hard sciences, e.g., physics and chemistry, conclusions or results could normally be confirmed by physical experiments, whereas in the soft sciences, e.g., economics and sociology, this is normally not possible, and postulations, conclusions, are normally "open-ended", more subjective.

Our very own life experiences, though useful, apparently have a limiting effect on our mind. Our mind is apparently so used to and familiar with the concept of limits, e.g., that the apparent infinitude of the cosmos seems difficult to get used to. For example, scientists seem unable to come to terms with an infinite cosmos which has infinite dimensions, where traveling in a straight path in any direction would have a beginning but no ending and would be an eternal journey, i.e., the cosmos is infinite in all directions. A study of the literature on astrophysics would show that scientists have apparently interpreted the infinitude of the cosmos by using familiar geometrical objects such as the circle and the Mobius Strip, e.g., the possibility of traveling round and round in a circuitous path through worm-holes, which are tunnels, each tunnel connecting two regions of space-time distant from one another (the worm-hole represents the shortcut for reaching another universe which could otherwise be reached only by traveling in a circuitous route through normal space-time), with no beginning and no end, i.e., infinitely or traveling round and round a Mobius Strip-like space with no beginning and no end (the Mobius Strip, which is a well-known geometrical curiosity, has some very interesting results if it were cut lengthwise and, e.g., one-quarter, one-third, and half-way along its width – twisted and interlaced strips of different lengths and widths, reminiscent of parallel universes, would result). They apparently liken traveling in infinite space in the cosmos to traveling round and round in circles or circuits; the circle (or circuit) is a very familiar geometrical object which has no beginning and no end, unlike the straight line which has a beginning and an end, and one could practically go round and round the circle (or circuit) non-stop and without end, i.e., eternally, if one wishes and could live that long. Why couldn't travel in infinite space in the cosmos be interpreted as traveling endlessly or eternally in outer space in a curved line, which has a beginning but no end unlike the circle (or circuit) or the Mobius Strip which has no beginning and no end, through curved space-time, which could be carried out in infinite directions? This might be difficult to "swallow", as here on this planet we know that there is a serious limitation on how far we could travel and we could hardly imagine ourselves or anyone undertaking such a journey, though we could find it easy enough to go round and round in circles or circuits. We should not let our experiences limit or bias our mind.

A really logical or intelligent mind should be egoistically, emotionally, and experientially detached. In other words, a truly logical or intelligent mind is a truly objective mind.

6 Godel's Proof and Mathematical Logic

According to Godel's proof, which has had far-reaching effects on logic and mathematics, any formal axiomatic system, e.g., arithmetic, contains undecidable propositions; in other words, it contains sentences S such that neither S nor the negation of S can be proved. This is Godel's first incompleteness theorem. Its corollary, Godel's second incompleteness theorem states that a formal system's consistency, e.g., that of arithmetic, cannot be proved by means using the formalization of the system itself; the proof can only be derived by using a stronger system. In short, some statements or propositions cannot be proved, e.g., the statement, "This statement is false."

What Godel meant was that the undecidable propositions are the result of an inherent weakness in the axioms of mathematics – a formal system's consistency can only be proved by using a more powerful set of axioms. This implies that the axioms of mathematics should be replaced by a stronger set of axioms if there were to be no inconsistency in mathematics. The question is whether there would be mathematicians or logicians clever enough to conceive such a strong set of axioms. In the words of computer programming, if we input garbage the computer would churn out garbage – garbage in, garbage out. Mathematicians therefore ought to be cautious of what they put into mathematics if they do not want garbage results.

The author would like to raise a few points of curiosity here. If, by a quirk of fate, if Earth were to be remade and human beings were to be reborn, would mathematics, mathematical proof, and logic ever be the same? Are mathematics, mathematical proof, and logic accidents of nature or fate? If civilizations or life exist in other galaxies, what would mathematics, mathematical proof, and logic be like there?

Apparently because we human beings have 10 fingers and could count with our 10 fingers our number system is "tens-based", e.g., 10, 100, 1,000, 10,000, 100,000, and 1,000,000. If we had, e.g., 13 fingers, 20 fingers, or, 35 fingers, imagine what our number system would have become, and what would our mathematics, mathematical proof, and logic then be? Can you imagine a "thirteens-based", "twenties-based", or "thirty-fives-based" number system? What would then be the odd numbers and the even numbers? Would the concepts of "oddness" and "evenness" then have to be modified? Is the nature of mathematics, mathematical proof, and logic predetermined by the physical form and nature of the human being and the form and nature of his environment? If human beings had been unable to count, if they had no fingers, or if they had no sight, would there ever be mathematics (which could be regarded as the science of numbers), and if mathematics were still able to "co-exist" with human beings, what would it be like? Yet, many mathematicians and logicians would view mathematics and logic as a reality which is independent of the existence of the human being. However, on this last point, the author harbors some doubt; the author wonders whether mathematics could have existed at all if all human beings had been born without fingers and eyes and had been unable to count!

https://doi.org/10.1515/9783112214367-006

And, yet, there have been savants, even idiot savants, who have been able to compute very large numbers in their heads at unbelievable speeds. This seems to have been some miracle of nature. Therefore, couldn't there be savants (or geniuses) who could perform similarly impressive and miraculous feats of mathematical reasoning or logical reasoning? Such a person, if he had existed, could have out-thought or out-reasoned a genius such as Godel.

We could only hope that such a savant or genius would materialize to reform mathematics and mathematical reasoning so that there would be no more inconsistencies in mathematics and logic so that there would really be a solid foundation for mathematics and logic. This genius could perhaps reinvent mathematics and produce many great theorems and axioms.

7 Some Ideas Relating to Godel's Undecidable Theorem and Mathematical Logic

Godel postulated that there are statements or propositions that could neither be proved nor disproved to be true and created a sort of revolution in mathematics, i.e., not all the formalism of mathematics could be proved or disproved rigorously – there are always some paradoxes to bewilder the mathematician.

The following is one such paradox:

a) The writing on one side of a sign board says, "The statement on the other side of this board is false."
b) The writing on the other side of the sign board says, "The statement on the other side of this board is true."

If (a) above is a true statement, (b) contradicts it, and if (b) is a true statement, (a) contradicts it.

The following statement is also paradoxical:

This statement is false.

If this statement is false, this statement is true, and if this statement is true, then this statement is false.

Then, there is also the paradox concerning classes or sets. For example, the class of all classes is a class. Similarly, the class of all catalogues, to take another example, is a catalogue. But, here goes the paradox: The class of all dogs is not a dog.

Zeno's paradox is another well-known paradox. This paradox concerns a race between Achilles and a tortoise. The tortoise, being a slow creature, is given a head start over Achilles. But, Achilles would never be able to catch up with the tortoise, let alone overtake it. If we keep on halving the distance between Achilles and the tortoise to indicate how Achilles would soon catch up with the much slower tortoise, we would find that even if we go on halving the distance between Achilles and the tortoise till infinity the tortoise would always be ahead of Achilles though the distance between the tortoise and Achilles would get smaller and smaller with each halving. Thus, Achilles would never be able to catch up with the tortoise, let alone overtake it. This runs counter to common sense and experience. We know that Achilles, moving at a greater speed than the tortoise, would overtake the tortoise at some point in time. We have no doubt about this. We would be able to observe this happening. But, by the above-mentioned logic, this is impossible. Here, the author would like to pose the question: Should we trust logic or our physical senses more?

There is little doubt that the mathematician or logician is likely to trust logic more than the non-mathematician or non-logician, for logic is the tool he uses to create new mathematics and abstractions. The man in the street might leave logic to the ivory tower logician and be content with just being able to see Achilles overtaking the tortoise.

https://doi.org/10.1515/9783112214367-007

Mathematics is a subject that is concerned with proofs or evidences that certain mathematical statements are true. But Godel's undecidable theorem states that such proofs, or dis-proofs, are not achievable all the time. At the metaphysical level, or more fundamental level, we might not even be able to prove our very own existence or actuality (Could one accept Descartes' proof of his own existence, which is as follows: I think, therefore I am?), our sanity or the reality of our senses; sometimes, we could not even tell whether an event or occurrence had been a dream or had been real. There also seems to be some inconsistency in mathematical proofs. Mathematical proofs rely on axioms, which are obvious, unproven statements, lemmas, which are proven statements leading to a more important statement or statements, and theorems, which are proven statements. Mathematicians demand rigor in mathematical proofs, but they could accept unproven statements which are obvious to them, i.e., axioms. However, what appears as obviously true to one mathematician might not be obviously true to another mathematician. Furthermore, proofs acceptable to one group of mathematicians might not be acceptable to another group, e.g., the proof by contradiction which is commonly used in mathematical proofs is not acceptable to the group who call themselves intuitionists. The twin primes conjecture, which postulates that there is an infinitude of twin primes or primes separated by 2, e.g., had not been proven or disproved so far despite many attempts by mathematicians, though practically all mathematicians believe that the conjecture is true. The twin primes conjecture may be one of those statements in mathematics which are true but whose truth is not provable, in accordance with Godel's undecidable theorem. Since thousands, millions, or billions, of twin primes (this quantitative evidence, though it might be acceptable as proof of the infinitude of the twin primes by scientists, is not acceptable to the mathematicians) have been found and the conjecture is so obviously or apparently true though its (complete) proof is lacking (perhaps, the proof or disproof is impossible, vide Godel's undecidable theorem), the twin primes conjecture could perhaps be treated as an axiom instead, i.e., it could be regarded as an obvious but unproven statement. Perhaps, to avoid such inconsistency, mathematics should do away with axioms, which may not be possible. For example, Boolean algebra has provided the axioms for a rigorous application of mathematical abstraction; it gives us the "and", "or", and "not" statements, which are fundamental in logic, and whose equivalent in mathematics are "times", "add", and "subtract".

With the inconsistency and paradoxes encountered it is a little wonder that mathematicians feel rather nervous about their field of work. Godel's theorems have cast a doubt on the soundness of mathematical logic, to the dismay of many mathematicians who swore by it, e.g., the great mathematician, David Hilbert.

8 Mathematical Logic and Proof

Whether mathematical proofs and arguments are "rigorous" enough to be accepted by the mathematical community is a matter of opinion and culture. How do we measure objectively or quantify the "rigor" of a mathematical proof? As a matter of fact, how ought a mathematical proof be like? Take a hypothetical example here: Would mathematical thought exist in the mind of an extra-terrestrial? How would an extra-terrestrial's mind view our mathematics? Would mathematical proofs make any sense to him?

In the middle of the nineteenth century, the famous logician and philosopher, C. S. Peirce, announced that "mathematics is the science of making necessary conclusions". Mathematics could be "about" anything as long as it is a subject that exhibits the pattern of assumption-deduction-conclusion. The ideal mathematician rests his faith on rigorous proof. He believes that the difference between a correct proof and an incorrect one is an unmistakable and decisive difference. He could think of no condemnation more damning than to say of a fellow mathematician, "He does not even know what a proof is." Yet, ironically, he is unable to supply a coherent explanation of what is meant by rigor, or, what is required to make a proof rigorous. In his own work, the line between complete and incomplete proof is always somewhat fuzzy, and often controversial. Cantor's radical ideas on infinite sets had been ridiculed, and he had to suffer nervous breakdowns for it. Now they are regarded as "correct" mathematics. This is not surprising. But it shows that the subject of "mathematical rigor" itself is controversial, even opinionative and prejudicial. It depends also on the status of the person who puts forward the mathematical proof.

According to an international authority in Engineering Science, who often had to apply mathematics in his work, Professor William F. Taylor, mathematical truth is reasoning that leads to correct physical relationships, where empirical demonstrations are possible. Belief in a non-material reality removes the paradox from the problem of mathematical existence, whether in the mind of God or in some more abstract and less personalized mode. If there is a realm of non-material reality, then there is no difficulty in accepting the reality of mathematical objects, which are simply one particular kind of non-material object. Hence, a mathematical proof could be regarded as a procedure by which a proposition about the unseen reality could be established with finality and accepted by all adherents. It could be observed that if a mathematical question has a definite answer, then different mathematicians, using different methods, working in different centuries, would find the same answers. A well-known mathematician even ventured to call mathematics a form of religion. To him, mathematics ought to be regarded as the true religion.

In the opinion of some, the name of the mathematics game is proof – no proof, no mathematics. In the opinion of others, this is nonsense – there are many games in mathematics. Mathematics, then, is the subject in which there are proofs. Tradition-

https://doi.org/10.1515/9783112214367-008

ally, proof was first met in Euclid. Millions of hours have been spent in class after class, in country after country, in generation after generation, proving and reproving the theorems in Euclid. After the introduction of the "new math" in the mid-1950s, proof spread to other high school mathematics such as algebra, and subjects such as set theory were deliberately introduced so as to be a vehicle for the axiomatic method and proof. In college, a typical lecture in advanced mathematics, especially one given by an instructor with "pure" interests, consists entirely of definition, theorem, proof, definition, theorem, proof, in solemn and unrelieved concatenation. Why is this so? If, as claimed, proof is validation and certification, then one might think that once a proof has been accepted by a competent group of scholars, the rest of the scholarly world would be glad to take their word for it and to go on. Why do mathematicians and their students find it worthwhile to prove again and yet again the Pythagorean theorem or the theorems of Lebesque, Wiener, or Kolmogoroff? Proof serves many purposes simultaneously. In being exposed to the scrutiny and judgment of a new audience, the proof is subject to a constant process of criticism and revalidation. Errors, ambiguities, and misunderstandings are cleared up by constant exposure. In its best instances, a proof increases understanding by revealing the heart of the matter. It also suggests new mathematics. The novice who studies proofs, even incomplete proofs, gets closer to the creation of new mathematics. A proof is ritual and a celebration of the power of pure reason. Such an exercise in reassurance may be very necessary in view of all the messes that clear thinking clearly gets us into.

The logical analysis of mathematics, which reduces a proof to a (in principle) mechanizable procedure, is a hypothetical possibility, which is never realized in full. Mathematics is a human activity, and the formal-logical account of mathematics is only a fiction – mathematics itself is to be found in the actual practice of mathematicians. An interesting phenomenon should be noted in connection with the difficulties of proof comprehension. A mathematical theorem is called "deep" if its proof is difficult. Some of the elements that contribute to depth are non-intuitiveness of statement or argument, novelty of ideas, complexity, or length of proof-material measured from some origin which itself is not deep. The opposite of deep is "trivial" – this word is often used in the sense of a put-down. However, it does not follow that what is trivial is uninteresting, useless, or unimportant. Despite this hierarchical ordering, what is deep is, in a sense, undesirable, for there is a constant effort toward simplification, toward the finding of alternative ways of looking at the matter which trivializes what is deep. We all feel better when we have moved from the analytic or abstract toward the analog or practical portion of the experiential spectrum.

According to Imre Lakatos, in his *Proofs and Refutations*, mathematics, too, like the natural sciences, is fallible, not indubitable. It, too, grows by the criticism and correction of theories, which are never entirely free of ambiguity or the possibility of error or oversight. Starting from a problem or a conjecture, there is a simultaneous search for proofs and counter-examples. New proofs explain old counter-examples. New counter-examples undermine old proofs. To Lakatos, "proof" in this context of informal mathematics does not mean a mechanical procedure which carries truth in an unbreakable chain from assumptions to conclusions. Rather, it means explanations, justifications, and elaborations that make the conjecture more plausible and convincing, while it is being made more detailed and accurate under the pressure of counter-examples. Each step of the proof is itself subject to criticism, which may be mere skepticism or may be the production of a counter-example to a particular argument. A counter-example which challenges one step in the argument is called by Lakatos a "local counter-example", while a counter-example which challenges, not the argument, but the conclusion itself, is called a "global counter-example". *Proofs and Refutations* is an overwhelming work. The effect of its polemical brilliance, its complexity of argument and self-conscious sophistication, and its sheer weight of historical learning all dazzle the reader. In fact, Lakatos had been idolized.

On the one hand, we have real mathematics, with proofs which are established by "consensus of the qualified". A real proof is not checkable by a machine or computer, or even by any mathematician not privy to the gestalt, the mode of thought of the particular field of mathematics in which the proof is located. Even to the "qualified reader", there are normally differences of opinion as to whether a real proof (that is, one which is actually spoken or written down) is complete or correct. These doubts are resolved by communication and explanation, never by transcribing the proof into first-order predicate calculus. Once a proof is "accepted", the results of the proof are regarded as true (with very high probability). It may take generations to detect an error in a proof. If a theorem is widely known and used, its proof frequently studied, if alternative proofs are invented, if it has known applications and generalizations and is analogous to known results in related areas, then it comes to be regarded as "rock bottom". In this sense, all of arithmetic and Euclidean geometry are "rock bottom".

On the other hand, to be distinguished from real mathematics, we have "metamathematics" or "first-order logic". As an activity, this is indeed part of real mathematics. But, as to its content, it shows a structure of proofs which are indeed infallible "in principle". We are thus able to study mathematically the consequences of an imagined ability to construct infallible proofs. We could, e.g., give constructivist variations on the rules of proof and see what the consequences of such variations are.

There are various schools of mathematical thought as well. For example, there are the logicist and the constructivist schools of thought. If mathematics were indeed neutral and objective, why is this so? Even the foundation of mathematics, which is set theory, was found to be shaky by Godel, whose Incompleteness Theorems caused deep cracks in the important monumental works of such mathematicians as Frege and Russell. Is mathematics a part of nature that is waiting to be discovered, or is it an invention of the human mind? The author would consider mathematics a bit of both, for mathematical proofs are subject to the interpretation and imagination of the human mind. For instance, how could we find an example of $\sqrt{-1}$ in reality (What is $\sqrt{-1}$ actually equal to?), though it is commonly used in mathematics?

In conclusion, the author would like to state that a mathematical proof ought to be closely scrutinized and ought to be regarded as true (only to a high probability). It ought to lead mathematicians into thinking more deeply into the problem and possibly coming up with a better or more interesting proof or proofs. Most mathematical papers which are published in the journals are actually regurgitations and clarifications of established mathematical proofs. Mathematical papers with original and fresh ideas that arouse considerable interest and thought are rather rare. Whether a mathematical proof is indeed correct or not is a matter of disputation. Take the twin primes problem, for instance, where some of the best brains in mathematics have been trying, without success, to prove the infinity of the twin primes for centuries. There is perhaps one way to indubitably convince a top-flight mathematician of the infinity of the twin primes. That is, if God Himself brings him before a huge screen upon which all the twin primes are infinitely projected in running and ascending order of magnitude. Not surprisingly, some mathematicians may expect a cleverer proof than this. How should the proof be, then?

9 The Philosophy and Principles of Logic

Introduction

What is logic? Is it only humans who are capable of conceiving logic? Are animals, insects, or even plants and trees capable of logical thought? If man has evolved from animals, who in turn have evolved from simple cells, then animals, insects, plants, trees, and other living things should also have the capacity for logic, though their kind of logic could be different from human logic. Here, the author would attempt to put human logic "under the microscope".

Philosophy and Principles of Logic

A person would be at the height of naivety, if not folly, if he were to put all his hopes and confidence on the efficacy of logic, if he were to swear by logic. What is this thing that we call "logic"? We can simply define it as the valid relationship or connection between ideas or concepts. How is it that human minds can see this relationship or connection between ideas, and the capacity for doing so varies from person to person? This is a complex question. The logical capacity of the human mind has yet to offer a satisfactory explanation of the relationship of the human mind with logical thought, and, how can we be sure that the perceived relationship or connection between ideas or concepts is indeed logic?

Two famous mathematicians and intellectual gadflies, Godel and Russell, had dissected the intricacies of logic and subjected it to a thorough examination that is reminiscent of a scientist closely examining the entrails of a dead animal. And, what did they find in the entrails of logic? Plainly, paradoxes and contradictions. Which goes to prove that logic is not worthy of the faith of the logician and the man of intellect, as every thinking person thinks it is. Godel's proofs of the inconsistency of logic had whittled the efforts of mathematicians such as Frege, Russell, and Whitehead in reducing mathematics to logic to zero.

So, is logic for real? Analogous to the manner in which the human being has been designed or fated to breathe (through his nose and lungs), eat (through his mouth), and hear (through his ears), the human brain or mind is equipped by Mother Nature to "behave" in certain patterns – hardwired to "behave" as such, you may say. Thus, logic constitutes the behavioral patterns of the human brain or mind. It is the intrinsic nature of the human brain. It is part of the way life is designed (by Nature) to be lived. All this implies that, had the human brain been of a different construction or make, its "behavior", or, powers or manner of logical behavior, sense of beauty, sense of smell, hearing, touch, or taste, or even the sense of justice, could have taken on different forms and dimensions. What kind of logic would we be speaking of then?

https://doi.org/10.1515/9783112214367-009

Thus, the "logic" of the mad man is as real to him as the "logic" of the man of genius is to the latter. Would the "logic" of the man of genius enable him to get things done or survive in a colony populated by mad people who live by a different kind of "logic"? In fact, in the absence of consciousness, the real is meaningless. Does it matter to a dead person (we suppose he has no consciousness, but we cannot be sure about this) if he were chopped into a thousand pieces and disposed into the sewers? The peculiarity of logic seems to be a function of the brain.

Consider the following simple chain of logic:

A is bigger than B
B is bigger than C
Therefore, A is bigger than C

This makes sense when it could be confirmed empirically. To a person who is blind as a bat this is difficult, if not impossible, to register in his mind. In like vein, the apparent reasonableness of many scientific theories are received with skepticism in the absence of affirmative empirical data. However, the same does not apply to the supposedly more exact science, mathematics, where mathematical proofs are not based on or supported by empirical data (except for that branch of mathematics known as arithmetic, where the results of addition, subtraction, multiplication, and division could be checked, confirmed or proven as correct by a physical count). A scientific theory backed by empirical data (which is tangible or observable proof) should thus be regarded as more reliable (more exact) than a mathematical theory backed by a proof that is abstract (that is, based on the imagination or creativity of the mind but that has no tangible or physical substance as its basis), and, not the other way around.

It would not be much of an exaggeration to say that logical principles, such as syllogism and the like, are artifacts of the human psyche. They are, simply put, man-made laws, made by man and could be unmade by man. We should be more concerned about what is the real, the tangible than mere logic. We should let our physical senses (sight, hearing, touch, smell, and taste) guide our logic and not allow logic to influence our physical senses. It should be accepted that what man could experience with his senses, i.e., sight, smell, hearing, touch, and taste, are real. There is a simple proof that the senses are more real than logic – it is the fact that we mortals tend to argue about ideas (the logic of ideas as such) but not about the data we receive through our five senses, e.g., we do not normally argue whether a color is green or not, a noise is loud or not, a surface is smooth or not, a foodstuff tastes salty or not, a vapor smells foul or not, for such things are obvious to our senses. This is not the case for ideas and logic, whose "consistency" may not be that obvious. Here, the skeptic might vouch that the senses could be deceptive. (Yes, sometimes only.) That man could not be certain of or prove his own existence, and so forth. What, then, are reality, existence, and proof? They are a function of the mind or consciousness, they are things which are abstract, intangible, and thus not as obvious as physical matter is to the mind, which is supplied with data about physical matter by the aforesaid five senses. (Unfortunately, these five senses could not supply any data about reality, existence, and proof to the mind.) The mind could choose to ascertain or doubt. If it chooses to certify, then it is "certain," and if it chooses to doubt, then plenty of questions would be forwarded. This is because the mind has the quality of prejudice. A person could choose to see or experience the very reality that is at the core of his existence, or he could literally shut his eyes, shut off his senses, and bar the incursion of reality. The choice is entirely his. A person could struggle to live and face life's realities, no matter how unpleasant they are, or he could run away from realities – especially the very abhorrent realities, by taking his own life even, for example. Logic has to be preceded by the existence of life or consciousness, which are real. Logic, which is the result of life or consciousness, is not real if it is not relevant and if it is not the basis of a correct conclusion, prediction, or outcome. It is similar to our five senses of sight, hearing, touch, smell, and taste because, like these senses, it is a function of consciousness. In fact, the power of logical reasoning could be regarded as a sixth sense, which could see the abstract beauty in the connection of ideas and thoughts, just as the eye could see beauty in physical objects, the ear could appreciate beautiful music, the skin could experience pleasant sensations, the nose could enjoy pleasant aromas, and the tongue could relish delicious cuisine.

We have spoken of proof above. What is its connection with logic, especially mathematical logic? Mathematicians are fond of demanding proofs for mathematical propositions, and these proofs are expected to exhibit incontrovertible logic. But when asked to define what makes or constitutes a valid mathematical proof, or how a valid mathematical proof should be, they would probably be stumped. Mathematical proofs normally comprise axioms, lemmas, and theorems, which are the accepted truths as the bases of their logic or correctness. An axiom is a fact or statement which is definitely true and accepted as a principle or rule. A theorem is something that has been or must be proven to be true by careful reasoning, e.g., a geometrical theorem. A lemma is a smaller theorem that precedes the main theorem. Instead of lengthy sentences, mathematicians utilize their "short-hand," which is an array of symbols, and all their mathematical reasoning is compressed tersely into these symbols, which could be as arcane as Greek is to the uninitiated. These symbols are supposed to aid mathematicians in their thinking, and they could rightly be regarded as the language of mathematics as much as English, German, or French is a language respectively to the English, German, or French literary man. Not surprisingly, many people find mathematics difficult. Similarly, and not surprisingly too, many people find German, French, Greek, Tamil, and other arcane languages difficult. In fact, any unfamiliar language, the mathematical language included, is difficult. Why do some people, who have the apparent lack of capability for mathematics, fear mathematics so much that they are overwhelmed by it, attributing their weakness in the subject to some inferiority in their mental capacity, a kind of fear or feeling of inferiority which they are not likely to experience when they fail to master the intricacies of a foreign language? The author postulates that the ideas of mathematics, when presented in a language that the mathematical student or novice is familiar with, would be much more comprehensible to the latter. So, it is not rational or logical to fear mathematics but not fear an arcane, unfamiliar language. Perhaps, weak mathematical students have been made to feel that way by their mathematics teachers.

The mathematician and logician make use of symbols and equations to synthesize or analyze and arrive at certain conclusions or truths with certain procedures or algorithms. These symbols and equations, like the words and grammar of the English tongue and the syntax of any other language, are artificial, invented. They are not part of Nature but simply the invention of the mind. They are a form of short-hand, a form of media of communication, whereby ideas are compressed into as few letters or words as possible in order to aid reasoning; in other words, they facilitate reasoning, especially complex reasoning, which would otherwise be very difficult. The reasoned conclusions derived from mathematical procedures or algorithms should normally have axioms, lemmas, and theorems, or accepted truths, as their bases, which is expected by the mathematical fraternity. A correct conclusion based on several theorems could itself be regarded as a theorem. An exercise in mathematical logic could be regarded as a mode of expression or communication used by mathematicians. Take, for instance, the statement: If I have a successful career and money, I would have happiness. This message could be translated into the following mathematical form or equation:

$$x + y = z$$

where x represents a successful career, y represents money, and z represents happiness.

With some resourcefulness, the logician could transform the above message into the following:

$$xy \rightarrow z$$

As long as the recipient of the above message grasps the meaning of the message without having to experience any difficulty, the “logic” of the mathematical equation is inevitable.

It is generally felt that the logic in mathematics, the mathematical language itself, is difficult to grasp. Is mathematics or mathematical logic any more difficult than any of the arcane languages such as Erdu, Tamil, and Greek? The symbols of mathematics and the symbols of logic have some parallels with the letters and words of languages – they are messengers of meanings and ideas. The elaborate symbols of calculus might intimidate, but at least they make more sense to us than the convoluted twists and turns of the letters of a foreign language such as Tamil or Greek. Incidentally, the mathematical symbol for summation ($\sum$) in calculus is a Greek letter. In fact, the symbols of mathematics could take any form, depending on the mathematician’s inventiveness and imagination. Is it then any surprise that behind the abstruse symbolic logic of mathematics and its impenetrable symbols and figures are hidden concepts that are simple and easily comprehended? For example, the constancy of the total electrical energy, potential and actual, in the world could be expressed through the following rather forbidding partial differential equation:

$$\partial^2 U/\partial x^2 + \partial^2 U/\partial y^2 + \partial^2 U/\partial z^2 = 0$$

Unless the mathematician invoking the above-mentioned equation explains or makes clear what the various symbols represent and what the equation stands for, no one would understand what he is trying to convey. A mathematical symbol might represent several words, a sentence, or several sentences which describe an object (the mathematical symbol could be said to represent this object) – the symbols thus act as a thinking aid for the mathematician. Though mathematics is the specialized "language" mathematicians use in their tasks, the actual language of communication, such as English, French, or Latin, e.g., is still a necessity for mathematics, for the mathematical symbols representing the English, French, or Latin words or terms normally have to be re-interpreted in these words or terms to be understood. Hence, the importance of language as well in mathematics and logic. In fact, thinking or reasoning is not possible at all without language – language is thus fundamental to logic and mathematics. To think or reason is to use the words or terms of a language, though in the case of mathematics, much of these words or terms would be converted to convenient symbols to facilitate thinking.

Mathematics has always been deemed a logically rigorous subject. It has of recent times seemed to have descended a rung or two down the ladder of credibility. It did not help to improve the image of mathematics as a logically precise subject when two mathematicians resorted to having a high-speed computer run innumerable computations and regarding the result of these computations as a proof of the Four-Color Theorem, a kind of mathematical proof that a number of mathematicians object to, the first mathematical proof to rely on a computer for checking its validity. Mathematical proofs are normally checked by humans, and this is a deviation from the norm. The question on the skeptics' mind is: how are they to know if the computer has made some errors?

Pi, e.g., is regarded as transcendental. How could we tell that the continual computation of its supposedly transcendental value would not end at some point toward infinity, thus proving it is not really transcendental? So what if pi is transcendental? It is not going to make any difference to the well-being of civilization. The renowned British mathematician, G. H. Hardy, was probably right when he said that he had never done any useful work in his life as a pure mathematician.

The image of mathematicians received another tarnishing when one of their kind got the letter "i" to represent the square root of −1. Logically, −1 could never have a square root. But there is nothing to stop a mathematician from being illogical as such, especially when he is an authority in his subject. This casts some aspersion on the rigor of mathematical logic.

What is the importance of logic? Everyone seems to think that he is logical. You hear people saying this makes sense, that does not make sense, and so on and so forth, sometimes arguing till they become very agitated. When we try to make sense out of something, we are exercising our power of logical thinking. If logic is so sure and steady, why do people argue at all? Disputes, arguments, the proceedings in the courts of law, the debates of attorneys, men of science, politicians, logicians, mathematicians, and the men in the street suggest that logic is fluid. Logic is reason, and people are concerned about the reasons for many things they encounter. We expect to find logic in thoughts and ideas. But how often we find that this is not the case. Logic is thought to be important for human survival. Guided by logic, civilization shows decorum and moral righteousness. Logic, commonness of thought, or common sense, implies set patterns of thinking, predictability, a sense of security, and understanding – as we expect others to be capable of thinking and acting logically as we do we could predict how others would act under varying circumstances, we could see the "patterns" of others' reactions and behaviors. Logic enables us to understand and solve our problems. Because there is logic in others' minds, it is possible to persuade, influence, or negotiate with others through the use of logic. But is logic really trustworthy, and would it always help solve our problems? Of course, applying logic in our dealings with people who really think and behave logically would be a breeze. But often, people are biased, prejudiced, egoistic, emotional, or opinionated, and logical persuasion might not work with them, whereas emotional persuasion might – it might be better to work on the emotions of such people, rather than their rationality or logical faculty, to get things done with them. Hence, the importance of logic in human affairs is sometimes questionable. Logic or objective thinking would be more important in human endeavors involving the hard sciences, such as engineering, medicine, and computer programming.

Two well-known mathematicians, Frege and Russell, had tried reducing mathematics to logic and had run into paradoxes and inconsistencies. Proofs in that branch of mathematics known as arithmetic are based on set theory, and set theory has been found to be riddled with paradoxes and inconsistencies. That is, the foundation of mathematics is shaky. This fretful turn of events is attributed to the Incompleteness Theorems of Kurt Godel, which state that there are certain mathematical statements that are true but not provable. This means that some fundamental parts of mathematics cannot be proven, though they may be true. This is a problem for mathematicians who demand proofs for everything. Mathematics suddenly found itself in troubled waters.

What is proof? Evidence, logical argument, something that affirms a point – all these can be regarded as proof. A logician might claim that a proof is a reason or argument that renders the truth of a statement or conclusion indisputable. To a mathematician, a proof is a logical chain of arguments involving axioms, lemmas, and theorems that leads to a valid conclusion. The question here is how to check the validity of a conclusion involving really abstract mathematical concepts, concepts with hardly any physical basis, such as those found in pure mathematics (as opposed to applied mathematics) – the problem lies in the difficulty of visualizing these concepts, e.g., concepts such as infinity, transcendence, and divergence. When it comes to real objects such as doors, pens, and cups (which are tangible), we could easily check on them by looking at them, touching them, or manhandling them (by using our physical senses). For example, to check or prove whether a door opens inward or outward, we just have to open the door to check. To check or prove whether an idea or statement, which is not tangible but abstract, is true or logical, we surely cannot rely on any of our physical senses (sight, hearing, smell, touch, and taste) but on our mind or consciousness – an idea or statement is only conveyed through the written word or the spoken word, which we can respectively read or hear with our aforementioned physical senses, but only our mind or consciousness (another intangible entity) is able to consider, evaluate, or prove whether an idea or statement is true or not. To prove an intangible, abstract entity such as an idea or statement with an intangible entity such as the mind is obviously a difficult task as compared to checking or proving whether a door (which is tangible) opens inward or outward (with our hand, which is tangible), for example. How could this difficulty be overcome? The author suggests that the mathematician or logician working with abstract ideas find analogies of these ideas in real-life examples, real cases that have occurred in the past, and/or cases that are still present. For the mathematician, he could examine the mathematical ideas (axioms, lemmas, theorems, etc.) used currently and in the past and see whether any of them are similar in logic or reasoning to the idea or statement under scrutiny. He could also try to convert the idea or statement into a physical model (which is tangible), or at least a diagram or drawing (which he could see), if possible. Such activities would facilitate the mathematician's or logician's thinking. If the idea or statement has a parallel some time in the past or present, or if it could be physically confirmed with a physical model or a diagram or drawing at least, then it could be regarded as valid or true. A proof, such as a mathematical proof, has to have a chain of reasoning or evidence that invokes a feeling of *certainty* that the conclusion it supports is valid or true. Without this *certainty*, a proof fails in its objective and is considered "wrong". How do we prove that a proof is correct or valid? Don't we need the proof of a proof, or, the proof of the proof of a proof, the proof of the proof of the proof of a proof, and so on? This would be impractical and ludicrous. How should a proof be? For example, what are the requirements for a valid mathematical proof? The mathematician would have a difficult time answering that. Similarly, the following would be hard to answer: What are the requirements for a valid logical train of reasoning? One of the requirements is apparently: It must invoke a feeling of *certainty*, an intuitive feeling, that the conclusion is correct. "feeling of certainty" implies that there is some subjectivity in logic – examining the same proof or reasoning, different people might feel different degrees of "feeling of *certainty*," probably with some not feeling any "feeling of *certainty*" at all. So, if a "feeling of *certainty*" that a conclusion is valid or correct is achieved without a so-called valid proof, could this be accepted? In any case, there is some subjectivity in proofs. Though having a proof, no matter how subjective it is, is better than having no proof at all, a person should not be so dogmatic about proofs as they are apparently not “fool-proof”.

Mathematicians have a number of ways of proving a mathematical statement. For example, they could resort to the indirect (*reductio ad absurdum*) method, whereby they assume that the opposite of the statement that they want to prove is true and then find a contradiction in this opposing statement, hence proving that the original statement is true. They could also use induction, whereby more similar cases are assumed to follow several cases. In deduction, a valid conclusion could be drawn from several premises. Is it possible for a very powerful mind, possibly a very highly intuitive one, to derive a true mathematical statement without having to resort to proofs or reasoning? This appears possible. Some of the great mathematical minds have had the capability of intuitively creating great mathematical theorems without having had to resort much to a great reasoning effort. What is the concept of a mathematical proof, or even mathematics, to an extra-terrestrial intelligence, if one exists? According to mathematics professor and author Roger Penrose, mathematical truths are arrived at by the human mind independent of algorithms (formal procedures) – they seem to have been arrived at by a “leap of intuition”, a “creative leap”. That means no man-made computer programmed to prove a mathematical theorem could ever emulate the way the human mind could perform the task. Hence, the importance of intuition to logic (mathematical logic in this case); we could say that logical reasoning is a function of intuition.

The other question is whether mathematics is real and exists independently of the human mind or whether it is not real but just a creation of the human mind. Obviously, some parts of mathematics, e.g., arithmetic, are real, while some, such as calculus, are creations of the mind – for example, in arithmetic, we are concerned with quantities which we could physically check or confirm, whereas in calculus, we are talking about methods involving approximation (which could be arbitrary) in the calculation of areas, volumes, and rates of changes. It would be practically impossible for a non-mathematician not familiar with mathematical methods such as algebra, geometry, or calculus to understand the mathematical truths conveyed by them. Like the person who needs to know English or French to understand what the English-speaking or French-speaking person is saying, a person has to know the “language” or methods of mathematics to understand what the mathematician is saying through his mathematics. Abstract mathematics is as real as abstract art. Wherefore is abstract art without the creative talent of the artist? Undoubtedly, there would be many who would look upon the works of the abstract artist as utter rubbish rather than meaningful works of the creative mind. There would be many as well, even highly intelligent people, who would regard mathematics as rubbish, and even some mathematicians themselves, e.g., G. H. Hardy and Morris Kline, had bemoaned the lack of consistency and practical utility of mathematics. The bases of mathematics and mathematical truths are assumptions or principles, and the assumption that these assumptions or principles would remain constant. These assumptions could be so abstract as to be detached from reality. Due to this, the logic of mathematics might not be that easy to understand.

However, what normally passes for logic is often some explanation for an observed phenomenon, an explanation that has to "make sense" (some subjectivity is implied here). The question is: how it would "make sense". This "making sense" here is intuitive and is hence subjective, arbitrary, and disputable. That is the reason why there are different schools of thought in science, philosophy, even mathematics, and other intellectual pursuits. It is also because no one could say for certain which school of thought is the really correct one, one that is supported by true logic. On the other hand, all the different schools of thought could be partially correct, and each could be so to varying degrees, or, they could be all wrong. As stated earlier, correct logic is difficult to define. We only "sense" that the logic is correct when we encounter it, and different people might "sense" it differently, as mentioned earlier. Of course, one way to settle this problem is to obtain a consensus on how everyone thinks, to obtain the majority viewpoint.

History has shown that theories, the result of logical reasoning, are not immortal. Ideas have been faulted or debunked. New theories have been known to replace older ones. This implies that logic is not as objective or real as it is supposed to be. A logical posing, a hypothesis, is impermanent. For example, scientists have always been formulating theories to suit the empirical data. The problem is that often more than one theory, even many theories, could be supported by the physical data. So, what is the reality or the truth in this case? This is a great problem for the logician. Consider the example here. It is considered a fact (or assumed?) that electricity or voltage is the result of the meeting of a positive charge and a negative charge. When an electrical instrument short-circuited, it is considered a fact (or assumed?) that a positive charge and a negative charge have met when they should not have done so. A clever, intelligent mind (say, a famous scientist whose views would carry weight) could have interpreted these phenomena in any of the following ways, for example:

1) Energy has flowed from a stronger point to a weaker point.
2) Molecules at two points have changed their composition.
3) Energy has escaped from one point to another point.
4) An external invisible force caused the phenomena.
5) Some magnetic forces caused the phenomena.
6) The forces in the electrical circuit outbalanced the forces outside the electrical circuit.
7) A foreign body entered the electrical circuit and caused the phenomena.
8) The temperature in the electrical circuit exceeded the temperature outside the electrical circuit, and so on.

To any person with an open mind, any of the above-mentioned theories makes some sense. If fate had taken a different course, and if a well-regarded scientist had formulated any of the above-mentioned theories to explain electricity, the physics textbooks could have been differently written. This shows there could be different interpretations of the same phenomena. These interpretations, once universally accepted, become the so-called physical laws or laws of nature. The physical laws are the "cause," and the phenomena of nature represent the "effect".

When a reason or argument makes sense, when the logic of an idea is apparent, what is actually taking place in the mind? It could be stated that logic is related to the perception of "cause and effect". Any person has in his mind many experiences of causes and effects. For instance, the following are common experiences:

1) A heavy object, falling from an elevated plane, crashes to the ground due to gravity.
2) An airplane flies due to the effect of air pressure.
3) Wood floats on water, while stones sink in it, as the density of the former is less, while the density of the latter is greater than that of water, and so on.

The mind has a data-bank of such experiences, of causes and effects. When new data or new experiences are encountered, they are compared with the contents of this data-bank. If there are similarities, the new experiences are understood, i.e., the "cause and effect" of the new data is understood. Otherwise, the new experiences may bring about puzzlement or wonder. Therefore, what makes sense to a person might have the reverse effect on another person whose experiences might be much different. For instance, being kind to colleagues is logical or sensible behavior to one person (as it has got him the good will and cooperation of colleagues in the past), but to another person, it is not so (as it has resulted in him being exploited by colleagues in the past). Hence, what is logical, what makes sense, could be rather subjective.

There could be at least two ways of reasoning or logical deduction. One, from the causes, we could deduce the effects. Or, two, from the effects, we could deduce the causes. Logical deduction results in understanding or knowledge. Unless deduction is performed with empirical or real data (as opposed to "fictitious" or imagined data), all kinds of inconsistencies and paradoxes might result. Logicians, mathematicians, and scientists are all concerned with "beautiful" ideas. The problem is that the beauty of ideas (as opposed to the beauty of tangible objects) could be so abstract that not everyone is capable of appreciating it. Take the beauty of mathematical ideas. There is purportedly no idea more precise than a mathematical theorem, which has a rigorous proof as its basis. However, a mathematical theorem does not need to be substantiated by empirical data, whereas a scientific theory has to be so. An unproven mathematical idea is known as a conjecture. Mathematicians find great challenges in proving or disproving conjectures. It is always difficult to tell whether a mathematical proof is valid or not. Normally, mathematical proofs are based on the work of other mathematicians. Hence, the importance of research, discussion, and conferences. Mathematical work could be done with a pen or pencil, paper, and a library, whereas scientific work requires much more than these; a well-equipped laboratory is also needed. Mathematicians, and even scientists, have to “stand on the shoulders of giants”, i.e., rely on the work of their predecessors and peers. Mathematicians have, in the past, been divided into two camps: the intuitionists and the non-intuitionists. The non-intuitionists support the law of the excluded middle, whereby a mathematical statement is considered either true or not true. Not so for the intuitionists; for them, there are statements that are neither true nor not true, especially statements involving infinity. For example, to the non-intuitionists, – –p = p, while to the intuitionists, – –p is not equal to p. Since Godel entered the scene, chinks have been found in the armor of mathematics. Suddenly, paradoxes were found to pop up with higher frequency. Take, for instance, the following statement:

This statement is false.

If this statement is false, it is true – a self-contradiction.

Or, the statement:

I am lying.

Here, “I am lying” implies “I am speaking the truth” – another self-contradiction.

Thus, as Godel had put it, there are statements that are true that could not be proven and such statements are incomplete, and, there are statements which are false which could be proven, and these statements are inconsistent.

These paradoxes and contradictions have been a great concern to mathematicians and logicians who have firmly believed in the solid foundation of mathematics and logic, a number of whom even believed that the two subjects are one and the same. Why is this so? To the author, the “causes and effects” of empirical or physical data are apparently paradox-free. These “causes and effects” could be physically checked and confirmed. As you can see, Nature has been constant all along. Take the following examples: The sun always rises in the east. (When has it ever risen in the west?) The equatorial countries are normally hot and are snow-free. (When has snow ever been found in an equatorial country?) All living things eat and drink to survive. (Can one find a living thing anywhere in the world that survives without eating and drinking?) And so on. As for the above-mentioned self-referential statements, “This statement is false” and “I am lying”, they are “fictitious” or made-up, i.e., they are not realistic – in fact, in real life, no one would ever make such statements to contradict oneself unless one is playing the fool. In real life, e.g., there is no way for a person to be on hunger strike and be enjoying a sumptuous feast at the same time, to be asleep and awake at the same time, to walk and sit at the same time, to talk and remain silent at the same time, and so on, unless he could split himself up into two persons, with each performing opposing tasks, which everyone knows is impossible. Thus, statements based on real-life settings would be paradox-free.

Let us look at the well-known Zeno’s paradox. According to this paradox, Achilles and a tortoise compete with one another in a race. The tortoise, being a slow creature, is given a head start. But, according to Zeno, Achilles would never be able to overtake the tortoise at all. For, if one keeps halving the distance between Achilles and the tortoise, the halving would continue to infinity, i.e., go on forever, and there would always be an interval between Achilles and the tortoise, though this interval would decrease to infinity (or, decrease forever). This is only based on Zeno's theory. Everyone knows it is not a fact and will never be a fact. (That is, if Achilles had really wanted to overtake the tortoise, he would definitely have overtaken the latter effortlessly, though Zeno's paradox proves otherwise. Such a paradox seems no more than a mere "play of ideas". It appears that Godel's incompleteness theorems present an analogous situation – the theorems point to some paradox concerning the foundation of mathematics, thus implying that it is faulty. Could we not regard this "faultiness" in the foundation of mathematics (as in the case between Achilles and the tortoise in Zeno's paradox) as more apparent than real?)

So, are we to take such paradoxes seriously, no matter how logical they appear to be? Here, we have to choose between reality and the logical unreal, between real life and the life of the imagination. Of course, if one lets the imagination run wild, there could be problems. Therefore, we should plant our feet firmly in the real-life world and work our logic with real-life situations to avoid disappointing ourselves with paradoxes.

Paradoxes, or logical self-contradictions, are a manifestation of man's powerful imagination, coupled with a curious and strong intellect. They are a mental challenge to the intelligent and the logician, to anyone who is interested in logic.

A paradox is somewhat like the following self-contradicting statements:

1) Black is white.
2) When I am talking, I am silent.
3) When I am happy, I am sad.
4) To be honest is to be dishonest.

Would any sober and sane person ever make such interesting, though laughable, statements unless he is joking? Everyone knows that such self-contradicting things could never be real and could only be figments of the imagination.

Logic and proof are just mental entities, abstractions, and could never have a life of their own independent of the human mind. The mind creates, intuits, or senses logic and proof, and not the other way around. If logic or proof were made of matter, its existence would have been indubitable, and there would not have been disputes or problems relating to them. It seems that people tend to over-estimate the importance of logic. People tend to erroneously think that if they observe things carefully and apply the rules of logic rigorously to explain them, they would be able to unravel many of the mysteries of nature. Though this has been the case for ages, man has never been able to fully understand the fundamentals of nature – for example, life's origin still seems a "gray" area, and, consciousness, which is an integral part of life, is still not fully understood yet. Though we have been rather successful in fathoming the things that happen or exist outside us, we could not even truly understand ourselves, our origin, our consciousness, our ego, and our intelligence. Apparently, the application of logic does not always bring the expected results. The most logical or the most intelligent are often not the ones achieving the most desirable results. There are a myriad of events taking place out there in the actual world, the world of the senses, the real, and not the world of the imagination, that greatly influence one's actions and destiny, e.g., luck, opportunity, coincidences, physical attraction, favoritism, environmental factors, relationships, and so on.

Scientists are supposedly highly intelligent people with logical minds. They study physical phenomena, apply logical thinking, and form hypotheses about them, the hypotheses to be checked and confirmed by experimentation and observation. Any physical phenomena that cannot be explained or accounted for or in defiance of logic or common sense can be considered supernatural or occult. The supernatural or occult can cause fear. People tend to fear things they could not understand. Sometimes, people try to impose their own interpretation or explanation of such things which could only be considered their personal opinion and not something backed by solid facts or logic. There are much happenings in the physical world that are beyond logical interpretation and are relegated to the realm of the supernatural or the occult. The intellect could only form hypotheses about these unexplainable phenomena, being unable to confirm these hypotheses as yet with empirical facts (still under investigation). Claiming more importance for logic than it deserves is tantamount to exaggerating, though logic obviously plays an important role in our lives (because we allow it to). It is rather common to hear protestations such as "It is not logical!" or "It is stupid!" (even by fools, as even fools apparently think they are logical beings). The physical senses, which derive empirical data from the natural world (and not the world of the imagination – many apparently intelligent people like to live in this world), should take priority over the logical faculty, which has quite a tendency to deal with stuffs from the world of the imagination. It should be the case of matter over mind. For the scientist, supportive empirical facts should always supersede pure abstract logic, while for the pure mathematician, the reverse is apparently the case. Godel, the logician, might had had conceived of a man claiming that the self-referential statement he made about himself lying was true. Such a man, a sane and serious one too, has yet to make an actual appearance in this world (though he might already exist in some other world or worlds, e.g., a parallel world which might exist – such a person could definitely be found in the world of the imagination or the world of make-believe).

A person could do better if he allows his physical senses and experiences (historical or experiential facts) to guide him instead of relying too much on logic. Over-reliance on the application of logic often results in solutions with open-ended conclusions and circular reasoning, which may invite endless debates and disputes. Logic, beauty, meaning (languages), justice, feelings, emotions, and so on, they are all concepts or imagery. They have no existence independent of the brain, which is solid matter. On the other hand, the tangible natural world has an apparent existence that is independent of the mind. A person, for instance, might own a house, a car, and a posh furniture set. It is apparent that such things would be there even if their owner were not around. The physical senses hardly fool us, but the mind often does (as is especially the case with the mind of the fool). In the actual world, we would always see "black" as "black", e.g., though we could have assigned a different name to it, for instance, "kcalb", "albck", or, even "white". Thus, even if we call an object "white", we could understand it as "black". (Therefore, "white" could mean "black") Our physical senses have an apparent "commonness", e.g., our senses of colors, heat/cold, brightness/darkness, and softness/hardness, – we do not normally dispute over whether an object is green, blue, or black, whether an object is hot or cold, whether it is bright or dark, whether it is soft or hard, and so on. This implies the sure reality of sensing through the physical senses.

There may be an abundant flow of ideas and concepts all around. But ideas that are brilliantly logical are of no consequence if they are not realistic, based on real things or facts. Such ideas serve little or no useful purpose at all. A thing that is real exists independently of consciousness or the mind. Is reality real? Reality means nothing to a person anyway if he is devoid of the physical senses and consciousness. Without physical senses and consciousness, not only would reality be out of consciousness, out of mind, but logic would be too.

Is the logic of a set of propositions true or valid? How do we prove this logic? For example, how do we prove that 1 + 1 = 2? A few well-known mathematicians used set theory to prove that 1 + 1 = 2, their proof running to 97 pages of complex symbols. Anyway, this proof was found to be incomprehensible by practically all mathematicians, as they were unfamiliar with the symbolism used. So, here is a proof which was supposed to make the proposition 1 + 1 = 2 obvious to everyone but ended up obscuring this fact because practically no one understood it. Hence, the proof, on account of its not being comprehensible, contradicts its objective of making the truth of the proposition 1 + 1 = 2 obvious to everyone and should have been thrown out. Worse still, since the proof did not succeed in making it clear that the proposition 1 + 1 = 2 is true, its incomprehensibility could have had the adverse effect of making the mathematician perusing it feel that 1 + 1 = 2 might not be true after all. By Ockham's razor (which stipulates that the shortest proof is likely to be the best or most correct proof), this 97-page proof fails the test of logical elegance. The author could even prove that 1 + 1 = 2 might not be true. We will come to that later. As mentioned earlier, after we provided a proof for a proposition (such as, in this case, 1 + 1 = 2), how do we know that this proof is correct or logically consistent? Don't we need a proof for this proof? After a proof for this proof is obtained, then how about a proof for the proof for this proof, and so forth ad infinitum? Of course, in mathematics, which is supposed to be a logically precise subject, the bug stops at the proof for the proposition because obviously no one could live forever supplying such proofs ad infinitum. But if mathematicians strictly demand proofs for any proposition to be accepted as logical, then it is similarly logical to demand proof for the proof of the proposition, proof for the proof for the proof of the proposition, proof for the proof for the proof for the proof of the proposition, and so on ad infinitum. This really seems absurd and is impractical, though logical. We might as well prove that our mind is sane, working well, logical, and capable of supplying the proof for a proposition. We might as well also prove that our existence is real. Why don't mathematicians think that it is logical and sensible to prove the efficacy of the mind supplying the proof or prove the existence of the mind supplying the proof? Logic thus appears to be subject to double or more standards. In mathematics, certain axioms, e.g., the axioms of Euclid, are fundamental and are accepted as obviously true or correct (without a need for a proof). These form the bases for other mathematical truths or theorems. So, is a 97-page proof involving set theory to prove $1 + 1 = 2$ really necessary? Mathematicians such as Russell and Frege had placed set theory as the foundation of arithmetic, and set theory had been found to be riddled with paradoxes, resulting in a shaky foundation. Perhaps one might argue that this is unnecessary, that this is logical hair-splitting and over-reacting. It must be very, very obvious, and, the feeling of *certainty* must be very, very great, that $1 + 1 = 2$. We could count and add one to one, we could get others to count and add one to one to confirm what we have confirmed, and the result would always be two. There is no doubt at all about this. Probably, mathematicians such as Russell and Frege had wanted to be "ultra-certain" about the operations and results of arithmetic. Or, they could had had tried to be too smart or too clever by half. But the results of arithmetic, such as $1 + 1 = 2$, could be easily checked and confirmed and are clearly and undoubtedly true. They could have been regarded as fundamental axioms of mathematics (without the need for a proof), like Euclid's axioms. Strictly demanding a proof for everything implies that the proving process should go on forever, without stopping. Of course, this would never do, as no one could live forever. So the proving procedure has to halt at some point. Thus, the proving normally ends at the proof for the proposition for all practical purposes.

The author said above that $1 + 1 = 2$ might not be true. You might wonder how this could be the case, as to you, one plus one is obviously two. The author senses an inconsistency in this proposition. For example:

a) 1 (apple) + 1 (orange) = 2 (apples), or 2 (oranges) is apparently an illogical proposition.
b) 1 (big apple) + 1 (small apple) = 2 (big apples), or 2 (small apples) is also apparently an illogical proposition.
c) 1 apple (equal to the one added to in all respects) + 1 apple (equal to the one added to in all respects) = 2 apples (equal to one another in all respects) is apparently a logical proposition.

So, in arithmetic, for propositions to be strictly or truly logical, we should specify what the quantities such as 1 and 2 represent, for certainly, 1 apple cannot be equal to 1 orange.

The physical senses seem more consistent than the logical faculty. The physical pertains to matter, while the logical pertains to ideas. Matter is tangible, while ideas are not. We should trust or rely more on what is tangible than on what is intangible (such as ideas and logical consistency). You may say that matter predominates over the mind – without the brain, which is matter, there will be no mind – a damaged brain results in a defective mind. Does this not prove that matter is more important than the mind or consciousness (read: logic)? The causes and effects, or the laws of nature, are apparently consistent, while inconsistencies are manifested in the ramifications of logic. In the natural world, for instance, a river or stream always flows down-hill but never up-hill. But, in the world of the imagination or make-belief a river could flow up-hill or in any other direction. It is little wonder that there are paradoxes when logical reasoning is based on unreal and impractical situations. However, we must not forget that there have been physical phenomena which defy understanding or logic; these are normally classified as paranormal, occult, or supernatural events.

If there are cracks in logic, how should those who rely on it to get work done act, e.g., lawyers who apply logic in arguing court cases, doctors who apply observation and logical reasoning in diagnosing illnesses, scientists who use logic to analyze physical data, and so on? Precisely because of the "gray areas", imperfection, or fuzziness of logic, lawyers, e.g., are needed to argue cases in court, to iron out things and get at the truths. In an argument or debate between two parties, each party tends to think that he is right or logical while the other party is wrong, with neither party wishing to admit defeat or give in. Each of the parties might have valid points or reasons. If there is no absolutely right or wrong conclusion (i.e., there is no excluded middle, but there is fuzzy logic), then it is obvious that neither party could be completely right or completely wrong – there could be plenty of "gray" areas (the propositions are correct under certain conditions and wrong under others). The two parties or debaters are thus likely to end up in a circular argument, with no definite conclusion (neither completely right nor completely wrong, but right under certain conditions and wrong under others). The two parties could get a third neutral party to arbitrate or decide who is right and should accept the neutral arbitrator's decision as final (If the matter under dispute is of sufficient importance the services of a professional arbitrator could be engaged). Alternatively, the following techniques of resolving the dispute or conflict could be resorted to, provided the subject under dispute is of sufficient importance:

Alternative A (The More Scientific Method)

1) State the objective(s) of the exercise.
2) Determine and specify the criterion or criteria upon which the right or correct conclusion or decision is to be based. If there is disagreement, go to (9).
3) Each of the two opponents is to agree that they will be objective in resolving the dispute.

4) Each of them is to objectively research the pros and cons of their respective stands. The research could be carried out in research libraries or other institutions.
5) Each pro and con the two opponents bring up is to be substantiated by documentary proofs, e.g., research findings, published papers, articles, books, and so on, which have been accepted as truths or facts by the community of scholars, as is normally the case.
6) The two opponents are to meet again after an agreed-upon reasonable period of time.
7) Their substantiated findings (with documentary proofs) are then compared. If there is no disagreement, the total numbers of pros and cons obtained by them are computed. Should there be disagreement, proceed to (9).
8) If the number of pros exceeds the number of cons, one of their propositions is adopted. If the result is the reverse, the other proposition is adopted. The conclusion or decision made here is to be in accordance with (2) above. The solution should be accepted by both parties at this stage if the whole exercise has been objectively carried out.
9) Should there be disagreement pertaining to the lists of pros and cons, or should there be a tie in the number of pros and cons, and an agreeable solution fails to materialize as a result, both parties are to find a third neutral party to referee, whose conclusion is to be final and accepted by both parties.

Alternative B (The Quicker Method)

1) State the objective(s) of the exercise.
2) Determine and specify the criterion or criteria upon which the right or correct conclusion or decision is to be based. If there is disagreement, go to (8).
3) Both parties agree to be objective in resolving the dispute.
4) Each party is to write down, objectively, the pros and cons of their respective propositions. Each may carry out their own research.
5) Each party is to meet after an agreed period of time and objectively compare their lists of pros and cons.
6) If there is no disagreement on the lists, the total numbers of pros and cons are added up. Should there be disagreement, proceed to (8).
7) One of their propositions is adopted depending on whether the number of pros exceeds the number of cons, or vice versa. The conclusion or decision made here is to be in accordance with (2) above.
8) Should there be disagreement pertaining to the lists of pros and cons, or should there be a tie between the number of pros and the number of cons, both parties are to get a third neutral party to referee, whose conclusion is to be final and accepted by both parties.

The above-advocated methods of resolving logical disputes of a serious nature represent the more practical and scientific approaches. Disputants need not go round and round arguing their cases and reaching no acceptable conclusion, feeling angrier and angrier with the other party for being stubborn in his views. Time and energy are thus not wasted over a dispute which is circular and has no ending. More importantly, enemies are not made. These methods of resolving logical disputation should also apply to organizations, interest groups, and nations. The exercise of logic is hardly an easy one due to the imperfect state and fuzziness of logic. The principle behind the above-mentioned methods of resolving the problems of logical dispute is the same as that behind our courts of law and parliaments, where attorneys and politicians, respectively, could have their arguments heard and resolved in an acceptable, civilized, and logical manner.

We should allow experience (experiences of past causes and effects) and observation to guide us more than logic. Bad experiences of the past, i.e., causes in the past that have resulted in undesirable effects, should teach us to avoid them in the future. The reverse of this should apply as well. But we have, at times, failed to learn from our experiences; that is when history repeats itself.

Logic is the common denominator in all kinds of ideas and all fields of knowledge – mathematics, science, sociology, economics, psychology, engineering, medicine, information technology, and so on. Mathematics is understandably considered an exact science as it deals with countable quantities and measurable weights, dimensions, distances, speeds, areas, and volumes, all having more or less precise results (this area of mathematics is known as applied mathematics and normally involves quantitative data) – it is in pure mathematics, which deals more with abstract concepts where ideas are qualitative rather than quantitative, that logical inconsistency flourishes. The hard sciences, such as physics and chemistry, involve quantitative data, utilize applied mathematics, which is fine, and are relatively free of logical inconsistencies. It is in the soft sciences, such as psychology, economics, and sociology, which deal with qualitative data, that logical inconsistencies could be found to abound; as a matter of fact, the data in this field tend to be more subjective and hence disputable. Mathematics could be utilized by the soft sciences as well; e.g., statistical analysis (applied mathematics) is used in economics and sociology. Logic could be defined as the valid connection between ideas; for example, there could be a valid connection (logic) between a mathematical idea and an idea in sociology, between an idea in economics and an idea in psychology, between an idea in information technology and an idea in mathematics, between an idea in biology and an idea in medicine, and so on. Here, it is more or less apparent that quantitative data (which are more tangible and hence they could be more easily checked or confirmed) are more "logic friendly" than qualitative data (which are subjective, intangible, disputable, and not easily checked or confirmed). So, when the mathematicians Russell and Whitehead, attempted to reduce mathematics to logic in their monumental work, *Principia Mathe-*

matica, they were in effect downgrading the status of mathematics as a specialized field to a non-specialized, general field (i.e., logic).

Moreover, there is a peculiarly elusive quality about logic. Logic could not exist independently of words and symbols (diagrams, pictures, mathematical symbols, etc.). Could a person ever reason or think without words, symbols, or ideas? Definitely not. What is the implication of this? This is very important. Please note this carefully: *The implication of this is that the more adept a person is with words and symbols, the better he can think – in other words, to improve one's ability to think logically, one has to improve one's ability to use words (in languages) and symbols (in mathematics) – so, being good in the mastery of words and symbols (or, in languages and mathematics) equates with being good in thinking logically, words and symbols being the tools or aids to thinking.* Logic is a quality of ideas. Ideas are evoked by words or symbols. We say that ideas are logical or describe ideas as such. When we say that an act is logical, we mean that the concept or idea behind that act is logical. Words and symbols are human inventions. They are the tools or media for conveying ideas. And logic is only evident in ideas. Thus, it follows that had the human race not existed, there would not have been words and symbols; without words and symbols, there would not have been ideas or concepts; and, without ideas or concepts, there would not have been logic. The mute and deaf person communicates with sign language (symbols). The blind, mute, and deaf person communicates through braille (symbols). Could it ever be possible at all for a person to reason or think without words or symbols, to reason or think in a vacuum, with nothing, so to speak? Certainly not. For to reason or think is to engage words and symbols. Therefore, logic is not an objective reality that exists independently of human existence; it is the result of human existence.

Logicians and mathematicians who regard logic as an objective reality independent of the mind are, therefore, erroneous. When Russell and Whitehead attempted to reduce mathematics to logic in their scholarly tome, *Principia Mathematica,* they had, in effect, overvalued or exaggerated the role of logic. It seems a misconception to regard logical thinking or intelligence as a precondition for human survival. Analogous to the situation where an automaton or robot acts "intelligently" according to mechanistic rules (software or program), the human race could possibly survive by acting according to certain prescribed mechanistic rules, by "adapting" to the environment in "prescribed ways". The adaptive behaviors of some of the lower forms of life, such as plants, e.g., creepers and deciduous trees, the former creeping toward the direction of life-giving sunlight, the latter shedding their leaves in summer to avoid the loss of moisture through the leaf surface, exhibit an "intelligence" or "logic" of their own. There seems to be an apparent innate "logic" or "intelligence" in nature itself. The physical laws of nature, with its discernible patterns and symmetries, the mechanistic rules of the world of nature, seem to represent the logic of the universe. To attempt to understand these physical laws is an exercise in logic or logical thinking. Thus, by implication, if the physical laws had been different, human logic would have been different too. This means that logic is neither an objective truth, an absolute truth, nor a constant. (Witness, for instance, the fluctuation of scientific theories through the ages.) If the human brain were constructed or "hard-wired" in a manner different from its present state, what would the mind or logic be like then? In all probability, a different kind of mind or logic. Perhaps logic could be classified into at least two types: human logic as it is now, and non-human logic (as evident in the animal and plant kingdom). In fact, the more logic is contemplated here, the less sense it seems to make.

It is often said that man is an emotional animal, that he is ruled more by emotion than by reason or logic. When this is the case, he would act illogically and might carry out rash acts which he might regret later. It is also often noted that sound logical deduction requires a calm, clear, and objective mind. So, logic and emotion are "enemies". It is either one that is in control of the person or the other.

Thus, the discussion on the nebulous nature of logic ends here with some reservation on the part of the author with regard to its actual importance. Logic should be engaged to enable a person to arrive at a correct or valid conclusion, which could be the basis for a correct decision or action. Theoretically, logic should help a person to be successful in his endeavors, to get the results he is aiming at. But, as explained earlier, this might not take place due to some intervening factors such as luck and opportunity. A logical proposition could be so difficult to grasp, so perplexing, that it serves little or no purpose. On the other hand, a logical proposition could be difficult to comprehend when presented in a certain form, e.g., the calculus, and could be simplicity itself when stated in, for instance, simple English. A dish could be presented in a palatable or unpalatable form depending on the chef. The same rule applies to logical propositions. The author scorns those ivory tower or pseudo-intellectuals who hide the logical quality, or lack of logical quality, of their propositions behind an intellectual fog or obscurity so that they would appear smarter than they are. Unfortunately, quite a number of philosophers, mathematicians, and logicians are guilty of such a petty act. What they actually achieve amounts to much ado about nothing.

A truth, any truth, e.g., a logical proposition, should be made as clear and obvious to everyone as possible. If it is obscure, it ought to be regarded as doubtful. The author subscribes to Ockham's razor, which stipulates that the simplest and shortest solution to a problem is likely to be the most correct solution.

Finally, coming back to the question of how real and objective logic is: like breathing, heartbeat, and the physical senses, logical thought is a manifestation that a person is alive. Breathing pertains to the nose, the air-passage, and the lungs; heartbeat pertains to the heart; the physical senses of sight, hearing, smell, touch, and taste pertain to the eyes, the ears, the nose, the skin, and the tongue respectively; and logical thought pertains to the brain. Logic, the result of the brain cells at work, is the equivalent of the physical senses and emotions; like the physical senses and emotions, logic is an aspect of consciousness. All of them depend on the various parts of the functioning brain for their existence – the various centers of the brain interpret signals from the various parts of the body, such as the eyes and ears, giving rise to sensations, feelings, and conscious thought. It has been established by neurologists that how capable a person is of logical or abstract thinking depends on the number of glial cells in the brain. The brain cells that give rise to logical thinking are real. Logic is a function of these brain cells and is secondary. Supposing these brain cells are constructed differently and behaved differently, would human logic exist, and if it does, would it be as it is now? So, could we really say that logic is real and objective? Perhaps we could state that the reality and objectivity of logic are relative, depending on the construction of the brain. There is a further characteristic of thought. It is possible to arrive at a logical conclusion without having to go through the reasoning process step by step but by a sweep of intuitive thought, a process known as heuristic thinking. Like creative thinking, such as in the case of inventing a machine, heuristic thinking depends on some unconscious processes that take place in the brain. The brain has a mysterious mode of operation. Instead of trying to analyze logic, we ought to try analyzing the functional aspects of the brain that give birth to logic.

In conclusion, the author would like to mention that a logical mind should be objective, open, realistic, flexible, and undogmatic as well. As logic is apparently fluid, such mental qualities are important for adaptation to varying situations. In a non-human world, human logic would probably be meaningless and useless.

10 Is a New Kind of Logic or Reason Possible?

We, *Homo sapiens*, pride ourselves on our intelligence, our logic, our reason – we consider ourselves the most intelligent life-form on this planet, if not in the universe. However, many scientists believe that there is an infinite number of alternative universes in higher dimensions beyond those we know, or could know, and, there is a possibility that beings who are much more intelligent than we are could be found in a number of these alternative universes. On this score, some scientists even postulate that for every human being on this planet, there is an exact replica of them in one, or more, of these infinite alternative universes; the alternative universe in which this exact replica exists is known as a parallel universe. What is the implication of all this? The implication appears to be that intelligence, hence, logic, is infinite, limitless, boundless; intelligence and logic, if they could be found in any of these infinite alternative universes, could be vastly, even totally, different from those of our earthly ones, perhaps so different that they could escape our notice or recognition should we encounter them.

What is intelligence? Practically everyone would agree that it is that mental ability which enables us to comprehend, reason, and form conclusions. It is that ability which enables us to overcome the problems of the environment, to survive, and to progress. With it, we are able to learn and acquire knowledge. With knowledge, we progress, and with more knowledge, we progress further. This thing, intelligence, is really wonderful, though its origin is as yet not really understood. It is something abstract, intangible, and yet it could bring tangible results. It involves the play of reason or logic.

What is the actual nature of intelligence? Could intelligence be characterized by an entirely different "form of reason"? In the exercising of intelligence, in reasoning, our thoughts rely on the principle of causality and the logical link (we talk about a priori, a posteriori, and syllogistic reasoning). For example, Event A gives rise to another event, say, Event B; Event B gives rise to Event C; reason would conclude, thus, that Event A would give rise to Event C in this instance. Reason (which is that thing in the brain which makes us accept a chain of events as logically true and sensible) tells us that this sequence of events is "logical", that this "cause and effect" makes sense. Could reason not be regarded as a "sense", like the "sense" of beauty, the "sense" of guilt, or, the "sense" of pity? It appears that the very structure and make of the human brain (or is it the DNA, which is the genetic code?) enable every person to universally experience the same "senses" in the same way, e.g., what looks logical (or beautiful) to one person would generally appear so to others (excepting perhaps the mentally defective and the insane). Is reason or logic really something which is God-given, created (in which case, it could have been created differently, though a number of scientists would probably like to believe that God had little or no choice but to create it the way it has always been), or, is it something which exists spontaneously?

https://doi.org/10.1515/9783112214367-010

Moreover, reason or logic has to co-exist with ideas, and ideas could be obtained through "experiencing" the environment (our surroundings). Could reason exist without ideas? Could ideas exist without living beings, consciousness, or matter? Certainly, reason or logic could never exist in the void (where there are no consciousness or ideas, no matter, no events, no happenings, no experiences, no senses, e.g., sight, hearing, taste, touch, and smell).

Would reason or logic still be the same if the inherent nature or structure of our brains (or our DNAs?) is entirely different, and if the inherent characteristic of Nature is totally different? Could reason or logic even exist at all then, or, reason, if it exists at all then, be having another form? In this situation, there could be other possibilities. How about, instead of intelligent beings, the existence of non-thinking, non-feeling, and non-conscious beings whose behaviors are purely mechanical, being affected by an entirely different set of physical or natural laws, e.g., a scenario where gravity, sunlight, air, water, and so on, do not exist? Could such a society of zombies ever exist? The die-hard might argue that this would never happen; pointing out that such an environment would never support life. Though such an environment would not support the kind of life-forms which are now found on our planet, there could be other types of life-forms, life-forms which could adapt to this kind of environment. We should not assume that Nature has no other choices, except supporting the kinds of life-forms which we are familiar with. In such a situation, what would really be the likes of reason or logic, if it exists? It could perhaps be reason or logic which is characterized by something absolutely against our common sense, e.g., different laws of causality, spontaneous events, i.e., events that occur without cause, magic, levitation, telepathic control of objects, and so on. How about a different kind of consciousness, e.g., supra-consciousness, a consciousness capable of extra-sensory perception or capable of foreseeing or foretelling the future? In such a setting, would reason or logic exist, and, would it be needed? For example, living things such as trees and plants are without apparent intelligence or reason, but they could survive for many years, even centuries. Couldn't pure consciousness, reason or logic exist by itself without being embodied (in human or animal bodies)? How about the existence of ghosts or spirits, or ghost-like or spirit-like beings? All these are not mere figments of the imagination but are apparent logical possibilities. Practically all of us must be familiar with such topics as spirits, ghosts, and levitation, e.g., which are controversial subjects. Not surprisingly, many among us are firm believers in spirits and ghosts (some might even claim that they have actually encountered such "beings"). That a human being could ever think of such apparently absurd objects implies that there is a possibility that such objects could exist or materialize, no matter how remote this possibility seems to be.

The human mind is apparently incapable of comprehending the mind-boggling possibilities described above, as they are detached from our everyday experiences, and we could not logically link any of them to what we already know, nor could we confirm them by experimentation. It is perhaps this limitation of our mind which makes us incapable of comprehending the origin of life or the universe. It seems much easier to accept the notion that life was created by a Supreme Being. Spontaneous life, i.e., life which had appeared out of nowhere or without any cause, seems more difficult to accept. This appears to be due to the fact that we have been much exposed and used to causality, so much so that we feel that any occurrence, happening, or event has to have a cause.

If a different set of physical laws (a set of physical laws which had been different from our physical laws) had been responsible for the creation of the first life-forms on earth, it would be difficult to deduce how life on earth originated; it is evident that if that were the case we would not be able to figure out how the first life-forms on earth originated based on our present physical laws, and, we would have to be able to think "out of the box" if we were to get the answer. Some scientists have been carrying out experiments in their laboratories to simulate the creation of life in the early universe. If the creation of life in the early universe had indeed been the result of some earlier, unknown physical laws, these experiments and their results would have been invalid if they had been based on our known physical laws, which are different.

The faculty of reasoning resides in a particular area of the brain. If, by a stroke of fate, Nature (and for the believers, God) were to re-construct that part of the brain, would reason or logic be the same anymore? If, by some twist of fate, a new race of intelligent beings were to replace the human race, which had been destroyed by, e.g., a cosmic explosion, what kind of intelligence or logic would then have materialized?

It would be interesting here to consider the case of mathematics, which is considered a logically precise science. Even so, there have been controversies over mathematical proofs. (What is acceptable as a proof or reasoning to one mathematician might not be acceptable to another mathematician.) What does this imply? Does this not imply that logic is not as precise or certain as we would like it to be? The important question is whether in a universe where there are intelligent beings who are different from *Homo sapiens*, would mathematics be around. If mathematics have a role to play in this universe, what would mathematical proofs or mathematical reasoning be like? Would they be similar to ours or different? If different, whose mathematical proofs are better? And, by "better" what do we mean? Is it higher precision or certainty, is it conclusions which could be more easily checked or confirmed, e.g., through physical, tangible experiments, is it more precise language or symbolism being used, is it easier, greater comprehensibility, and greater accessibility?

Could reason or logic be relative? If the human brain (or the DNA) had been different, and if the universe had been different, could there have been an altogether different kind of reason or logic (which, to us, might be incomprehensible, against common sense, absurd, and unbelievable)? We need not think that far. At the mundane level, we could consider whether there are any physical differences between the brain of the sane person and the brain of the insane person which account for their different intellects. These important questions deserve serious thought. In relative terms, if, hypothetically, the few billions of people inhabiting this planet were insane and you were the only person who was sane, would your logic still be logic, or considered logic, anymore? What use would your logic, the logic of the sane person, be in this insane world?

Logic may be imperfect, even faulty. But this should not be any cause for concern. What is important is that we keep an open, objective mind where logic is concerned and be intellectually honest. We should be objective and honest enough to admit that our human logic may not be perfect and may be faulty, and we should always be prepared to consider alternative views and compromise. In the final analysis, the truest and purest logic is true objectivity and flexibility; the truly objective and flexible mind should be able to accept, adapt to, and make use of any kind of logic, whether earthly or extra-terrestrial, in its quest for the desired results. Such a mind should also be able to be brutally honest with itself, such as being able to accept any personal shortcomings or faults without the need to deceive or excuse itself. Unfortunately, true objectivity and flexibility appear to be rare, and most people are apparently egoistic, biased, and emotional, resulting in much crooked thinking and logic.

11 The Basis and Reality of Logic

It appears to the human mind that logic is objective and real. Would a non-human, say, animalistic mind, conceive of such a thing as logic? At least, animals seem to apply logic, i.e., they have intelligence. It is man's ability to use logic well, i.e., his ability to reason well, which puts him in a class way above the other life-forms; at least, this is what man thinks of himself.

One should ask oneself, "What is the utility of logic? Has it truly any reality? What is its basis?" Logic could be regarded as the set of rules which governs thinking. By presuming that a person is logical, we could somewhat predict his behavior. Man, through his upbringing and education, has been more or less trained to think or look at things in a certain way. This "way" is the logic which we are here talking about. By studying how others behave (and think), it would be apparent that people behave (and think) more or less in the same "way"; we may refer to this human tendency of acting (and thinking) alike as the herd instinct. The problem is that, though people may act (and think) alike in many things, there are a number of "controversial" areas, e.g., religion and politics, where, unfortunately, this may not be the case, where the reverse may be true.

Human beings apparently have a tendency to prove things or demand proofs for things. A proof could be regarded as evidence or information which supports a postulation or stand. It is as though we have a sense of insecurity, a lack of confidence in our thinking, that we frequently look for proofs, which would give us confidence in a postulation or stand. We need a proof because we need to feel certain, confident, about the validity of an idea or logic. If we, e.g., wanted to determine whether the statement, $1 + 1 = 2$, is logical, or, valid, we could prove or affirm the logical soundness of this statement by physically counting one twice; it is apparent that this would be how people would normally check the logical rectitude of this statement, though some well-known mathematicians in the past had resorted to 100 pages or so of esoteric mathematics to prove that this statement is true. In other words, we could experientially or experimentally see for ourselves that the statement, $1 + 1 = 2$, is indeed true, and we would be certain that the statement is true. Whether it is a 100-page mathematical proof or a proof by physically counting, the objective has been to ascertain or make certain, to convince oneself as well as others, that the statement, $1 + 1 = 2$, is indeed true. We could say that logic has a physical (and therefore real) basis, though it may not be so at times as when we are dealing with really abstract entities which are not amenable to physical measurement or experimentation. In the above case, it is evident that the hard, physical evidence, which is the process of physically counting one twice to obtain two, is the more obvious (and much simpler) proof compared to the 100-page mathematical proof, which is abstruse and difficult; it also complies with Ockham's razor, which stipulates that the simpler proof is likely to be the more correct proof. For example, to check, confirm, or prove whether "$15 \times 10 = 150$"

https://doi.org/10.1515/9783112214367-011

is correct and not erroneous, we could simply get a calculator to multiply 15 by 10 to see whether we would indeed get 150 (which is the simplest and most direct way of doing so, though mathematicians would probably not accept it as a proof), divide 150 by 10 to see whether we would indeed get the answer 15, divide 150 by 15 to see whether we would indeed get the answer 10, add 15 to 15 nine times to see whether we would indeed get the answer 150, or, add 10 to 10 fourteen times to see whether we would indeed get the answer 150; we do not resort to a mathematical way using the esoteric theory of sets and what-not in doing so, though the mathematicians would probably prefer it. Imagine, if in the business world, e.g., in the banks, if people do not trust or have no confidence in their calculators or computers and have to rely on 100-page, 200-page, 300-page, and so on, mathematical proofs to affirm the logical validity or soundness of the often complicated numerical computations produced by the computers/calculators, what would have become of the business world? Worse, such abstruse and complex mathematical proofs are very likely to cause fear, confusion, lack of confidence, and serious problems, which would defeat their purpose. This shows that it is much easier to check, confirm, or prove the validity of the logic of a statement by a method which is reliant on something that is tangible, that is, physical (that involves the physical senses – sight, hearing, touch, smell, and taste); as they say, seeing something tangible or physical is believing, as compared to just viewing it in the abstract during the process of logical reasoning.

Logic could be said to be based on the sum total of our physical environment, our needs, and our sensitivities. It is not possible to divorce logic from this, to separate the two. In other words, logic cannot exist in a vacuum. Neither can it exist without the means of carrying a message or idea, be they words, symbols, signs, or pictures. How do we reason without words, symbols, signs, or pictures? How can we reason if we cannot see, hear, feel, read, and articulate? How can we reason if we do not have the necessary data or are unable to obtain or access the necessary data? How real is logic? Or, is it just a learnt behavior or "way"? Can it be that without the outside or environmental influence, logic does not exist? Can logic be regarded as such as the response to an outside stimulus only, without which logic would not exist and would not be real?

It should be remembered that logic has great utility for us. It enables us to get the desired results, to survive, and to progress. This applies also to the other forms of life, though logic may not be as successfully applied by them. Man's (and other life-forms') needs, emotions, feelings, and sensitivities determine how they would behave, how they would apply logic, or formulate the rules of logic or reasoning.

Logic is practically a means to an end. Perhaps, really pure, truly objective logic is non-existent, i.e., logic seems to be variable. Even for such relatively logically precise subjects such as mathematics and science, there appears to be some subjective element, which is manifested in the apparent elegance, beauty, and credibility of a piece of mathematical or scientific reasoning. In practice, logic or logical reasoning has to have the characteristic of elegance or beauty about it or some plausibility or credibility about it, which is rather subjective, in order to be worthy of our attention. For the relatively logically precise subjects like mathematics and science, which belong to the hard sciences, unlike, e.g., the soft sciences, the validity of the logic is more amenable to confirmation, by physical measurement or physical experimentation. All human beings are apparently capable of appreciating a fine piece of logical reasoning, as they do a fine piece of art.

Logic appears to be a creation of the human intellect, with no independent existence or reality of its own. It is immaterial, intangible, and a sort of a "ghost" of an idea. The human race places much importance on it, while it represents the set of rules which governs how the mind or intellect shall "behave", e.g., the mathematician and the scientist have to be familiar with and work with many rules, principles, and formulae, the mathematician and the scientist apparently have to build on the work of their predecessors and peers, to work with the "conventional wisdom", in order that the recognition or credit for their work is more forthcoming, i.e., they are not expected to be heretical; in the case of the mathematician, e.g., his work, which is published in a mathematics journal, is expected to be presented in a canonical manner. If their work had been much different from, even to the extent of contradicting, those of their peers, wouldn't it be very difficult to gain recognition or credit from their peers? It implies that great original thinking, even if the logic were infallible, would have some problem getting recognition or credit. It also implies that logic could be rather subjective, and, in the above case, rather subject to peer influence. Perhaps, we could state that logic is not real, only surreal, and not truly objective, only relatively objective. There are also a number of paradoxes found within logic itself, which is a manifestation of some weakness in logic. Those who rely too much on it may be disappointed, as things may not turn out as expected or desired.

12 Reason and Logic in Human Affairs

The human race is an intelligent species. Man, through his superior intelligence and his sound reason, has been able to dominate his environment and determine his future.

Take, for example now, the wonders of man's scientific achievements. It is now possible to travel or move at a fantastic pace. It is now possible to transplant vital organs. It is possible to make machines do all the unpleasant work for us. The list goes on.

It should be natural for man to act with reason, as he is reputedly the smartest creature that roams the Earth. But often, man behaves in such a way that is incomprehensible, such that it makes one wonder where his reason is. Man is apparently always motivated by selfishness. The reasons in his behavior are thus apparently selfish. Sometimes, and quite often, his behavior, instead of furthering his own selfish intentions, is even detrimental to his own well-being. Take, for instance, the drug addict who robs to pay for his expensive drug habits. He may get high and get a kick from drugs, but if he were arrested for robbery (and worse, murder, if he had murdered his victim), he is likely to spend years behind bars, in which case he is likely to suffer mentally, physically, and spiritually. It is plain common sense that crime does not pay, yet some people, even some from well-to-do families, still resort to it by reason of greed, lack of self-discipline, and, perhaps, short-sighted foolishness.

In short, man has an ego. If he is insulted and snubbed, he is likely to act unreasonably. He tends to take offense at criticisms and act defensively. It is not easy to live with a fellow human being. You have an ego, and so have others. When egos clash, you get arguments, quarrels, and fights. The world now may have practically all the amenities for the good life. But the good life is often marred by the so-called "evil" of man, the bad things that they think and do. Scientists may aim to improve the lot of mankind, and seek truths. But, perhaps, the most important task facing them is to make it possible for everyone to live together harmoniously, regardless of race, language, religion, and creed. It is rather pointless having a luxurious life when there is plenty of disharmony and discord. Would one be able to peacefully enjoy life's luxuries thus? One may ask oneself what leisure time means when there is leisure to pick quarrels or fights with a fellowman. Think of the unpleasant scenario. There is unrest. There are demonstrations and riots. There are conflicts, fighting, and wars. There are legal wrangles and court cases. There are innumerable disputes over all and sundry, even over the most petty things. There are instances of people being murdered because they had stared (they had probably looked admiringly only) at their own "murderers" without knowing that it would be their last stare (for the person-being-stared-at had taken offense at being stared at). If all the beautiful women in the world take offense at being stared at, many of their admirers might have been murdered. Staring could be taken for a state of admiration or curiosity and need not be a manifestation

https://doi.org/10.1515/9783112214367-012

of anger or hatred. Some people might wear a scowl all the time, though they might not feel anger or harbor any evil intentions. Moreover, appearances can be misleading, and many a man have apparently been wrongly taken in by appearances. For example, we are all familiar with the person who dresses well but is penniless. Thieves, for instance, have been known to dress smartly to parlay suspicion. In all this, one may wonder what has happened to reason or logic.

The main point here about human reason is that it gives understanding. Knowing and understanding the reasons behind the things which go on around one, our fellowmen, and other important things make one surer of oneself and able to maintain one's calmness and poise, help to ensure that one does not misjudge things happening around one and lose control of oneself as when one gives way to one's emotions, and, give one the ability to carry out activities in the right direction. Often, it is apparently the lack of understanding, or, not knowing the reasons that makes enemies of men.

A person may ask himself why a fellowman behaves in such an abhorring manner, and hits out at him with words or even physical force. If he were more rational, he would resort to words of reason or persuasion. If words of reason do not help, it is doubtful if angry words or force would be of any assistance at all. As the saying goes: One cannot force a horse to drink water if it refuses to do so. It is naturally difficult to behave in a sensible manner when practically everybody around one seems to behave like brutes devoid of common sense. But one should retain one's rationality and attempt to change society for the better with it. There are many ways of going about this. One way is to write, express one's views publicly, and to win over others to one's points of view. Another way is to attempt to sway people to accept certain ideas through political or social activities, through the manifestation of one's sincerity, confidence, and intelligence. Though many human beings seem to be brutes, they are apparently basically intelligent and can understand a sound reason. High intelligence is manifested in many areas of life; robots, computers, and drugs that cure deadly diseases are but a few examples of what man can achieve with his fine intelligence. If man can control his emotions and his ego, and lets his intelligence take control all the time the world would be a very much better place to live in.

Harmony and peace ought to be sought and maintained. Otherwise, life is meaningless. What is the point of life when one's life is always in danger, when one cannot trust one's fellowmen, when there is the fear of harm in one all the time? One may go insane with fears and suspicions. Why does man accept the status quo and not do something about ridding society of greed and evil? This is hard to understand. Something should be done about this. Somebody, at least, should do something about it. Some may pray to God to salvage the souls of mankind, as they have sinned. This is action, albeit a passive one. More activeness is needed. For instance, a person with enough concern for his fellowmen may set up an association known as the Society for Harmonious Relationship Between Men. An intellectual or philosopher may set up a sort of philosophical club to introduce philosophy to his fellowmen, to attempt to coerce his fellowmen to think rationally and to seek the truth all the time. A person may lobby for certain rights or lobby against something baneful to society. There have been only a few intellectuals with the courage to fight for human rights in some countries where they are lacking in recent times. A few such intellectuals that come to mind are Andrei Sakharov of Russia and Fang Li Xhi of China.

At this point, the reader may ask, "What is reason, and why is it so important?" This is not an easy question to answer. In fact, what is reasonable to one person may be absurd to another. Reasons should be ideas that spring from a mind that is free from all self-bias and prejudice. What happens if two minds, equally free from self-bias and prejudice, hold contrary ideas or opinions? Who is to judge whose ideas make more sense or are more logical? Well, the sensible person does not, in the first place, always think he is right. He is prepared to consider another person's views, and if they make more sense, he is prepared to accept them. This is intellectual honesty and is the hallmark of a true intellectual. The really reasonable person is always realistic. He does not run away from unpleasant truths, nor does he delude himself. But, apparently, many people could not face harsh facts and therefore could not really be reasonable. There appears to be some truth in the saying: One man's meat is another man's poison. It appears that many people cannot think rationally. This stark fact is apparent in the social unrest which is so common all over the world – shocking crimes, student riots, military takeovers, oil crises, slanders, assassinations, and so on. It seems that we will never be certain whether what we regard as facts are really true. Even in the most objective branch of knowledge, viz., mathematics, Kurt Godel had proven that mathematical truths are not absolutely certain or free from contradictions. So, how can other branches of knowledge, which are relatively less objective, be proved true with any greater certainty? Here, the most important thing is perhaps to maintain an agility and flexibility of thought – the main point is perhaps to have subtlety of thought and to be able to "double-think". The person who is too sure of his ideas and thoughts is, in a sense, not rational. It would appear that many people are less rational than they think they are. It would be wise if everyone were able to learn all the time, even if he were into advanced old age; one should always keep an open and broad mind. Confucius was not wrong when he said that, at old age, he still had much knowledge to acquire – and Confucius was already considered a very wise man by his peers!

Reason evidently has a very important role to play in human society. Man is apparently different from all other forms of life because he is capable of rational thought. Who knows? Even the lower animals and plants may have rational thinking capabilities of their own – their logic may be rather different from human logic. For instance, we have hydroponics, the science of plant communication. If, in all our actions, we are more guided by self-interests and greed than any other thing, and stop at nothing at self-serving, what would become of society? Fortunately, there are vestiges of conscience and rationality in all human beings, which make it possible to talk things over by using reasons rather than resorting to brute force and violence. We are apparently proud that we are unlike the wild animals of the jungle. We are apparently proud that we are civilized, polite, and sophisticated. Apparently, we tend to think that we are rational and aim to be so, though quite often the rationality is only assumed, make-believe.

Everyone should be rational – guided more by sound reason than by emotion. We all reckon that there is reasonableness in the hearts of our fellowmen. To deal with a fellowman is straightforward – just appeal to his reason (or his emotions). Try appealing to the reason of the wild tiger or the ferocious lion and see what will happen? There are degrees of reasonableness, depending on the times and situations. It is reasonable to tell a lie, if it could help avert disaster, or save a life, for example. It is not reasonable to lie if that lie leads a person to carry out a harmful act. It could be considered reasonable to steal if the thief were a Robin Hood (stealing from the rich to help the underprivileged). It is not reasonable to steal something because you rather not buy it at a high cost. It is relative.

To the human mind, everything has its reasons. Thus, we have scientists and philosophers who look for reasons and truths. An interesting question is: Must everything have a reason for existing? We should ask ourselves: What is the reason for having reasons? We are curious. We want to know, to find out, even if finding out something does not always naturally lead to our well-being. In actuality, it is sometimes better not to know the truth. For example, if one were to know the truth that one were below average in intelligence, would one not be a worse person, a less striving and less ambitious person who would tend to seek the help of more capable people? On the other hand, if one felt that one were smart and attractive, even if that was not what one actually were, one would be able to act with confidence and courage and would want to have a part in many things in life. Man apparently needs to know the reasons for many things, for that knowledge gives him a sense of security (but knowing the reasons may sometimes lead to a sense of insecurity).

It can probably be said that as long as there is life, there is reason. In this respect, animals can be considered to have reasoning power in them. Observe the animal, or even the insect, and you will probably be able to see reason in their behavior. Reason, of course, should be comprehensible in order to be considered reason. Whatever is not comprehensible, however good its reason or logic actually is, is likely to be regarded with suspicion and may even be discarded as nonsense. There are many things in the universe which are incomprehensible. For example, how did life on the Earth come about, is there extra-terrestrial life, is there a soul in all human beings, and so forth? What is disappointing is that quite often, on deeper scrutiny, things may look rather contradictory. For example, one may explain that, according to the "steady state" theory, the universe just exists by itself and is forever steadily expanding, with no beginning and no end – this is related to the concept of infinity, which is rather difficult to comprehend, for we are rather used to the concept that everything has a beginning and an end. On the other hand, the "big bang" theory states that the universe is the result of an explosion of gases which became matter, i.e., the universe has a beginning.

The author believes that he is a person of reason. He is here writing about human reason, and it requires reasoning power to analyze human reason. If the reader agrees with what the author has said so far, he is likely to regard the author as a person of sound reason. If the reader regards the author's ideas here as trash, he may be right, though the author is "reasonably" confident that they are not so. If the author tells the reader that he is absolutely certain that what is written here is the absolute truth, then the author may not be in possession of such a reasonable mind after all. Everything apparently has its reason or logic. Even a thief or a murderer has a reason for stealing or killing – the thief steals because he apparently needs to survive, the murderer kills because he apparently needs to get rid of an uneasy feeling, e.g., hatred or envy, and to achieve some peace of mind (which might only be achievable by a murderous act). We tend not to condone such criminal acts because we tend to sympathize with the victim. Have we ever stopped to wonder why such criminal acts have to take place? Have we ever tried to see things from the criminal's point of view? Theoretically, crime is not always wrong. Robin Hood, for instance, robbed the oppressive rich to help the poor. Would anyone with a sense of fair-play frown on his kindhearted activities? The very famous scientist, Albert Einstein, had said it: Everything is relative. It depends on how one looks at things, what yardsticks for comparison one is using. Let us put it this way. To survive, to feel happiness and peace of mind, and to be free from worries, are apparently always good enough, or strong enough, reasons for our actions.

Can reason be other than what it always has been? Can there be a reason or logic that is vastly different from our normal concept of reason? For example, can society accept lying, stealing, killing, and other corrupted, evil actions as reasonable, logical? Can there not be life without reason or logic? Can there not be plant-like, vegetable-like life, wherein everyone apparently lives as he likes or wishes without concern about rhyme or reason? Can a life devoid of emotions or feelings be possible if the universe were to be remade (by God)? What purpose does logic or reason really serve? At best, logic or reason seems to contribute to the betterment of man. But not really all the time. Logic or reason can sometimes be self-defeating. For example, a person who thinks of too many reasons why a thing should not be carried out will fear too much to want to carry it out. He will thus "chicken out" too quickly.

13 Humanity and Logic

Humanity, in the plenitude of its abilities and the certainty of its logical faculty, stands conspicuously apart from the lower species of beings. Apart from human beings, who else would claim to possess spirituality or sensitivity of the soul? The human psyche constitutes an aspect of Nature which humanity has yet to come to terms with. The perpetual search for an explanation for the origin of life, the constant groping with the workings of Nature, all manifest a typical characteristic of human nature, viz., the sense of wonder, puzzlement, and the urge to know. Since time immemorial, many theories, which are always in a state of flux, have been forwarded for the sake of coming to terms with the mysteries of Nature. A full understanding of some of the most important fundamentals in our lives has so far been denied to us. It is as though God, whose hands seem to have been responsible for creating the cosmos, has posed a great conundrum to mankind to solve. In lieu of hard evidence for the myriad theories which explain our origins, there are only inductions, deductions, and perceptions. Man has a brain, a mind, and is therefore able to think and reason, or so he thinks. Given a brain or mind of a different disposition by its Creator, whoever He may be, what good or bad reasons, ideas, and perceptions are not possible? If, by some stroke of fate, man is equipped with a mind which is only capable of evil thoughts, evil by our present standards, wherefore would be humanity, conscience, and logic? It is apparent that humanity quite often displays more folly than wisdom when lost in the intellectual wilderness upon being confronted by the conundrums presented by the unfolding of Nature. As when many scientists derided Darwin when the latter postulated that we descended from the apes, which had apparently been deemed an insult to humanity, while scientists today seem open to this theory of evolution; man in the past had accepted that the world was flat and that the sun revolved around the Earth, and would have had laughed at the modern man who dares to believe that the Earth is round and that it revolves around the sun.

Anyone who brings up matters relating to the spiritual realm or the netherworld is likely to draw the scorn of the modern-day scientist. It seems that the latter is skeptical about such things, though those who have had reported having encountered spiritual beings or ghosts could swear that their encounters were real. After all, seeing is believing. And, there have been some reported cases of people who apparently had after-death experiences, i.e., these people had apparently died before, had a glimpse or an experience of the afterlife or the netherworld, but had come back to life again; there have also been a number of people who apparently had near-death or out-of-body experiences. The spiritual might defy logic or common sense, but it should not be ignored. It is indeed strange that the scientific community does not seem interested in spiritual matters. Scientists could have subjected those who apparently had experiences with the spiritual or the afterlife to some kind of tests to confirm the authenticity of their experiences, e.g., the more scientific lie-detector tests to affirm that their

https://doi.org/10.1515/9783112214367-013

information is not lies or the result of an overworked imagination. They could also carry out investigative work to determine the authenticity of the information. Scientists should be having an open, objective view about such things instead of keeping them out of their minds. It seems that the scientific community has been reluctant to embrace any scientific theory which implies the workings of the hand of God or the spiritual, though this trend may be changing a little now. Einstein, the most famous scientist of the twentieth century, had apparently believed that there is a God, and he had often been quoted for the remark, "God does not play dice with the Universe".

All this seems to manifest that there is a limitation to the human intellect. What harm is there in keeping an open mind on ghosts, spirits, Supreme Creators, and the like? It is as though these scientists already had in mind the whys and wherefores of the physical world and were only interested in certain theories, especially theories that could be embellished with beautiful mathematical equations. For researchers in the various sciences, though they are apparently expected to be creative and produce new ideas or results, it seems that they are bounded by some unwritten law not to stray too much from the conventional wisdom. What good is it for humanity if their bright sparks succumb to peer pressure and conform to certain acceptable modes of thought, and dare not to take the challenge of adopting a radical mode of thought for fear of being ridiculed or ostracized by their peers? What good is there for humanity if their intellectuals are more interested in politicizing than in truths? It is as though humanity is living in a world of make-believe. Out of all this morass, one thing seems evident. Mankind, through its ability to conceptualize and dictate reasons and laws, has an inherent capability of creating the world-conditions it wants, including modifying and controlling the workings of nature. In fact, man seems capable of playing God. And man can be God, if he wishes to be regarded so. To demonstrate this point, a few examples here would suffice – the great achievements of cloning and artificial intelligence. Human intelligence seems to be a double-edged sword. It could advance and benefit society, or it could lead to harm, even self-destruction. The fate of mankind lies squarely in the hands of the powers that be. The control of the equipment of peace and the weapons of destruction lies with the political leaders and their governments. They are the ones who are going to dictate how science and technology are to be used – for the advancement of society or for the destruction of other men. The hierarchical nature of society today is not unlike that of past feudal societies. The modern-day bosses could be regarded as the equivalents of the feudal lords of the past. At the top of the social hierarchy today is the president or prime minister of the state, who could be regarded as the equivalent of the emperor or ruler of the past. Just when we thought we have been rid of the last vestige of feudalism, a kind of neo-feudalism now seems to starkly stare at us. The feudal lords of the past took away the livelihoods or the lives of their unwanted subjects by ordering their execution. The modern-day lords dispense with unsatisfactory or unwanted underlings by booting them out of employment or breaking their rice-bowls, as we are fond of saying. Humanity does not seem to have improved drastically with the more advanced scientific methodologies which are now available but seems to have shifted the points of their problems elsewhere. Improved scientific techniques that help to stamp out hardships have apparently given birth to new difficulties. For instance, the technologies which make working a breeze today apparently demand the sweat and toil of the so-called techies whose job it is to learn the new technologies and apply them. New technologies may take many years to master and tend to require continual training and education. Learning today apparently does not, and could not, stop after the day one collects one's college diploma, and has to continue with the seemingly exponential progress of technology. Life-long education is apparently the order of the day, and technology apparently progresses at a much greater rate than our ability to absorb it. The irony is that while advanced technology is aimed at making life easier for all, it is making life harder at the same time, negating practically all the advantages it brings. Is it not a nerve-wracking, if not miserable, life when one has to spend one's spare time, e.g., the evenings, reading up or studying, or attending computer, engineering, or management classes to upgrade one's knowledge and skills so that one could perform one's hi-tech job well, when one could be relaxing at home or somewhere else and having a whale of a time?

Humanity now apparently live their lives amid the tumult and strains of a complex society. Our post-modern society witnesses an erosion of past values and a great surge of liberalism. Once well-received ideologies such as communism and socialism now seem to have tottered, if not collapsed. In their places now seem to be modified forms of these ideologies, e.g., democratic socialism, or something else. We have within each of us the ability to speculate on what the future holds for each of us. Some would envision a future where everyone works from their homes, with their homes linked to their offices by vast computer networks, which is, in fact, practiced by some companies now. The other possible future is that our important decisions would be made by highly intelligent computers – government by computers, so to speak.

The essence of humanity is its soul or spirit, coupled with a conscience which dictates how it ought to behave toward others. Humanity likes to think itself as different from and superior to the other forms of life in that it is able to distinguish right from wrong, moral from immoral, and is civilized, so to speak, whereas the other forms of life are lower and wild. This mode of thinking is, in fact, rather exaggerated. For instance, loyalty is a human virtue, but does humanity have more loyalty than a well-trained, domesticated pet dog (toward its master)? Humans could think any thoughts they wish. Their brains are theirs alone. And they apparently hardly realize their folly while they indulge in their vanities. Their apparently large egos seem to color their thoughts, which are often tainted with prejudices, misconceptions, misunderstandings, and self-centeredness. The saint-like human is apparently not only a dying breed; he is probably extinct now, his innocence or saintliness probably buried deep down in some unfathomable grave. Where humans exhibit some hints of saintliness, it seems only to be so by default. Yes, they are so as long as they could benefit, it seems. There is always an ulterior motive, it seems. It seems that nobody does anything for nothing. It is good that we could scratch one another's back. If not, at least you should scratch my back. It is perhaps flawed thinking to suppose that a person, because he is human, would have other people's welfare in mind. Perhaps he understands greatly that others' welfare implies his own welfare. For example, one would not want the fire in one's neighbor's house to grow any bigger and rush to extinguish it before it spreads to one's house. Even the charitable act is not to be viewed as a truly kind one, according to some philosophers, who had way back rationalized that a kind or charitable act was a selfish one. According to them, the philanthropist donates huge sums in order to assuage his conscience – he feels bad if he does not do so. It could be said that all human actions are more or less guided by a deep-seated desire for self-preservation. Besides the mere do-gooder whose kindly acts are for the sake of avoiding an uneasy conscience, there are those with altruistic motives who just do good for its own sake. Could the latter not be construed as selfish, as they care only for the pleasure of doing good? For them, though the expressed aim is to perform a kind deed, to help, the ulterior motive (which is self-satisfaction) is selfish. Only the dead are truly selfless, it seems.

The condition of humanity is manifested in emotions. Where poverty, sufferings, and misfortunes abound, we humans experience depression, sorrow, and unhappiness. Successes, breakthroughs, overcoming obstacles, and harmonious companionship are the bringers of happiness and satisfaction. Sometimes, our feelings fall neither here nor there and are somewhat indescribable; the emotions are mixed, and there is some confusion; the person feels happy and is yet sad, or bad and is yet glad. Emotions apparently represent a distinctive quality of humaneness, and they seem to be displayed more by humans than by the lower forms of life, though dogs have been known to mourn their deceased masters and die heart-broken. Indeed, people who are skeptical that animals could be more sensitive and attached in their relations with their own kind or their human masters think too highly of themselves. They seem to think that the human race is so great that every other species is inconsequential. The origin of emotions is mysterious, but they are believed to have been born in the soul. The play of words, rhythms, and music could stir up the emotions or evoke them. The brilliant speaker or orator could work up his audience's emotions like a magician conjuring some tricks. There are apparently people who do not have much emotions. The cold, poker-faced person gives the appearance of apparent calm and lack of excitement. He seems to experience but one constant emotion – calm detachment. However, animals apparently display their emotions in ways different from humans, e.g., by wagging their tails, making quick, loud noises, and frolicking on the ground. Animals have been known to cry but do not seem to laugh at all, and even if they do laugh out of happiness, we humans apparently are not aware of it. They apparently have their physical and mental limitations. Yet they appear to have some superior traits, e.g., the ability of dogs to detect objects by sniffing and the capability of apes to quickly and gracefully clamber up trees. The apparent trouble with humans is that they normally choose to view things in a way which makes them feel and look good. They apparently have plenty of pretenses but could yet claim to be intelligent, logical, and clever. They may be excellent actors in the world stage, but they seem to suffer more from a swollen head than a brainy one. Like Lucifer, they sometimes tear at the fabrics of society, causing damage here and there. With their evil thoughts and destructive moods, they sometimes resemble the devil. They are apparently equally adept at playing God or Satan. They create their own world, the living environment.

Humanity has apparently been born into a world of consciousness and imagination. These have their origin in the brain. A man may see an object which is either actually there or it is not there at all but a figment of the imagination. Either way, the object is as real and meaningful to him as he is conscious of its "existence". With his consciousness and imagination, he could live in and be conscious of the actual world he lives in or he could dissociate himself from the actual world and live in the inner world of the imagination. He may not be able to control the progress of what takes place in the actual world, but he is capable of giving rise to what takes place in and controlling the events of this inner world. In fact, this inner world is private and belongs solely to him. When the external, real world is stressful and unbearable, he could retreat into the sanctity of this inner world. Novels, the cinema, artistic works, and other creative works of fiction could transport a person into this fantastic inner realm. The inner world and creative imagination of the psyche apparently know no bounds. The myriad inventions and creations of humanity – great works of art, technological and scientific innovations, and masterpieces of fiction and poetry – apparently testify to this. It is apparently in this respect that humanity differs vastly from other forms of life, unless, of course, if animals such as dogs and cats write their own kind of novels, produce their own kind of poems and works of art, and make their own kind of music and something more, which would apparently all be incomprehensible to humans.

Humanity could be construed as representing an abstraction, an intangible something, the sum-total of which apparently knows no limit. Apparently unbounded in scope, humanity, and hence, humanism, represents everything and all things that the human mind is capable of comprehending and being conscious of. Humanism apparently represents that distinct quality of humanity which sets it apart from all other forms of life. And what exactly is this quality which makes human beings so distinctive? Human beings, being the apparently wonderfully clever and imaginative people they are, apparently regard themselves as humane, kind, civilized, moral, and unbarbaric and unanimal-like. At least, this is how they seemingly tend to regard themselves. It is a subjective, arbitrary thing. The narcissism is apparent. The ego-play is evident. It is only the normal, your-fellow-next-door human who prides himself on his humaneness but who yet unconsciously or subconsciously behaves like a despicable weasel. The truly objective-minded intellect would probably regard humanity as not more superior than other forms of life. There are also beauty and intelligence in other forms of life, though it is a beauty and intelligence of apparently quite a different sort. Vis-à-vis the other forms of life, the human race appears more intelligent with regard to its ability to adapt to an adverse environment and overcome obstacles. In some other respects, their apparent folly detracts from their humaneness. One of man's greatest follies is perhaps vanity. Status symbols, plastic surgery for added beauty, branded fashion goods, cosmetics, hair styles, fancy dressings, fancy cars, and many, many more. All his posing and vainglorious activities. All these strutting. All for what, if not for pampering an apparently oversized ego? And the extent to which humanity indulges its vanities is often extreme, even absurd, beyond reason. Suppose a person wears cosmetics or make-up which is beautiful, but which hides an ugly face. Yet this person could be adored for her beauty. Natural beauty seems rare nowadays, with people opting for cosmetics, plastic surgery, and what-not. Perhaps the same could be said about intelligence or logic. With so much education and training courses and a diffusion of knowledge everywhere – through schools, the media, publications, journals, forums, and seminars – what is now great intelligence is probably pseudo-intelligence or only apparent intelligence. In other words, man may appear more intelligent or more logical than he actually is. Man's raw intelligence or true intelligence may not be that great. In some cases, it may be even false intelligence – foolishness in the guise of cleverness, as when a man uses obscure terminology to hide his ignorance and lack of comprehension, which is effectually an empty parade of erudition. That a man could ascend to such a height of folly or plumb to such a dire predicament speaks much of his wanting intellectual caliber, his not very strong sense of propriety, his lack of decorum, his imperfection. Man has always prided himself in his capability for abstract thinking, and reasoning could therefore be regarded as an act of human vanity, as though other species are incapable of reasoning. This capability might be even more pronounced in other forms of life, though it could be unfathomable to the human being, as it is perhaps on a different "wave-length". Unbelievable as it might seem, animal abstract thinking, if it exists, could be more abstract than human abstract thinking, and who is to tell this could not be so? If man has an intuition and thus have an intuitive understanding, there is evidence of this in other life-forms too. We should not be blind to the fact that man could not perform, or perform as well, many activities which other species are capable of. Not surprisingly, because of this, the other species might think that man has a more inferior intelligence. Does this signify that man is less intelligent, at least in some respects? Sad to say, even the concept of intelligence is arbitrary and not fully comprehended. Intelligence tests developed by psychologists evaluate, at most, a few types of intelligence. By consensus, it is now believed that there are many more, much more, types of intelligence. We have "invented", if not discovered, a new type of intelligence – emotional intelligence.

Supposedly, the human being is not only flesh and blood; he has a soul, which many believe lives on after a person's demise. He is characterized by having a spiritual awareness, which results in the worship of God and other spiritual beings. This quality does not seem to be present in other forms of life. Presumably, the soul resides in the brain. There is synonymous-ness between the soul and the mind, which could be regarded as two terms describing the same thing. The seeming lack of spiritual awareness in other life-forms, however, does not mean that they have no souls. All living things, perhaps including plants and trees, should have souls. The life of man could be interpreted as the life of the soul. The solidity and tangible-ness of man and all other forms of life are apparently translations of the soul into its material form. Soulfulness is apparently consciousness, awareness, feeling, sensing, and emoting. As long as a person has these experiences, his soul is apparently alive in the material world. The absence of such experiences could mean either of two things – the person is asleep or he is dead. However, death does not mean that the soul is dead. In any case, the soul is a controversial subject. Though the body of a deceased person might be dead and not functioning, the soul could have left the dead body for another realm and could be very much alive. Man generally savors life and fears death. The rare exceptions to this are the suicide cases.

The human being is apparently most proud of his ability to exercise logic in his thinking. While other life-forms apparently do not exhibit such complexity of thought, man seems to be always trying to make the abstract even more abstract, so much so that some others might find it rather incomprehensible. But, from the author's experience, a very abstract concept, e.g., a complex mathematical equation, when expressed in simple layman's language or terms, could be simplicity itself. Mathematics, e.g., is apparently difficult. It is likely that it is the arcane symbols and terms used in mathematics which are the obstacle or difficulty, while the mathematical ideas themselves may be relatively simple and comprehensible to most people. The fact is that intellectuals or the professions are fond of communicating with jargon or technical terms, apparently in order to protect their enclaves from the encroachment of the outsiders, the uninitiated, who, when they listen or look in, tend to have little or no inkling of what they are talking about, thus possibly making simple ideas sound or look more profound, more impressive, than they actually are. We humans may think that being capable of abstract thought makes us more "human" and less animal-like. But, tongue-in-cheek, animals could have abstractions which are so "abstract", so different and so uncommunicable, that they are unfathomable, unknowable, to their human counter-parts. This may be why we humans apparently lack a full comprehension of the psychologies or thoughts of other life-forms, though, as scientists, we may study their behaviors and habits. Similarly, if one day extra-territorial beings from other universes, who may be very intelligent, visit our planet, how would they be able to make themselves understood by earthlings? At the earthly level, a person communicating in an arcane language, e.g., Erdu, would be somewhat like an extra-territorial being, as he would be incomprehensible to the non-Erdu speaking people, just as the mathematician with his arcane mathematical language would be incomprehensible to the non-mathematician. Human beings should not underestimate the mental capacity of other life-forms, whose modes of thought could indeed be very different from ours. For example, animals staying in the deep jungles might have their own animal logic and reasoning, and if they could survive with these in the wild, which if a human being were stranded there, would have difficulty surviving, they could not be stupid. It is perhaps man who is stupid in assuming that other life-forms lack intelligence and intellect. Man is probably mistaken when he believes that logic is objective. Logic depends on interpretation, how we view or look at things. For example, the following simple equation would appear undoubtedly logical to any human being, based on a commonly accepted or well-accepted interpretation:

$$1+1=2$$

Then, how about:

$$1+1=3?$$

Though many would say that this equation is wrong, non-sensible, or illogical, to the author it could be interpreted in another way and still be logical. We could interpret this equation as one and one result in three, with the third being reproduced, i.e., a married couple gives birth to a child. This equation could be regarded as a representation of the following statement: A man and a woman marry and reproduce a child, resulting in three persons in the family. What, then, is logic if it is not something which could be arbitrarily, subjectively interpreted? The latter appears to be the great problem with our logic. And there are so many kinds of logic – Aristotelian logic, Boole's logic or laws of thought, Frege's logic of arithmetic, and so on. If logic were really objective, it should be universal in form. Logic or reasonableness is but apparently just one of the many creations or inventions of the human intellect, a quality which we normally do not attribute to other life-forms. For example, we do not reason or argue with a dog or a horse. Logic or reasonableness may be regarded as the interpretation of the relationship between two or more objects, events, or situations, whether one object or situation leads to another, and whether the latter leads to yet another, and so on. Logical reasoning thus results in a chain of related thoughts. This interpretation or perception of the relationship between two or more objects, events, or situations depends on the person's past experiences, attitude, open-mindedness, and mental acuity, and is anything but objective, though for practical purpose it could be regarded as relatively objective, provided that there is agreement among a sufficient number of people pertaining to its logic. Supposing, four or less out of ten persons agree that there is logic, could that thing be classified as logic? What if five persons out of ten agree that there is logic? What if six or more persons out of ten agree to the logic? What if two out of two persons agree that it is logical? Who is right and who is wrong? Even if we are certain that there is the logical connection, we could perhaps never be able to prove it or prove otherwise. Even if we have the proof one way or the other, we could probably not prove that it is the perfect, absolute proof. It all boils down to how the logic or relationship is interpreted, and interpretation implies subjectivity. Why worship logic so much and overplay its importance? If there were such a thing as objective logic, there would never be disputes, debates, arguments, fights, wars, conflicts, and controversies. To avoid any difficulties with logic, man should first change himself, especially his attitude. He should learn not to be stubborn in the face of controversies but be compromising, calm, unemotional, and flexible, with the ultimate objective of achieving amicable agreement and harmonious personal relations. Why be foolish by insisting on a point which could never be satisfactorily proved? Would it not be best to incorporate the merits of each different viewpoint into a grand viewpoint with which every party is most satisfied with? Philosophers had taught men to question and seek the truths and apparently knew that there is no such thing as absolute truth. Men should learn to be philosophers. If all men could have the mind-set of philosophers, human society would definitely be a happier, more harmonious, and conflict-free place. The trouble with man now is that he is often so certain of his logic to the extent that he becomes arrogant, whereby he misses the point that there is no absolute logic or absolute truth. He has apparently been traveling through the tunnel of folly, confusion, and ignorance all this while. He should divert his travel path to that of intellectual enlightenment and wisdom. He should be awakened from his apparently fool-hardy dream-world and made to take a perusal of his faulty mode of thought. All the world's problems and difficulties imply that men have been thinking the wrong thoughts and adopting the wrong systems. There is a place for a new logicism.

Apparently, little does the hard-headed human being realize that he has all along been living in the arbitrary, make-believe world of LOGIC. Consider the blindly accepted logic of the following basic equations in mathematics:

$$0 \times 0 = 0 \tag{1}$$

$$1 \times 0 = 0 \tag{2}$$

$$1 \times 1 = 1 \tag{3}$$

$$-1 \times -1 = 1 \tag{4}$$

Don't you think that eq. (4) does not make logical sense? -1 is less than 0. If so, how could -1 multiplied by -1, or an object which is less than 0 (in this case, -1) multiplied by another object which is also less than 0 (in this case, also -1), be 1? If $0 \times 0 = 0$, and if -1 is less than 0, then -1×-1 ought to be less than 0 instead of 1. To the author, -1 seems to be a more logically proper or acceptable answer. However, we should not worry unduly about this unsatisfactory state of things. Such rules of logic are made by man and could, of course, be unmade by man. To rectify the problematic nature of human society, there ought to be a new logicism, which the author shall call Flexism. "What is Flexism?" one might ask. This could be regarded as a mode of thought which is derived from a mind-set that is characterized by broad-mindedness, open-mindedness, flexibility, and subtlety. Human society could forge this kind of mind-set through suitable training, education, and intellectual stimulation. In such an intellectual culture or climate, all religions, ideologies, philosophies, and theories would be tolerated, their merits (and demerits) noted. There would be real freedom of thought and exchange of ideas. With intellectual tolerance and honesty, conflicts and strives would be abated. With these abated, would it be necessary to resort to wars and other forms of destructive behavior?

Our understanding of the natural world is apparently based on some commonly accepted rules or logic. How about those paranormal or apparently supernatural phenomena, then, which could not be explained or understood in terms of these commonly accepted rules or logic? Perhaps, for such phenomena, there are different rules or logic. Our scientists are apparently reluctant to work on the paranormal and the apparently supernatural, often dismissing them derisively as the province of cranks. It appears that since the scientists feel that they could never understand or explain these strange, far-out, and abnormal phenomena, they should give up trying to understand them. On the other hand, they should not give up but take such things up as a great intellectual challenge, a huge and challenging puzzle. What does all this imply? As our standard physical laws or rules of logic apparently do not apply to the paranormal and the supernatural, we should seek some other principles or rules of logic which explain them. Thus, all this implies that there could be other kinds of logic. It means that logic is limitless, boundless, and infinite.

What is the usefulness or importance of logic? To the author, logic or reasonableness is only useful when dealing with people who are logical, people who are capable of appreciating logic or a reasonable argument. Logic is useless when dealing with illogical people, people whom we may regard as brutes or animals. The problem or headache is that the apparently illogical people may think that they are logical and regard others who are truly logical as illogical. In other words, logic may be nonsense to the illogical, while to the logical nonsense will be nonsense. Consider this interesting situation: You are living in a hypothetical society with a population of, say, 20,000 people. You are the only person there who could think logically, while the other 19,999 people there are all fools, being incapable of logical thinking. Do you think that there would be any use for your powers of logical thinking in this society? Would you find living in this society challenging, interesting, or a bore? Logic could only be appreciated by people who are capable of logical thinking. Logic could only be used to persuade, influence, or convince someone who could appreciate the logic, who is capable of logical thinking. Though these 19,999 people, or fools, are not capable of logical thinking, each of them might have their own particular mode of thought, which, to them, is "logic", and not surprisingly, these 19,999 individuals might regard the only logical-thinking person as a "fool". Now, who is really the fool, the real fools or the logical-thinking person? The truly logical person should not be so foolish as to use true logic to persuade or convince a real fool, a person really incapable of logical thinking, but might use other (possibly illogical) modes of thought to convince the real fool, e.g., by playing on his emotions, such as fear, desire, and happiness. To the author, logic is relative. It depends on who views the true logic – to the truly logical-minded, the logic is indeed logic, while to a fool or a person incapable of logical thinking, the logic is probably "nonsense," and the only logic to this fool is probably their illogic, their nonsense. So, if I were a logical person, the true logic would indeed be logic, and the fool's "logic" would be nonsense; if I were a fool, I would probably agree that a fellow fool's nonsense is "logic" and the true logic is "nonsense". Therefore, what logic is to a person depends on whether he is really logical-thinking or really a fool who is incapable of logical thinking. Therefore, logic, as described above, might, after all, be useless or of no importance in certain situations (such as those described above). On the other hand, if all these 19,999 people are highly intelligent and well capable of logical thinking, how would you find living in this society now? Wouldn't the logical-minded person find it a breeze?

At the metaphysical level, logic is itself mysterious. Is logic really a necessity in life? Life-like objects such as robots and automatons, who "survive" on artificial intelligence, may be practically more intelligent and more capable than many humans. Could they be regarded as logical objects? Could they be regarded as conscious objects? After all, they appear as intelligent, or more intelligent, and life-like, or lively, as a human being. Where does logic, or consciousness, originate from? Must logic necessarily be what it is now? Could logic have had been different if the conditions supporting life had been different? Is logic, or, consciousness, real, and not just an illusion? How do we prove or confirm it is real? How do we prove, confirm, and convince others that a logic is true and correct, for if we could really achieve this all the time, disputes and conflicts would have been unnecessary and would have been avoided? But proving, confirming, and convincing others of the validity of a logic is apparently a great problem, as it is apparent, and grievously so, that no two minds think or reason totally alike, with the egos, motives, and emotions of the parties concerned apparently adding to the problem, resulting in disputes and conflicts galore. The way around this daunting problem would be explained elsewhere in this tome.

14 More on Mathematical Logic and Proof

A mathematical proof could be controversial and might have some not-so-obvious loop-holes. A proof is information which shows that a statement or a proposition is true. The quantitative aspects of mathematics and reasoning are actually self-evident and require no further proofs than the fact that their self-evidence could be simply checked or confirmed by counting, weighing, or measuring. (Is it really necessary to resort to set theory, which is qualitative, in-concrete, not-so-obvious, abstract, and rather subjective evidence, to prove the consistency or logical soundness of arithmetic?) It is the qualitative part, which is intangible, incapable of being counted, weighed, or measured, and lacking in obviousness, that requires something more concrete (more obvious) as evidence of its soundness or correctness.

The concept of infinity is often used in mathematics. Infinity is an abstraction and cannot be quantified. Since this is so, the proof of the infinity of something, e.g., the primes, would be inherently qualitative and, therefore, controversial. This is perhaps the reason why mathematical problems involving infinity are difficult.

Mathematicians expect mathematical proofs to be infused with rigorous logic. But, it is known that logic is not always consistent, as paradoxes have been discovered (by Godel). A person might analyze the premises of mathematical, scientific, commercial, or sociological reasoning. The contents of the premises might vary for these different subjects, but they all might have a common denominator which makes the premises relate with each other, leading to a valid conclusion. This common denominator is “logic”. But, the problem is that it is difficult to ascertain the “logic” of abstract principles, whereas the “logic” of quantities (which are tangible) or, arithmetic, could be easily confirmed or checked by counting, measuring, weighing, or comparing. Our mind is able to straightaway see “logic” in the above-mentioned subjects if the reasoning involves quantifiable objects. Grasping the logical links between qualitative concepts is an apparently more arduous task. Since quantitative reasoning is obvious and indubitable, it is curious that mathematicians would want to use axioms and set theory to prove the consistency of arithmetic (logical correctness concerning quantities). It is truly the validity or correctness of qualitative reasoning (reasoning with unquantifiable, abstract objects, such as infinity) that is hard to ascertain, for how does one measure, weigh, count, or compare unquantifiable objects or qualities?

The Intuitionist school of thought, championed by Dutch mathematician, Brouwer, had wanted to dispense with the law of the excluded middle, wherein a proposition is either right or wrong (though Brouwer had finally been coerced into accepting its validity for finite sets only). For a finite sequence, e.g., 1 to 10 (1, 2, 3, 4, 5, 6, 7, 8, 9, 10), if it is not 1, then it is any number from 2 to 10. For an infinite sequence (such as 1, 2, 3, 4, 5, 6, 7, 8, 9, 10, . . .), it is not so straightforward to infer thus. However, for qualities such as tall and short, or, right and wrong, there could be a “middle” state where a thing or person is neither tall nor short, nor, right nor wrong – a person could be of average

https://doi.org/10.1515/9783112214367-014

height, and an action could be deemed right under a set of circumstances and wrong under another. Therefore, the principle of "reductio ad absurdum" – that is, to prove that a proposition is true by finding a contradiction in the assumption that it is false – is not consistent. A proposition that is not false might not be true either, i.e., it could be neither true nor false. To conclude, e.g., that a man who is not clever is certainly stupid, is stupid in itself. What if the man is of average intelligence?

The logic of quantities is obvious (as they could be checked or confirmed by counting, weighing, measuring, or comparing). The problem lies with counting, measuring, or weighing qualities, if it could be carried out at all. The following equation, which displays "quantitative" logic, might be modified to display "qualitative" logic, as follows:

$$1 + 1 = 2 \text{ ("quantitative" logic)} \tag{1}$$

$$1\,\text{apple} + 1\,\text{pear} = 2\,\text{apples, if } 1\,\text{apple} = 1\,\text{pear} \tag{2}$$

or,

$$a + b = 2a, \text{ if } a = b \tag{3}$$

The "quantitative" logic of eq. (1) is self-evident or obvious. The difficulty of "qualitative" logic is represented by the following queries or doubts:

In what ways is the one apple above the same as or equal to the one pear, or in what ways is a = b?

1) Is it the weight?
2) Is it the dimensions?
3) Is it the color?
4) Is it the surface texture?
5) Is it the taste and so on?

There is indeed no apparent or obvious similarity between the apple and the pear, or between a and b – their similarities were just assumed – in fact, their similarities, if any, would be controversial.

Godel's proof pertaining to the incompleteness of a formal system of logic, or the logic of arithmetic, which is not completely accepted by mathematicians, is a proof by contradiction. It is far from obvious and very difficult to comprehend. Its reasoning is "qualitative". Euclid's proof of the infinity of the primes falls into the same mold; it is as follows:

Assume P is the largest existing prime in the following sequence of primes:
2, 3, 5, 7, 11, 13, 17, 19, 23, . . ., P

There would always be a prime, Q, or, a prime factor of Q, that is larger than P, whatever the value of P is, comprising the composite of all the primes from 2 up to P, plus 1, which is as follows:

$$Q = (2 \times 3 \times 5 \times 7 \times 11 \times 13 \times 17 \times \cdots \times P) + 1$$

This contradicts the proposition that P is the largest existing prime number, and, therefore, the number of primes is infinite.

There is a quirk in the above-mentioned sequence of prime numbers. Whereas all the other prime numbers are each an odd number divisible only by 1 and itself, 2 is the only prime number which is even and divisible only by 1 and itself. Probably, it ought to be replaced by 1, which, perhaps, could be regarded as an odd number that is divisible only by itself, but (probably for the reason that it is also the divisor of all the numbers) it is not regarded as a prime number. This may look arbitrary. Euclid's proof is proof by "reductio ad absurdum" and is hence inconsistent, and, controversial, according to the Intuitionist school of thought (which posits that a proposition is either correct, incorrect, or undecidable).

The application of logical reasoning could sometimes lead to some preposterous results regarding infinity. An example is Zeno's famous paradox regarding Achilles racing the tortoise, whereby it is posited that Achilles would never overtake the tortoise, as there would always be an "infinity of intervals" between them. In actual fact, Achilles, on account of his much greater speed, would always be able to overtake the tortoise, outrunning the latter effortlessly. Infinity is either an invention or a discovery of the human mind. It is a controversial, mind-blowing concept, for we are all used to finite systems.

The solving of the Four-Color problem in 1976 by Haken and Appel was not without some controversy, though its validity has now been largely accepted by the mathematical community. Haken and Appel solved the problem by reducing it to a "building block" configuration of 1,482 maps, whereby proving that these 1,482 "building block" maps are four-colorable would imply that all maps are four-colorable. They worked on these 1,482 "building block" maps with powerful computers for 1,200 hours of computing time. They showed that these 1,482 "building block" maps are indeed four-colorable, thus proving that all maps are four-colorable. The snag here is that mathematical proofs have been normally checked by humans. In this case, it was not possible for a human to check the proof. What if the computer had indeed made a computing error? There was no way for anyone to check and confirm that the computer had not made any computing error, and this had to be accepted on faith. This proof was the first of its kind and was a precedent for the mathematical community.

Though e and pi represent transcendental numbers, we might never be certain that their transcendence might not end at some point toward infinity. But for practical purposes, we accept their transcendence.

If the Intuitionist school is right, mathematical proofs embracing the law of the excluded middle might not be acceptable. We have yet to come to terms with our understanding of mathematical proofs. "Quantitative" mathematics or arithmetic is obvious, self-evident – we could easily count, weigh, measure, or compare quantities – we therefore need not offer a more elaborate proof for an arithmetic. Likewise, we do not need to prove our existence by logical inference, for it is obvious or self-evident that we exist. (No question about it. A person just needs to look at the reflection of himself in the mirror and would find that his reflection would have a one-to-one correspondence with his actual self.) But the French mathematician and philosopher, Descartes, inferred his own existence by the simple argument, "I think, therefore I exist." He, and Godel as well, had resorted to a mythical kind of logical reasoning to prove the existence of God. If man could actually see God, does he still need to resort to logic to prove His existence? As they say, "Seeing is believing". In a court of law, witnesses are normally called to give evidence – witnesses who have seen or witnessed something related to the case. Why don't the judges use pure logical reasoning instead of calling witnesses to help determine a case, if logical reasoning were so reliable? In mathematics, it should be what is not obvious that requires a formal proof, and it is actually the "qualitative" or "judgmental" aspects of mathematics that are not so obvious.

What is mathematics really? There is no complete agreement on what mathematics actually is. Some mathematicians and philosophers regard it as the invention of the human mind, while others attribute it to human discovery (which means that mathematics exists even if the human mind does not exist). Whichever it is, no one could be certain. However, the author thinks it is a combination of both. It is, therefore, important that one keeps an open mind on all mathematical ideas.

15 Are We Rational?

Human beings are distinguished from animals in that they have the ability to reason. If that is really so, if we really have the power to reason, it seems only to be the power to reason correctly on infrequent occasions, while the rest of the time we reason wrongly, are oblivious to reason, or do not reason at all. If we reason frequently and reason correctly too, there should be no reason why the world would have to have so much problems, such as quarrels, disputes, and wars.

To be intelligent is one thing, but, to be rational, is quite a different thing. To be rational involves a number of personality traits, such as honesty, especially intellectual honesty, moral uprightness, emotional control, discipline, care, and objectiveness. An intelligent man may not be rational, unless he possesses these qualities. Rationality could be regarded as an attribute of the character, while intelligence is an attribute of the mind.

When we look around, we could probably see more irrational selfishness or self-centeredness in people than rationality. When a driver, e.g., horns madly at the driver in front, he is not rational; it is not that the driver in front wants to block his way but that he too is being blocked by the vehicle in front of him, which has probably broken down or is involved in an accident, an event which is beyond his control and which he certainly abhors. When one person accuses another person without substantial proof or finds fault with others, he is also not rational, for we all know that no one is perfect, and for that reason, we should expect others to make mistakes at times, and if others make mistakes we should be understanding toward them instead of faulting or blaming them – certainly, nobody likes to be reprimanded for making mistakes, not even those who carry out the reprimand themselves.

When our country lost the 1978 Regional Soccer Cup, our national soccer coach was openly castigated as a "terrible" coach, but if our country had won the Cup, there is little doubt that our national coach would have been given acclaim not quite unlike that given to a national hero. Many of us who are proud citizens had, of course, liked to see our country retain the Regional Soccer Cup championship title, not to mention our national coach and our national soccer team. But soccer coaches are not gods and therefore could not be expected to guarantee success in any game – we could only expect them to increase the chances of success or make success more likely. How irrational, how unfair to our national coach had many of our soccer fans been? And his frustration, it could be imagined, was aggravated, when soccer fans went to the extent of making abusive phone calls to his home.

Sadly, there are apparently many intelligent people, even brilliant people, around who jump to conclusions too fast or jump to the wrong conclusions. They misinterpret other people's actions and other people's motives. They misunderstand. They sometimes even let their emotions control them. Many of them seem to think that they are good judges of character, and many, too, seem to be inflexible in their opinions, be-

https://doi.org/10.1515/9783112214367-015

lieving that they are smarter than others, and are not prepared to let others tell them what to do or to listen to others. However, even fools could sometimes be right and hence deserve a hearing. If a criminal is entitled to be heard and defended, why shouldn't a fool be entitled to a hearing by others?

With the apparently higher standard of education now, people should be more rational. But this does not seem to be so. People appear to be better equipped with knowledge, with the advancement of knowledge – education appears to fill their minds with knowledge and intelligence – but it does not seem to improve their ability to think and act rationally. It is little wonder that the greatest philosopher of this century, Bertrand Russell, had lamented the irrationality of man, and he believed that it is this irrationality that would probably lead to world catastrophe one day.

The author believes that the first step now toward being more rational is for us to realize how irrational we have been and how we could all have been happier if we had adopted a more rational attitude, a more tolerant, a less self-centered, and a more detached attitude. The author hopes that this article helps to fulfill this first step.

16 The Conduct of Rational Argument

The author thinks it is unnecessary for him to define what argument is, as everyone, including the reader, must have argued at one time or another and therefore knows what it is about, and, also because he wants to avoid having the reader arguing about his definition (sic). The fact that argument is so common is telling. It implies that our human logic or reason is far from perfect and riddled with problems. This is the reason why the author wrote this tome; he has felt a strong need to address and resolve this problem and is hoping that all readers would join him in achieving this important goal. Writing this tome is the first positive step taken by the author. But, perhaps more important is that the author, together with the readers, walks the talk, i.e., carry out the proposed ideas, set the example. It is thus highly important that argument, which is inevitable since our human logic is full of imperfections, is properly and rationally conducted. The big question is how.

The author could see at least two types of arguments. One is an argument for the sake of argument. The other is an argument with an objective. Those who engage in the first type of argument may enjoy themselves greatly if their arguments are well-received, but if their arguments are not well-received, they may not only be an irritant to the target audience but may also suffer considerable personal embarrassment. As for the second type of argument, it is part and parcel of progress in life and is commonly used in all strata of society. It is used amongst members of the family. It is used amongst friends. It is used by executives and managers. It is used by writers and philosophers. It is used by scientists. It is used by politicians. Lawyers thrive on it. It may be employed even by animals – who knows?

It is perhaps true that people tend to view arguments with distaste. It is common, e.g., to hear people disdainfully remark, “He likes to argue”, or, “He argues too much”. But it should be alright and natural to argue, provided it is not overdone, it is not done in an overbearing manner, the opinions are not forced down people’s throats, and the argument is rational.

When two or more persons have differences in opinion, there is bound to be an argument amongst them to decide whose opinion is right or reasonable, and the reasonableness of the argument is all the more important if it concerns an important decision or action.

However, we should be more concerned with the question, “How should people argue fairly and rationally?” The author is aware that there are honest and dishonest ways of arguing and would like to refer to a book, Straight And Crooked Thinking (the author could not recall its author’s name), which presented about 20 or more dishonest tricks in argument. The author is of the view that all arguments which are aimed purely at appealing to a person’s emotions or sense of pride and prejudice to gain his acceptance are dishonest. The author could think of at least one instance of such dishonest argument, which would be enough to show what harm dishonest arguments

https://doi.org/10.1515/9783112214367-016

could do, especially when they are employed by people who run our lives. Hitler, e.g., had been able to impose his irrational theory of the German "master race" on his people by "emotional" argument and had succeeded in stirring up their anti-Semitic feelings. The gross result had been the extermination of many innocent Jews, and the world had been thrown into a disastrous world war, viz., the Second World War. It seems that now the world is more wary about who their leaders are and are determined to avoid such a disastrous large-scale war.

Coming back to the point on how we should argue fairly and rationally, the author is of the view that we should, firstly, be intellectually honest – we should be honest and realistic enough to acknowledge to ourselves that we are not perfect and, therefore, would not be right all the time – we should have flexibility in our opinions. If, at first, we hold opinion A to be right and if Mr. X then tells us that opinion A is wrong and opinion B is right, we should give Mr. X a chance to expound fully his argument, and if his argument is rational and proves opinion A wrong, we should be selfless enough to discard opinion A and adopt opinion B. But the trouble, it seems, is that people tend to be either too egoistic to admit that they are wrong, or they like to deceive themselves that they could not be wrong, that they are very smart.

Secondly, it is important that the exponent of arguments be emotionally detached. He should not lose his temper over arguments. There are, in fact, many instances where arguments have turned into quarrels or even fights. It is a well-known fact that when a person is angry, he cannot think rationally. Therefore, it is important to remain calm and cool.

Thirdly, the debater should be objective. He should forget about the self or the ego and concentrate on cold, hard facts, even if the facts are repugnant to his ego, e.g., if proven, he should be able to accept the fact that he is narrow-minded, naïve, foolish, and so on. The author is of the view that it is wise for one to accept, e.g., the fact that one is foolish, for one would then try not to be foolish; otherwise, one would always remain a fool.

Finally, we should bear in mind that nothing is absolutely certain, not even knowledge. For example, Newton's theories had been proven inadequate by Einstein, and, who knows, one day someone might find faults in Einstein's theories. (In fact, a scientist has recently postulated that the speed of light is not really constant, as held by Einstein – the speed of light had been greater at the time of the early universe.) Bertrand Russell had found flaws in Hegel's philosophy. Therefore, we should be on the look-out for better reasons or arguments, and we should be prepared to listen further and ask more questions.

Perhaps all these are easier said than done. The author suspects that these have not occurred to the minds of many. But once these are accepted as being quite rational, it is perhaps fair to expect people to be more reasonable in their outlook or more rational in their arguments.

17 The Logic of Moral Behavior

We live for happiness and the moral principle of right and wrong is the guideline by which we live our uncertain and short life toward achieving that evasive happiness.

Though life is short and uncertain, some of us are so certain that they are right that they are prepared to quarrel with others over a principle and in so quarrelling adds even more uncertainty to life and if the quarrel turns into a big fight, life may not only be more uncertain but may even be gone.

Everyday you hear people blaming others, "He is wrong!", "He is stupid!". You see that some people do not want to mix with a certain group of people because of differences. You even see atheists, as well as devoutly religious people, antisocial people, women-haters, even recluses, even suicidal people, and so on.

Everyone behaves the way he thinks right. If only they know the right way to behave. But there is no definitely right way. Even the right ways of older sages and establishments such as Confucius and the Bible do not seem right or applicable anymore. How many of us for instance would offer the left cheek if slapped on the right cheek? The author can imagine that anyone who is slapped on the right cheek would reply with a volley of solid punches not incomparable with those thrown by a professional boxer.

Everyone's sense of right and wrong depends on the influence of his environment – the immediate influence of his upbringing and education, the influence of friends, peer groups, and so on, and the general influence of society. The murder of human beings was right and a necessity in certain societies such as that of the cannibals and that of the head-hunters in Borneo. Bigamy is wrong for us but right for the Islamic people. The burning of widows was proper for the Indians but is unthinkable to us. Today, old practices or mal-practices are gone. We are more educated, supposedly more intelligent. But perhaps foolishly intelligent. We are now more sophisticated, less blindly religious, less blindly obedient, more questioning, harder to convince, more complicated, more materialistic, more intelligent, more secure, more self-centered (self-centeredness being the main result of competition) and so on, but less wise and less moral, and even more contradictory, it seems.

All things that are right or appear right seem to have a selfish basis. To be selfish is natural and necessary for survival and is understandable or justifiable. To be too selfish is not right because it is offensive to other people, which mean loss of one's respectability. To be too unselfish is not right too because one is then likely to lose out to others and therefore this is actually not fair to oneself and one may even be regarded as a fool.

Many moralists and deeply religious people avoid meat as they abhor cruelty to animals who are feeling beings, like human beings. They survive on vegetables and fruits, which are also living things of a different form than animals and human beings. The lower form of life, whether life is of a lower form or a higher form, is

https://doi.org/10.1515/9783112214367-017

basically life and all life should be respected. How can one be so snobbish as to respect the higher life or the life with brains and disregard the lower life because it has no brains? Would human beings totally disregard and treat idiots or "vegetable" human beings like animals or plants? No, they would care for them and probably register them with an institution such as a mental institution. In such unequal attitude, we are actually prejudiced against all other forms of life and prejudice is not right. But our survival evidently depends on the death of some of the lower forms of life. The argument is that if we must survive we must cause the demise of some lower forms of life, i.e., eat these lower forms of life, which some regard as a sin. Perhaps, by sinning we generally mean doing wrong to our fellow-beings, while it seems that this term cannot be applied to the lower life, i.e., we cannot sin against the lower life. So, sinning is a matter of how one regards it, i.e., it is a subjective issue. Just because we have to survive, a selfish instinct, we cannot be considered sinners if we cause the death of some lower forms of life. To some, drinking some nutrients such as milk seems to be the solution to this problem which would not involve the loss of life. This is not true. To live and to produce milk a cow has to feed on grass, a form of life, and drinking the cow's milk is like abetting a sin (you would not spend the money of a criminal, which had been obtained criminally, if you really disapproved his crime). So, the solution to this problem seems to be simply either to drink plain water, or do not eat or drink at all, or, to drop dead.

We could conceive or invent another possible reason in favor of the slaughter of animals. If animals could live on other animals or plants, why couldn't we live on them? After all, we would be helping the victims of these animals to take revenge and at the same time we could help to reduce the "ferocious" threat these animals bring to the members of their victims' families.

According to the bible adultery is a sin and the Romans had stoned adulterers to death. But today adultery seems common and is not a crime though it tends to lead to marital break-ups if discovered by the partner – after all, adultery has the consent of both the involved parties, unlike rape, which is criminal.

The norms and the standards of society also determine right and wrong. If the majority of us social animals feel that an action or opinion is wrong, then it is wrong, and vice versa. If society considers all forms of criticism of the state as wrong, unpatriotic or treasonous then outspoken criticisms regarding the affairs of the state are wrong or even criminal, even if they are justified.

The whole of the Third Reich had considered the murder of millions of Jews justified because the charismatic influence of a great leader had persuaded the whole hierarchy that it was justified. Yet the whole world has viewed the whole episode with horror. So what is right for one person or one culture may not be right, or, may be wrong, for another.

Is there an absolute determinant for right and wrong? There appears to be none. And yet some people are so arrogantly sure that they are right. Even truths are not absolute. Theories had been proven wrong and theories will be proved wrong, though new theories will materialize. The preciseness of mathematics and science is not perfect and mathematicians are sometimes even skeptical of the exactness of mathematics.

The moral principles of right and wrong are invented by man; man has created religions, theories and many other things. Man has invented lies. He has invented nonsense too which he quite often regards unwittingly as truths. A perfectly right or perfectly wrong action or behavior must be the result of a perfect, infallible mind. And since nobody is perfect moral principles are also not perfect.

When we say lying is bad, we mean lying to us is bad. When we say arrogance is wrong, we mean arrogance toward us is wrong. When we say war is bad, we mean war which endangers our lives is bad. When we say robbery is criminal, we mean we do not want to be robbed. When we say nuclear war is insane, we mean we do not want to face catastrophe. And so on and so forth. So what is wrong or right is determined by one's selfish considerations. But is selfishness right? It appears that spontaneous selfishness is instinctive and should be regarded as neither right nor wrong. But when it is the result of rational consideration and results, wittingly on the part of the selfish individual, in somebody being put to a disadvantage, then it is morally wrong. In this sense, animals cannot be sinners since they apparently lack the power of rational thought. In fact, many humans also seem to lack this power and perhaps the "sins" of such stupid humans should not be considered sins or should be overlooked or forgiven. People of unsound or insane mind are exonerated for committing a criminal act. But there are many people who are not of unsound mind but are stupid or who are so impulsive that they do not think at all before they act. Can these people be blamed if they commit sins? However, it appears that they are blamed and even hated, sometimes. Perhaps, they should not be blamed. There are of course highly intelligent sinners and these people should be punished because of their possibly great evil influence and the possibly disastrous outcome of their actions on their fellow-men by virtue of their high, calculative intelligence.

When we say honesty is good, we mean others should be honest with us. When we say hard work is good and admirable, we mean we would like those who work under us and those who work for us to be hard-working but admirers of hard work could quite often be seen shirking or shrinking from work. When we say generosity is good, admirable, we expect others to be generous toward us. When we praise patriotism we expect others to care for the country so that as constituents of the country we would be taken care of at the same time.

It appears that everyone, every behavior, is selfish. The kamikaze pilot, though manifesting patriotism and selflessness, can be construed as a selfish individual because he believed that it was worth dying for the country because of the honors and respect he would have had earned, even if he was dead after winning the honors. A martyr who willingly died for his country would have had died happy and pleased with himself and everyone in the country would be pleased to see him die because he died for them and had tried to save and had probably succeeded in saving the day for the country, though national tears would be shed, outwardly. Today, getting involved in a duel would be irrational, even insane. But in the days of old, getting involved in a duel was natural and failure to accept the challenge to a duel would be despicable and the person who turned down the challenge was regarded as a coward and would lose his honor.

Moral principles were thought out, conveyed, taught and acted on apparently because it was necessary to do so. Imagine what would have happened if there were no such things as moral principles or rules, or laws which were actually based on moral principles? Under such a condition, if a person felt like taking your wallet, he would just grab it without feeling bad about it, without any consideration at all except perhaps for the consideration that he might fail to reach the wallet and might end up getting a wallop from you, i.e., if he could be as thoughtful of the consequences as that, though it is quite doubtful that we could find more than a handful of such thoughtful people. If he did not like you, he would probably cause you physical injury, depending on his whims and fancies, and if he hated you he would probably murder you. And, likewise, he might end up being murdered by you. He might not feel the need to work because he could take what he wanted but did not have by persuasion or by force, depending on his whims and fancies, from his well-off neighbors. He could actually do whatever he liked. There would definitely be chaos if everyone did whatever he liked, such as killing, robbing, raping, quarreling, fighting, loving, befriending, walking around naked, shouting, cursing, swearing, torturing, and carrying out other inhuman, animal-like activities. With such a lack of constraints on human behavior, it would not be a surprise that the human race would wipe itself out sooner or later, especially with the availability of nuclear weapons. No one then would be able to live in peace. Just as your neighbors would be fearful that you would strike at them or rob them, you would be fearful of your neighbors in like manner. Everyone would practically be living under high tension, high stress, having to be alert all the time. Human behavior would be wild and bizarre.

Today, in our advanced civilization we have well-defined laws and codes of ethics. We are living in a comparatively more ordered, though more complex, world. For instance, many countries have succeeded in lowering their crime rates, not only through better organized police forces, but also through better education, which means more people are exposed to the inculcation of moral values and trained in rational thought. We are generally expected to display moral scruples, though quite often selfish considerations outweigh them, and this makes our behavior more and more predictable, the apparent result of which is that there is more understanding amongst the human race, more humanity, as compared to the days when there was great unlawfulness and disorder, when an emperor's or a monarch's edict, which could be whimsical and absurd, was the unquestionable law, and heads could be chopped off because the emperor or his courtiers had grudges against some people.

However, it should be remembered that moral principles are made and can be unmade by man. At times, it seems, people are getting less humane because of the over-competitiveness of the society at large. Because of the keen competition for the trappings of success, such as wealth, properties and high posts, it is perhaps not possible or practical to be too humane or too considerate, as it would probably result in one losing out in the rat-race. Out of selfish considerations, people tend to behave in a manner which makes them appear less humane, and it is rather surprising that the government and the society at large not only seem to ignore or take for granted such not so humane behavior but seem to be encouraging them, indifferent to humane considerations, as in the instance where meritocracy and elitism go hand in hand. Perhaps, we have reached that level of sophistication in humanity that we could afford not to try so hard at being humane, whence our humane tendencies began to slip. There will be considerable alarm to everyone when the slip is too great and it becomes too obvious, whence everyone, especially the government, will feel the threats of a degenerating society acutely and wish that things will be put right again.

Hence, it is for the selfish survival of the human race that moral principles evolve. We may experience various emotions when we see our fellow-beings adhering to or contravening the moral principles which they are expected to abide by. For instance, when we see two persons fight, we may be disgusted because we feel that it is uncivilized to fight. But when we see two dogs fighting we may not feel any emotion about the incident at all, except perhaps for the feeling of amusement, and we do not care so much at all, if we do care at all, whether any of them will be killed; this is because, so far as we are concerned, dogs, or, animals, are not capable of rational thought and do not live by moral principles. Perhaps, animals also have their moral principles. Who knows? Because animals are different from us and cannot communicate in the way we do, we apparently have a problem understanding them and it is harder for us to read their minds or predict their behavior. It is probably true that because we cannot understand animals well or share the same sentiments with them we generally choose to ignore them, unless they are our pets, or, a nuisance to us. However, it is sometimes regrettable that human beings are becoming more and more like animals. Even animals sometimes display more kindness and humanity.

Absurd though it may seem, the author had been once quite conscientious, felt morally not quite right, about damaging or destroying the paraphernalia of nature such as twigs, plants, stones and rocks. The author had then been in one of those philosophical states of mind. He had then often entertained the idea that such objects might have feelings or a life of their own. The author could not prove to himself that such objects definitely did not have feelings of their own and neither could he prove otherwise. But he did recall that stone-age men worshipped rocks (idolatry), trees, lightning and thunder, and he also did hear others admonishing against tempering with old trees, claiming that they have spirits. Such things did cast some doubt in the author's mind as to the certainty that inanimate objects have no feelings of their own. Under such circumstances, the author feels that we should err on the safe side and try not to damage or destroy such inanimate subjects unduly. Perhaps, such objects have a residing spirit or soul in them. Perhaps, there is a life in them so different from ours that it is impossible for us to comprehend it or it misses our attention altogether. Perhaps, to be morally sound, one has to think not only in regards to animate objects but inanimate objects as well, i.e., one should consider everything under the sun. In a sense, it is not right to damage, ravage or destroy inanimate objects such as plants, trees and rocks, which may be regarded as the gifts of nature or, as some religious persons may wish to regard them thus, the gifts of God. Such destructiveness may cause environmental or ecological damage, which may in turn adversely affect us.

There should be rationality and moderation in moral behavior. We should not kill quite harmless organisms such as ants and grasshoppers because they have apparent intelligence. But if they are harmful and carry diseases, e.g., flies and mosquitoes, it is perhaps justifiable to destroy them, for it is a matter of either they destroying us or we destroying them, and to prevent ourselves being harmed by them we destroy them, and this should be carried out only if absolutely necessary. Of course, we should all subscribe to the principle of "live and let live", but within reasonable limits.

Moral principles are apparently the result of rational thought. People who sin and carry out evil deeds must also be intelligent enough to do so; such deeds are apparently the result of selfish, inconsiderate, intelligent consideration. In this respect, animals cannot sin as they are apparently lacking in intelligence while only human beings are capable of sinning as they are relatively intelligent. For the sake of everyone in society human beings should use their intelligence to perform good deeds and make this world a pleasant place for everyone to live in, in other words, moral principles should have a high place in society. Though selfishness is the essence of living, we should all be but selfish in an enlightened manner and should always resort to reason and moderation.

However, in today's dog-eat-dog, competitive world, where many are desperately striving for survival and success, consistently staying on the moral high ground may not be possible. People in their desperation for survival, success or status may resort to cheating, sabotaging, intrigue, politicking, theft, even murder, and other immoral or criminal acts. There is a Chinese saying, "Honesty will result in one ending up begging". Therefore, it may not be practical to be moral or too moral. Would a person rather choose to be honest and be unsuccessful or poor, or, a little dishonest and be well-off and successful? Nowadays, there seems to be more and more white collar crimes, e.g., siphoning off company funds, fraud, insider trading, cheating, and so on. Thus, even if one had wanted to stay on the moral high ground practical considerations may make this a less favorable option. The decrease in moral standard may simply reflect that there is some defect in society, e.g., the over-competitiveness and ruthlessness of its members. On the other hand, a harmonious, cooperative, caring society is conducive to the maintenance of high moral standards. In a competitive society, it is "either you survive or I survive" (a win-lose situation), but in a harmonious, caring society, it is "you survive and I survive and both of us survive together" (there is teamwork and cooperation – a win-win situation). Isn't it obvious that a harmonious, cooperative, caring society is better, more desirable than a dog-eat-dog, over-competitive society? Isn't it better to have everyone a winner, whereby everyone is happy, rather than only some winners, who will be happy, while the rest are unhappy losers? It is only in a win-win kind of society that there will be general harmony, teamwork, cooperation, peace, happiness, and, thereby, high moral standards.

18 The Logic of Evil

How is evil defined? An action that results in harm or pain to another, without regards to the feelings or well-being of this other being, should be regarded as an evil act. An evil act could be regarded as a selfish act as it is normally carried out for selfish or egoistic reasons. For example, a dictator who gets his opponents murdered to remain in power is regarded as evil.

In actuality, life is not that simple whereby acts by human beings could be easily classified as "evil" or "good". Take a simple case here. Man eats the animals such as pigs, cows or fishes, generally without any sense of guilt, generally knowing that such "innocent lives" have to be lost to satisfy man's hunger. However, this ought to have made him feel guilty, as indeed, in some rare cases, e.g., in the case of the vegetarian, this fact apparently does make the vegetarian feel guilty enough to want to refrain from eating meat (but, who knows, even vegetables may have "lives" of their own). So, sometimes, as shown above, "evil" acts are unavoidable.

Evilness is hence a question of relativeness. When a tiger, e.g., pounces on a man or another animal and kills it and devours it, it is not regarded as evil. When a man kills another man, if it is not done out of self-defense, it is normally regarded as evil. Can we not say that evilness is a standard of behaviors invented by man to prevent behaviors that are detrimental to society from being perpetuated by any of its members, an evil act or acts being regarded as criminal by society and punishable by the law?

As all men are selfish by nature and evil acts are selfish acts, evil acts are apparently acts perpetuated by the relatively more selfish beings that have the courage to take the risk of facing punishment, even death, by flaunting the law in the process of executing the evil or criminal act.

So, in a sense, everyone is evil, the question is "How evil?". A certain degree of ruthlessness in man is acceptable. But being too ruthless or too evil is normally frowned upon. Indeed, evilness is a relative thing. Sometimes, as propounded by the famous political philosopher, Machiavelli, evil acts are even justified if they result in more or greater good, i.e., the ends justify the means.

https://doi.org/10.1515/9783112214367-018

19 The Logic of Cause and Effect or Causal Effect

As mentioned in Introduction, the central idea of logic is that true premises must lead to a true conclusion. An action or event that takes place may lead to a result. When this happens the action or event could be said to have "caused" the result. This is what we mean by "cause and effect" or "causal effect". Thus, in future, if we witness (experience) the same action or event (cause) we could deduce or conclude the same result (effect). Also, in future, if we witness (experience) the same result or happening (effect) we could deduce or conclude the same action or event (cause).

Being able to experience or to use the physical senses is thus important to logical reasoning. For example, in the chain of reasoning, where A is taller than B and B is taller than C, whereby A is also taller than C, it would be difficult for a person who has no sense of sight (a blind person), whereby he could not actually see and compare the heights of A, B, and C, to understand this chain of logical deduction. For a normal person, who has the sense of sight, he could physically (and mentally) see straightaway that A would be taller than C. The great British philosopher, David Hume, had emphasized the importance of sensation or seeing in the causal effect of logical deduction. For example, if we saw ball A moving forth and hitting ball B, and, ball B in turn moving forth and hitting ball C, whereupon ball C moves forth, we could conclude right away that the action of ball C moving forth was also caused by ball A. What happens if, for the above chain of actions, we only saw ball B hitting ball C whereupon ball C moves forth but missed out on (missed seeing) ball A hitting ball B? Wouldn't we then conclude that ball B was the sole cause of ball C moving forth, a mistaken concept? Then, what happens if, for the above chain of actions, we missed out on (missed seeing) both ball A hitting ball B and ball B hitting ball C, but only saw ball C moving forth? Wouldn't we then be wondering why ball C was moving forth, what has caused it to move forth? Hence, it is clear that physical sensations such as seeing are important to logical deduction, the science which involves the cause and effect, or, causal effect.

However, couldn't the above chain of actions, if one has seen the complete chain in action, be interpreted in some other ways as well? Let us see. How about the following interpretations, which seem plausible and not unreasonable?:-

1) The chain of actions is the result of invisible gravitational and/or other forces acting on the three balls, A, B, and C.
2) The chain of actions is the result of some person's or persons' manipulations.
3) The chain of actions is the result of a remote control technology.
4) Ball B attracts and pulls ball A to itself (by electromagnetic, or, some other force?), resulting in ball A hitting it. Then ball C similarly attracts and pulls ball B to itself, resulting in ball B hitting it, whereupon it moves forth.
5) The chain of actions has been pre-planned and pre-programmed; it is the result of an advanced, powerful wireless computer-related technology involving some

https://doi.org/10.1515/9783112214367-019

electromotive force. The actions of the three balls, A, B, and C, have been as exactly planned.

6) The chain of actions is the result of changes in the environmental conditions, e.g., changes in temperature and/or changes in air conditions.
7) There are some live objects residing or hidden inside the three balls that cause them to move and/or hit each other. Hence, the chain of actions.
8) The three balls have life and consciousness, could themselves decide their own actions and could discuss with and influence each other on what actions to take. The chain of actions is a result of this.
9) The chain of actions is a result of the intervention of some higher, powerful, divine force or forces.
10) Other possible causes.

The author would like to pose a question on a hypothetical situation here. Supposing our physical laws had been different and the construction of our brains had been different, and hence our thinking processes and powers had been different, what could have caused the above-mentioned chain of actions in this new physical environment, in what possible ways could this chain of actions be interpreted by our now differently constructed brains and could this chain of actions have taken place at all in this new physical environment? For example, what if this new physical environment has no gravity, temperature and air? Perhaps, this chain of actions could never have occurred in this new physical environment. And, even if this chain of actions could and did occur in this new physical environment its interpretation by our now differently constructed brains could have been an altogether different one, so different that perhaps our current human brains could not imagine, our current human logic could not apply or connect.

The author brings up all this to show that the cause and effect or causal effect of logic is not foolproof. Besides the ten ways or so of interpreting the chain actions of balls A, B, and C described above, there could be many other ways of interpretation. Which is actually the correct interpretation, or, the best interpretation? Perhaps, no one could say for sure (for how could one be sure that one's physical senses have not missed some details even if one could be sure one's physical senses have not deceived one?). For example, how would the mind of a person suffering from a mental disorder such as schizophrenia or insanity interpret this chain of actions? A schizophrenic's interpretation, for instance, could be expected to be different from that of a person of sound mind. But to the schizophrenic, or, his fellow schizophrenics, his interpretation could be the "truth", the "logical" one (while the interpretation of the person who is of sound mind could be regarded or interpreted by the schizophrenics as "nonsense" or simply ignored). So, now what would become of logic if everyone in the world were now schizophrenic or insane? All this shows that logic is relative; it depends on the quality or construction of our brains, our past experiences of causes and effects (including the physical laws we are familiar with such as Newton's laws, other facts we are familiar with, et al.), our physical senses (sight, hearing, smell, taste, and touch – what happens if we have additional senses, e.g., being able to see through solid objects or see in the dark without light, extra-sensory perception or telepathic powers, and so on?) and whether our physical senses have witnessed or experienced all the details of an action or event (and have not fooled us).

Therefore, as rational people, since it is possible for our physical senses to miss out some physical details (or even deceive us) resulting in an incorrect interpretation of physical data and our past experiences (experiences of causes and effects and knowledge of physical laws and other facts) might be insufficient to enable us to interpret an action or event correctly, we should always regard our interpretation of an action or event as tentative (subject to future possible confirmation of the contrary). A rational person should thus often qualify his propositions or conclusions with terms such as “probably”, “most probably”, “likely”, or “most likely”, and should be prepared to consider “contradictory” proofs or evidences and modify or change his propositions or conclusions if necessary.

We should be careful with the handling of logic.

20 The Theory and Logic of Numbers

Numbers are meant to be finite, counted, and for describing quantities. Though numbers are finite, they do lead to infinity, i.e., if one were to start counting numbers one could count to no end if one wanted to. Bear in mind that quantity is a fixed amount. Hence, any number must be a fixed amount.

Numbers take various forms; they take the form of rational or irrational numbers, prime, even, or odd numbers, and so on. These symbolize the property, or, "quality" of numbers. Their charm is something that appeals to the intellect and the imagination.

Numbers have no beginning and no end. They need not begin with "1" or end with, e.g., "1,000,000". They can begin with "minus something", e.g., −10, and end with something toward infinity. Numbers can form series or sequences of definite order or indefinite order. Patterns are discernible in numbers. Series or sequences of numbers can be of ascending order or descending order, arithmetic progression or geometric progression. For example, 3 + 6 + 9 + 12 + 15 + 18 + 21 + 24 + 27 . . . is an infinite series with an arithmetic progression and 1 + 1/2 + 1/4 + 1/8 + 1/16 + 1/32 + 1/64 + 1/128 + 1/256 . . . is an infinite series with a geometric progression, and, 3, 6, 9, 12, 15, 18, 21, 24, 27, . . . is an infinite sequence whose numbers increase in an arithmetic order and 1, 2, 4, 8, 16, 32, 64, 128, 256, . . . is an infinite sequence with a geometric progression.

Numbers are rational if they are whole numbers or integers, or if they take the forms of fractions, e.g., 3/8, 7/9, and 11/201. Only rational numbers can be classified as "even" or "odd". Even numbers are sums of two even numbers or two odd numbers. Odd numbers are indivisible by even numbers such as 2, 4, and 10, except by odd numbers, such as 3, 5, and 7; they are each a composition of an odd and an even number added together, e.g., 5 is actually 3 + 2, and, 9, 7 + 2, where 2 is even and 3 and 7 are odd numbers. An irrational number may be an algebraic number (as are all rational numbers) or a transcendental number (e.g., e). A real number is either a rational number or an irrational number. A complex number is the sum of a real number and an imaginary number, the latter being equal to the product of a real number with i (i represents the square root of −1). Cardinal and ordinal numbers are the more abstract entities pertaining to sets.

Prime numbers are odd numbers with a special quality, viz., divisibility by itself and by 1 only, and nothing else, e.g., 41 is only divisible by 1 and 41, which gives results of 41 and 1, respectively. The prime numbers are also regarded as the "atoms" or building-blocks of the integers or whole numbers.

The study of operating with numbers is known as arithmetic. However, in higher mathematics, e.g., calculus, symbols are also important. Numbers represent quantities and they can signify big or small quantities. They, together with symbols, are the objects of logical reasoning in mathematics. Arithmetic merely involves addition, subtraction, multiplication and division, but in higher mathematics substitution involving

https://doi.org/10.1515/9783112214367-020

symbols comes in. For example, in calculus, symbols such as "∂" or "Δ" stands for change in rate or variance, "∞" stands for infinity and "∑" stands for summation, and, in algebra or geometry, alphabets such as, e.g., "x", "y", "a", and "b" are used to denote unknown quantities or distances.

In mathematics numbers and alphabets are closely akin. For example, the infinite sequences, 1, 2, 3, 4, 5, 6, 7, 8, . . . and a, b, c, d, e, f, g, h, . . ., each represents a certain order, a consecutive order. "1" can be regarded as equivalent to "a", "2" to "b", "3" to "c", and so on, indefinitely (one-to-one correspondence). The theory of sets deals with such equivalences. In set theory, which was created by Georg Cantor, a set within which is the sequence, 1, 2, 3, 4, 5, 6, 7, 8, . . ., and a set within which is found the other sequence, a, b, c, d, e, f, g, h, . . ., are both equivalent sets. The set theory is studied in modern mathematics and is a fundamental theory in the theory of numbers. In the above case, it is possible to substitute for "1" the letter "a", for both are equivalent. If, however, $1 + 2 = 3$, then $a + b = c$, an equivalence. Intersection sets, and joint and disjoint sets are other sets which are studied in set theory. Intersection sets give us a clearer understanding of subtraction, as shown below:

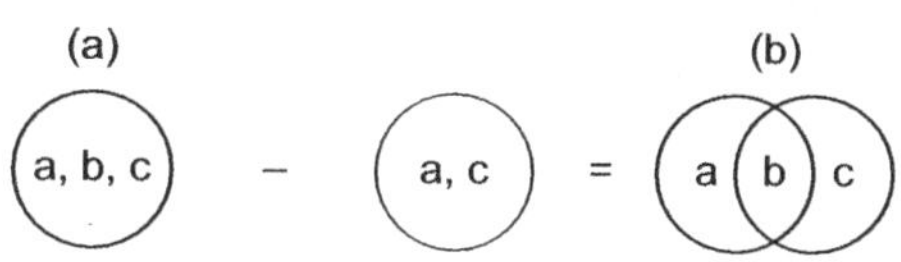

As shown above, (b) is the result of intersection, or actually, subtraction.

Joint sets are shown below:

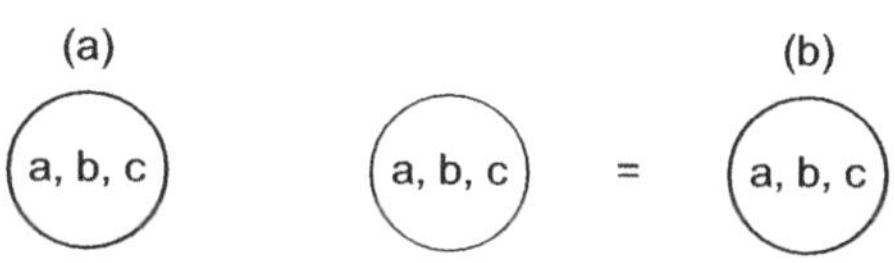

The two sets in (a) are joint together to give (b). This gives the impression of equality or sameness for indeed the letters, a, b, and c, in the two sets in (a) are the same. Disjoint sets give the opposite impression, the impression of inequality or difference.

The study of the theory of numbers cannot be complete without the study of set theory. In the sciences, e.g., physics, electronics, and chemistry, numbers play an important role, though such subjects are mainly concerned with practical theories. But some of these theories could be worked out mathematically; that is when numbers come into play. In theoretical physics, e.g., technical symbols incorporated with numbers make the subject highly abstruse, which requires knowledge and experience to tackle. In calculus, in differentiation, the quantity without a root at the end of the equation, e.g., +7, or, –36, is omitted, and the coefficients and roots in the equation are varied in value and power respectively; in any case, the coefficient is increased in value and the root is decreased in power, e.g., $2x^8$ (2 here is the coefficient), on differentiation, becomes $16x^7$ (the coefficient now becomes 16, obtained by multiplying 2, the coefficient, by 8, the power of the root, x, and the root, x, is decreased in power from 8 to 7, by 1); in some cases, the root is omitted, as in, e.g., 5x (x here is the root) which becomes 5. In integration, the effect is the reversed of differentiation; the equation would be lengthened, unlike in differentiation. These are some of the rules in calculus which mathematicians have to abide by.

Differentiation is used to calculate variation in something, e.g., rate of increase, or, decrease, in speed, while integration is used to calculate such things as area and volume.

Symbols which precede numbers, e.g., the "+" symbol which signifies "plus" and the "–" symbol which signifies "minus", are important. The "+" symbol is normally left out, if functionless; e.g., we know when we look at the number "8" that it means "+8" as well. But the "–" symbol would always be denoted. In calculus, in the computation of distance or variance, a "+" sign before a number signifies a positive value and a "–" sign before a number signifies that the number has a negative value. For example, if the distance from A to B, being positive, is 8 miles (or, + 8 miles), the negative value, which is the distance measured from B to A, would be −8 miles, the positive and negative signs in this case symbolize opposite directions of measurement. In vector calculus, e.g., the plus and minus signs indicate opposite directions, such as the directions of fluid flows.

If one looks at the diagram below, one would understand this point:

```
                   I
                   I
  <---------       I   --------->
-I----I----I----I---------I----I----I----I-
-4  -3  -2  -1  I0    1    2    3    4
                   I
                   I
```

Notice the direction of increase of the positive values, 1, 2, 3, 4, and the direction of increase of the negative values, −1, −2, −3, −4. They are opposite, the positive values increase to the right, as is traditional, the negative values increase to the left.

In the diagrams below, A is assumed to be 0 in (1) and B is to the right and is 10 miles (or, +10 miles) away from 0 or A. In (2) B is now assumed to be 0 and A is to the left of B or 0. As any value to the left of 0 is normally regarded as negative, the distance of A from B is to be regarded as −10 miles. In this case, −10 miles are 10 miles, in terms of distance. Only the directions of measurement and starting points of measurement are different. But strictly speaking, i.e., arithmetically speaking, 10 cannot be equal to −10. 10 can only be equal to −10 if the conditions stated above are present.

```
     |---------------------------------------|
(1)  A        ----------------------→        B
                    10 miles
```

```
     |---------------------------------------|
(2)  A        ←----------------------        B
                   -10 miles
```

Numbers and symbols cannot be dissociated from one another; any number is understood to have either a positive or a negative value, hence, a positive or a negative symbol. All numbers which are not negative are understood to have a positive or "+" sign preceding them; it is unnecessary to put the "+" sign before the numbers with positive values unless addition or summation is involved.

The theory of numbers like any other theory evolves from other systems. It should thus be treated in relation to the systems of mathematics evolved by the various great mathematicians such as Isaac Newton, Georg Cantor, Jules Henri Poincare, Georg Friedrich Bernhard Riemann, Niels Henrik Abel, and Evariste Galois.

21 The Logic of Mathematical Equations

What is an equation? $A = B$. $C = y + x$. $S = x^2 + 2yx + y^2$. These are equations. The first two equations are linear equations. The third equation, where x and y each has a power of 2, is a nonlinear equation. Notice the “is equal to” symbol (i.e., =). A, e.g., is still equal to B after an increment in value provided that B’s value is also increased by a margin equal to that of A, e.g., x; hence, $A + x = B + x$. On the other hand, $A - x = B - x$. Notice that when you decrease the value of A you have to decrease the value of B as well by the same margin, x, in order that A in its new form equals B.

Look at this equation, $A = B = C$. Is it not obvious that $A + B = B + C$ is true? Is it not true then that $A + C = 2B$, $A + B = 2C$, $B + C = 2A$? Look at the proof. Substitute for A, B, since B is equal to A, and C, also B, since B is also equal to C. Hence, $B + B = 2B$. Hence, $A + C = 2B$ is true. Equations are meant to be solved. Simple equations can be solved by practically anyone who is acquainted with algebra. It is perhaps the quadratic equations which present the difficulty. The roots (variables or unknowns, typified, e.g., by the symbols x and y) of the equations however can be found by anyone who is acquainted with the methods of finding the roots of equations and who has gone through the woes of solving equations.

Now, equations are of different degrees. There are, e.g., first-degree, second-degree, third-degree, and fourth-degree equations (the degree of an algebraic equation or polynomial is the highest power or sum of powers in any term of the equation, e.g., $x^4 + 2x^2 - x$ and xy^2z are both of the fourth degree). It is possible to solve the first-degree, second-degree, third-degree, and fourth-degree equations, but not considered possible to solve the fifth-degree (quintic) or higher degree equations in general. One might be able to solve any of these equations by finding the square or cube root of the whole equation, or, if two (or more) equations were given, one could multiply one or both of the equations by a certain number or numbers and subtract one equation from the other, or, add both equations, so as to bring about, finally, a less complex equation with only one root (where there is only one symbol, e.g., x, or, y, which is one of the roots of the equation), which could hence normally be solved without much difficulty.

Let us go further. An equation could be expressed, e.g., as “the sum/subtraction of something and something and something is equal to something”, e.g., $x^2 + 2xy - 4y^2 = 6$, where x and y, the roots of the equation, could each be seen to have the power of two. To find the roots or in other words the values of these symbols (i.e., x and y) is the problem of algebraists. In many cases, a few equations are given, the values of the roots of each of these equations must be assumed to be the same. Look at the following equations (we may call them simultaneous equations):

$$x^2 + 4x^2y + y^2 = 8 \tag{1}$$

https://doi.org/10.1515/9783112214367-021

$$2x^2 + 8x^2y + 4y^2 = 30 \tag{2}$$

The way we should go about finding the roots of the above equations is to multiply one of the equations by a certain number or both the equations by two different numbers (each equation by one of the two numbers), and, after that, either subtract one equation from the other or add up both the equations so as to bring about a new equation with only one root (or symbol/variable/unknown) in the form as follows: $2y^2 = 14$ (therefore: $y = \sqrt{14 \div 2} = \sqrt{7}$). Once this reduction of the two equations to a simple equation with only one root has been carried out, the problem of solving the equations is almost complete. Let us carry out this operation here. Let us multiply eq. (1) by 2 ($(x^2 + 4x^2y + y^2 = 8) \times 2 = 2x^2 + 8x^2y + 2y^2 = 16$) to get the following equation:

$$2x^2 + 8x^2y + 2y^2 = 16 \tag{3}$$

Now, subtract eq. (3) above from eq. (2). After this operation we get the following equation with only one root (y):

$$2y^2 = 14 \tag{4}$$

Therefore: $y = \sqrt{14 \div 2} = \sqrt{7}$.

In this way, we obtain the value of the root, y, which is $\sqrt{7}$. After finding the value of one root (in the above case, y) of the equations (eqs. (1) and (2)), the value of the other root (in the above case, x) of the equations (eqs. (1) and (2)) could be found by substituting the root (y, in the above case) in one (anyone) of the two equations (eqs. (1) or (2)) with its newly found value ($\sqrt{7}$, in the above case) and carrying out the necessary computation. If we, e.g., substitute y in eq. (1) with $\sqrt{7}$, we would get the following:

$$x^2 + 4x^2\sqrt{7} + 7 = 8$$

$$x^2 + (4x\sqrt{7})x^2 = 8 - 7 = 1$$

$$x^2(1 + (4x\sqrt{7})) = 1$$

Therefore: $x = \sqrt{1 \div (1 + (4x\sqrt{7}))}$.

Boolean algebra makes use of the theory of sets. The equations consist of sets of symbols (normally in the form of alphabets). Look at the following:

$$A(B + C) = A(C + B)$$

$$A(B + C) = AB + AC$$

$$BA + CA = (B + C)A$$

Notice the inter-positioning of letters.

In any case, one of the axioms concerning such algebra is that the product of two symbols, e.g., A and B, in a certain order is not necessarily equal to the product of the two same symbols in another order. AB is not necessarily equal to BA. In this case it should be borne in mind that A stands for anything, except a number, e.g., an action, such as acting (A being the initial for acting), and B for anything, except a number, e.g., boating. Why is AB not necessarily equal to BA? See the following explanation:

Eating comes before digestion, for we have to take in food before digestion, which is the decomposition of food into simpler and more soluble elements, can occur. It is therefore absurd to say that digestion comes before eating. Let eating = E and digestion = D. So it could be seen that ED = DE is not true, for ED is in the right order as it represents the order that eating must come before digestion, while DE is not so, as it represents the reverse. Such equations could rightly be called equations of orders; one order is not necessarily equal to another. To solve or understand such equations, the symbols should be carefully noted and understood as representing whatever things they represent, and their orders should be realized and understood as either right or wrong – a right order cannot be equal to a wrong order.

In geometry, AD = DA is correct. AD represents the distance between points A and D measured from A and DA represents the same distance measured from D.

The symbols cannot be the symbols representing numbers, for if "A" represents the number "1" and "D" the number "2", then AD = DA is incorrect for then it means 12 = 21 which is obviously not true. The integer, 12, is never quantitatively equal to the integer, 21, for 21 is quantitatively greater than 12 by 9. But if by 12 = 21 one means that 12 is qualitatively equal to 21 then one may be right; 12 may be equal to 21 in that they are comprised of the same integers (1 and 2), i.e., integers quantitatively the same.

Hence, in the study of equations one should find out in what sense something is equal to something. Is it quantitative equality or qualitative equality? Normally in equations the symbols so used stand for numerical values; therefore, quantitative equality is involved. But this is not so in other forms of algebra, such as Boolean Algebra, where symbols may stand for anything apart from numbers. In Boolean Algebra, the symbols may stand for certain classes of things. The axiom is such that the class of an object belonging to a certain class of objects is equal to the class of this certain class of objects. Equations in different branches of mathematics should be differently understood and dealt with. In the new mathematics, which involves the theory of sets, in geometry, and in elementary algebra, the equations should be dealt with accordingly. Thus, we could separate equations into classes or types. Algebraic equations and equations of that branch of mathematics known as New Mathematics, or Boolean algebra (which could be regarded as the algebra of classes), should not be confused, and should not be solved in the same manner.

The study of equations comes under algebra, which is the branch of elementary mathematics that makes use of symbols representing unknown quantities or variables in order to determine their values by the elementary operations of arithmetic. Equations may involve numerical values, properties of objects, or order of events, things. The study of order itself could in a way be regarded as the study of the properties of numbers. For example, the numbers or integers, 1, 2, 3, 4, 5, 6, 7, . . ., in this infinite sequence increase in a certain order known as a consecutive order, wherein each consecutive number is always of a value greater than that of the number just before by 1. The numbers of a sequence can increase, e.g., in an arithmetic order or geometric order. The sequence 3, 6, 9, 12, 15, 18, 21, . . . is infinite whose numbers increase in an arithmetic order with a marginal difference of 3. The marginal differences can be turned into another infinite sequence, 3, 3, 3, 3, 3, 3, ... The sequence 1, 2, 4, 8, 16, 32, 64, . . . is infinite with a geometric order or progression. The marginal differences can be turned into the infinite sequence, 1, 2, 4, 8, 16, 32, ... The numbers of a sequence can also increase without any order or pattern, e.g., 1, 7, 21, 68, 73, 104, 175, . . ., which is an infinite sequence with numbers that increase without any order, whose marginal differences can be turned into another infinite sequence, 6, 14, 47, 5, 31, 71, . . ., whose numbers increase/decrease without any order.

Equations could, hence, be described as the logical aggregation of symbols and numbers in a certain pattern or order. To juggle with these symbols and numbers and so create new equations which are logical is the main task of the mathematician doing work with equations.

22 Logical Deduction

The systematic study of logic first took place two millennia earlier in Greece where Aristotle had developed a system of logic involving the division of concepts into fixed "categories".

Leibniz, one of the greatest mathematicians of all time, who had been fascinated by Aristotle's system of logic, invented a special alphabet whose elements represented not sounds but concepts. A language with such an alphabet makes it possible to find out by symbolic calculation which sentences written in the language were true and what their logical relationships were. As an initial step toward his idea of an alphabet of concepts, he foresaw the necessity to be able to enumerate the various ways of combining such concepts. This resulted in him carrying out a systematic study of the problem of counting complex arrangements of basic elements in his monographs, Habilitationsschrift and Dissertatio de Arte Combinatoria. The subject of his dissertation for a doctorate in law was the use of logic to resolve cases in law thought too hard for resolution by the normal methods, this dissertation having been received with acclaim by the University of Altdorf, which awarded him the doctorate.

He had also designed a calculating machine capable of addition, subtraction, multiplication, and division. He described a machine that was capable of solving algebraic equations in 1674. One year later he made a comparison of logical reasoning with a mechanism. He thought it might be possible to reduce reasoning to a kind of calculation and build a machine that is capable of performing such calculations. But he knew that unaided this goal would be practically impossible, though he had thought and written about it all through his life. He was aware that the special characters used in arithmetic and algebra, the symbols he had introduced for the differential and integral calculus, and the symbols used in astronomy and chemistry all indicated how important a really appropriate symbolism could be. Leibniz called such a system of characters a "characteristic". To him these symbols, unlike the alphabetic symbols that had no meaning, represented a "real characteristic" wherein each symbol represented each definite idea in an appropriate and natural way. He claimed that a "universal characteristic", a system of symbols that was "real" and that encompassed the full scope of human thought, was needed. He consistently emphasized that a system of properly chosen symbols is useful and indispensable for deductive thought, e.g., in arithmetic and algebra.

He thought it was first of all necessary to create an "encyclopedia" or compendium encompassing the whole range of human knowledge. After accomplishing this it would be possible to choose the important underlying notions and to allocate appropriate symbols to each of them. Finally, it would be possible to carry out the manipulations of these symbols, which represented the rules of deduction; these manipulations he called a "calculus ratiocinator", the algebra of logic, which today might be regarded as symbolic logic.

https://doi.org/10.1515/9783112214367-022

Leibniz had proposed an algebra of logic. This algebra specified the rules for manipulating logical concepts in the way that ordinary algebra specifies the rules for manipulating numbers. A special new symbol ⊕ was introduced by him to represent the combining of quite arbitrary pluralities of terms. The idea was to combine two collections of items into a single collection, which contains all of the items in either one. The circle around the plus sign indicates that it is unlike ordinary addition as it is not numbers being added. However, to some degree, the same rules apply for logical concepts as for numbers. But there are also rules that are very different from those for numbers. In Leibniz's Axiom 2, $A \oplus A = A$ – this indicates that combining a plurality of terms with itself would yield nothing new – combining all the items belonging to a given collection with that same collection of items would simply produce that same collection all over again. As for the addition of numbers it would be quite different – $2 + 2 = 4$ and not 2.

Almost a century and a half after Leibniz's death a self-taught British mathematician, George Boole, had also carried out what Leibniz had set out to do. Boole realized that the power of algebra depends on the fact that the symbols representing quantities and operations obeyed a small number of basic rules or principles, which implied that this similar power could be applied to objects and operations of the most diverse kind as long as they obeyed some of these same rules. In his early work, he applied algebraic methods to the objects that mathematicians call "operators", which "operate" on expressions of ordinary algebra to create new expressions. He was especially interested in "differential operators", which involve the differentiation operation of the calculus. These operators were regarded as important because many of the fundamental laws of nature take the form of differential equations or equations involving differential operators. He showed how certain differential equations could be solved by utilizing methods of ordinary algebra applied to differential operators.

Boole realized that what is important in logical reasoning about words such as "people", "alive", or "horses" is the "class" or "collection" of all individuals described by the word in question – the "class" of people, the "class" of living things, or the "class" of horses. He tried to express this kind of reasoning in terms of an algebra of such classes. He used letters to represent classes in the same way that letters had previously been used to represent numbers or operators. If the letters a and b stand for two particular classes, then he wrote ab (in contemporary terminology, ab is called the intersection of a and b) for the class of things that are both in a and in b. Thus, if a stands for "white objects" and b for "horses", then ab stands for "white horses"; further, if we let c stand for "large objects", then cab would represent "large white horses." To Boole this operation applied to classes in some ways similar to the operation of multiplication applied to numbers. But he wondered what bb would represent if b is the class of horses. It must be the class of things that are horses and are also horses; therefore, $bb = b$. He had based his whole system of logic on the fact that when x stands for a class, the equation $xx = x$ would always be true. This led to a further question: In ordinary algebra, where x stands for a number, when would the equation $xx = x$ be true? This equation would be true only when x is 0 or 1 and for no other numbers. Thus the algebra of logic was precisely what ordinary algebra would become if it were restricted to the two values, 0 and 1. For this to make sense we have to look at the rules of multiplication involving 0 and 1, i.e., 0 times any number is 0 and 1 times any number is that very number, which are represented by the following symbols:

$$0x = 0, \quad 1x = x$$

$0x$ would be identical to 0 for every x, if we regard 0 as a class to which nothing belongs (in modern terms, 0 represents the empty set). $1x$ would be identical to x for every x, if 1 contains every object under consideration (1 representing the universe of discourse in modern terms). Since ordinary algebra deals with addition and subtraction besides multiplication, Boole in his algebra of logic had to provide an interpretation for + and –. Therefore, according him, if x and y represent two classes, $x+y$ would represent the class of all things that are to be found either in x or in y (called the union of x and y in modern terms). Thus, if x is the class of horses and y is the class of cows, then $x+y$ is the class consisting of all horses and cows. To him $x-y$ represents the class of things in x that are not in y. Thus, if x represents the class of all people and y represents the class of all children, then $x-y$ represents the class of adults. $1-x$ would represent the class of things not in x, so that:

$$x + (1 - x) = 1.$$

The following shows how Boole's algebra works. We could substitute x^2 for xx, using ordinary algebraic notation. So, his basic rule could be written as $x^2 = x$ or $x - x^2 = 0$. Factoring this equation according to the usual rules of algebra leads to the following:

$$x(1 - x) = 0.$$

It means that: Nothing can both belong and fail to belong to a given class x.

This exciting result helped to convince Boole that he was on the right track as it concurs with Aristotle's principle of contradiction.

The system of logic that Aristotle had introduced deals with inferences of a very special and restricted kind called syllogisms. The following is an example of a valid syllogism:

All dogs bark.
All things that bark are animals.
Therefore, all dogs are animals.

The first two statements are premises while the third statement is the conclusion. In the above syllogism whatever properties are substituted for "dogs", "barking things" and "animals", so long as the two given premises are true, the conclusion would be true as well.

Let us look at the following syllogism and consider how Boole's algebraic methods could easily be used to demonstrate that this syllogism is valid:

All x are y.
All y are z.
Therefore, all x are z.

To say that everything in x also belongs to y is similar to saying that there is nothing that belongs to x but not to y; in other words, $x(1 - y) = 0$, or, $x = xy$. Similarly, the second premise could be written $y = yz$. Using these equations we thus arrive at the following desired conclusion:

$$x = xy = x(yz) = (xy)z = xz.$$

However, not every proposed syllogism is valid. An invalid syllogism, e.g., could be obtained by interchanging the second premise with the conclusion in the earlier example:

All x are y.
All x are z.
Therefore, all y are z.

There is no way this time to use the premises $x = xy$ and $x = yz$ to arrive at the supposed conclusion $y = yz$.

Boole could not agree with the widespread belief that syllogistic reasoning constituted the whole of logic. He indicated that much ordinary reasoning involves what he regarded as secondary propositions, i.e., propositions that express relations between other propositions. This kind of reasoning is not syllogistic.

Let us look at the following propositions (each labeled with a letter):
A = Robert left his check-book at the shop.
B = Robert's check-book was found at the shop.
C = Robert wrote a check at the electronic goods store yesterday afternoon.
D = After writing the check yesterday afternoon, Robert put his check-book in his trouser pocket.
E = Robert has not used his check-book since yesterday afternoon.
F = Robert's check-book is still in his trouser pocket.

Our logical reasoning would proceed as follows:

Premises

If A, then B.
Not B.
C & D.
If C & D & E, then F.
E.

Conclusions

Not A.
F.

Similar to Aristotle's syllogisms, the above represents a valid inference. In order for an inference to be valid, the truth of sentences known as conclusions has to be inferred from the truth of other sentences called premises. According to Boole, the same algebra that worked for classes would also work for such inferences. He used an equation such as $x = 1$ to indicate that the proposition x is true and the equation $x = 0$ to show that x is false. Thus, "not x" could be represented by the equation $x = 0$. "x & y" could be represented by the equation $xy = 1$; x & y is true when x and y are both true. Algebraically, $xy = 1$ if $x = y = 1$, but $xy = 0$ if either $x = 0$ or $y = 0$ or both equal zero. The statement "If x, then y" could be represented by the following equation:

$$x(1-y)=0.$$

In other words:

$$\text{If } x = 1, \text{ then } y = 1.$$

In fact, substituting $x = 1$ in the above equation leads to $1 - y = 0$, which means, $y = 1$.

Using these ideas, the above-mentioned propositions or premises, A to F, could be expressed by the following equations:

$$\begin{aligned} A(1 - B) &= 0, \\ B &= 0, \\ CD &= 1, \\ CDE(1 - F) &= 0, \\ E &= 1. \end{aligned}$$

Substituting the second equation in the first we arrive at $A = 0$, which is the first desired conclusion. Substituting the third and fifth equations in the fourth we arrive at $1 - F = 0$, in other words, $F = 1$, which is the other desired conclusion.

The fact that Boole's algebra could capture the kind of reasoning that takes place informally and implicitly in ordinary human interactions gave rise to the hope that more complex reasoning could also be captured. Mathematics might be regarded as a system of highly complex logical inferences. Thus, a theory of logic might encompass all mathematical reasoning.

The following shows Boole's method of reasoning pertaining to the existence of God:

The premises are as follows:

1st· Something is.
2nd· If something is, either something always was, or the things that now are have risen out of nothing.
3rd· If something is, either it exists in the necessity of its own nature, or it exists by the will of another being.
4th· If it exists in the necessity of its own nature, something always was.
5th· If it exists by the will of another being, then the hypothesis that the things, which now are, have risen out of nothing, is false.

The above propositions could be expressed symbolically as follows:

x = Something is.
y = Something always was.
z = The things that now are have risen out of nothing.
v = It exists in the necessity of its own nature (i.e., the something spoken of above).
w = It exists by the will of another being.

Boole then obtained the following equations from the above premises:

$$1-x=0,$$

$$x\{yz+(1-y)(1-z)\}=0,$$

$$x\{vw+(1-v)(1-w)\}=0,$$

$$v(1-y)=0,$$

$$wz=0.$$

His system of logic included Aristotle's but went far beyond it. However it fell far short of what was required to fulfill Leibniz's dream. As Boole's algebra represents a system of rules for calculating it could in some way be regarded as the calculus ratiocinator that Leibniz had sought.

When Boole was 32 his first revolutionary monograph on logic as a form of mathematics was published. His more polished tract, *The Laws of Thought,* was published seven years later. His great achievement was to show that logical deduction could be developed as a branch of mathematics. Mathematical logic has seen uninterrupted development since Boole's time. Computers today rely on Boolean algebra for their computational work.

In 1879, Gottlob Frege published a booklet of less than 100 pages, which was titled *Begriffsschrift,* and, which had the subtitle *A Formula Language, Modeled Upon That of Arithmetic, for Pure Thought.* This work was probably the most important single work ever written in logic. He had striven for a system of logic that included all of the deductive inferences in mathematical practice. Like Boole who took ordinary algebra as his starting point and used algebraic symbols to represent logical relations, he wanted algebra, like the other parts of mathematics, to be built as a superstructure with his logic as a foundation. To avoid confusion he introduced his own special symbols for logical relationships. Whereas Boole had regarded propositions, which express relations between other propositions as secondary propositions, he saw that the same relations that connected propositions could also be used to analyze the structure of individual propositions. He made these relations the foundation of his logic. This crucial insight, which has gained general acceptance, is the basis of modern logic. For instance, Frege would analyze the sentence "All horses are mammals" using the logical relationship "if . . ., then . . . ":

If x is a horse, then x is a mammal.

Similarly, he would analyze the sentence "Some horses are pure-bred" using the logical relationship " . . . and . . . ":

x is a horse and x is pure-bred.

The letter x has been used differently in these two examples. In the first example what is asserted is true whatever x might be, i.e., for every x. In the second example the assertion is only for some x. In the symbolism that is in current use "for every" is written as "$\forall$" and "for some" is written as "$\exists$". Thus, the two sentences could be written as follows:

$(\forall x)$(if x is a horse, then x is a mammal),
$(\exists x)$(x is a horse and x is pure-bred).

The symbol $\forall$, which is an upside-down A, indicates the word "all" and is known as an universal quantifier. The symbol $\exists$, which is a backwards E, indicates the word "exists" and is called an existential quantifier. Thus, the first sentence could be read as follows:

For all x, if x is a horse, then x is a mammal.

While the second sentence could be read as:

There exists x such that x is a horse and x is pure-bred.

The logical relation "if . . ., then . . . " is symbolized by $\supset$, and the relation " . . . and . . . " is symbolized by $\wedge$. Using these symbols the above sentences become:

$(\forall x)(x \text{ is a horse} \supset x \text{ is a mammal})$,
$(\exists x)(x \text{ is a horse} \wedge x \text{ is pure-bred})$.

They could be abbreviated as follows:

$(\forall x)(\text{horse}(x) \supset \text{mammal}(x))$,
$(\exists\, x)(\text{horse}(x) \wedge \text{pure-bred}(x))$.

On further abbreviation they become:

$(\forall x)(\text{h}(x) \supset \text{m}(x))$,
$(\exists x)(\text{h}(x) \wedge \text{p}(x))$.

Let us revert to the earlier example of Robert and his check-book whereby we have the following premises and conclusions:

Premises

If A, then B.
Not B.
C & D.
If C & D & E, then F.
E.

Conclusions

Not A.
F.

Using the symbol ¬ to represent "not" and the other symbols we have earlier introduced, this now becomes:

$$A \supset B$$
$$\neg B$$
$$C \wedge D$$
$$C \wedge D \wedge E \supset F$$
$$E$$

$$\neg A$$
$$F$$

One last symbol should be introduced here, namely V, which represents " . . . or . . . ". The table below gives a summary of the symbols, which have been introduced:

¬ not . . .
v . . . or . . .
Λ . . . and . . .
⊃ if . . ., then . . .
∀ all
∃ some

Though the following sentence has a logical structure that would be missed by Boole's analysis it would be easy to express it in Frege's logic:

All failing students are either stupid or lazy.
Using the following symbols:

F(*x*) for *x* is a failing student,
S(*x*) for *x* is stupid,
L(*x*) for *x* is lazy,

the sentence could be expressed as:

$$(\forall x)(F(x) \supset S(x) \vee L(x)).$$

It is obvious that Frege was not just developing a mathematical treatment of logic. He was in fact creating a new language and was guided in this by Leibniz's notion of a universal language, which would gain its power from a judicious choice of symbols. The following examples, where *L*(*x*, *y*) represents "*x* loves *y*", show the expressiveness of this language:

Everyone loves someone.	$(\forall x)(\exists y)$ *x* loves *y*	$(\forall x)(\exists y)L(x, y)$
Someone loves everyone.	$(\exists x)(\forall y)$ *x* loves *y*	$(\exists x)(\forall y)L(x, y)$
Everyone is loved by someone.	$(\forall y)(\exists x)$ *x* loves *y*	$(\forall y)(\exists x)L(x, y)$
Someone is loved by everyone.	$(\exists y)(\forall x)$ *x* loves *y*	$(\exists y)(\forall x)L(x, y)$

The following is a further example:

Everyone loves a lover.

This statement could be expressed as follows:

$$(\forall x)(\forall y)[y \text{ is a lover} \supset L(x,y)].$$

If we interpret being a lover as simply loving someone we could replace "y is a lover" with (∃z)*L*(y, z), finally getting the following:

$$(\forall x)(\forall y)[(\exists z)L(y,z) \supset L(x,y)].$$

Boole had used ordinary mathematical methods to develop logic, which he simply considered another branch of mathematics. There is something circular about using logic to develop logic, which was unacceptable to Frege, who wanted to show how all of mathematics could be based on logic. To be convincing he had to find a way to develop his logic without using logic in the process. As a solution he developed his Begriffsschrift as an artificial language with very precise rules of grammar, or, syntax. This made it possible to show logical inferences as strictly mechanical operations, the so-called rules of inference, referring only to the patterns in which symbols are arranged. This was also the first example of a formal artificial language, which was constructed with a precise syntax. We could thus regard the Begriffsschrift as the ancestor of all computer programming languages that are in common use today.

According to Frege's rules of inference if $\Diamond$ and Δ are any two sentences of the Begriffsschrift, then if $\Diamond$ and $(\Diamond \supset \Delta)$ are both asserted, then one is allowed to also assert the sentence Δ. To carry out this operation it is not necessary to understand what $\supset$ means. This rule cannot lead to error as it only enables one to proceed from $\Diamond$ and (If $\Diamond$, then Δ) to Δ. To actually employ the rule it is necessary to match up the individual symbols constituting the sentence $\Diamond$ with symbols in the first part of the longer sentence one by one. In the example of Robert and his check-book we had the following premise:

$$C \wedge D \wedge E \supset F.$$

If we were to assert $C \wedge D \wedge E$, the rule would allow us to also assert one of the desired conclusions, i.e., F. The following shows how the matchup would be like:

$$C \wedge D \wedge E \supset F,$$

$$C \wedge D \wedge E.$$

The logic of Frege has become the standard logic taught to undergraduates in logic courses in mathematics, computer science, and philosophy departments and the basis for a large body of research. It had indirectly led Alan Turing to formulate the concept of an all-purpose computer. His logic was a great advancement over Boole's. It pioneered as an exact system of mathematical logic, which encompassed, at least in principle, all the reasoning normally used by mathematicians. Commencing from some premises in Frege's logic, Frege's rules could be deployed in an attempt to reach a desired conclusion. But if the attempt did not succeed there was no way of finding out whether this was due to a lack of cleverness or persistence or whether the valid conclusion just did not follow from the given premises. This weakness in Frege's rules of logic meant that Leibniz's dream of a calculating machine was still unfulfilled.

So far we have seen the development of logic from the time of Aristotle to Frege. But the rules of logic or algorithms that have been developed have apparently been aimed at numerical computation or arithmetic and reasoning with classes or sets. Logical deduction is of course much more than these. Our present day computers could blindly follow these rules or algorithms for routine numerical computations. Unlike the human mind, they lack the intuition, creativity or the heuristic ability to perform more advanced forms of logical deduction, such as, e.g., solving advanced mathematical problems such as the Bieberbach conjecture and Fermat's last theorem. The human mind could carry out logical deduction with a leap of intuition, which a machine or computer would never be able to. Even the computer, which has beaten a world-class chess-master at chess has to have a chess-playing program or algorithm encompassing the techniques or ideas of another chess-master. The rules developed by Frege, Boole, Leibniz, and Aristotle act as a guide for logical deduction but even as a guide they are apparently far from sufficient. Their rules deal with right or wrong, yes or no answers, whereas in real life situations things requiring logical deduction are often much more complex (and variable too) and do not fit neatly into such classifications. We have to also consider this troublesome aspect of logic known as fuzzy logic. As a model for logical deduction, the author would like to suggest the famous fictional private detective and logician Sherlock Holmes, who had been modeled after a medical doctor, Dr. Joseph Bell, by his creator, Arthur Conan Doyle. Holmes was able to solve all his cases through his great powers of observation or attention to details (form premises) and his superb ability at logical deduction (form conclusions). So, in logical deduction facts are required and these could be obtained through observation, interrogation, and research. Then we have to link the data or facts collected – we have to figure out the logical connection between them – and arrive at a valid logical conclusion. Even if there were a little doubt about the validity of the conclusion we could still proceed to act upon it based on the justifiable understanding that "no matter how improbable the solution might seem it must be the correct one after all else have failed".

Logical deduction is a process that might involve trial and error or trial and elimination. There might be several possible theories or conclusions based on the evidence at hand. You consider each theory carefully, weighing its pros and cons, discarding all the unlikely theories one by one, till only a few likely ones might remain. Then you pick the most likely correct theory or conclusion from amongst these few. (Often, you could not be absolutely certain as to whether the theory or conclusion you think is correct is indeed correct. This is one of the setbacks in life. So, an objective and flexible mind-set is important.)

Another effective method of logical deduction is reasoning by analogies. Here we compare the current situation (premises) with a similar situation (premises) with its result in the past, and if both the situations (or premises) are indeed similar then we adopt the result of this past similar situation as our conclusion and act on it.

Logical deduction should involve objective, unprejudiced thought, sound judgment, patient and careful gathering of facts and formulation of premises, and thoughtfully and carefully arriving at a valid conclusion.

23 The Author's Encounters with Logic

The author's encounters with logic are mostly through the written word or books. You will find that what most people say would seem more logical or sensible when they put them down on paper. This seems to indicate that when a person has to put down ideas on paper, he would have to focus his mind or concentrate more, thus resulting in a more coherent, more logical piece of communication (the written word, in this case), as compared to his spoken communication. The author knows that this is the case with him. However, his face-to-face meetings with people who show great powers of logical reasoning seem relatively few.

The author's first serious meeting with logic per se occurred during his schooldays. He was browsing through a shelf of books and saw a set of books in three volumes or so titled *Encyclopaedia of Philosophy* and became addicted to philosophy after perusing their contents. The volumes opened his eyes, expanded his mind and kept him thinking about things in life. For example, there was this comment that we were born with a blank mind (tabula rusa, which means blank slate, the author guesses), and in the process of life, this blank mind or slate becomes filled with impressions and ideas. This means that everyone was born innocent and picked up evil only along the process of life. This suggests that one's living environment is important in determining one's future in life.

There was also this interesting stuff about Zeno's arrow. An arrow might be flying or zooming toward its target at great speed, but at every point in time, it is stationary. There was also this philosophical wisecrack "It is better to be an unhappy Aristotle (famous Greek philosopher) than a happy fool" and I was forced to ponder whether this is indeed true. The author has also read the famous French philosopher, Jean Jacques Rousseau's tract, *The Social Contract,* which is a thin book, and came away very impressed by its reasonableness or logic. He learnt that our present-day democracies are based on the principles advocated by this tract. He has bought a rather thick, verbose philosophical book written by the British philosopher, David Hume, titled *A Treatise of Human Nature,* but unfortunately he found it dry and obscure and gave up before he could finish it. Hume had stated that the awareness of cause and effect (cause and effect is an important aspect of logic) would not be possible without seeing. The author read Paine's *Common Sense* and quite liked it. He also tried to digest Kant's *Critique of Pure Reason* but found the book challenging to read but very obscure. He has since been wondering what Kant had been driving at about reasoning or logic. He vaguely recalls that its central theme is something about the quality of the mind or brain being responsible for the kind of mathematical ideas that we have. Kant should have had come straight to the point instead of rambling through hundreds of pages of text.

The author likes the *Analects of Confucius* very much. This scholarly work is a social do's and don'ts. It is common-sensible and very easy to digest. Confucius spoke

https://doi.org/10.1515/9783112214367-023

about family units, emperor-subject relations, social behavior, his disciples, wise men, etc. The author likes his advocacy that the best government is the government that governs least. But he also wonders whether it is better to have no government at all. He also read the famous Chinese novel, *Romance of the Three Kingdoms,* and was very impressed by the exploits and intellect of the famous prime minister of that period, Chuko Liang. His cleverness and cunningness remind the author of Sherlock Holmes, the famous fictional British private detective noted for his great intellect and powers of observation. He was very impressed by one episode of Chuko Liang's exploits. At one time, Chuko Liang and a very small group of men in a city were surrounded by their opponents who much outnumbered them. There was no way for Chuko and his men to out-battle their opponents. The only way out was to outwit them. Chuko, with his great intellect, reasoned that his opponents knew that he was a cautious person and would not carry out any rash acts and that if he were seen carrying out a rash act, they would become suspicious that he was up to something and would not have dared to attack him and his men. Relying on this logic, Chuko got his men to fling open the gates of the city and carry on as usual. Then, he went about nonchalantly playing the zither in full view of his enemies. The enemies got worried, just as Chuko had predicted, and dared not carry out the attack. They finally retreated. Such clever people like Chuko and Holmes give the impression that logic and the intellect have a very important role to play in life. Holmes is the author's favorite, and his role model. The author would like to bring up a little of his exploits. He has read all the Holmes stories. There was this case, where the murder victim's dog did not bark at the time of its master's murder and Holmes deduced that the murderer must be someone known to the victim. But couldn't it be the result of the dog having had a bad cold at that time and had then temporarily lost its sense of smell? There was also this situation when Holmes was at the end of his wits. He had considered all the hypotheses earlier but none of them seemed right at all. Now, he was suddenly faced with a new hypothesis, which only seemed partially right. Holmes was in doubt, but he still went ahead to act on this hypothesis because he reasoned that "no matter how improbable this solution seems, it must be the correct one after all else have failed". The author had information that the Holmes books are standard texts on police detection work adopted by the British police force.

The author has also enjoyed reading Russell's philosophical works. He finds that they have been written in clear, easy-to-understand English. The exception is his joint-work with Whitehead, his and Whitehead's magnum opus, *Principia Mathematica,* full of incomprehensible symbols (Are they Greek or Latin symbols? – the author is not sure) and totally out of the author's (and other mathematicians') mental range. His *Principles of Mathematics,* which shares the same theme with Principia, is much more comprehensible. Russell had a liberal philosophical outlook and had been a good example of a philosopher who took an active interest in public affairs. The author remembers he bemoaned the lack of common sense or logic in people, in general. He had been an intermediary in the Cuban Missile Crisis. Russell had been a Wrangler, a high ranker in the mathematics tripos examinations. He was an example of a man of high intellect, who was genuinely interested in the affairs of his fellow men, attempting to influence public affairs with his high intellect, especially through his writings and talks. He was witty too. When asked to define matter, his reply was "Never mind". When he was irked by a famous mathematical genius, whose name was Wiener, who came to England to study mathematics under him, he commented sarcastically "Wiener good for mathematics? Then mathematics must be no good". He was a person who had the courage of his convictions and had even been penalized for his beliefs. He really walked his talk. The author thinks more intellectuals should be like him instead of hiding in their ivory towers. He wrote prolifically and amongst his books are *The Conquest of Happiness, Marriage and Morals, Logic and Knowledge, Why I Am Not a Christian,* and *My Philosophical Development.*

The author has also read Machiavelli's, *The Prince,* which is a treatise on the art of ruling. Machiavelli was also a man of high intellect. To him, the end justifies the means. That means any act of evil is justifiable if it helps to bring about good to all the people. There was a genius, though an eccentric one, called Wittgenstein, who was much admired by Russell and Keynes, the famous British economist. To him, logic depends on the use of language. Wittgenstein had a great influence on Russell. His famous tract, *Tractatus Logico-Philosophicus,* emphasized that language only has meaning as a result of the way it could "picture" the world; in other words, the structure of a meaningful sentence has somehow to "mirror" or represent the way reality is structured. For him, logic was just a linguistic activity, a mere study of the structure of "empty" tautologies. There was also the German philosopher, Hegel, much of whose language and thought had been used by Marx; Hegel originated a dialectic process of reasoning within an idealistic philosophy, whereas Marx accepted the initial ideas of his view of history and fully applied the dialectic to a materialistic philosophy. Hegel thought that all changes take place through antagonisms and struggle but did not believe that violence or revolution by the masses is the desirable way. As for Keynes, he revolutionized economics, he was the Einstein of economics. Against the current train of economic thought (which was: The economy would pick up on its own), he advocated that the government intervenes in the economic affairs of the country, e.g., by creating public works, et al., emphasizing that the economy would never pick up on its own, without governmental intervention. Keynes was a brilliant and quick-thinking man who wrote a number of books on economics and finance and had been a powerful influence in the financial aspect of the British government. He also authored a tome on statistics, which was titled *The Theory of Probability.* His magnum opus was *The General Theory of Employment, Interest and Money.* He was also a very sharp logician, who could out-argue or outsmart most people and he also had the power of quick comprehension. Russell regarded him as the sharpest person he had ever met and said that when he argued with Keynes, he seldom emerged without feeling like a fool. He had wide interests, became rich through prudent investments, and was regarded as a genius by some. He was the scientific brain in economics.

We could see that logic and theories are applied in all areas of life, economics, politics, government, science, and so on. In politics, Gandhi of India, and in his footsteps, Martin Luther King Jr. of the U.S., had preached and practiced passive (nonviolent) resistance in achieving their political goals. Scientists today are searching for a unified field theory that explains the mechanism of the whole universe, and superstrings, which are theoretical string-like sub-atomic particles (the vibrations of these string-like particles representing the different levels of energy), seem to present the best hope for a general field equation that sums up the workings of the whole universe, something which Einstein had tried but failed to achieve. Einstein, of course, had revolutionized physics in the twentieth century with his theory of relativity. His theory of relativity states that nothing could travel faster than the speed of light because any object that approaches the speed of light would contract in length and have an infinite mass, which causes it to reduce speed. Also, anyone approaching the speed of light would experience a slowing down of the passage of time and would not grow old so quickly. His theory also predicted that a person who travels faster than the speed of light could go back in time, i.e., he could go into the past and meet his deceased ancestors. All these are counterintuitive and science-fiction-like. Is time travel really possible? His theory on nuclear energy resulted in the creation of the atomic bomb. Marx had provided the theoretical framework for socialism and communism, social systems where the proletariat, the working-class, dominates, through his magnum opus, Capital. This is in contrast to the capitalist system, which is dominated by the business class.

In psychology, we hear about such abstract terms as the ego, the id, the conscious mind, the unconscious mind, and sublimation, all very abstract and controversial too. Psychology is the science of the mind and since the mind is intangible and abstract, how do you conduct experiments on it? So, psychological theories are difficult to prove. According to Freud, any person's behavior is always influenced by the desire to please the opposite sex. A revolution in psychology involves the introduction of intelligence tests, whose objective, when first introduced, was to spot the mentally defective. Of course, such tests are widely used to spot intelligent and talented people today. The author had tried taking a number of such tests in his younger days as he was curious about his level of intelligence. He did very well in some of them and not so well in others. Such tests also purportedly find out whether a person is capable of logical thinking. These tests have been found to be culturally biased. They claim to measure a person's speed of thought and a number of items have to be completed by the one taking the test within a certain time limit. Familiarity with the test items, of course, helps a person to score. There is the element of luck involved as well. Many of the items require the person to figure out relationships or "patterns", and, unless the person is already familiar with the test items and knows how to go about it straight away, there is going to be some trial and error involved, which could waste a lot or a little of the test-taker's time, depending on his luck – take this example: there are say five possible ways of approaching the problem and the test-taker has to try out all the five ways; a lucky test-taker might hit the solution after trying the first method, while an unlucky one might have to make four false starts before he hits the fifth, which is the correct method, wasting much time as a result. So, intelligence tests are not foolproof.

A tome that almost resembles a Sherlock Holmes story is *The Theory of Evolution,* which had been authored by the naturalist, Charles Darwin. He was purportedly a person of mediocre intelligence and even he himself considered himself so. But *The Theory of Evolution* is a great work of observation of details and logical deduction. It explains how living things change their features or characteristics in adapting to a changing environment, where only the fittest survives. The author has also perused Galton's *Hereditary Genius,* which is a study that shows that ability is in-born. He also read *Discourse on the Method,* which had been authored by the famous French mathematical genius and philosopher, Rene Descartes. Descartes proved his own existence by stating, "I think. Therefore, I am". He proved the existence of God with the following argument, "The concept of perfection exists. God represents the concept of perfection. Therefore, God exists."

The British philosopher, Francis Bacon, had said that reading nourishes the mind as eating nourishes the body. Reading, in the author's opinion, would improve one's ability to analyze things and to think logically. Of course, it also depends on what one reads. If one reads comics or pornography, it is not going to help much. The author has also read John Stuart Mill's *Utilitarianism* and *On Liberty.* Mill had been a child prodigy who later suffered a nervous breakdown. His Utilitarianism philosophy, like Bentham's, is about actions that bring the greatest good to the most people. Schopenhaur, a German, is a philosopher of pessimism. To him, civilization is on the decline. According to Jacques Derrida's deconstruction philosophy, all truths are relative and there are no absolute truths – that is, we live in a world of tentative truths and abstractions – that means logic is tentative.

As you can see, the author likes philosophy. Philosophy is the science of the search for truths and must include logic, as the application of logic is needed in the search for truths. Though philosophy has no direct practical value, having a philosophical outlook or bent in life is important. Being able to think logically is one of the qualities of the philosopher, the other qualities should be objectivity, the love of truth and knowledge, and curiosity. But, a person would never be bothered to search for the truths or be bothered with logic if he is not naturally curious.

Take the author for example. He has been curious about many things. That is why he reads very widely. For example, he is not a mathematician by training but he is very interested in mathematics, spending quite an inordinate amount of time poring over its literature. For, he is one of those who could appreciate the beauty of mathematics and mathematical logic. In fact, he generally appreciates the beauty of ideas, especially clever ideas or ideas with clever or unusual logic. One of these ideas is the physicist's idea that time travel is possible. Is time travel really possible? The author would say that it is not physically possible. But mentally, it is possible – thinking of the past is traveling back to the past, thinking of the future is traveling forward to the future, though our physical self is firmly planted in the present. In fact, time is a thing of the consciousness. Nobody can gift wrap time and give it to another person as a birthday present. As a matter of fact, without the watch or the clock, a person would be pretty lost about time. Time is just an abstraction, an aspect of consciousness, and it is not logical for scientists to speak about it as though it is physically real, as though it is a tangible, solid object – the fact is that time has no physical or tangible reality. That is why, theories of time travel are characterized by paradox and absurdity (one famous paradox is that of a person traveling to the past to the time before he was born to kill his mother, which means that he would never have been born – this upsets the causal effect and is of course absurd); they have more to do with matters of the imagination than real stuffs. Yet, you find scientists spending an inordinate amount of time writing papers and books on time travel. Even Einstein, whose theory of relativity postulated the possibility of time travel, finally had doubts about it.

There is a saying that the universe is mathematical in nature. Mathematics has been used to forecast natural phenomena, and even the economy. This is known as mathematical modeling, whereby with certain mathematical formulae or equations and certain assumptions or inputs, outcomes could be predicted. With equations, it is possible to describe the orbits of planets, the behavior of subatomic particles, the movement of share prices, etc. A scientist has coined the term "the unreasonable effectiveness of mathematics" to describe the surprising practical importance of mathematics, including pure mathematics, which is normally thought to be devoid of practical value. This seems to imply that mathematical logic and the laws of nature have something in common. As a result of this close link between mathematics and physics, which is the study of the laws of nature, there is now a new branch of mathematics known as mathematical physics. In fact, some mathematicians have had become involved in physics, e.g., Godel had communicated with Einstein about time travel, Roger Penrose has immersed himself into the study of black holes, and so on.

Man's existence has always been logic defying. Scientists have tried simulating the creation of life in the laboratory and have found that even given all the right conditions for life to form, the chances were still very very slim for life to form. They could only offer the explanation that there must be a higher Being who created life. In the past, scientists had spoken about the primeval soup, the ingredients, which had helped to bring about life.

As mentioned earlier, the author's first involvement with logic has been through philosophy. Then, he became interested in physics and, later, mathematics. Because physics uses plenty of mathematics, sooner or later, one has to pick up mathematics to understand physics. What is most frightening about mathematics is the symbolism. Mathematicians could create any symbol to represent something and the symbols could be confusing. Even mathematicians often have problems understanding the work of their peers, especially if they are not specialized in that particular field of work, not to mention the nonmathematician. Another reason that mathematics is difficult is the fact that a person normally needs to have a foundation or background in some more basic mathematics before he can understand the more advanced mathematics, as mathematical principles generally involve an interconnected chain of reasoning from the simpler or more basic principles to the more advanced principles, not to mention the jargon or unfamiliar terms some time being used. So, a mathematics book is normally read slowly as compared to reading a novel as the brains require more time to digest its contents. The reason why many give up mathematics might be because they lack the patience to slowly digest the contents of a mathematics book; they might rather prefer to race through a novel. Mathematics might be too slow for them. They might be too fast for mathematics. So, if they do not bother to spend time on mathematics and they do not understand mathematics, they are not stupid. What about those who take the trouble and have the patience to learn mathematics, but cannot understand mathematics, are they stupid? Either they are stupid or the mathematics book is poorly written, or, both. The other difficulty is that mathematics, especially pure mathematics, often deals with concepts that are abstract and have little or no bearing with real-life situations or objects, giving rise to a visualization problem. What is difficult to visualize would be difficult to understand.

The field of mathematics has expanded so much now that no mathematician can expect to master the whole of mathematics, which means that specialization is required. Though mathematics is a challenging subject, many people prefer to take up the more practical subjects such as engineering, accountancy or law. Perhaps, mathematics is best taken up by those who are interested in the rigor and precision of mathematical logic for its own sake. You would notice that mathematicians like using such terms as "QED" (which means "easily demonstrated"), "It is obvious that" or "It is trivial" in their mathematical reasoning, when in fact their mathematical logic could be very difficult to follow and is not obvious. This is the irony of the situation. In fact, the logic could even be wrong, and since no one understands the symbolism, no one knows it is wrong. Worse still, some one might think that the mathematician's solution is apparently correct although his symbolism (and, hence, his logic, which is actually wrong) is incomprehensible because his symbolism looks deep and impressive. Just as bad, this poor guy might feel stupid (even suffer an inferiority complex) because he could not understand the mathematician's stupid logic. In mathematics, as in all arcane subjects, one could easily hide one's ignorance behind an appearance of profundity. The author is not saying that all mathematicians are like that. What he is saying is that a mathematician could choose to be like that and could quite easily get away with it. Of course, there are many really good mathematicians around who would not behave so unprofessionally. The author brings up the case of the mathematician here because he is supposed to talk about logic, and the mathematician is an exemplary practitioner of logic. The same could be applied to the other professions such as the motor mechanic, the doctor, or the electronic engineer, whose fields of work involve knowledge that is arcane – any of them could easily dupe the uninitiated. We should thus be wary about logic, especially the abuse of logic.

Mathematics also involves the study of patterns in numbers as well as in objects. For example, in topology, it might be possible to transform an object into other shapes, with all the transformed objects sharing a common feature with the original object – if that is the case, we could generalize that a point (or feature) of the object would always remain unchanged after transformation. In fractal geometry, we could prove that from all the irregular patterns or shapes gathered together, there would emerge a regular pattern (which implies that symmetry is a common feature of nature) and this new mathematical subject apparently has wide practical applications. (This subject has been pioneered by the French mathematician, Benoit Mandelbrot.)

The well-known British mathematician, George Boole, had authored a book on the principles of logic, where basic logical principles such as the operations of arithmetic had been described; he had also introduced what is called Boolean algebra. There are also the Venn diagrams introduced by John Venn, in an attempt to put logical reasoning on a firmer footing with graphics. The operations of arithmetic and algebra actually represent an algorithm or procedure for reasoning with numbers. But, is an algorithm or procedure for reasoning with abstract, qualitative objects such as cooperation, dissatisfaction, or hunger possible? It would be quite difficult unless one could equate each of these abstract, qualitative objects with a quantity or number, whence one could then resort to the operations of arithmetic and algebra. In *The Philosophy and Principles of Logic*, the author has given two procedures or algorithms for settling disputes or arguments. The same procedures or algorithms could be more or less utilized for reasoning with abstract, qualitative objects, though some modification would be needed. Here, you could imagine yourself disputing or arguing with yourself (instead of disputing or arguing with another party). The algorithm or procedure, after some modification, could be as follows:

1) State the objective(s) of the exercise.
2) Determine and specify the criterion or criteria upon which the right or correct conclusion or decision is to be based. If there is really a problem in this, go to (8).
3) Tell yourself that you have to be objective.
4) Give yourself a reasonable period of time to think about and/or research (get documentary proofs on) the matter.
5) At the end of this period, sit down, collate the documentary proofs if there are any, write down what you think and/or what you found out through your research are the alternatives or options, the pros and cons of each of them, etc.
6) Tally the pros and cons of all the options.
7) The option where the pros exceed the cons by the greatest amount should be the correct option. The conclusion or decision made here is to be in accordance with (2) above.
8) In the case where there is a tie between two or more options, whereby (7) above could not be implemented, you are to seek the opinion or counsel of a third party (or parties) and follow the third party's decision (or the consensus decision of the parties).

The above method represents a more scientific and objective way of arriving at a decision or judgment as compared to just simply turning the matter over in one's mind haphazardly and without a plan of thought.

Logic, right logic or spurious logic, has often been used in propaganda and advertisements to sway people's minds. A classic example is Germany during the Second World War. The chancellor of Germany then, a dictator by the name of Hitler, and his propaganda minister, Goebbels, were known to be great orators who could sway the crowd and cause mass hysteria. The masses apparently believed their spurious logic that the German represents the Master Race and is of course superior to the other races – the German is thus entitled to rule over the other races. With this logic, Germany practically launched itself into a disastrous war with other nations, in what could be regarded as the battle for territorial conquests. Has any one ever wondered what makes a superior race? What is a superior race superior in? Is it intelligence, physical characteristics such as height and appearance, athletic prowess, creative talents, business acumen, moral character, cultural heritage, or a combination of all or some of these qualities? How are these qualities to be judged or measured? Don't you think all these are subjective? Is it sensible to call yourself, or, your race, superior if you are outstanding in a few areas? Is an evil genius thus a superior person? Then, there are some advertisements that convey spurious messages. Some cigarette advertisements apparently convey the message that smoking their brand of cigarettes is a sign of manliness. It might be more a sign of recklessness than manliness, as smoking causes health problems. There are also the liquor advertisements that associate drinking their liquors with maleness. Of course, drinking might result in social problems as when someone got himself drunk, drove, and killed himself or someone in a road accident or got himself into a drunken brawl. Is this what being male is all about? The author has once been amazed by a liquor advertisement that associated its drink with soccer. Of course, heavy drinking would probably end up in a drunk person who could hardly walk straight, let alone being able to deftly dribble the soccer ball past opponents for a fantastic score near the goal line. Imagine what would have happened if our national soccer team were to be heavy drinkers? Propaganda and advertisements convey messages that have some apparent logic, right logic or spurious logic, which is aimed at influencing a person's thoughts. There are of course such media of communication that have noble objectives. For example, there are advertisements by the health authorities that warn members of the public about the hazards of smoking and there are those that plead for donations for the underprivileged.

Logic could be a double-edged sword. It is good to be able to apply the rules of logic and obtain fruitful results. It is good to be able to persuade an important person or a group of people to carry out an action that is worthwhile with fine logic. But displaying too much of such powers just might incur the envy of some small-minded people, who could not stand someone who is apparently smarter than they and these people could go around backstabbing the brilliant guy, causing problems for the poor guy. So, the brilliant guy should have some wisdom in the use of logic and should not be a loose cannon indiscriminately firing logic balls. Sometimes, it is logical to be stupid or act stupid. We should only apply logic and apply it well when the situation really calls for it. We should not overwhelm others so much with our logic that they would feel humiliated and inferior. Sometimes, it would have been smarter, and more logical, to pass our brilliant ideas to others and let them get the credit for them so that we could get or retain their goodwill and cooperation. Remember that the famous sage Confucius had taught us to be humble. So, we should be humble (but we could quietly, only quietly and never ostentatiously, be proud of our humility, which Confucius said is a virtue, as well as our intellect). In fact, logically, if one is really smart and really has something to boast about, one should do so without hesitation if by doing so, one gets rewarded (and not harmed of course). It is only those who boast about having a quality or qualities that are nonexistent who deserve to be chided. In sales and marketing, the salesman unstintingly boasts about his product and his company; he even displays his product knowledge or his technical knowledge. A salesman who talks humbly about his product or company and displays humble product or technical knowledge should be scolded by the customer for trying to sell him something mediocre or inferior, which he is trying to avoid; a salesman would not be a salesman anymore if he chooses to be a humble man; he would end up being a really humble man (whom Confucius would have liked very much) and a poor one too because he would not get any sales and would lose his job. You are your own salesman. You have to sell yourself to others, especially to your boss, who might expect you to be capable and smart. Certainly, you are not going to act humble (because Confucius said that one should be humble) and not show off your strengths in front of your bosses and give the impression that you are only mediocre, if not below average, so that you end up being bypassed for promotions, or, worse, losing your livelihood. Of course, you have to assess the situation carefully first. Some bosses expect you to be as smart as possible and it is alright to them if you are smarter than they. Some expect you to be smart but not as smart or smarter than they. Some do not expect you to be smart at all but be a puppet or yes-man. Thus, a person has to be wise in the exercise of logical brilliance. Otherwise, as they say, he might win the battle but lose the war. There is a poser here. Who do you prefer to be: a person who is not smart but could give the impression that he is smart (has high IQ, illumination quotient, in this case; possibly, a good actor), or, a person who is smart (has high IQ, intelligence quotient in this case) but could only give the impression that he is not smart? Of course, in life, impression and appearance play an important part. In any case, qualities such as intelligence, honesty, and drive are intangible and hard to evaluate, and it takes time to assess such qualities in a person through his past and present achievements and record, a process that might be difficult. The author would think that he prefers to be the first person. What is the use of having great intelligence when that intelligence is not recognized, appreciated, and given a chance to really contribute to one's fellow men?

Differing viewpoints representing differing logic have resulted in social problems, arguments, quarrels, segregation, conflicts, and wars. We may call them ideological differences. They could be religious differences, political differences, differences in outlook in any other thing, or simply personal differences. But the problem is that logic tends to be fuzzy and there are often no clear-cut yes or no answers to things. Relying too much on logic might make us more argumentative, and this is not conducive to social harmony. The best policy is apparently to keep an open mind about things. That is, it is best not to be dogmatic. It is best to have the quality of objectivity. We should place more importance on objectivity than logic. An objective mind is truly above logic and would not fall victim to the inconsistencies and disadvantages of logic. Objectivity should be the one and only true "logic". All other forms of logic appear artificial and "make-believe" – they seem to serve to imprison our mind in "chains", stifling its creativity and originality, not to mention objectivity, in the process. The question is how to attain objectivity, the mental Nirvana. The key to attaining objectivity of the mind is to divorce it from the ego, a no mean feat for an ordinary mortal.

It would be appropriate here to say something about the doyen of logic, Kurt Godel's, ideas. Godel's proof pertaining to the incompleteness of a formal system of logic or the logic of arithmetic is based more on qualitative logic than quantitative – he resorted to applying the principle of "reductio ad absurdum", a qualitative concept, hence the paradox or confusion. What he had done was to have integers within integers (quantities within quantities) or arithmetic within arithmetic, using the counting and comparing operations of arithmetic itself to prove the inconsistency of arithmetic; he was to show that there is paradox within the logic of arithmetic itself – there is a self-contradiction of the form "This statement is unprovable" or "This statement is false." His theorem is as follows:

> "There are universal truth machines (UTMs) that know that the statement, "There are UTMs that know the truth of a premise but cannot tell the truth", is true, but can never tell or utter the truth." This statement is true because if the UTMs say that this statement is true they will contradict themselves, and, therefore, in order not to contradict themselves they have to refrain from making this statement, i.e., keep silent, as the following would demonstrate:
>
> G = UTM will never say G is true.
>
> a) If UTM says G is true, then "UTM will never say G is true" is false.
> b) If "UTM will never say G is true" is false, then G is false (since G = UTM will never say G is true.)
> c) "UTM will never say G is true" is in fact a true sentence.

Finally, though we always try to be logical about many things, we should not forget to be logical about logic.

24 Conclusion

It is not uncommon to find apparently intelligent, logical people disputing over some matters or affairs, some logical points. In fact, it is rather common. This is manifested in the differing schools of thought, philosophies, ideologies in all areas of human affairs such as politics, science and technology, mathematics, religion, medicine, commerce, sports, entertainment, the arts, literature, social matters, legal matters, and other mundane matters. These differences have resulted in social problems, legal disputes, conflicts, violence, and wars; logical controversies have made many people feel frustrated, outraged, or angered, with some even feeling that their intelligence has been insulted, resulting in arguments, quarrels, fights, violence, loss of friendship, enmity, injuries, and even loss of lives. The problem is that each conflicting or opposing party firmly thinks or believes that his view is correct. Each has been dogmatic, or, to put it more bluntly, stubborn.

Now, wouldn't it be ideal if the rules, principles, or sureness of logic were as indisputable as black is from white? Unfortunately, this is far from being the case. If the reader has gone through all the preceding materials in this book, he would have realized by now that logic is far from perfect and laden with paradoxes, "multiple standards," and circularity. As logic is dependent on words and symbols and words and symbols might represent different subtle shades of meaning to different people and are thus not perfect, it (logic) would tend to appear differently to different people. As the philosophical genius Wittgenstein had posited, logic could be regarded as a purely linguistic activity, which means that the weaknesses in the language that we use could result in weak logic. So, in our logical deductions we should pay special attention to the language or symbols that we use. We should ensure clarity of language or symbols (no double or multiple meanings or interpretations, no nuances, no undertones, no emotions). In this respect, the "language" of numbers would be fine, as Godel had recommended; for example, we could encode our logical sentences (reasoning or argument) in numbers, as suggested by Godel. Moreover, logic could be difficult to understand, especially if the argument is a very long one (involving many sentences and unfamiliar terms). In fact, it appears that some logicians would prefer their reasoning or argument to be that way, so that they would look profound or intelligent – their view is that they would appear more superior or intelligent if fewer people understood what they had to say – in mathematical terms, their reasoning or argument would have "depth" (instead of being "trivial").

But, in many cases the same lengthy and obscure reasoning could be made clear and obvious by the expediency of a more careful and selective use of language and the leaving out of technical or unfamiliar terms and jargon, by the use of plain, simple language, by the sincere effort of the expositor to make himself as clearly understood as possible.

https://doi.org/10.1515/9783112214367-024

The author is certain that any idea or reasoning, no matter how difficult or complex, could be understood by an ordinary person if the expositor knows how to simplify and explain that idea – that is, to achieve this communication the expositor should be a good teacher or instructor, and not everyone could be a good teacher. The author could vouch for this from his own experience. For example, after perusing a complex mathematical reasoning, the author might have some doubts or questions as to certain terms, meanings, or implications. But, through some coincidence the author subsequently chanced upon another text on the similar subject whereby the expositor expresses himself in simpler terms, even supplying analogies to make himself better understood, and showing some crucial steps (which are missing in the above-mentioned mathematical reasoning), and (voila!) the logic or reasoning suddenly became clear to the author.

Thus, one of the means of avoiding logical inconsistencies (or, logical disputes) is to use the simplest and clearest language possible and be sincere in making oneself really understood by the other party, as well as being sincere in trying to understand what the other party is really saying. In other words, one should be a clear expositor or good teacher and a good listener.

As stated earlier on, if a mathematician or logician were asked to define what constitutes a correct mathematical proof he would probably be stumped, though he might say that he is sure that the mathematical proof is valid. Therefore, there are no hard and fast rules where logic (in this case mathematical logic) is concerned. Mathematical research papers submitted to a mathematical journal are sent for peer review by at least one, or, more specialists in the respective areas of research. If the reviewer or reviewers recommend publication then the paper would be published in that journal. After publication, if there are no objections to the paper by the mathematical community and if there are favorable comments on it by the established people in the field, then the ideas in the papers are considered correct. In other words, the logic of a mathematical paper is considered correct by consensus. Likewise, in many courts of law the validity of legal arguments is determined by the consensus of the members of the jury. Similarly, unsettled disputes between two parties could be settled more expeditiously and cheaply through arbitration or mediation, whereby a neutral third party would hear both sides and recommend his decision. Of course, it would be fine if all the powers that be throughout the world could gather together and draw up the logical principles or rules that universally govern all human affairs, legislate these rules (whereby they become laws), and ensure that everyone adheres to them when conducting their affairs, be it politics, commerce, religion, or other mundane matters. Today, every country has laws governing moral behavior. When could we have laws governing logical behavior? Perhaps, to start with, a law could be passed to the effect that all serious disputes be referred to mediation or arbitration by qualified, professionally trained mediators. This measure would certainly greatly reduce human conflicts or conflicts due to logical differences.

There would probably be disagreement as to what logic simply is, as to the definition of logic or what constitutes logic. How could we expect there to be easy agreement on logical arguments especially the more complex logical arguments? Logic also appears to be relative; as stated earlier an action could be regarded as logical under one set of circumstances and illogical under a different set of circumstances. That is, logic could be fuzzy. This further complicates the matter of logical deduction. Come to think of it practically all the problems in our society seem to boil down to differences in logical concepts (ideological differences). Having logical (or, cultural) differences is one problem, but the greater problem seems to be that people are generally quite inflexible in their attitude toward logical arguments – they tend to be stubborn and dogmatic in their thought and to be egotistical – many even seem to think they are always right, and this is a "most dangerous" attitude. As logic is a function of the quality of our brains, and, if by some stroke of fate, our brains had been "hardwired" differently, then our logic, if it would exist at all in this instance, could be of a different sort. The author would not be surprised that a vastly superior mind due to the much superior quality of the brains would be able to arrive at the truths or know the truths without the need for logical deduction or reasoning. In any case, logic is dependent on intuition or what we call the sixth sense. The much superior, differently constructed brains could give rise to a mind that is so exceedingly sharp that it is able to know or see the truths instantaneously (without having to waste time "thinking about it") – the intuitive power of this mind could be so great that it enables it to perform such a feat (by a great leap of intuition). What is more, the truths perceived by this relatively much more powerful mind could be different from those perceived by our more limited minds. If this is so how could anyone tell whether this "super-genius's" perceived truths or our ordinary mortal's perceived truths are the correct truths? Could there be two different types of truths, which all might seem contradictory? Here, what would be the criterion or criteria for truths? As mentioned earlier, logic appears to be relative; this implies that truths, which contain logic, could be regarded as relative – this means all truths could be regarded as tentative (and not absolute, as posited by Derrida in his deconstruction philosophy). Here, the author would like to define the validity of a truth or logic as the accuracy with which it enables the person to realize his forecast, prediction, or, conclusion; the more accurate the forecast or prediction achieved the better the truth or logic (relativity of truth or logic). We should always be aware of the limitations of our minds, e.g., our limited memory, our frequent slowness of comprehension, and our frequent mental fatigue. Therefore, we should always be mindful of the logic and truths that we espouse.

Considering all these would it not be logical or reasonable to have an objective, realistic, fair and open mind, good judgment, and carefulness when dealing with matters involving logic? Arguing with someone over some finer points in logic till the cows come home would thus not be such a sensible thing to do. Hence, when it comes to logical disputations, which could be expected, knowing the nature of logic, there should be "give and take", or compromise, by the parties concerned; the parties should not be so dogmatic or firm about their respective stands.

Thus, if society as a whole adopts the stance or attitude toward logic or truth propositioned here, harmony and peace could be expected to prevail.

25 Epilogue: A Constructive and Important Proposal for Logic

Now that the author has taken the reader through an "eagle's eye" view of logic, he feels impelled to show the reader the "way" to utilize logic for the maximum possible result in the most efficient (harmonious and conflict-free) manner possible, for he does not want to be guilty of not offering a more constructive position on the subject.

What is striking is that if humanity were so intelligent (and so implicitly logical), why do we end up having so much troubles in our lives – squabbles, disputes, controversies, violence, terrorism, wars – bad news? Why are apparently highly intelligent human beings so bent on abusing and destroying one another, creating some kind of living hell on earth, while on the other hand they, e.g., the Christians, pray to the Lord Almighty to forgive their sins and give them a chance to enter the paradise of Heaven in the afterlife (creating hell on earth but hoping for heaven in the afterlife – a contradiction)? Why not use their intelligence or apply their logical capacity to turn Earth into Heaven instead? It might seem as though we are living in a fool's paradise instead. Indeed, human beings have often made fools of themselves, whether they realize it or not. For example, they enact laws to make violent acts such as causing injury and death illegal but conducting wars, which could result in lots of injuries and deaths, including that of innocent people, is not illegitimate, and, rather rampant too. They abhor dictatorships or autocracies and generally wish to see dictators or autocrats destroyed, favoring the noble principles of democracy instead, but would often turn out to be virtual dictators to those who work under their authority. They consider greed in human beings an undesirable trait but preach capitalism, whose working is based on the profit motive, or, the greed for gain or money. Even the education systems, whose noble objective is to produce intellectually and morally sound people, have often ended up with graduands who are only concerned with scoring A's in their examinations and being able to secure good jobs after graduation, with hardly any concern about the true quality of their minds. Gangsterism is considered undesirable and illegal, but politicians such as members of parliament, congressmen and senators, who are legislators or lawmakers, join political parties or "gangs". (Indeed, one may consider them legalized gangsters.) If one looks around some more, it is possible to witness more of such logical contradictions by apparently highly intelligent or highly logical people. What is wrong? What is the cause of the "lapses" in logic in such instances? Is it the result of twisted logic at work? Is it the human brain that is faulty? (As a matter of fact, a psychologist had claimed that the less intelligent a person is, the more likely it is that he would contradict himself. In the light of the above-mentioned examples, does it mean that there is a serious limitation in the general intelligence of human beings?) Or, is it logic itself that is faulty?

Despite all the contradictions, there is some logic in all the above-mentioned "madnesses". You see, violence in any form, absolute powers of rulers, illiteracy,

https://doi.org/10.1515/9783112214367-025

greed, and forming cliques or gangs are not considered good for society or civilization as a whole and the powers that be, the governments or the rulers who have to take charge of their respective constituencies, pass laws or legislate accordingly, and put a cap on such undesirable elements to facilitate their governing, but they do exempt themselves from the same rules if it suits them for doing so. (As a matter of fact, some top government leaders have been known to be immune to certain laws.) So, there it is – the logic of governing and controlling and the logic of self-preservation or self-interest. What must be disturbing now is the degree of violence perpetrated by criminals, terrorists, and war mongers. It only takes one aggressive, war-like country (person) to start a war (fight). The other country (person) might want peace very badly but would have no choice but to fight back and defend itself (himself), if attacked. Great deterrence for wars (or fights) is required – this is a logical measure. The question is what kind of deterrence. Logical deduction suggests that the troubles encountered in our sophisticated society reflect a defective social system or defective government. Brilliant scientists and engineers have given us fantastic gadgets and fantastic inventions. But is that all there is to it? A sociologist had once said that what is now most needed is the invention of a new social system that ensures that there are peace, harmony, fair and equal distribution of wealth, tolerance, happiness, satisfaction, and other utopian characteristics. How do we apply logic and intelligence now to get society to change from one where tension, terrorism, and violence are the order of the day to one where there is really harmony and graciousness? The author would like to challenge all logical, sensible people to think about it and try or help to invent such a social system, or, at least, try to help bring about such a harmonious social environment, e.g., by personal example.

To the author at least, our squabbles and conflicts are the result of our apparently low "tolerance level". Most people seem unable to tolerate opposing viewpoints and like to surround themselves with yes-men, people who are mirror images of themselves, so to speak. This indicates a sense of insecurity and distrust in most people. This is probably the fault of education. Students have been generally trained to memorize textbook facts and regurgitate them in examinations. When have they ever been encouraged or taught to question their textbooks or the logic of their textbook facts? Also, in examinations, they are expected to present their answers in a "certain manner or format" if they are to do well, i.e., there is only one expected answer to the question, one "logic". (This had been the case of the imperial examinations in ancient China, which had encouraged rote learning of prescribed texts and which had helped produce generations of government bureaucrats who had been generally subservient to the monarch and intolerant of unconventional ideas.) However, the author thinks that this situation can be remedied. Students should be encouraged to read very widely, apart from their recommended textbooks, even other textbooks (better still, do away with their recommended textbooks and let them select their own textbooks, subject to the given syllabi), whereby they would be exposed to other viewpoints, other "logics". Students should be taught that there could be more than one right answer for many aspects of life, so that they would become more tolerant of opposing viewpoints.

Now, many people regard mathematics as the most exact, most objective science. Is this really true? Nay! There are at least three schools of thought in mathematics, namely, the intuitionist school, the formalist school, and the logicist school – each of these three schools views mathematical reasoning or logic, or, mathematical proof, differently. As mentioned earlier on, the symbols of mathematics are just convenient tools adopted by mathematicians to convey mathematical ideas; the mathematical ideas could just as well be conveyed with plain English words or the words of any other languages, though this might prove more cumbersome than using mathematical symbols, but this would certainly not detract from the rigor of the logic of the mathematical ideas. In any case, compared to the other sciences, mathematics is generally regarded as objective, though mathematical ideas have not been entirely free of controversy. Why is this the case? The author fathoms that this is due to the fact that mathematics relies on the minimal use of words. Words could convey feelings and emotions (as in poems) – you could use strong words to convey strong feelings and emotions or tone down your words accordingly, but you could never use mathematical symbols (and numbers, the mathematical operations of which such as addition, subtraction, multiplication and division are definitely objective and indubitable) to convey any feelings or emotions. Anything that conveys feelings and emotions, in this case, words, implies subjectivity. Therefore, mathematics, being a subject that involves the minimal use of words, becomes by default a relatively less subjective (or, more objective) science. The author would like to point out that the rigor of the logic of a logical economical idea, a logical psychological idea or a logical political idea is no less than the rigor of the logic of a logical mathematical idea. Yet, people generally view mathematics as a more superior or more rigorous science as compared to the other sciences. The symbols used by mathematicians represent the "short-hand" language of the mathematician that is comparable to any arcane foreign language. However, there is a possibility that mathematics, or mathematical ideas as such, are more difficult to grasp than the social sciences such as economics or psychology, because they seem to be more divorced from our everyday experiences, as compared to the social sciences. It should now be clear why the use of logic in human affairs is so beset with problems, especially if it is in the form of emotional words. Apparently, different "logics" could be found to be at work and they could be found to "battle" for supremacy. Controversies appear inherent in logic. Hence, conflicts.

How to remedy this dire condition? Conflicts have been looked upon with abhorrence and disgust, but how could they be avoided? The author has an important proposal here. Since there are so many schools of thought or philosophies, so many standards (especially moral standards, wherein complaints of "double standards" are common), so many systems of doing things, with the perpetrators of these each respectively considering their own thing better than their opponents', how could there not be conflicts? The obvious solution to this dilemma is to have only one philosophy, standard, or system. But how could this be achieved? It is generally accepted that there are always pros and cons, advantages and disadvantages, good and bad, in every situation. The most logical thing to carry out here is thus to, as far as possible, adopt the best parts of the various schools of thought, standards, or systems to create only one school of thought, standard, or system, and subject it to future modifications, should better ideas arise. (Good, impartial arbitrators, negotiators, or, discussion leaders or conference chairpersons would thus be needed.) The author is aware that such a project is not going to be easy to implement as egos are involved and logic itself is manifestly imperfect and elusive. However, it is better to make an attempt to implement it than to let our conflicts simmer, persist and, worse, deteriorate to the point-of-no-return into violent ends and destruction.

Let us have a peek at a few scenarios pertaining to this important standardization. First, take the case of mathematics, a science respected for its rigor and objectivity. The community of mathematicians could try to eradicate the divisiveness of the science through the formation of a common or standard school of mathematical thought, which incorporates the best from the intuitionist school, the formalist school and the logicist school (subject to future modifications). In this way, we could expect mathematicians of these different hues to have their disparities diminished or eradicated by default. Now, no intuitionist, formalist, or logicist could insist that theirs is the better way of doing mathematics – their different ways of doing mathematics are likely to be right or good in some respects and not so right or good in others – combining the strengths of the three different schools gives mathematics the best that could be obtained, as compared to just the strengths of any one of the schools.

Next, we look at the sensitive issue of religions – Christianity, Islam, Hinduism, Buddhism, Taoism, Shintoism, Judaism, Zoroastrianism, and so on. Religion has always been one of the chief causes of divisiveness in society. It would be ideal to have the best facets of all the world's religions combined together as a whole as one common religion for the spiritual and moral enlightenment of all humanity – and thereby be rid of all the troublesome holy wars, jihads, and what-have-you once and for all.

Then, there is the serious divisiveness caused by political differences. We have politicians proclaiming themselves democrats, communists, socialists, autocrats, monarchists, rightists, leftists, moderates, and so on. Political differences could lead to serious consequences such as rebellions, coups, revolutions, civil wars, and bloodshed. Why not have an amalgamation of the best political principles and aim for one standard political system, combining the best of the many political systems, and, avoid any possibly internecine political clashes as a result?

All this may appear idealistic and difficult, if not impossible, to achieve. However, without ideals, aims, or plans, would not society be heading for more trouble? Humanity should aim at minimizing diversification of ideas and logical concepts. This diversity has already led to overly much conflicts in our lives. There should now be a push for the curtailment of this diversity and an increase in standardization and unification. To achieve this, it is imperative for us to cultivate the qualities of objectivity, graciousness, tolerance of opposing viewpoints, and a "give-and-take" mind-set – as stated earlier, education has an important role to play in this. Humanity should try to understand each other's cultures better through wide reading, travels, and discussions, with emphasis on wide reading. Food nourishes the body. However, reading nourishes the mind. (So claimed the British philosopher, Francis Bacon.)

Though some might be skeptical of the practicality of such an ambitious plan proposed here, these people ought to realize that at least someone has the gumption to try to find a way out of this "dilemma of logic" – the situation where people do not think alike, where there appears to be different "logics" at work. If we were to give up hope because the problems seem insurmountable, then we could expect the worst. Earlier on, the author has stated that a dogmatic, stubborn attitude toward ideas and logic could result in conflicts. If people with differing ideas were objective, flexible, and open-minded enough to consider each other's ideas and seek the middle ground and compromise rather than say, "I am right and you are wrong!", then a large proportion of our conflicts would have been eliminated or reduced. In fact, this very attempt at discussion, negotiation, and compromise is part of the solution – for, the parties involved to be willing to sit down and discuss is already part of the problem solved. (Let us hope that none of the parties would stop doing this and become inflexible or stubborn. If they could not settle their differences through discussions, they could always resort to arbitration later. There is no need to quarrel or fight.) If the leaders of society themselves such as educators, trainers, motivators, opinion leaders, and government leaders could adopt this enlightened, liberal intellectual attitude and coerce their followers to do likewise, there would be progress toward a more harmonious, graceful society.

The educational system has to foster such a mindset, together with other influential social institutions such as the media and the government. The leaders of society and the powers that be should initiate and sustain such programs of mental cultivation for the common good and harmony, setting examples themselves, instead of giving leeway to disputes, disharmony, violence, and wars in society, and instead of letting the status quo continue. For instance, enlightened political or government leaders could set a good example by allowing opposition leaders to participate in the government instead of ostracizing them.

An initiation or campaign on harmonious interaction could be carried out by somebody with some influence or respect. This person or group of persons might be political leaders, professionals, authors, film stars, and so on. – in fact, anyone whose shining example would influence others to follow suit.

We should not let logic, i.e., differing "logics", make casualties or victims out of us. We should ride this "tiger" safely and efficiently. It could be supposed that the world is now generally a "mess" because of the "failure of logic" – because there are many ideologies and areas of conflict. Why allow this status quo to persist when it is clearly a cause for concern? We should find a way out of this conundrum, one of which has been mentioned above. There are no absolute rights or wrongs in many aspects of life. So, why be egotistical and always insist that one's ideas are right or better, instead of giving other viewpoints or ideas a fair hearing and adopting them whenever they are worthy? There should always be intellectual honesty in our dealings with our fellow men. Otherwise, we could end up as dishonest people with crooked logic. Besides, not being able to appreciate the good ideas and fine logic of others is tantamount to being stupid.

To the author, logic is also concerned with thinking about possibilities. In fact, in life, there are many possibilities, and to be able to realize all the potentials or possibilities requires some mental discipline, creativity, and flexibility. One should never just jump to a conclusion and never budge from it. One should be more cautious than that and keep wondering whether there are other possibilities, other equally valid or better conclusions.

There must be an aim to logic. If the application of logic helps to achieve what one aims to achieve with it, then one could regard it as valid; otherwise, the validity of the logic should be doubtful.

It is very important for logic to be used wisely, realistically, with caution and care, and, without prejudice, in order to achieve the maximum possible result and avoid unnecessary conflict or friction.

However, life is not always a bed of roses and often, we have to deal with people who are apparently unreasonable or illogical? The author himself has had many such encounters before and they have indeed been very frustrating experiences. Indeed, trying to talk sense with such people is worse than talking to the wall. It appears that such people have an attitudinal problem. It also appears that our society is now so beset with problems and conflicts because of the presence of such people, especially those in influential positions or positions of authority and power. The situation seems dreadful and intractable. But, as mentioned earlier, we should never give up hope, for to do so, is to allow the worst to happen. Something has to be done to remedy this adverse situation. The author has been mulling over the ideas in this book for many years and has taken great pains to write this book, hoping that the ideas could be passed on and help bring about the changes for the better. The author here entreats readers to think of and suggest ways of dealing harmoniously and effectively with such irrational people. To the author, such people seem to have an attitudinal problem, but this might not be the case. There might be some defect in their intellectual equipment or brains, and, if this is the case, the situation is indeed serious. But not to do anything to change the status quo is to allow the worst to take place. We all have to do something. For the author, he would like to present copies of this book to these irrational people and persuade, convince, or motivate them to go through the book. As mentioned earlier, education has to play a role here. Such irrational people are possibly a reflection of the failure or inefficacy of the educational system. (A revamp of the educational system could be in order here.) Suitable re-education for such people through formal training or informally through publicity campaigns might help, but its implementation might pose a problem. Nevertheless, we should try and must not give up hope. Another possibility is to use the "logic" of fear, rewards, and punishments (motivation) in dealing with these irrational people. This is also not easy and is tricky.

Having said so much about logic and its impact on human affairs, the author would like to propose that society finds a more foolproof way of selecting the most logical, ethical, and sane persons for leadership and responsible positions, for having gone through the educational system and attaining the highest academic degrees is no guarantee that the candidate has these desirable qualities, as history has proven, especially if the educational system is flawed.

It is bad if members of our society are hardly or not aware of the possibility of their leaders lacking in logical capacity, ethical quality, and soundness of mind, despite their high academic qualifications and professional achievements, and, it is apparent that quite often, society elects its leaders more with its heart than its brains, as the elections of professional wrestlers and actors into positions of authority, where they subsequently fumbled or turned out to be ineffective, have proven. The least that the author could do here is to make the readers aware of this dangerous possibility and to exhort them to be cautious when choosing their leaders.

The author feels that this book would have achieved something important if it could get its readers to treat logic fairly, realistically, and effectively, if it could get them to have the right mindset, if it could get them to expand their mind, which all could be regarded as the preconditions for harmony and progress. He hopes that the readers would be good examples for others to follow.

It is ardently hoped that through this book something constructive and positive would be achieved.

Part Two

1 What if Ideas, the Logic of Ideas, Are Such That . . . ? Further Thoughts . . .

Ideas, Colors, and Quantities

When it comes to ideas, there are always "gray" areas and it is often difficult to tell whether an idea is logical or not. Inevitably, this often leads to argument or dispute. For instance, someone might argue that communism is good for society and somebody else might disagree, and, such disputes or arguments are not uncommon in everyday affairs.

On the other hand, people apparently do not argue over or dispute the colors they perceive. Red would always be perceived as red, blue as blue, green as green, and so on. If this were not the case could you imagine the resultant chaos? Imagine yourself at a traffic light junction, for example. The traffic light just turned red. But a driver perceives the red to be green and continues to drive on. You would soon hear a loud "Bump!" and witness a vehicle smashup. Unless the driver is really color-blind, this kind of incident does not normally happen.

The same applies to numbers and other quantities such as weights and lengths. Three would always be three, four would be four, five would be five, and so on. Have you ever encountered the situation whereby the quantity "three" is being construed as seven by someone, twelve by another, ten by yet another, and nine by someone else, for example? Ten kilograms would always be 10 kilograms. Twenty meters would be 20 meters. And so on. Nobody disputes these ideas.

If only other ideas were as "exact" as colors and quantities. Why are the other ideas, especially the more abstract ones, not endowed with this desirable quality, having a tendency to give rise to disputes instead? Would the world not otherwise be a more harmonious and happier place, instead of being torn by differences and conflicts?

Or, is it the brains that conceive those ideas that are at fault, that are defective? Why does nature endow human beings with such seemingly defective brains?

Some apologists might argue that that is nature's way of maintaining the "ecological balance" in the human world, nature's way of limiting the growth of human population, by self-attrition such as quarrels, skirmishes, fights, and wars. How do they know? Are they sure that that is the design of nature? Is not birth control the more sensible answer to the problem of overpopulation? Why does nature endow human beings with the ability to procreate and prevent their population from growing too much by creating some self-attrition process for them, making them suffer as a result? Is nature that wicked? Is it not more sensible for nature to prevent overpopulation by endowing human beings with a more limited ability to procreate? There goes the danger of speculation not based on irrefutable logic. There goes the failure of logic, the controversy, which could result in conflicts galore.

https://doi.org/10.1515/9783112214367-026

Perhaps, it is not as simple as that. Unlike colors and quantities (which could also be considered as ideas, e.g., the idea of "red" or the idea of the number "3"), ideas are generally abstract, i.e., they are not seeable, measurable, countable, or touchable (capable of being physically checked or affirmed); they are not tangible. What one could not see, measure, count, or touch (could not physically check or affirm) could hardly be regarded as real or true. In this respect, ideas could not be regarded as real or true if they are devoid of manifestations, which are within the reach of our physical senses, e.g., sight or touch – we could only regard the ideas as real or true if the ideas cause, give rise to, or result in a physical phenomenon or phenomena, which could be witnessed or felt by any of our physical senses (that is, sight, hearing, smell, taste, and touch) – in other words, ideas, the logic of ideas, could only be confirmed as realities or truths through their manifestation in some physical or tangible phenomenon or phenomena, especially if these phenomena are predicted or forecasted outcomes.

Otherwise, ideas and their logic could at most be regarded as hypotheses or conjectures; unless the ideas or logic could be supported by proofs or evidence, especially physical, tangible evidence, as described above, they could not be regarded as real or true. In other words, ideas and logic, to be valid, real or true, should have the quality of "cause and effect" or "causal effect" – an idea or logical relation that causes something to happen could be regarded as valid, real, or true, especially if this something that happened is tangible (obvious and indisputable). That is, logic should produce results. If not, ideas or logic are no better than imagined objects, make-believe or fiction.

An interesting question here. If you were to play the part of God and remake human beings, how would you make the human brain and shape human thought? Couldn't the laws of human thought be as immutable as the laws of nature such as the law of gravity, the laws of thermodynamics, or the laws governing planetary motions, for example? How do we "create" a logic, or, rule of thought or reasoning that would never present any problem? Is it possible to create or invent a universal set of equations for all our practical logical reasoning or deduction, a universal set of laws for practical reasoning, which could be universally accepted, as surely as anyone would accept that one plus one equals two? It could even be something that is akin to a law or moral principle, which has to be abided by all law-abiding, moral people, the citizens of the world, the whole human race. Leibniz had created such a "calculus" of reasoning but the result fell short, in fact, very much short.

Perhaps, our everyday languages for communication purposes are not the right media for the execution of exact, rigorous logical deduction. Our languages are known to convey multiple meanings, nuances, emotional undertones, innuendoes, vagueness, and obscurity at various times. If we could use a language of logical deduction that is as precise and exact as numbers or arithmetic that would be fine. We could possibly use symbols or even numbers or codes to represent our ideas instead of words and play around with or manipulate these symbols, numbers, or codes instead of words in our logical reasoning, for example (as mathematicians have been doing so for mathematical ideas) – we could avoid the use of words as much as possible, if possible, be completely free of its usage, in order to achieve rigorous logical deduction, if reasoning with words, especially imprecise words, causes problems. We could develop or invent a specialized, universal language (just as Newton and Leibniz had invented the calculus) with the intent of achieving rigorous and precise reasoning with ideas, general ideas, any ideas, and not just mathematical ideas or ideas concerning quantities, dimensions, or other measures. We could call this language Logicguage (the language of logic) or by any other appropriate name. It might be something as symbolic as a computer language, something cold, hard, and unemotional, or, better. Perhaps, because of the abstruseness, tentative nature, ephemeral quality, and inexactness of ideas in general (with the exception of ideas concerning numbers and objects which could be counted, measured or weighed, which could be physically affirmed or checked), the realization of such a language might not be possible at all. Otherwise, by now some logicians or linguists would probably have invented such a language. (This does not seem to have been for want of trying. Mathematicians and logicians have, e.g., invented the predicate calculus, which is a language that could be used to analyze sentences containing quantifiers; the propositional calculus, which was designed to analyze arguments whose only logical constants are truth-functional connectives such as "&" (and) and "⊃" (implication) but the language is not refined enough to capture all those arguments that we intuitively recognize as valid; modal logic, a language designed for analyzing more complex types of argument; proof theory, a branch of logic used for the study of formal languages independently of any content the symbols might have but there is no way of finding out whether or not a rule of inference would allow us to pass from true premises to a false conclusion; and, model theory, a branch of logic, which is concerned with the interpretations of formal languages and is used to assess the adequacy of a formal language as a tool for reasoning. Hence, the importance of a formal language to logical reasoning.) Perhaps, we could aim for a language used specifically for logical deduction that is as close to the above-described ideal language as could be. What should the qualities of such a language that facilitates precision of thought and logical reasoning be? The author thinks it should at least have the following qualities:

1) The words or symbols should be precise and clear in meaning, i.e., there should be no other meanings (implied or otherwise), there should be no other interpretations and should not be prone to misinterpretation. (The English language has a weakness here in that a number of words has more than one meaning.)
2) The sentences should be simple, not complicated, and meanings and messages should be clear, devoid of ambiguity or vagueness, so that anyone could understand the meanings and messages easily, without any misinterpretation or misunderstanding.
3) The layout or juxtaposition of the sentences should be simple and straightforward.

4) There should not be jargon or technical terms, which the uninitiated might have difficulty in understanding and might misinterpret.
5) The ideas and logical relations of the ideas thus conveyed should be easily understood, clear, not subject to ambiguity, vagueness, doubt, or misinterpretation.
6) There should not be emotional undertones, nuances, innuendoes, and other shades of meaning.
7) The language should be easy to master and use.
8) The rules and grammar of the language, including the terms to be used, should be capable of being incorporated into a universally accepted and adopted standard textbook or reference text for the benefit of the human race in their everyday intellectual activities. We might name this reference text, the special language of logic, the special language of thought, the special language of reasoning, logic guage, logic language, thought language, logic tongue, logic lingua, thought lingua, reasoning lingua, thought tongue, or whatever other appropriate name. (That is, the language should have universal acceptance and usage.)
9) The language should thus be an international language.

Linguists and logicians should seriously think about developing such a specialized language that would really facilitate precision of thought and reasoning. Mathematicians and logicians have had done that for mathematical reasoning; the Italian mathematician and logician, Peano, had, for instance, developed a clear notation for the new discipline of mathematical logic and had proposed five simple axioms for number theory.

A bane faced by people reasoning with ideas is the circularity of reasoning that is often encountered, something that leads to no end in an argument and no solution, e.g., debates such as whether socialism is desirable for a nation, whether it is better for an organization to decentralize, or whether John is a better person than Jim. The author thinks that such debates could be easily settled if the disputants exercise some patience and common sense and take some trouble. If you were to get the computer to execute some action, what would you do? Yes, write a program, prepare an algorithm, or, procedure for the computer to follow. Likewise, for the disputants to arrive at the right conclusion, they should follow a certain logical procedure instead of just simply exchanging arguments that lead nowhere. Such an algorithm or procedure, which could be regarded as the rules or laws of debate or reasoning, could be as follows:

1) State the objective(s) of the debate or reasoning.
2) Determine and specify the criterion or criteria upon which the right or correct conclusion or decision is to be based. If there is disagreement go to (8).
3) Agree to be objective and apply facts only.
4) Get writing-papers and pens.
5) Each disputant is to list the advantages and the disadvantages, the good and the bad, or, the pros and the cons of the options, presumably two options, though the

options could be much more – is it better to implement the system or better to implement another system, for example? (The disputants might adjourn the meeting after a period of time to do their research and arrange to meet again some time later.)

6) Then each disputant is to objectively compare their lists and cooperate and together objectively compile the final list of advantages and disadvantages, pros and cons, or good and bad of each option. Where there are "gray" areas or areas, which are not clearly advantages or disadvantages, both parties are to compromise and arrive at a joint conclusion or joint decision. If there is any disagreement proceed to (8).
7) Both parties are to then jointly adopt the option that enjoys the most advantages and suffers the least disadvantages. The conclusion or decision made here is to be in accordance with (2) above.
8) Only when both parties tried and exhausted all means to reach a settlement, should a neutral third party or arbitrator be called in to offer a judgment or decision, which is to be abided to by both parties.

Thus, if a proper procedure or system is adopted for considering various viewpoints, rather than just having the disputing parties talk and argue "till the cows come home" and trying to forcefully persuade one another to accept his own views, matters of intellectual concern could be resolved in a more systematic, scientific, and efficient way, and the decisions that are thus made with so much attention and patient effort are likely to be better than those enforced through mere persuasive powers or strength of personality.

The author would like to see a law sanctioning or enforcing such a procedure or algorithm for resolving matters of intellectual concern. The above is only a rough guide for this procedure. Perhaps, a more formal, elaborate, and legal procedure would be appropriate. Our legislators should work together with the logicians to develop such procedures. This should enhance peace and harmony and would be a boon to society.

The author would like to make some daring, even radical, suggestions here. He would like to suggest that we have the following laws pertaining to the settling of differences and disputes (with regard to logical principles):-

1) Make logical deduction a compulsory subject in our school and tertiary curricula.
2) Have a standard text on logical deduction adopted as a "bible" or "reference book" on logical deduction – legalize it thus.
3) Have a law to outlaw personal insults and vehement arguments.
4) Have a law to make it compulsory for disputing parties to settle their serious differences through arbitration if they could not settle them amicably amongst themselves.
5) Have a law to make the decisions of the arbitrators binding on the disputing parties.

6) Have penalties for those who infringe on the arbitrators' decisions.

Next, we could provide some motivators for logical, sensible, and decent behavior. For instance, we could institute annual awards or recognition for those people who have been nominated and voted as "most logical persons", "most sensible persons", "most reasonable", "most intellectual", or even, "model citizens". It would appear that having the above-mentioned laws and motivators would be another "logical" way out of the problems caused by the "failure of logic". The author is confident that with them in place, peace and harmony would prevail.

Perhaps, in the not-too-distant future the above-described procedures for resolving differences and disputes would be obsolete. The author could envision that with the great progress of computer technology a day could come when the computer would be universally utilized to help us resolve disputes and make decisions. A universally accepted computer program could be created by an international team of, e.g., computer scientists, psychologists, behavioral experts, subject specialists, and logicians for this purpose. This program should have a very huge databank of past actual cases and scenarios and related outcomes and decisions, which is being updated all the time. A person could input his requirements into the computer and the latter could churn out the necessary advice (which is, of course, based on the information stored in the computer's databank). This program is comparable to our current simulation software, e.g., the simulation software for electronic circuit design, mechanical parts design, and aircraft design. It could be regarded as the simulation software for decision-making and problem-solving and would probably come in various subject categories, e.g., business, social, sports, legal and religious. The computer represents a neutral, impartial, unemotional, efficient, and smart advisor and friend who could help us solve problems and make decisions quickly. Hopefully, disputes, quarrels, and wars would then not be necessary, as the computer could be depended on to quickly, efficiently, and objectively help us settle our differences and problems. Only in areas where the computer could not be of help, which would probably be few, would we have to revert to the earlier mentioned procedures. When this day arrives, many decision-makers and leaders could become redundant. Perhaps, for this very reason decision-makers and leaders would try to ensure that this day would never come. It would be selfish of them to block the advent of such a powerful computer, which could generally help eradicate or reduce controversies and conflicts and bring about great peace and harmony.

Here, the author would like to enumerate the axioms of logic, which are as follows:-

1) Logic is relative, not absolute; logic is a matter of interpretation.
2) Logic is characterized by "cause and effect" or "causal effect".
3) Logic is a function of language and symbols, i.e., it could not exist without language and symbols.

4) Logic is the relevant relationship between ideas, even disparate ideas; it is a quality of ideas.
5) Logic does not exist in ideas that contradict each other.
6) Logic could produce results, i.e., its principles could be applied to get tangible results; it leads to successful predictions or forecasts.
7) Logic is abstract, invisible, but its reality, truth, or validity could be manifested in predicted or forecasted physical, tangible phenomena; the validity or truth of logic could be proved or substantiated by experiment.
8) Logic could be precise (as in arithmetic or other quantitative concepts) or vague (as in nonquantitative, qualitative concepts).
9) Logic depends on the quality of the brain; if the brain were differently constructed or hardwired, logic could be different.
10) Logic is intuitive; it is the sense of certainty of the correctness or validity of ideas; it is the sense of the beauty of ideas.
11) Logic is not immutable, unlike the physical laws of nature.
12) Logic affects the way people live their lives.
13) Life or consciousness is the pre-condition of logic; logic depends on the physical senses (sight, hearing, taste, smell, and touch) for its input; hence, logic depends on the physical senses for its existence.
14) Logic also depends on education or training, i.e., a person could be trained to think more logically.
15) Logic could help society to improve but could also detract from society's well-being because of its arbitrary nature.
16) Logic could cause controversies and disputes and is imperfect.
17) Logic could be subtle and elusive.
18) Logic could be effective when utilized with objectivity and other positive mental attitudes such as tolerance of opposing viewpoints and patience.
19) Logic is concerned with the fact that there could be other possibilities or reasons.
20) Paradoxes are a manifestation of the imperfection of logic.
21) Logic is also implication; if proposition one implies proposition two, the two propositions have a logical relation; if scenario *x* implies scenario *y*, the two scenarios have a logical relation.
22) Logic could be characterized by pattern or predictability.
23) For evidences and proofs, it is better to rely more on the physical senses (sight, hearing, smell, taste, and touch) than just logic, that is, the physical senses are more reliable than logic; even logic has to depend on the physical senses for its input and its very own existence, as stated in axiom (13) above. (Seeing is believing. Physical sensing is believing. Wouldn't you prefer to physically see something in order to accept its reality or truth than just accept its reality or truth based on logical argument alone?)
24) Logic could be absurd, seemingly illogical, unexpected, even bizarre; however, as long as it produces the expected results it could still be regarded as logic. (Exam-

ple: A lie, untruth or absurd, nonsensible proposition aimed at convincing and successful at convincing a person, perhaps a person of unsound, illogical mind or insane person, to carry out an act, for instance, could be regarded as logical, even though it is apparently illogical under normal circumstances, as long as it satisfies axiom (6) above.)

25) Logic is possibly artificial and not real.

We should always have the above-mentioned axioms in mind when utilizing logic. Otherwise, we could be bewildered by it.

2 Something to Ponder About

The author would like to invite the reader to think about the following pithy anonymous quote:

> **IN THE PRESENCE OF FOOLS, IT IS NOT FOOLISHNESS WHICH MAKES A FOOL OF YOU BUT CLEVERNESS** (read **LOGIC**).

The question here for everyone to ponder, especially those who swear by logic, is whether human logic could be of use in a society which is totally different from ours, e.g., an extra-terrestrial civilization, which could possibly regard human logic as nonsense.

The author believes that, in the last analysis, it is the truly objective and realistic mind that knows when to take logic seriously and when not to take it seriously that would triumph over the obstacles and problems of the inconsistency of logic and use it most effectively.

https://doi.org/10.1515/9783112214367-027

3 Some Interesting Bizarre Scenarios That Would Seriously Test the Power and Reliability of Logic

So far, logic has been fortunate in that it only has to deal with the constants of nature. For example, Newton's laws and Einstein's theory of nuclear energy still pretty much apply today. We have been lucky so far.

What would happen to logic and to the human race if all the laws of nature vary from time to time or fluctuate? Take, for instance, if one day there is hardly any oxygen in the atmosphere, the next day the oxygen returns, another day it is gone again, and so on; the force of gravity varies from hour to hour or minute to minute; and Earth revolves round the sun one day, deviates and revolves around the Moon the next day, round elsewhere on yet another day, and so on, what would have been the consequences? How would we now interpret this new nature, this pattern-less disorder, this chaos? How would human logic now react to all this? It is unimaginable. Probably, the human mind would have been overwhelmed and gone mad. Moreover, the human being as it is now would, in all likelihood, not be able to survive such a natural habitat. Just imagine what the constant changes in the gravitational force alone would have done to the human body – one moment feet firmly on the ground, the next moment thrown about due to the serious decrease of the gravitational force, then on the ground again (perhaps dead), then thrown about again, and so on. Could any person with a logical mind predict what would happen next? Could any logical person think of a possible way out of this chaotic natural habitat? Unless nature now endows the human body with the constant ability to evolve and adapt to the constantly changing environment, there does not appear to be any way at all for the present human race to survive this horrendous ordeal. The whole human race would, in all probability, be wiped out, become extinct, gone the way of the dinosaurs. Who dares vouch that such a scenario, or something close to it, or resembling it, would never occur? If the whole human race were to be wiped out like this, there would be no more logic, no more use for logic, no more need to talk about logic, no more use for this book.

Let us now look at another interesting but not-so-drastic scenario. The physical laws of nature might remain constant, unchanged. But the structure of the human brain is constantly evolving, changing. How would the new human brain now react? Interesting question, isn't it? If the human brain were now to evolve into something less powerful, woe betide the consequences. If it were to become more powerful, then wouldn't its interpretation of nature be better and more accurate, wouldn't it understand nature better and solve the problems presented by nature better, wouldn't it also understand logic better, be able to refine it, and, use it better? What if the evolution of the brain results in it being alternatively more superior and more inferior? The consequences are hard to imagine. There would, in all probability, be great problems.

https://doi.org/10.1515/9783112214367-028

Logic abhors chaos (nonlinearity), it seems. (As a matter of fact, scientists and mathematicians are still trying to come to terms with chaos or turbulence. In fluid dynamics, e.g., the Navier-Stokes equations could predict, from initial conditions, the motions of viscous fluids which are of steady flow and constant velocity (with a low Reynolds number, say, below a few hundred), and incompressible, in two dimensions, with the velocity field independent of time. However, the equations fail when turbulence or chaos sets in (when the Reynolds number exceeds, say, 3,000). The Navier-Stokes equations had been formulated for predictions relating to regular fluid flows, but turbulent fluid flows imply irregularity. A one-million-dollar prize awaits the person who solves the Navier-Stokes equations.) Where there is regularity, pattern, or predictability (linearity), logic is likely to be found there.

The above-mentioned examples have been brought up to show the precarious status of logic. Therefore, logic could be regarded simply as something invented (or, is it discovered?) by the human mind to link separate objects, events, or actions. This means that if this link or relationship could not be invented (or discovered?), there could not be logic. This implies that logic is possibly artificial and not real – it depends on how one views the separate objects, events, or actions, which depends on one's past experiences, one's physical senses, and whether one's physical senses have not missed out any details.

4 A Tongue-in-Cheek, Even Humorous, Slant on Human Logic

Human beings have brains, grey matter, from which springs "logic". Logicians such as Godel and Russell had made profound studies of logic. Godel had found paradoxes in logic. Russell had reduced mathematics to logic in a joint effort with Whitehead. Why is there all this preoccupation with logic? Does logic mean that much to a person's life? To the author, logic is due to the intrinsic nature of the human brain and is part of the way life is designed (by Nature) to be lived, just like the ways the human being has been designed to breathe (through his nose and lungs), eat (through his mouth), and hear (through his ears). We could say that the human being lives according to reason or logic, upon which ethics and laws, e.g., are based. Reason or logic is the way the mind acts or behaves. As long as there is a brain, there is thinking, and there is logic. Logic is the way a human being thinks. Simple as that. The logic of the non-human beings, the lower forms of life, i.e., the way they think, might be vastly different from that of the human being. Human beings regard these forms of life as less intelligent. On the other hand, some, if not all, of these lower forms of life might regard human beings as the less intelligent species. The author thinks that in some ways these lower forms of life are more intelligent (more logical). Many of them are greatly adaptable to their environment (evidently endowed by nature to be so), which is some kind of a sign of intelligence. Here, tongue-in-cheek, the author would like to point out that much of these lower forms of life are "their own bosses" – self-employed – they could survive without having to work for, and be at the mercy of, some bosses (unlike the human being). Now, who is smarter?

https://doi.org/10.1515/9783112214367-029

5 More on the Logic of Moral Behavior

Our moral sense is self-centered. What we do not like done to ourselves is considered "wrong" and what is acceptable to ourselves, our well-being, is "right". Therefore, injury and harm to others, like injury and harm to ourselves, is "wrong" or "evil". By the same reasoning, helping and doing good to others is morally "right".

To carry the above-mentioned reasoning a little further, what we do not like done to ourselves, we should not do to others. The converse is also true.

When we have done "wrongs" to others we normally feel bad, i.e., troubled by our conscience. When we see the "wrongs" done by someone to somebody else, we could sympathize with the latter and abhor or detest the doer of these "wrongs".

The question is whether our sense of "right" and "wrong" is innate or learned. The author suspects it is learned rather than innate. For instance, when we devour another life-form, e.g., a chicken or beef, we generally do not feel the guilt. We know that the chicken and cow who have become willing or unwilling victims of our appetite have life, sense, and feeling, and would have suffered the throes of death when slaughtered, and yet most of us would have no feelings of guilt over this in the enjoyment of our chicken and beef. Animals have also been subject to all forms of suffering in the name of science and experimentation. This is all true, except for a small minority, who choose not to eat meat and adopt a vegetarian diet. Who is to say that vegetables do not have life and feelings, e.g., feelings of pain and sorrow? In fact, there have had been experiments conducted by scientists which show that plants do have "emotions".

Man seems to be facing a great moral dilemma. To the author, it seems that there is more than one moral standard – one standard for the human being and another standard or standards for non-humans. Who is to say that the ground one treads on or the chair one sits on, e.g., has no life, feelings, or emotion? How does one know for sure?

To be aware of this great moral dilemma, and to do one's best to prevent or minimize all forms of suffering – to humans and non-humans alike – should be the Tao (way) of life. Strictly speaking, and theoretically so, to prevent sufferings inflicted on others, we should all kill ourselves. As long as we live, we seem likely to bring suffering and unhappiness to others. For instance, when we are successful in securing a job, we deprive others of one, others who may need the job more, and when we beat our competitors in business, studies, sports, and other areas of life, we inevitably bring failure, and, hence, unhappiness and pain, to those whom we have "vanquished". So, we could not avoid hurting or harming others. Unless, we kill ourselves and fail to exist. In which case, we harm only ourselves. Theoretically, at least it should be thought so, all forms of life, whether our life or others' lives, should be treated equally. By killing ourselves are we being fair to ourselves? And, if we con-

https://doi.org/10.1515/9783112214367-030

tinue to live, we would continue to bring pain and unhappiness to others – humans and non-humans.

Is there a right Tao (way) of life whereby we do not bring pain and unhappiness to others? It seems this could only be so if we do not need to depend on other things and other people for our own survival – if we do not need to eat, or even breathe, if we do not need to compete, if we do not need to work. But, all this is not possible.

We all face this great moral dilemma. There appears to be no way out of it.

6 The Logic of Good Versus Evil

The human mind is inventive in the sense that it is capable of creating all kinds of concepts. Perhaps, the most important concepts are those of "good" and "evil". Typically, "good" is normally associated with "white" and "evil" with "black"; this is perhaps the result of cultural influence.

The "good" deeds or "good" character are characterized by kindness, happiness, pleasure, comfort, and sense of security, the "evil" deeds, by pain, disgust, discomfort, and desire for revenge or retribution. It appears that these concepts are based on selfishness. Simply, what is beneficial to one is "good", what is harmful or what detracts from one's well-being is "evil" or "bad". Let us look at some parallels in life. When a criminal kills an innocent person to rob him he is "evil". When a person kills a fly or a bee, for instance, nobody says he is "evil". Perhaps, the person sympathizes with the murder/robbery victim in the first instance (the person could imagine himself in the victim's position) and would not like to suffer the same fate. But, who would be able to sympathize with the fly or the bee (who would be able to imagine himself a fly or a bee)?

In history, German dictator Adolf Hitler had ordered the Jews in Germany exterminated, an order which had been executed by his minions, many of whom had been trialed and sentenced to death by the allied forces for war crimes after the Second World War. And Hitler has been thoroughly denounced an "evil" genius ever since. On the contrary, a pest control operator who exterminates a whole colony of termites or bees is rewarded (paid) for his "evil" deed – nobody bats an eyelid, as long as the termites or the bees pose a nuisance to the human being or the public they deserve extermination. On the other hand, if a thief or a robber were caught and sentenced to death for a relatively minor crime (barring murder), many would probably adjudge the punishment too harsh or unnecessary.

The person who donates to charity, who comes to the aid of the underprivileged, is given the nod of approval by society. He is "good". The person who vandalizes, steals or robs, cheats, or does anything that detracts from the well-being of society is branded "bad" or "evil".

Yet, evil deeds could sometimes be justified. For instance, if evil deeds were resorted to as a means to an end, a good ending where there is peace, calm, and prosperity, some might condone the evil deeds and some might even say they are "good". This is the Machiavellian philosophy.

What could be concluded from all these? There is probably no such thing as absolute good or absolute bad or evil. The human mind could always "invent" justifications for anything, including "evil" deeds. "Good" and "evil" are basically the "creations" of the human intellect. For instance, Iraq's invasion of Kuwait in the Gulf crisis, which had resulted in the loss of innocent human lives, was greeted with approval by some Muslim quarters for religious reasons, while the Western allies thoroughly con-

https://doi.org/10.1515/9783112214367-031

demned this "evil" act of aggression. When Abraham wanted to sacrifice his son (an innocent life) for God (to prove his faith to God), nobody denounces him a would-be murderer (and therefore an "evil" man). Does it mean that "evil" deeds are justifiable, and perhaps not to be regarded as "evil", depending on what reason(s) and whom they are carried out for? Is it not true that the human race has double, or even multiple, moral standards?

7 More on the Logic of Evil

To be able to think and act morally apparently requires a certain level of intelligence. As compared to man, animals are much less intelligent and are therefore considered as beings without morals or ethics, though on occasions some humans are considered similar to or worse than animals where morals are concerned.

If, e.g., a person's level of intelligence is so low that he is not aware of the consequences of his actions, i.e., their effects on others, and if he thus causes harm to others, should he be regarded as evil? If an idiot or moron causes harm, is he evil? An idiot or moron, being mentally challenged, should not be considered evil for harm done, and would probably not be capable of harming others, though they might do so due to their foolishness and lack of understanding. If an insane person, a person totally incapable of logical thinking, causes harm, is he evil? In the eyes of the law, an insane person or a person of unsound mind is not culpable for any acts of harm or evil; he could even be spared for murder. It is those who are not highly intelligent but intelligent and cunning enough to interact with others who are difficult to classify. If such persons cause harm to others, are they evil? If they are aware of the harmful effects of their actions and yet carry them out, then they could be considered evil. If they are not aware of the harmful effects and carry out the harmful actions, they should not be regarded as evil.

To know what are the really right things to be done evidently requires a certain level of intelligence, i.e., to understand morals or ethics requires a level of intelligence. On the other hand, to understand what are harmful also requires some level of intelligence, and a person who has this understanding and still carries out a harmful deed is to be considered evil. However, there are some who are so lacking in intelligence, so lacking in understanding, even misunderstanding, that their actions take the form of evil acts (without their understanding of the evil). But these "evil" acts should not be regarded as evil acts and should instead be regarded as foolish acts or acts of stupidity.

The point here is that an evil act might not really be so, but is just an act of foolishness or stupidity.

https://doi.org/10.1515/9783112214367-032

8 The Difficult Logic of Morals

Is it true that only humans have morals? This may not necessarily be true. Animals seem to have their own rituals, which are perhaps their morals. And, what exactly are morals? They are the rules for behaving or acting in a manner that is not offensive to other fellow beings. And, why are some behaviors or actions so abhorrent? Abhorrent behaviors are selfish, inconsiderate, or harmful. Any action or behavior that detracts from the welfare of our fellow beings are immoral or not moral. Human beings are self-centered or selfish. They expect others to have consideration for their well-being or welfare. Any behavior that takes into consideration their welfare or well-being would be moral. Any action or behavior that detracts from or robs them of their welfare or well-being is immoral. So, ironically, a moral behavior would be, in this sense, a behavior that encourages self-centeredness or selfishness, which are not desirable qualities but could perhaps be regarded as immoral qualities. Thus, strangely and in a contradictory way, a moral behavior could be regarded as a behavior that encourages immoral thought or behavior such as self-centeredness, or, selfishness, which are undesirable qualities, as moral behavior takes into consideration the well-being or welfare of fellow beings who are generally obsessed or very concerned with their own welfare or well-being. Is this not contradictory?

What are the aims and objectives of morals? Their aim appears to be one of preserving order in society. The rules and guides for behaving act as guides for everyone in society so that there would be uniformity of behavior, so that there is predictability, so that everyone could be expected to behave in the same way, so that there would be no mavericks to rock the boat, so that there would be smooth sailing, order, efficiency.

In a society ruled by the laws of evil, where murder, injury, hurt, swearing, vulgarities, and the bad and crass, are accepted norms, acts of evil could be accepted as standard practice or morally or conventionally right or good behavior. Here, what is normally regarded as good in our society would be unacceptable and regarded as "evil". So, now, the bad (considered bad by our normal, conventional society) would become good and the good would become bad. Is this not reminiscent of the underworld, and the world of crooks and gangsters, where criminal, immoral acts such as violence and robbery are "in" things, the normal things to do?

So, what are really good and bad, kind and evil? It depends on what our objectives are and what stance we adopt. This is the difficult and true problem of morals. Morals are the do's and don'ts of society, of behavior. These do's and don'ts could be arbitrary and subjective. Some of these do's might be don'ts in some societies and some of these don'ts might be do's in some societies. A safe rule of thumb to follow is: What we do not like others to do to us, we should not do to others. But, man in his self-centeredness and selfishness, often practices something quite different – he would not like others to do to him what he could be found doing to others. Double

https://doi.org/10.1515/9783112214367-033

standards here, so to speak. But, of course, many behave morally out of fear and/or selfish motive, e.g., fear of retribution or bad luck, or, having a bad afterlife, for the immoral or evil deeds they had done, and, the desire for good luck, a happy afterlife or being able to go to heaven in the afterlife, for the good deeds they had done. However, there is a problem faced perhaps by many, i.e., it is difficult to categorize a deed or action as good or bad, kind or evil. Good or bad, kind or evil could be subjective and arbitrary judgments. Deeds could, e.g., be considered good when considered from a certain angle and bad when viewed from another angle. Therefore, how could a person be blamed for an apparently immoral act, sometimes when that person did not consider his act immoral at all, while some others only (and not all others) considered his act immoral? Predictably, most people in their lifetime could probably be categorized as having been somewhere between good and evil, the good and the evil being the extreme cases (which is in compliance with the famous "bell curve" of statistics). Thus, the Almighty God might have a problem deciding on who should go to Heaven and who should go to Hell in the afterlife.

We could conclude here that the logic and principles of morals could be arbitrary, subjective, and even contradictory. This is the real difficulty of the logic of morals. It could indeed be an unstable, wishy-washy logic. This does not really reflect well on the logical faculties of the human mind. Perhaps, there is no such thing as absolutely valid or true logic. Perhaps, there is only logic, whose validity is relative, valid in some circumstances but invalid in others.

Finally, how are we really to behave, and what moral standards are to be adhered to? First, we should behave in such a way that we are happy. Second, since making others happy generally make us happy, we should make others happy. Therefore, we should all generally behave or act in a way that would make ourselves and others happy at the same time (win-win situation). We should not make ourselves happy at the expense of others (win-lose situation), who might take revenge on us and make us unhappy or harm us (resulting in a lose-lose situation). In other words, moral behavior should have utility value – it should be able to bring the most happiness to the most people, including the self. Only then would there be order and peace in society. If people have no morals and think only of themselves and carry out all kinds of evil, unkind acts on each other, can you imagine the result? Deaths, injuries, bloodshed, quarrels, violence, pandemonium, disorder, chaos, lack or loss of control, and hardships and difficulties would be the sorry outcome, and society would not be able to function smoothly at all and would be at a standstill.

Hence, the great and undeniable importance of moral standards or ethics in society.

9 The Logic of Human Relations

The author has long been aware of human sufferings inflicted by the hands of other human beings and has experienced quite a lot of them himself. A person who is wrongfully or unfairly treated by another person often finds himself in a helpless or hopeless situation, a situation that is beyond his control. Whether in the family, the workplace, or social circles, a person's relationship with others makes or breaks him. The art of human relations is very much an art, an inexact science, and a very important one. All the luxuries in the world would not have much meaning if one's relationship with others were poor – no man is an island – comradeship, companionship, or friendship is very important. But, comradeship, companionship, or friendship has taken a "back-seat" role nowadays, it seems, and many people have become "mercenary". Obviously, there is something wrong somewhere when there is so much unrest around us.

Few if any seem to realize what the causes of all the unrest are. The world seems to be full of animosity and dissent, and things often look so frighteningly out of control. It does not seem a place for saints to thrive in. There appear to be many devil incarnates around. We have had seen human conflicts escalating into two World Wars and many lesser battles. A third world war is unimaginable and could well be internecine. Man kills his fellow men in wars. Must he kill at all? Why is he so desperate as to kill? Is there really no other way out? The author feels that all these meaningless deaths could have been avoided . . . if society could develop the law of human relations. Put it this way: Wars and battles are the result of failures at human relations – disputes over territorial rights, personal dislikes (between races and/or between political leaders), cultural differences, religious differences, ideological differences, greed, power craze, and so on. Wars, battles, and violence take place generally when peace talks and negotiations fail (we could call this situation failed human relations).

The importance of good human relations must be emphasized. Human relations ought to be a compulsory subject in schools, institutes of learning, and professional institutes. Good human relations ought to be the ultimate goal of human life, though unfortunately this seems far from being the case. The author feels that freedom from conflicts is the most important ingredient for an ideal society.

Human relations are apparently based on some logic, reason, or principles, mixed with some feelings and emotions. As it does not exclusively involve logic but involves the emotions, likes, and dislikes as well, human relations is in effect highly complex and difficult. And, human beings are often more influenced by the emotions than by logic. We could persuade or convince a logical person with logic but how could we do so with an emotionally charged person such as an angry person or an agitated person? It is known that an emotionally charged person is not that capable of thinking logically or objectively. Also, egos often clash; some people would often try to

https://doi.org/10.1515/9783112214367-034

be smarter than their fellow beings (sometimes to the extent of making the latter look and feel stupid) and others would often put themselves above others. Human relations thus end up being messy. Before going further, it is appropriate to take a quick look at the apparent basics of human behavior, the so-called axioms of human behavior, which are as follows:

1) A person looks after his basic needs as a first priority. (bn_1)
2) He looks after the interests of his closest kin as the second priority, e.g., his parents, spouse, and children. (ck_2)
3) He looks after the interests of friends as it suits his own interests. (if_3)
4) He would defend and protect his own interests if others encroach on them. (poi_4)

Finally, a person's behavior (B_h) could be summed up as follows:

$$B_h = \sum_{1}^{4} bn_1 + ck_2 + if_3 + poi_4$$

Troubles apparently arise when people talk too much, especially when what leaves their voice boxes are not the result of sufficient reflection and consideration. As a result, there are frequent misunderstandings and hurt feelings, which lead to disharmonious human relations. What is the solution? The author would like to make a suggestion here. Talking or conversing is normally carried on spontaneously and quickly, and many people simply shoot off from their mouths the very first thoughts or ideas that strike them, though the more prudent and cautious might take their time to think and consider first before shooting off their mouths. Unfortunately, what leaves the mouth might hurt the listener's feelings and/or leave a poor impression of the speaker if it is dispensed rashly or with insufficient reflection, and, it is impossible to delete or cancel what had been said though the offender could offer his apologies for having said the offensive or hurtful things. On the other hand, communicating or corresponding by writing has the greater chance of achieving the opposite effect; the writer has more time for reflection and could even afford to carry out some research first before writing (something that is not possible in a conversation) and what is written is normally expressed more logically, in a more organized way, and with more grammatically correct and precise words; and, of course what is written could be deleted, changed, edited, and polished for greater impact (while this is not possible in conversation). The author suggests that we all try to listen more and talk less, and, resort more to writing to each other through memoranda, letters, electronic mails, or facsimiles, whenever and wherever possible. Compared to the spoken word, the written word is much more likely to be well-considered or well-thought, more logical, more objective, less subjective, less emotional, more precise, and better organized. The spoken words per se might not convey negative emotions such as anger and hatred but the manner the spoken words are spoken or delivered, e.g., the volume and tone of the voice, might give the impression that the speaker has these negative emotions, which might not be true, whereas the written words do not face such a problem. Therefore, writing or corresponding is the better and more effective way of conducting human affairs than talking or conversing; talking or conversing, though unavoidable, should be kept minimal. With the advent of the internet today, writing or corresponding is more viable.

Human beings come in all shapes and sizes, physically and mentally. There are all kinds of people, e.g., people with the following conspicuous characteristics: intelligent, foolish, shrewd, gullible, simple-minded, even-tempered, hot-tempered, logical, reasonable, unreasonable, emotional, prejudiced, broad-minded, narrow-minded, jovial, serious, melancholic, introverted, extroverted, religious, god-fearing, atheistic, superstitious, fearful, worried, happy, deep-thinking, superficial, comical, humorous, liberal, conservative, fashionable, old-fashioned, cultured, rough, educated, illiterate, artistic, technically inclined, honest, dishonest, frank, reserved, insular, fair-minded, philosophical, practical, gentle, sophisticated, self-centered, selfless, charitable, merciful, merciless, brutal, violent, peace-loving, glamorous, altruistic, quick-thinking, slow-witted, patriotic, treacherous, loyal, disloyal, active, passive, cooperative, prima donna-like, friendly, unfriendly, blunt, critical, self-possessed, optimistic, pessimistic, faithful, unfaithful, business-minded, crooked, ambitious, favor-currying, mercenary, evil, cruel, demanding, aggressive, nonchalant, irritable, irritating, calculative, harsh, lenient, kind-hearted, noble, ignoble, patient, impatient, easily satisfied, greedy, calm, anxious, scholarly, athletic, adventurous, imaginative, risk-taking, likes to play safe, reckless, rash, cautious, considerate, inconsiderate, sympathetic, unfeeling, cunning, proud, egoistic, public spirited, private, hateful, snobbish, humble, pompous, intellectual, alert, witty, observant, talkative, quiet, secretive, courageous, meek, cowardly, materialistic, mentally troubled or ill, mentally challenged, sociable, gloomy, stubborn, spiritual, shy, romantic, sensitive, thick-skinned, shameless, weird, petty, fault-finding, crazy, tactful, tactless, average, brilliant, persuasive, flexible, inflexible, soft-hearted, hard-hearted, realistic, self-deceiving, having a combination of various personal characteristics (which appears to be the norm), etc. Should we deal with or handle all the various categories of people in the same way? It is apparent that the various kinds of people would respond to their environment and to stimuli, in their own different ways. Logical argument might not work for many of them; motivational talks or playing on the emotions might work better with many of them. To deal effectively and amicably with the various kinds of people should therefore involve understanding their characters and satisfying their most important needs or desires, and, logic might not have that prominent a role to play at all. Clearly, coexisting harmoniously with some of these people would be relatively easy while doing so with some others would be comparatively difficult, perhaps even impossible. And, in many cases if someone likes your face or your style he would support or help you, while the reverse is also apparently true. You may say that this is not fair or logical, but human nature is such. If one realizes that one is going to cause enmity or clash when dealing with some difficult people one should avoid dealing with them if possible, but sometimes it is not possible to avoid such people and one should try one's best to deal with them as effectively and amicably as possible and hope for the best.

The bible teaches one not to hit back when attacked but offer the other cheek to the aggressor. In our everyday relations with our fellow beings, should we behave thus toward our enemies or aggressors? It might work in some instances and we could offer the other cheek if we are reasonably confident that it would achieve the desired result. Otherwise, we could either flee or fight back. We should assess and decide on such situations logically and objectively. But, probably, the best is still to negotiate peace and settle things amicably. In the last analysis, do we all not want to, ideally, coexist harmoniously with our fellow beings?

The world is a confusing place. There is turmoil all over. Unrest. Riots. Coups. Revolutions. Assassinations. Murders. Conflicts. Many of these seem meaningless and avoidable. The whole problem boils down to the vast human differences – racial, language, ideology, education, culture, family background, intelligence, maturity, emotional stability, religion. Take the following instances as examples. A religious person thinks his religion is superior to other religions. One man thinks his idea of going about doing a thing is better than another's. One race thinks their race is superior to others. One ideologue thinks his ideology is better than another's. One person thinks he is better than another. And so on.

Disputes or conflicts normally arise when one party thinks it is right or superior as compared to the opposing party. The problem is that such supposed superiority or "betterness" is impossible, or very hard, to prove, thus making it difficult to be accepted or recognized by the opposite party. Hence, conflicts take place, with both parties trying to prove to one another their correctness or superiority. Since proving to the opposite party one's correctness or superiority is very hard or impossible, and logic or reason itself is complex and elusive, the conflicts usually take time to resolve and could even escalate. The fact that the concept of superiority is a subjective one and not clear-cut makes it obvious that conflicts are inevitable. For example, how could the author prove that he is smarter than another person and is more likely to be correct in his views as compared to this other person? If the author shows another person his I.Q. scores, which are higher than this person's, e.g., this person would probably not be able to accept them as I.Q. tests have been known to be culturally biased and test only relatively few types of intelligence – even psychological experts could not ascertain what I.Q. actually is except to call it "ability to do well in I.Q. tests", and there are likely to be other areas of intelligence where the person with the high I.Q. does not do well but the person with the lower I.Q. would excel. Another example: A person excellent in academic pursuit might perform poorly in sports while the star athlete might do badly in scholarship, and, both academic pursuit and sports require intelligence, though intelligence of different sorts.

As we cannot excel in all things, we concentrate on doing what we are best at and exchange the fruits of our labor with others who are better at other things. Hence, cooperation is very important. Conflicts should be avoided or resolved without trouble. But how? This question is much easier asked than answered.

People in all walks of life have agreements and disagreements in many things. Those who agree might form groups or gangs, e.g., political parties and associations. Those who disagree might fight it out, argue, or quarrel with their opponents, e.g., through disputations, riots, revolutions, or wars. It is normally difficult to persuade, convince, or prove a point to others as egos are involved. For example, how could one prove that one religion is better than another? How could one prove that democracy is better than communism? (Wouldn't it be clear that under certain conditions, communism does more good than democracy? And, conditions are never totally the same in any two places.) How could you prove that a government is a good government? (And not just a lucky government that happened to come into power when the economy was at its most prosperous and there were least crises to overcome.) Such abstract matters such as "better", "good", or "bad" are not like subjects such as height, length, breadth, and weight, which could be measured and are thus indisputable. Worse, how could one prove that one was better or more correct than one's counterpart without offending the latter, embarrassing the latter, or making the latter feel that one were boastful (though one was merely expressing what one sincerely thought was a fact and hoping to win over one's counterpart to one's point of view)? Human actions, motives, and intentions are easily misconstrued or misunderstood resulting in unhappiness, grudges, or revenge-seeking. The human being apparently tends to be egoistic, self-centered, competitive, sensitive, and mistrustful. The other factor is that human beings apparently have different levels of intelligence so that what is logical to the more intelligent might not be understood (and regarded so) by the less intelligent, i.e., the more intelligent and the less intelligent might not see eye-to-eye. (It should be easier for the intelligent to deal with the equally intelligent as they would be more likely to see-to-eye; therefore, an intelligent person should seek out and deal with equally intelligent people or people having more or less the same personal character.) This is the apparent stumbling block for human relations. The obvious result: enmity, disputes, conflicts, wars. There are courts of law to allow attorneys to argue cases and prove cases. There are arbitrators to settle commercial disputes. There are parliaments for ideologues to argue and settle national policy problems. These are formal ways to settle formal problems. But, in our everyday life, and in our work, there are many problems that have no formal solutions, and perhaps no solution of any kind at all. Of course, if everyone concerned could compromise and could "give and take" instead of being stubborn there would not be problems. Unfortunately, stubbornness, prejudice, pride, and one-upmanship are apparently also some of the weaknesses of the human being. Antagonists often go for each other's throats and many have to win at all costs. This is poor human relations. This is the state when each party thinks that he is right and refuses to give way to the other. Experience shows that it is normally difficult to prove any party right or wrong in any disagreement or dispute. The other problem appears to be that parties having similar views or interests tend to gang together against parties having differing views. In certain situations, the line between right and wrong might be so thin that no solution might appear forthcoming. Theoretically, in such a dilemma, no rational beings should be at "each other's throats" – they should give and take and settle their disagreements amicably. To confound it all, other human weaknesses such as jealousy, greed, and pride all help to veer human conduct toward the irrational.

If we leave a couple of human beings to themselves, they are likely to argue, debate, quarrel and form gangs, parties or cliques. History has apparently proven this to be true. A simple staring incident might result in a fight or clash. A person who does not like another person's face or looks might find fault with or give trouble to the latter; this is bias or personal prejudice. This is far from logical, humane conduct, and is a manifestation of hooliganism. It is clear that laws have to be in place to maintain order in society, though the laws are normally not without their loopholes, i.e., they are not perfect. Look around and one could easily discern breakups, gripes, slurs, slanders, quarrels, disputes, fights, riots, wounding, murders, assassinations, coups, wars, etc. Because of jealousy, envy, greed, pride, fear, suspicion, and personal differences, cooperation and teamwork are often scattered to the winds, and, in their place are great competition, rivalry, and conflict. It should be emphasized that human cooperation and interdependence are of obvious vital importance, as no man is an island capable of producing all the goods and services to satisfy his own needs, i.e., no man could be a Robinson Crusoe, which is only a fictional character. Man has the following needs (ranging from the physiological needs at the lowest level to the self-actualization needs at the highest level) and would act or behave accordingly to satisfy them:

Self-Actualization Needs
(Need for challenge & self-improvement)

Esteem Needs
(Need for respect and admiration)

Social Needs
(Need for friendship and companionship)

Safety/Security Needs
(Need for physical safety, job security & retirement security,
even, in the form of pension, et al.)

Physiological Needs
(Need for the basics of life, i.e., food, clothing, and shelter)

Abraham H. Maslow's Hierarchy of Needs

Everyone, including the strong and the powerful, has to steadfastly note that someday he would become weak and would really require help. This point should drive home the fact that human cooperation is a vital fabric of society.

Man apparently classes himself way above other living things in terms of intellect and character. In the animal kingdom, e.g., it could be seen that the more powerful animals prey on the weaker ones; big fishes eat the small fishes. Is man any different in this respect? Throughout history it could be witnessed that the stronger and more powerful would normally intimidate, control, take advantage of, bully, exploit, even oppress the weaker people (like the stronger animals or creatures, which prey on the weaker ones). Class conflicts and snobbery are apparently quite common. Social harmony is apparently quite lacking; people could be found trying to outdo each other in status, wealth, and power. Competition and rivalry could be found wreaking havoc in our apparently turbulent times.

Despite the existence of all sorts of laws, there is much disorder in the world. All this implies that the existing laws are insufficient or ineffective. Is there a way out of this rut? Should there not be a law of human relations to ensure that people conduct themselves with decorum and social rectitude? For instance, when people could not agree or have disputes, there is an easily accessible legal procedure for quick and amicable settlement of the disputes, e.g., through enforced or compulsory arbitration. The author feels that the "arbitrator" should not be a human being, who could be partial and biased, but a super-smart universal computer, which, after all the parameters and statements pertaining to a dispute are fed in, would arrive speedily, fairly, and objectively at a conclusion or decision, which has to be accepted by the disputing parties as a "passed law".

It would be great if the lawmakers or legislators and the top scientific brains could develop such a super-smart universal computer. Computers have helped mankind to explore outer space, perform delicate tasks such as surgery, and carry out many complex tasks. Why can't they be roped in to help to eliminate our social or human relationship problems, which at present seem so complex and intractable? The author could envision very fast supercomputers, which could arrive at decisions very quickly and are very easy to operate (including perhaps voice recognition for those who could not write) effectively performing such an important task. These supercomputers, e.g., the pocket calculators or the personal computers, must have the implicit trust and recognition of everyone.

Imagine having super-smart computers that could speedily try cases, e.g., cheating and murder cases, quickly settle industrial disputes and hastily, but effectively, make decisions for us. Such super-smart computers would then rule our lives. And top decision-makers such as presidents, prime ministers, governors, and directors would in all likelihood be redundant, or be reduced to the role of "controllers of the super-smart computers" – all decision-makers would in all likelihood be redundant. In effect, these super-smart computers would be the de facto presidents, prime ministers, and other decision-makers of the world's nations and the de facto managing directors and managers of the corporations. Politics would in all likelihood be outmoded. Wars would in all likelihood be redundant. (Why fight when computers could settle the disputes between nations?)

Some computer scientists have predicted that some day the computer would rule the world; if allowed to do so, the chances are great that one day the computer would entirely control our lives. The only likely obstacle toward achieving this appears to be that politicians and managers would be reluctant to "share power" with the computer, let alone "hand over power". There is a detectable fear of computers. Many people are afraid of losing their jobs because of the computers; it is clear that computers could work very much faster and much more efficiently than human beings, posing a threat to the job security of the latter. There are some others who find computers too difficult to master.

Strictly speaking, we have more control over the computer than the latter has over us. If we wanted to stop the computer from interfering in our lives, all we had to do was to unplug it from the power point. It is really up to the powers that be to wisely utilize the computer in bringing about world peace and harmony. Let us hope that some wise leaders take this initiative.

10 The Logic of Competition Versus Cooperation or Teamwork in Economics

Man is an economic animal. The principles of economics dictate that business competition is good for the consumer. It is assumed that a business, in order to thrive over its competitors, would have to be better than the latter technically, financially, and management-wise. This means that it has to look around all the time and try to keep ahead of its competitors. Even if it is ahead of all its competitors, the problem is whether it could remain ahead all the time. A change of fortunes might just topple it from the pedestal. All these cuts and thrusts of the competition are supposed to benefit the customer who is likely to get better prices, better service, and better products. Is this assumption a reasonable one?

However, the author thinks that there is another side to the coin. Competition, it must be remembered, does result in tension, worry, and stress and too much of these is just not good for society. A society that clamors all the time after the "number one" position is going to be self-centered, inconsiderate, and distrustful. Besides this, is it not true that constantly trying to weaken and destroy the competition would weaken the business itself – extra costs, e.g., for research and development, advertising, and marketing would eat into company profits and impose unnecessary burdens on the business. Some would even adopt underhand tactics and attempt to deceive customers by claiming for their products or services some benefits that are either untrue or nebulous – technically, this is known as "passing off". In the end, the customer loses out.

The author thinks that so far economists have been blind to the fact that "unity is strength". In fact, when businesses team up, cooperate, and merge, they pool their resources and have a wider talent pool to rely on. Instead of thinning out their resources on advertising, marketing, and fighting the competition, they could now really concentrate on building up a better product or service and finding better, more efficient ways to manufacture a product or provide a service. All these could result in a cheaper and better product or service for the customer. The customer need not fear that lack of competition would result in exorbitant, skyrocketing prices. The government could always step in to regulate prices. Even the consumer representatives could impose checks on the pricings. The author thinks this is the reason why Japanese companies have been doing so well so far. For example, Japanese companies, which have new orders, have been known to share them with their competitors, which is unheard of in American or European companies. In other words, Japanese businesses are able to team together to serve their customers.

The author thinks that economists ought to rewrite their economics and emphasize on teamwork, cooperation, and mergers rather than competition. Like a good soccer or basketball team, society could function better when people team together and cooperate rather than eye, distrust, and outdo each other. The author thinks people

https://doi.org/10.1515/9783112214367-035

would then be living a much less stressful, happier, and healthier life. How do you account for Japan's rapid economic reconstruction after World War Two and its position as a world economic power today, if not for its capacity for teamwork, amongst other things? Surely, a soccer team of cooperating team members should score better than a team of prima donnas. The same principle should apply to the economics game. Economists should be well aware of the economies of scale that are available to the companies that team up or merge.

11 The General Logic of Cooperation, Teamwork, Harmony, Peace, and Happiness

Competition is bad, as it gets the worst out of people. People who are anxious and all out to get the upper hand and win over the competition tend to ride roughshod over the competition and benefit at the expense of the competition. This is a "win–lose" situation, where one party gains while the other party loses. People may resort to cheating, murder even, to win or gain the upper hand. People in competition may not be frank or honest with each another and may conceal things from each other. They may fear each other – fear of losing, fear of backstabbing, fear of being cheated, fear of being humiliated, fear of losing their livelihood, and many other fears, real or imagined. Thus, energy is wasted through fear and unnecessary activities, which are triggered by fear. A person's potential for great creativity and productivity is hence crippled by fear, wearisome conflict, and wasted energy.

The effect of cooperation and teamwork is just the reverse. It can be said that cooperation and teamwork get the best out of people. The synergy that results when people cooperate, help each other, go all out to oblige each other, and enhance, rather than hinder each other's performance, is simply great.

Those who say that competition keeps one on one's toes and is therefore desirable is just looking at the picture with a jaundiced view, a subjective mind, just so that competition could be justified. This is being rather silly. The author could recall that the cooperation and team spirit in a well-coordinated soccer team is simply morally and spiritually uplifting, leading to much apparent happiness and personal satisfaction on the part of the team members. Imagine if the team members were to backstab each other or quarrel amongst themselves. How terrible and saddening it would have been – both to the team members and to the spectator or observer.

What conclusion can be drawn from all these? The author dares say that all forms of competition, except for friendly competition as in friendly games, are bad. Business competition, economic competition, competitive examinations, competitive scholarships, competitive sports, international competition, and fights or wars can all be said to be bad and undesirable. Can they be avoided? Though their avoidance seems impossible, the author thinks it is really not impossible. It is really the competitive environment that has caused all the interpersonal, international, and even internecine conflicts. It is when one party gains at the expense of the other party or parties that causes so much personal dissatisfaction, personal envy, and hatred.

For there to be peace, harmony, and happiness, it is time now to be rid of the competitive environment. It is time to take radical action to turnaround the affairs of men and nations so that there would be harmony and world peace. First of all, the mindset of mankind has to be changed. All manner of men, from leaders to the man on the street, ought to be coerced into thinking that competition is undesirable and bad while COOPERATION and TEAMWORK are good for the society as a whole and

https://doi.org/10.1515/9783112214367-036

most desirable. Why be content with social unrests, conflicts, social ills such as crime and delinquency, stress and breakdowns, wars, and the like? Is it not obvious that the stresses and turmoil of life now in our modern society are the result of competition in schools, sports, work, and life? Take the radical steps. Abolish anything competitive in nature such as school examinations, competitive sports, business rivalries, rivalries between nations, and economic conflicts. Common sense ought to show that if mankind does not try to undermine each other in competition but instead *cooperate* and work together, the world would be a much more harmonious, peaceful and happy place. A central authority, e.g., the government, could take the lead and coerce people to cooperate and work together.

12 The Logic of Democratic Leadership

Society is presumptuous in that it thinks itself capable of judging brilliance, genius, intelligence, and even moral character. It is as if it is alright to expect a community of weak-minded people or morons to be able to appreciate the brilliant ideas of a genius. Is this not absurd?

Yet we have been so absurd indeed on many occasions, as when we have judged others on their ideas and actions when we have less experience, wisdom, or intelligence than they. We may think we know better than they. Yet, when we are in these people's shoes, we may end up worse off than they, a not uncommon thing altogether. This raises a pertinent question: What gives us the right or authority to judge another person's qualities, when we ourselves are lacking in those very qualities we are judging? It is as though instead of the more knowledgeable teacher teaching the student, the "greenhorn" student is trying to impart knowledge and wisdom to the teacher.

This is precisely the paradoxical situation society is saddled with. We judge and elect our leaders. Yet, how could we be sure we are qualified or have the wisdom to judge and elect our leaders? What happens, for instance, when men of evil amongst us succeed in electing into power a leader with similar evil qualities? A sociologist had stated that a decision of the masses tends to be more correct than a decision of a single individual. But there does not seem to be sound evidence to support this stand. Is our educational or examination system foolproof enough to produce really solid professionals, such as brilliant scientists and excellent administrators?

Unless the powers that be or those with authority are excellent in capability and morals, and, unless our political, educational, and selection systems are foolproof enough to vouch that those elected or selected for positions of authority or leadership are indeed of fine capability and morals, society is indeed running the risk of actually having the wrong leaders to manage it. Can you imagine the world's greatest power now, the United States, being led by an evil genius, a "Hitler"? A "playboy" president is bad enough (as an ex-U.S. president apparently was).

This brings to mind the many corrupted and treacherous officials or mandarins of feudal China, who had been supposedly brilliant individuals who had passed the highest imperial examination for the Chin-shih degree (the equivalent of the Ph.D.), the most difficult examination. And, what a sorry mess they had made of China then, with the common people suffering abject hardships and poverty, while these mandarins enjoyed wealth and authority at the expense of the commoners. That was why there were so much rebellions by the common people against those in power. Even those in power were fighting each other for more power. How chaotic!

Surely, we do not want a replay of this ugly scenario. That is why we have to be very careful about the election and selection of our leaders. To ensure that we elect or select the right leaders, we have to ensure that those who elect or select our leaders have the right qualities, such as excellent intelligence, judgment, and morals.

https://doi.org/10.1515/9783112214367-037

But, how do we ensure that the electorate has such proper qualities? How do we ensure that our educational system will throw up professionals and administrators with the right qualities? Can such qualities be judged or evaluated by certain fool-proof tests or examinations, e.g., an appropriate I.Q. and personality test? Perhaps, society should stipulate that only those who achieve a certain standard in such a test are eligible to elect, select, or appoint its leaders. We should think hard about this.

13 The Logic of Nature

Scientists have often made use of highly abstract mathematics to describe nature. For example, relativity theory relies much on Riemannian geometry and tensor calculus, and, quantum theory relies on group theory. There is a saying that nature is mathematical. Mathematics deals with patterns and much of these patterns studied by mathematicians are also evident in nature. In looking for a solution to the Riemann Hypothesis, which is considered the most important unsolved problem in mathematics, some mathematicians are working on the hypothesis that perhaps the distribution of the zeros of the Riemann zeta function (which are mathematical objects) is similar to the distribution of quantum particles in the microworld of particle physics. Group theory in mathematics is the study of the classification of members into groups based on their common properties and their characteristic possible permutations – it is the study of symmetries or patterns. Apparently, the symmetries or patterns that mathematicians deal with are also evident in nature, e.g., in the quantum world.

Is the similarity between some mathematical entities and some aspects of the natural world a mere coincidence? Or, is there a deeper meaning behind the similarity? It is apparent that mathematics is an effective instrument for the study of nature, and a well-known scientist had even dubbed this "the unreasonable effectiveness of mathematics", implying somewhat that the effectiveness is unexpected, surprising. Here perhaps, we could infer that how we think, how a mathematician thinks, depends on and is influenced by our environment, which includes nature. When we deal with quantities, e.g., in algebraic manipulations, in geometrical computations (involving space), and in simple arithmetical calculations such as addition and subtraction, what kind of objects are we concerned with, if not the objects of nature? We cannot just simply say that we are dealing only with imaginary objects (objects that do not exist), for even to think of imaginary objects, we need to experience nature first. It is nature, the environment, which provides the data for our thinking, our imagining, our mathematics. Our logical reasoning cannot exist in a vacuum or blank, where there is nothing, no feature, no data at all. The mathematician also cannot work in the blank.

Is nature really that simple? Does it indeed reveal itself so easily to the scientists and mathematicians? Are the apparent symmetries or patterns discerned in nature only a facade, beneath which lurks a deep, perhaps unfathomable, mystery? Scientists may, on the basis of such discernable patterns, jump to conclusions. They may experiment further and discern the same patterns. After that, they may confirm their conclusions. But, the conclusions may be the result of one way of interpretation. However, there could be some other plausible interpretation. Scientific theories have often been modified to suit new evidences.

Symmetries or patterns in nature imply order, predictability. Scientists who interpret nature may be comparable to two persons looking at a cup that is half filled with water; one person looking at the cup may say it is quite empty, half empty, the other

https://doi.org/10.1515/9783112214367-038

person may say it is quite full of water, half full of water. Thus, scientists may only be more conscious of the symmetries or orderliness of nature, unconsciously or consciously blocking out the asymmetries, the disorder, the chaotic. Apparently, the human mind prefers to see order, symmetry, which is normally equated with beauty; perhaps, the human mind has long been conditioned to react in this way; many of us have been taught to appreciate the beauty of nature since young. On the other hand, we should not forget the ugly aspects, the destructive power of nature, such as the harmful radiation of the sun's rays, the destructive lightning, the monstrous typhoons, the gigantic tidal waves, the terrible earthquakes, the terrifying volcanic eruptions, and the flash floods. Incidentally, all these natural disasters tend to be unpredictable or hard to predict; they are the opposite of the order that is predictable. Such disastrous natural phenomena are normally regarded as chaos or turbulence, which is unpredictable, the outcome of which is unpredictable, wherein a slight change gives rise to a more than proportionate increase in the outcome, which is a characteristic known as nonlinearity.

Of course, symmetry, set pattern, and order allow one to infer, arrive at logical conclusions, predict, forecast. On the contrary, with nonlinearity, disorder, disarray, chaos, or turbulence, whereby no set pattern is discernable, all this will be quite impossible. Incidentally, chaos or turbulence is also one of the important outstanding problems in mathematics; a mathematical theory or solution that is capable of minimizing, if not eradicating, the undesirable or harmful effects of chaos or turbulence, e.g., by making its force and movements somewhat predictable to some extent (please refer to the author's article, "The Navier-Stokes Equation and Turbulence", in the appendices, for a possible solution to this problem). It will benefit, e.g., the aircraft designer and the ship designer, who have to consider the possible effects of air turbulence and turbulence in the sea, respectively, while carrying out their design work.

Therefore, in looking at and appreciating the symmetries and the beautiful patterns of nature, we should not forget its darker side, which is characterized by disorder, lack of pattern, lack of predictability, chaos, or turbulence. To forget or ignore this darker side may be so at our peril.

14 The Paradoxical "Logic" of Time Travel in Theoretical Physics

To determine whether time travel is really possible, it is necessary to have an understanding of what time is. Time, it must be stated here, is something in the abstract, something belonging only to consciousness. It is a feeling, an awareness of the "passage of time", a consciousness that the "seconds, minutes, and hours, or, days, months and years, of time" are passing by. It is not an object per se. That is, it is not something we can catch hold of, physically feel with our sense of touch, or physically see with our eyes. It is an abstraction, a "quality" of objects that are moving, e.g., whether the objects we are observing are moving fast or slow, or faster or slower. So, time is an abstraction or "quality" associated with moving physical objects, e.g., the moving clock or watch hands, the moving automobile, and the flying aircraft.

According to Einstein, time is relative. For, according to him, if we are in a hurry, time would seem to pass faster or be too short; otherwise, time would seem to pass relatively slower. As stated here, time only "seems" to pass faster or slower. This means that time is only a matter of "feeling" or consciousness or awareness. It does not mean that when time "seems" to pass faster, our watch or clock would run faster, or, vice versa. As we all know, whether or not time "seems" to us to be passing faster or slower, the speed at which our watches and clocks run would remain the same, constant, regardless of our "sense of timing". This means that in actuality time is always constant and always moving forward. (The clock or watch always runs forward or clockwise.)

If time travel were indeed possible, a person could travel backwards in time and kill his parents so that they would never be able to meet, get married, and beget him. If this happened, he would never have existed, let alone being able to travel backwards in time to murder his parents. This is indeed a paradox. Can such a situation really be possible? The answer is "No!" In mathematics, such a self-contradictory argument is regarded as a proof by contradiction (*reductio ad absurdum*), whereby an assumed to be true statement is contradicted and proved to be false.

We can compare the above-mentioned paradox to Zeno's paradox. According to the latter paradox, if we let a tortoise and Achilles compete in a race but let the tortoise have a head start as it is a much slower creature, Achilles would never be able to overtake the tortoise at all, for if we keep halving the distance between the tortoise and Achilles, the halving could go on ad infinitum, without stop. This means that there would always be some distance between Achilles and the tortoise, no matter how infinitely small the distance is. This simply means that Achilles would never be able to overtake the tortoise. Now, ask, "Could such a thing ever happen in actuality?" Definitely not! One could carry out an actual experiment with a man and a tortoise and ten times out of ten, or, even, a million times out of a million, or, more, one would find that the tortoise would be overtaken by the man in very quick time.

https://doi.org/10.1515/9783112214367-039

One could safely conclude that Zeno's paradox exists in theory only, but never in practice – in fact, in this case, theory and practice do not mix. The same applies to the above-mentioned time travel paradox.

Though in the present framework of our physical laws and our personal being such as our consciousness and our intelligence, time travel appears impossible, its possibility should not be ruled out completely. Time travel could be a possibility in a world or universe where the physical laws and life-forms are different from those of this planet. Many of us have perhaps at one time or another heard of beings that have apparently been capable of passing through walls and appearing and disappearing from out of nowhere, seemingly capable of time travel or teleportation. Who are these beings? Have one never heard of spiritual beings, ghosts, or apparitions? Though it appears that scientists normally steer clear of research on such paranormal or supernatural subjects, as they are apparently not explainable by our physical laws and our everyday experiences but are puzzling and mysterious, they should present a very great challenge for the very great scientific mind. The very great scientific mind is expected to be a very curious mind, which is very keen to understand the unknown and the mysterious. Then why does such a great intellect normally eschew the paranormal and the supernatural, perhaps regarding someone who delves into such mysterious subjects as cranky or just having a wild imagination? Could any of these great intellects prove that such beings do not exist?

The way to greater understanding and intelligence should be to have an open and broad mind in all matters, whether worldly or other-worldly. Perhaps, the answer to whether time travel is indeed possible or not lies in the seemingly unfathomable physical laws, which may govern the apparently paranormal and supernatural. There is a great likelihood that the apparently paranormal and supernatural subsist on a different kind of logic or principles.

Coming back to the question of the possibility of time travel, it is indeed possible to travel in time, in our consciousness, as time exists only in our consciousness. To think of past events is to travel backward in time and to think of future events is to travel forward in time. This should settle all the controversies pertaining to time travel.

15 The Logic of Space-Time in Theoretical Physics

When Einstein first propounded the philosophy of the space-time continuum, there were critics who scoffed at this revolutionary concept, which predicts the existence of four dimensions in the universe: the three physical dimensions of length, breadth, and height, plus the fourth dimension of time; they could not conceive time as the fourth physical reality, they could not understand the relativity of time.

Einstein's theory seems to go against common sense. If time is the fourth dimension, could we travel in the time dimension as we could travel along the three physical dimensions of length, breadth, and height? Could there be more than four dimensions in the universe? Kaluza, e.g., had modeled a universe with 21 dimensions, reducible to 11 dimensions, a universe that could be described physically in five dimensions. And Einstein found his concept elegant and encouraged the publication of his theory.

Personally, the author finds Einstein's concept of time as the fourth dimension of the universe not that far-fetched, although some people could not accept this as sensible because time, to them, is abstract and lies outside our physical senses.

If a person, e.g., were to enter a completely dark room, whereby he could not see anything at all, and hence any movement, we could say that he could not see time (passing), and would lose his sense of timing. But when there is light, and he could see movement (even the movement of sunlight, moonlight, and day and night), e.g., he could see the hands of a clock moving, he could count on his fingers, he could count the points in space, he could see people and objects moving from point to point in space, he could see and is actually seeing time (passing).

Here, movement, any movement from point to point in space, in the author's opinion, is analogous or similar to the fourth dimension of time; it is the fourth dimension. Hence, we can have the four dimensions of length, breadth, height, and movement (along any of the length, breadth, and height dimensions) existing side by side in the universe, and this in the author's opinion, is similar to the "length, breadth, height, and time" model of the universe.

Time can thus be looked upon as a physical reality and not as a higher reality existing outside our senses, especially our sense of sight. When we are looking at the objects that are making their movement along points in space, the objects we are counting and the points in space we are enumerating we could not fail to be aware of time. If we touch these objects, we could feel the force of their movement, or, on the other hand, the force of our movement (along the objects we are counting or the points in space we are enumerating), and this would also give us a sense of timing.

If the author may put it simply, time could be seen by the naked eye and felt by physical touch.

How would you explain and describe time?

https://doi.org/10.1515/9783112214367-040

16 The Logic of the Existence of God

Since time immemorial, man has pondered on the question of the existence of God. Many believe in a God, a good God (as opposed to Satan, the evil One). Yet, a number call themselves "freethinkers", neither believing nor disbelieving in God. A number, however, do not believe there is a God and they are known as atheists.

There are numerous reasons believers give to support their belief in God (some of which might appear "circular" in their logic), for example:

1) Look at the beauty of nature all around us. Some Supreme Being must have created nature in all its beauty. He must be God.
2) I prayed to God for salvation. After that, my problems were solved.
3) My success in life is due to the grace of God.
4) I feel that God has spoken to me.

Man has also deified certain people or animals, e.g., Buddha is regarded as a God, and he was an aristocrat turned philosopher. Satanic cult members worship the Devil, Satan himself, who becomes their God. Primitive men even worshipped idols made of stones or wood. It seems Nature itself, the natural world, holds a deep mystery for man – man still could not understand his origin, the origin of all life, for example – scientists have tried to formulate theories to understand it better, but still have come no closer to understanding the origin of life, of the universe. So, how does man come to terms with the origin of life, including his origin. One way out is to simply assume a Supreme Creator, i.e., God, though this might seem to some, a "lazy short-cut out of the conundrum". In fact, this very assumption does lead to some controversy, as when, e.g., the Christians assume that those who do not believe in God would go to Hell after death.

Then, there is the concept of Heaven and Hell, a belief of the Christians, and the concept of retribution and blessing, which followers of other faiths may abide by. There is also the concept of reincarnation, rebirth after death, which is a belief held by some religious groups, e.g., Buddhists and Hindus, and is not really substantiated by any concrete evidence. The concept of Heaven and Hell simply stipulates that those who do good in life would go to Heaven, a paradise, after death, and those who sin would go to Hell or Purgatory after their demise. Whoever "created" this concept might have had a good reason. This concept does serve as a check on the behavior of human beings, coercing them to behave morally while they are alive. Yet, there are still so much evilness and vileness in man (something, which the Christians like to attribute to the work of Satan or the Devil). If only Heaven and Hell were a proven fact, how much better morally would humans become?

Do you really believe in God? If you do, how do you prove He exists? How do you prove you exist (a much simpler question to answer)? For these answers, the author would like to refer the reader to Rene Descartes, a famous French philosopher and

https://doi.org/10.1515/9783112214367-041

mathematical genius. His answers to these questions are interesting, and it is worth giving them some thought, at least since Descartes is reputed for his brilliant intellect. His answer to his own existence is as follows: "I think, therefore I am." His "proof" of the existence of God is found in his book, *Discourse on Method and the Meditations*. His deduction for the existence of God is as follows:

Descartes has a clear and distinct conception of the fact that he exists. He can therefore believe that whatever else he perceives with the same clarity and distinction is equally true. Moreover, he knows himself only as a thinking being; he is therefore assured that the soul and the body are entirely distinct. Since he has been able to understand his own being and essence without yet knowing anything about the world outside him, it follows that his self or soul is completely independent of the outside world, mind is distinct from and superior to matter. Next, by reflecting on the notion of doubt itself, he perceives that, as to know is a greater perfection than to doubt, he must be an imperfect being. But an imperfect being cannot produce the idea of perfection, which he nevertheless possesses. No other being, imperfect like himself, could have given it to him. Only a perfect being could have done so. Therefore, a perfect being, God, exists. Two other proofs of the existence of God follow: Descartes exists, possesses the idea of perfection, and is himself, imperfect. If he had created himself, he would have created himself perfect. He has therefore been created by another, who must, of necessity, be perfect since he (Descartes) has the idea of perfection. Lastly, there is the famous ontological proof, which is the most important. Absolute perfection is the sum total of all possible perfections. God, being absolute perfection, must necessarily exist, since existence is one of the perfections. To imagine that God does not exist is as absurd as to imagine a triangle that has not three angles. (Descartes was very pleased with this proof and wrote to Mersenne (a fellow mathematician): "I dare to boast that I have found a proof of this existence of God, which I find fully satisfactory and by which I know that God exists more certainly than I know the truth of any geometrical proposition." Indeed, this proof has, inside the framework of Cartesianism, a notable advantage, in so far that it is apprehended by *intuition* rather than by deductive reasoning, and therefore presents a greater degree of evidence.) Once the existence of God has been established, it is easy to show that, as a perfect being, he cannot deceive us and that consequently, we can place our confidence in the veracity of our clear and distinct ideas. Descartes had another concept: Human beings may be imperfect. But all human beings, with their imperfections, when viewed as a whole, may appear perfect.

In *A Brief History of Time*, a very popular book by well-known physicist, Stephen Hawking, two propositions are given pertaining to whether the Universe had been created by a God. The two theories are as follows:

1) Big Bang Theory – The Universe has a beginning at the time of the "big bang" (explosion). This implies the possibility of a Creator (God). (Edwin Hubble discovered that the universe is expanding – distances between stars are getting wider with the passage of time – possibly validating the Big Bang Theory.)

2) Steady-State Theory – The Universe has no beginning (in time), hence no creation of the Universe being carried out, and has always been as it is – the Creator or God does not come into the picture.

God is perhaps one of the "creations" of the human intellect. So far, whatever evidence of the existence of God, if any, is only circumstantial. To believe in the existence of God is an act of faith, and some people have more of this faith than others. Perhaps, we can say that we intuitively know God. One might argue here that it is not advisable to trust something based on strong intuition rather than sound deduction. But, nevertheless, the most intelligent people with highly deductive minds have believed in the existence of God – the world-renowned genius and physicist with the highly deductive intellect, Albert Einstein, was one of them. Some might even proffer the following as "proof" of the existence of God: "If so many highly deductive intellects, intellects of genius, believe in the existence of God, then God must exist."

There is a possibility that God only belongs to the realm of the abstract, the realm of thought. That is, if you think of God, He exists; if you do not think of Him, He does not exist. Perhaps, the totality of consciousness and intellect in all living things, especially humans (which has resulted so far in technological wonders such as computers and space travel) represents God, the supposedly evident powers of God. It is evident that the human being has mental or psychic energy that could be harnessed to influence the environment (there have even been cases reported where metal or other objects bend or levitate upon mental command). The totality of this mental or psychic energy or power could be extremely great, so great that it seems to emanate from a God. Some scientists have quite recently postulated that faith in God had been hardwired in the human brain, i.e., people who believe in God have the "God" gene in them.

If man is satisfied with the answer that God created all of us, the question of how God came about might be asked, especially by the atheists. In computer science, an object that emulates (or simulates) another object has to be more powerful (more complex) than the latter. Similarly, the Creator of Life should be more powerful, more complex than Life itself.

Charles Darwin's now scientifically embraced theory of evolution, which postulates that mankind had descended from the apes (though it was regarded as heretical and pooh-poohed when Darwin introduced it), is at cross ends with the Christian doctrine that man was created by God, who had made the first man, Adam, and the first woman, Eve. How do we reconcile this?

To the atheists, the nonbelievers, they would have no choice but to believe in the existence of God if they had experienced and confirmed firsthand His existence (by seeing and/or hearing Him), and, to assure that this sighting and/or hearing are no illusions of the physical senses, and more people should have had the same experiences.

Ultimately, whether one believes in any God or not is not so important, but one's intrinsic moral fiber is more important. There appears to be many God-fearing, pious people who preach the good of God but who practice rather evil deeds (though of course "good" and "evil" are also controversial, subjective interpretations). There are those who behave well and good because of fear, fear of retribution, or, fear of punishment by God (with many apparently seeking God's forgiveness for their shortcomings at various times). There are those who truly live by their moral principles or conscience and carry out only what is regarded as good or correct, immaterial of whether it is the dictate of religion or God, though such people of high moral fiber or moral principles might be few. The interesting question is who amongst the following is the more respectable and more worthy of emulation:

1) The pious, religious, God-fearing person, who carries out unethical, evil deeds.
2) The pious, religious, God-fearing person, who mechanically carries out good deeds and avoids evil deeds, as dictated by his religion/God.
3) The pious, religious, God-fearing person, who carries out good deeds and avoids evil deeds, out of desire for being blessed by God for performing the good deeds, and, fear of retribution or punishment by God if he carries out any evil deeds.
4) The atheist, who lives by his conscience and moral principles and carries out good deeds and avoids evil deeds.
5) The freethinker, who lives by his conscience and moral principles and carries out good deeds and avoids evil deeds.
6) The pious, religious, God-fearing person, who lives by his conscience and moral principles and carries out good deeds and avoids evil deeds.

To believe in God is amoral, i.e., neither good nor evil, and so is disbelieving. But to show disrespect, whether one believes or disbelieves, is not good. Can we say that a person who believes that apples bring fine health is good whereas one who disbelieves so is bad? Wouldn't this be silly? We should be concerned rather with the implications or effects of a person's behavior or beliefs (especially, if he has acted on them) on others, on the well-being of his fellow men. However, if belief (or, disbelief) in God causes problems or conflicts, then it is not good. On the other hand, if it helps to achieve peace and harmony, it is good and desirable.

It is perhaps because of the logical human mind being used to the logical principle of cause and effect (i.e., every event or happening in nature and life has a cause) that it supposes, assumes or believes in the existence of God, whom it interprets as the Cause of things in nature, things whose occurrences or existences are apparently not caused by any human or natural agencies. (This is ironical as in Chapter 17 of Part Two: it is stated that "Religion and God seem to appeal greatly to the not too rational, whereas the 'over-rational' or 'over-intellectual' tends to question God's existence and the religions.") As the human mind is so used to seeing things or reasoning things out as "causes and effects", anything that is mysterious or inexplicable is also believed to have a cause, though it is a mysterious or inexplicable cause; in this case, God seems to be the plausible Cause. Could things not exist spontaneously without a cause? (At the moment, God is apparently believed by theists to be the only Entity that exists without a cause.) However, if the mind is used to things happening without causes, then human logic might be different from what it is now (which is a logic entrenched in the principle of "cause and effect"), or, there might not be logic or understanding at all.

Before we conclude the discussion here, it is appropriate to bring up one more "proof" for the existence of God. This "ontological proof" is the brainchild of no less a person than the world-famous Godel, who was in fact the "Einstein" of logic. A similar proof of God's existence from pure reason, rather than from any evidence in the physical world, had been first attempted by Archbishop Anselm back in the eleventh century. Godel's proof was based on his use of Leibniz's concept of "positive" and "negative" properties and is as follows:

Axiom 1. A property is positive if and only if its negation is negative.
Axiom 2. A property is positive if it necessarily contains a positive property.
Theorem 1. A positive property is logically consistent (that is, possibly it has some instance).
Definition. Something is God-like if and only if it possesses all positive properties.
Axiom 3. Being God-like is a positive property.
Axiom 4. Being a positive property is logical and hence necessary.
Definition. A property P is the essence of x if and only if x has the property P and P is necessarily minimal.
Theorem 2. If x is God-like, then being God-like is the essence of x.
Axiom 5. Being necessarily existent is God-like.
Theorem 3. Necessarily, there is some x, such that x is God-like.

Godel seems to have had great faith in this proof of God's existence and had remarked that "the axiomatic method is very powerful". He was obviously in many ways a rather mystical thinker who believed that we participated in a Cosmic Mind, separate from all matter, in which we found the truths of, e.g., mathematics, when we searched in the right way. (To Godel, mathematics is a body of truths waiting to be discovered, i.e., mathematical concepts are real.) Could atheists or nonbelievers in God accept Godel's "ontological proof" that God necessarily exists?

17 The Logic of Religion and God

Practically every religion advocates a God or idol(s) of worship. Every religion professes to do good to its followers, to improve their lives, and each religion impinges upon the individual an authority that is not earthly in origin but heavenly. To ensure that no one questions or dares to question God and His authority, a religion, e.g., Christianity, may instill fear by stipulating that we ought not to question God's existence or His authority, for it is a sin to do so, a blasphemy.

That every religion is good in its purpose is undoubtedly true; the idea of God, that God protects, guides and forgives us for our sins, comforts and consoles us. Religion and God seem to appeal greatly to the not-too-rational, whereas the "over-rational" or "over-intellectual" tends to question God's existence and the religions. Bertrand Russell, to take but one example of a highly rational man, did not believe in Christianity, and even questioned the existence of God. Since God is more or less universally believed to exist, does it mean that people in general are not that rational? The answer appears positive. People are more guided by their emotions than by reason, it seems. It is the ego, the desire to satisfy one's own desires, rather than the desire to conform to reason that determines our actions. In practically all our actions, we consider our interests first before we consider other people's interests. This is only natural, for if by our actions we do not satisfy our needs and desires, then we alone are going to feel the pain within ourselves or suffer; when we feel hungry we eat, after which we feel happy and contented; if we do not satisfy this basic need for food, we would feel hungry, an uneasy feeling, an emotion, which, if prolonged, leads to misery. Since our feelings of uneasiness are ours and only ours to deal with, and not others', and because we have to eliminate these feelings of uneasiness as soon as possible, our actions naturally only bear on our self-interests, our desires and needs, first, and they only bear subordinately on others, after our immediate desires and needs are satisfied. Even in charity, we apparently think of ourselves first, without our own conscious knowledge; we feel an uneasy feeling, a pitiful feeling, for the less fortunate or the unfortunate; we want to eliminate this uneasy feeling or emotion and, therefore, we do charity, after which we feel at heart's ease, i.e., achieve a desired emotion.

This may explain why many people apparently try to coerce themselves into believing God and the ideas about God. Many people tell themselves, their friends, and sometimes, the world, that they believe in God. But subconsciously, they may doubt God. These people may sincerely want to believe in God and when they tell themselves and others that they believe in God, they may be actually trying to reinforce their belief (here, we could say that belief in God may be achieved through autosuggestion); deep in their hearts, the certainty about God may not be there, and by trying to reinforce their belief, they may wish to implant that certainty in their hearts. This seems impossible, for "pure" certainty may only be felt if concrete reasons or

https://doi.org/10.1515/9783112214367-042

proofs are available; mere striking of ideas or words on the mind, not based on strong reasons or proofs, may never create "pure" certainty. Since no one has given really concrete proofs about God and since strong evidences proving the existence of God apparently do not exist, the existence of God might be subject to doubt; entertaining such doubt is rational behavior, and this seems rare. There is still controversy over the subject of God. Practically every believer accepts God on faith.

Religion is apparently based on principles dictated by our self-interest mainly, and, secondarily, by our conscience, but to make our conscience at ease is, in a sense, to satisfy our self-interest. Religion too, like the theory of God, is not based on pure reason (it is apparently based more on self-interests). Therefore, we can either choose to follow a religion or not, depending on whether we "feel" strongly for a religion or not, and not on whether we "think" strongly about religion as a whole or about a particular religion.

The more insecure a person feels, the more problems he faces, the more likely he is to cling to religion and God. There are some who have prayed for and sought God's help or salvation but who ended up losing faith in God when they failed to get what they have prayed for. On the other hand, those who got what they prayed for are likely to have their faith in God strengthened. Let us assume, e.g., that we have a kind, rich, and generous friend (whom we might worship like a God). Should we always take advantage of his kindness and generosity or cause him inconvenience by always seeking free gifts or financial aids from him? We might be hesitant or feel shy to do so, and to be able to do so might require a thick skin. As educated and civilized human beings we ought to have some sense of decorum. Of course one could assume that this kind, rich, and generous friend would not mind being approached for help. There is however the probability that he might take offence at being bugged for assistance so frequently, when he finds that he could not live in peace. For the billions of people praying for and seeking God's help God might have to work extremely hard. Is it correct to burden the Great Helper with so many requests for help? It would be good to have faith in God's love for mankind and willingness to help mankind. This faith seems to be abundant (and some scientists think that this is an intrinsic quality of the human psyche – something that had been hardwired into the human brain).

A renowned psychologist had stated that it is those who have no religion or faith in God who are likely to suffer from mental illness. Without an all-powerful protector, without the sense of security that a Higher Power that is protective would protect one, one tends to have more fears and less mental stability. When a person talks about fate, destiny, or luck there is no role for logic. Intelligence and logic might help to guide a person's destiny in life to some extent. However, a person's life is pretty unpredictable on the whole. What one could not understand or achieve with logic, one leaves it to fate, destiny or luck, whatever it really means, implying some higher influence that is beyond our control. This may explain why God and religion are so important; they seem to be plausible answers to the incomprehensible and the mysterious.

18 The Inscrutable Logic of Nature

There are many things in life that still puzzle us. First and foremost, there is the question of consciousness. Consciousness is apparently the result of electrical messages passing through the brain. Moreover, to be conscious, a person has to be physically alive, i.e., his heart has to pump blood through his body, blood has to circulate through his body, and his lungs have to inhale and exhale. Consciousness is awareness or alertness, and is intangible. It could be regarded as the spirit or soul of a person and is mysterious. Consciousness or mind, apparently has influence or control over matter, over our environment. Where was a person's consciousness or spirit before he was born into the world and how did it come about when he was born into the world? Would consciousness or spirit be able to survive a person's death? Consciousness represents the self, humanity itself, but it is apparently still as intriguing and mysterious as ever. Though scientists might probe the mysteries of nature, they have, strangely, not been able to plumb the deep mysteries of consciousness, their own consciousness even, and the aforementioned questions remain unanswered, or, unanswerable.

Next, the subject of gravity is apparently still a puzzle to modern-day scientists. It is odd that though the effects of gravity are evident in the macroscopic cosmos, they appear to be absent in the microscopic quantum world (where quantum particles whiz about with great abandon without displaying the effects of gravity). It appears that the physical laws of the cosmos differ from those of the quantum world. How come? The scientists are apparently stumped. There are several theories relating to gravity, e.g., the theory that gravitons (particles) and gravity waves exist, which hopefully, would be detected in experiments. Gravity is now regarded as an attractive force, but it is not the attractive force of positive and negative particles, a phenomenon known as magnetism. Perhaps, it is an attractive or pulling force in reverse, a pushing force. The Sun apparently exerts a gravitational attraction on the Earth, as well as the other planets, which all orbit round it. This attraction is apparently stable, for otherwise, the planets would have been displaced from their respective orbits and collide with each other. Many apparently regard this efficient system of planetary movements as the work of God, the result of willful, intelligent design. Instead of regarding gravity as a force in the Earth's core that pulls everything toward it, how about considering it as an outward force that pushes everything toward the center of the Earth? Is this not also a plausible theory? This pushing force could originate from the space surrounding us, e.g., the air, gases, and light. Light and electromagnetism have been found to be the same thing. Perhaps, gravity may be similar to some other aspect or facet of nature.

https://doi.org/10.1515/9783112214367-043

Finally, the existence of perpetual motion machines has been debunked. But are not the various planets in our universe and other universes perpetual motion machines per se, as they apparently orbit in their various paths nonstop and perpetually for innumerable years? What is the driving force behind their apparent perpetual motions? Is this the handiwork of a higher Being or Intelligence?

It would be great to have all these matters sorted out, to have their logic revealed.

19 The Logic of Planetary Movements

A spinning ball rotates or spins round and moves in a more or less circular or elliptical path. A boomerang, spinning after a throw, also travels in a circular or elliptical path. If one throws a ball upwards, the ball would reach a maximum height and starts dropping till it hits the ground. Likewise, the force exerted on the boomerang is so great, greater than the surrounding forces of the air, that it travels away from the thrower, till the maximum distance is reached, when the surrounding forces of the air are greater than the force carried along by the boomerang; since the force exerted on the boomerang is "out-going", a "counter" force by the surrounding air has to be "in-going" to effect an opposite direction, whereby the boomerang would travel back to the thrower. This could explain the earth's circular, or rather, elliptical, path.

But, what is responsible for the earth's rotation, its movement? An invisible force, a magnetic force? Objects have been moved by magnetic force, by remote control. The north and south poles send out magnetic fluxes. The ocean currents are deflected due to the rotation of the earth. The Earth can rotate on itself, independent of external forces, due to the magnetic forces within itself, which spin it round its own axis. These magnetic forces vary in intensity at different stages in the earth's revolution, just like the force applied to and with a boomerang.

Finally, what makes the Earth follow the same path constantly, why does it not obviate from its normal path round the Sun? That is, what "suspends" it in position all the time? The Earth has to be kept in the same position by equal "opposite" forces that act upon it from all directions all over its spherical surface. The danger is that these "equal" forces might be upset and might become "unequal", resulting in the Earth straying out of its normal path, and crashing with or into another planet.

It is quite obvious that if the "equal" forces were upset, so would the other "equal" forces in the universe, resulting in a literally universal catastrophe of planets moving erratically in all directions. For, the universe is a conceivable "bounded" unity within an inconceivable unity, the apparently "boundless" outer space, and being a unity, a local disorder somewhere in the universe, would bring about a disorder that would involve practically the whole universe, or even, more.

Recent discoveries have more or less confirmed that magnetic fields do extend across the galaxies, a very vast expanse of space, and it is posited that these magnetic fields are carried by gravity waves, which are a theoretical substance, whose existence has yet to be confirmed. These magnetic fields probably have some bearing on the movements of the planets.

However, some might just simply credit all these wonders to the work of a supreme Creator, a great Being who could produce miracles. This simple reasoning might be more convincing to some and might make all the above theories with whatever doubts they arouse, seem redundant.

https://doi.org/10.1515/9783112214367-044

20 The Logic of Life, Nature, and the Supranatural in Our Universe: A No-Holds-Barred Analysis

Scientists have long been striving for a theory, a unified one that explains all the forces of nature. For example, the Superstring theory has been formalized to show that gravity, electromagnetism, the "weak" force, and the "strong" force are different aspects of the same thing in nature. It has been surmised that many things in nature, e.g., cells, grasses, and trees, resemble strings. We realize that many forms of nature take on a "string-like pattern". This is reminiscent of the classical quantum theory, wherein the microcosmos, the quantum world, is construed as similar in "pattern" to the macrocosmos, the planetary universe. Like the planets circling the Sun in orbits in our universe, subatomic particles, viz., electrons, travel in orbits round the nucleus in the atom, the quantum universe. Could such "similar patterns" be found in nature in other universes (that are as yet undiscovered)? If we believe that they could be found, as we believe that the "similar patterns" discernable in our universe should also be found in these other universes, we are not standing on solid, logical ground for doing so. Why should and why must the "similar patterns" discernable in our universe be found in these other universes? Life in these other universes, if it exists at all, could be different from or totally unlike life in our universe.

Men of thought, such as scientists and mathematicians, are always concerned about proving theories and conjectures. Scientists perform experiments to confirm their theories. Mathematicians apply logic or reasoning to affirm that conjectures are correct, they are concerned with proofs. A proof is reason, hard evidence. While scientists have to perform physical experiments to confirm results, mathematicians carry out their work entirely in their minds, in the abstract. Many scientists trying to gain a deeper understanding of the natural world seek patterns. They might then resort to mathematical formulas or symbols to describe or represent the patterns. In other words, mathematical beauty could be found in nature – that is, nature is mathematical. As many scientists have been mathematically trained and are mathematically inclined, it is not surprising that they could view nature through mathematical eyes. On the other hand, an artist or a person who is aesthetically inclined, could see nature in terms of beautiful shapes and colors. All this indicates that nature could be different things to different people.

Life is not so simple as many people seem to think. Life ought to be divided into two categories, at least: life after birth and "life" after death. We have apparently been preoccupied more with life after birth than "life" after death. The author is convinced that there is life after death on account of some actual, personal experiences. If there is indeed life after death, what are the physical laws of the Other World, the world where the deceased persons of this world apparently passed on to? A number of those who have lost their loved ones such as a parent or a life partner have apparently sensed the return of their departed ones to their homes, especially after dark, in

https://doi.org/10.1515/9783112214367-045

the night. They could have heard the departed ones' footsteps in the night and might be able to see them even, though many probably would not be able to see the departed ones. The author, on checking, realizes that these are the rather common experiences of the recently bereaved. Other phenomena include offerings placed at the altar apparently having been meddled with, doors closing apparently by themselves, familiar noises being heard (such as the beeping of the cardio equipment monitoring the departed one's heartbeats when he was dying in the hospital), noises of chairs apparently being moved, and noises of chairs apparently being sat on when it was clear that there was no person around at that time, mysterious sounds of furniture apparently being knocked into, physically feeling that an invisible object or force was touching one's face, hand, or other parts of the body. Now, the corpse of the departed one was either cremated or buried in the cemetery. How could he "come back to life"? How could "he" make his journey presumably from the cemetery, the columbarium, or, the Other World, to his former home (the bereaved's house)? How could "he", e.g., produce the cardioequipment's beeping sounds (to make his presence felt?)? The skeptic, who has not experienced any of the above-mentioned weird phenomena, might scoff and dismiss them as merely figments of imagination. But, to those who have apparently encountered or experienced them, they had been apparently real happenings and not the products of dreaming or an overactive imagination. All these boggle the mind and seem to defy all the physical laws we know. They are against common sense or logic and intuition. Are our physical laws, including the apparently counterintuitive laws of quantum physics, the only physical laws that are available? Could there be other physical laws in other universes, or elsewhere, which we do not know of yet? The apparent problem is that human beings tend not to be able to think "out of the box"; we tend to interpret what we observe, or reason, in terms of what we already know or understood, e.g., our known laws of nature. In all likelihood, at least to those who have actually had the above-mentioned weird experiences, there is an Other World or Spiritual World, though the scientific minded probably would like to have further, concrete proofs of its existence. There have been quite a number of publications and books on near-death experiences (NDEs), life after death, communications with the dead and what not. Are all of these not some kind of affirmation, no matter how lacking in substance or credibility, that there is an Other World? What good would turning a blind eye to all of these oddities and dismissing them as nonsense do? Many people, including intelligent ones, e.g., geniuses such as the physicist Einstein, apparently believe in God, a God who apparently has not made His presence directly felt, as the spirits of some of the departed ones apparently had (as described above). Where is the reasonableness or logic then of scoffing at, or deriding, those who claimed to have had the abovedescribed "weird" experiences of an apparently spiritual kind and who believe in the existence of an Other World, a spiritual world? Is it reasonable or logical to condone the act of a person who embraces a belief (in God) based apparently on pure faith and not concrete evidence, but scoff at or deride the person whose belief is apparently the result of some actual experiences, no matter how weird and against common sense they are? The scientific elite has apparently been trying to solve the many mysteries of nature and life, which is fine, but they should not neglect the supranatural or stuffs of the other worldly kind (for the departed's sake, at least), e.g., interesting, and, important, questions such as whether the human body has a soul or spirit that leaves it when the person dies, where this soul, if it exists, comes from when a person is born and goes to when a person dies. These are indeed challenging and intriguing questions that are somewhat linked to a morbid subject, viz., death, which perhaps few people, including scientists, like to deal with.

The soul or spirit could however be equated with consciousness. Consciousness itself is intangible, though it does manifest its presence through tangible or obvious results, e.g., speech and touch. The concept of death put forward by one physicist, who apparently believed in life after death, is interesting. He likened death to the soul discarding its old clothing, the aged or malfunctioning body. Then, after death, what new clothing would the soul get? Would there be reincarnation? This implies that though the human body is mortal, the human soul (or consciousness – we could equate consciousness with the soul or spirit) might be immortal. Consciousness is intangible and yet we know it exists. It appears as nothing (being intangible and in concrete), and yet it does exist, a seeming paradox. In any case, the human soul (or consciousness), after the demise of the human body, seems to continue to exist. Would it not be a more important challenge to the scientist to seek a theory for consciousness, the soul or spirit? Consciousness is apparently the most fundamental in life, in nature; without consciousness, we would not be able to know, to feel, to reason, to ask questions, to seek knowledge and truths; without consciousness, we would not be living (though we might be alive). A strong theory of consciousness that provides the answers pertaining to its origin, its possible continuity on the demise of the human body and reality (as perceived by our consciousness, which might vary from person to person), e.g., could well be the theory of everything, the ultimate theory, the ultimate truth, indeed the most fundamental and important truth. If we do not have such a complete understanding of our inner self, our consciousness or soul, to seek the truths about externalities, the extra-terrestrial and other distant objects would seem somewhat extraneous. To shun the great challenge of seeking this understanding is to shift the point of the problem elsewhere.

Perhaps, consciousness or the soul would survive the demise of the universe, should it ever happen. The following are some speculations regarding consciousness or the soul:

1) The soul or consciousness might survive the demise of this universe as it belongs to another space-time dimension, another universe, with differing physical laws, and, there might be more than one such universe; or, it might be immune to or unaffected by the physical laws, hence, destruction, of this universe.
2) This other universe or universes have access to our universe – their inhabitants could make visitations to this universe, while earthlings could not visit their universe or universes.

3) Could this explain the UFOs?
4) There is a parallel universe or universes for the soul or consciousness.
5) There is a more superior, or, supreme Being, a God, or, Gods belonging to another space-time dimension or universe, or, universes.
6) These Being or Beings have influence over humanity.
7) These Being or Beings are responsible for the birth of the universe and would be responsible for its demise, should it ever take place.
8) Human consciousness is divided into two kinds – conscious mind, and, unconscious mind, which have links with the Other World.
9) Dreams (which originate in the unconscious mind) might represent communication with the Other World.

Some years ago, an article in the New Scientist had described an experiment on the human soul. In this experiment, a dying person's weight was obtained with an atomic weighing scale three hours before demise. After demise, the person's weight was found to have dropped by three-quarter pound. Does this mean that the deceased's soul, which presumably weighs three-quarter pound, has left his dead body?

There are apparently a lot of mysterious phenomena staring right into our face, which have yet to receive our due, comprehensive attention. We expend relatively much effort in exploring the quantum world and the extra-terrestrial, for example. But, how about our inner world, our soul, spirit or consciousness? A definitive theory of this inner world should be just as important, if not more important, than the theories pertaining to the quantum world and extra-terrestrial life, and might prove to be the answer to everything, including the origin of the universe (which means the beginning of life or consciousness). Apparently, scientists normally regard the spiritual and the paranormal as being on the "fringes" and tend not to pay much attention to them. This is probably because the spiritual and the paranormal appear to defy logic or common sense, and, therefore, defy any meaningful explanation. In this respect, they are similar to magic, which is incomprehensible, and, mysterious. They would indeed be a very great intellectual challenge to the scientific world, which unfortunately, seems disinclined toward them; it not only appears that the scientific world lacks the confidence in tackling these "difficult" subjects but also regards them as impossible to comprehend, whereby research carried out on them would lead to a dead end and would be a sheer waste of effort and time. Is the supranatural really incomprehensible? Is not an understanding of it greatly important to humanity? Whether we wish to tackle or shun the challenges presented by the supranatural, the incomprehensible and the mysterious, it is difficult to deny that the supranatural has apparently been and is still a part of our lives, and hence, a part of our reality, though those who have never encountered or experienced any of the abovedescribed weird phenomena of an apparently other worldly or spiritual kind might beg to differ. We could expect the supranatural or the paranormal to operate on a different set of physical laws, which we have yet to understand.

21 The Possible Ultimate Logic Which Explains Everything in the Universe Including the Mysterious and the Unexplainable

In the twilight years of his life, Einstein had struggled in vain to discover a set of equations that would explain all of gravitation, the electric field, and the magnetic field. He conceived all the three in terms of Riemannian geometry and tensors, using different sets of equation to describe all three – with ten potentials for gravitation, three potentials for the electric field, and another three potentials for the magnetic field. Some felt that he was wasting his time in attempting to discover a Unified Field Theory, which would be, so to speak, a "theory of everything".

Now, physicists have produced a possible "theory of everything", the Superstring Theory. At CERN, e.g., scientists have been trying to get evidence of superstrings at work. A violin string vibrates at different frequencies and gives rise to different tones. In like manner, a superstring vibrates at different frequencies and gives rise to various forms, e.g., gravitation, the electric field, and the magnetic field. The idea sounds ingenious and scientists are still carrying out experimental work to prove that the theory is sound.

To the author, there is another possible "theory of everything", which is a plausible theory that explains all the phenomena of life. This theory may sound like determinism (as against free will). Within each living being, humans, animals and, possibly, plants, there is apparent intelligence, consciousness. We cannot quantify, touch, see, or smell intelligence or consciousness. We can only see *manifestations* of intelligence or consciousness, e.g., intelligent behavior and remarks. Intelligence or consciousness is similar to radio waves or electronic signals, which are invisible and can yet produce visible results, e.g., give sound and music, video displays, telecommunications, computerization, and so on; in fact, intelligence or consciousness is the apparent result of electrical impulses affecting the various areas in the brain. We can, e.g., view each living being as having a "transceiver" within itself, a kind of "equipment," which transmits and receives "radio waves" (for want of a better term) or "invisible messages." This is probably responsible for the so-called intuition, which all of us possess; this may explain why we can anticipate what others are going to do or say, why we can tell whether a person likes or dislikes us, is angry with us, and so on, without the other person having to do or say anything to us. It appears that we are all exchanging "radio waves" (or "brain waves" or "soul waves" if you wish) all the time, unconsciously, or semi-consciously. People of the same "frequency" or "wave length" get along well together, while those with different "frequencies" do not, it appears.

Is there a higher Being or Beings controlling our actions and thoughts? If there is or are such Beings we cannot see them. Such a Being or Beings, representing a higher intelligence, for instance, could be millions or billions of light-years away from us,

https://doi.org/10.1515/9783112214367-046

and exercise "remote control" over us (as though we are remote-controlled cars or TV sets). Such a Being or Beings could represent an extremely powerful "transmitter" that transmits "radio waves" (as well as receives "radio waves") over distances of millions or billions of light-years.

Such a Being or Beings could be around us even, though they might be invisible; they might be in the form of some kind of "radio waves", "electrons", or some other loose, subatomic forms, which have probably not been discovered yet. We know that radio waves, though unseen, are very powerful and effective, and could, for instance, give us pictures and sounds from outer space and enable us to communicate instantaneously over extremely great distances easily. (Some years ago, Nobel Prize winner A. Penzias and his colleague discovered some mysterious radio waves emitting from outer space. What could be the source of the radio waves?)

There is a possibility of such a Being or Beings invisibly controlling all lives and nature on earth and in the universe, an act similar to that of a person manipulating the control panel, which guides the operations of a contrivance by remote control, i.e., a person manipulating radio waves. If this were the case, a human being is somewhat comparable to, for example, a remote-controlled robot, airplane, or car.

Inanimate objects such as minerals, rocks, and liquid forms could also transmit and receive radio waves or signals, and could be, in theory, subject to such "radio controls" (by the higher Being or Beings).

The electric and magnetic fields, and perhaps gravitation, are some kind of energy waves or signals, which could be and are being controlled and harnessed by man (in fact, man makes much use of the electronic in modern technology), and could of course be subject to control by a higher Being or Beings as well, if they do exist. Human consciousness is the result of electrical impulses (electrons) acting on the various areas of the brain. What causes these electrical impulses to act the way they do? As human beings we apparently do not fully understand our own consciousness, e.g., what causes us to think or behave the way we do. If one could influence the way these electrical impulses act on the brain one could influence the way a person behaves or thinks. Is this not plausible? This is one interesting and important aspect of technology, which ought to be looked into, and, whoever possesses this technology would in all likelihood hold great power, to do good or evil. Who knows, some higher (or more intelligent) Being or Beings (whom many would apparently refer to as God) might be already in possession of this powerful "technology", and, human beings could well just be the pawns in a highly sophisticated chess game played by the Gods. And, who dares swear that this could never be the case? Imagine a chess game (or any other board game for that matter), imagine yourself being one of the pieces on the chess board and imagine that that chess piece (yourself) has consciousness. Your movement (as the chess piece) is now controlled by a higher, more intelligent being, the chess player. The downside of this theory is that human beings, unlike the so described chess pieces, are apparently capable of independent thinking and capable of questioning God (apparent free will, perhaps more apparent than real) whereas the chess pieces apparently could not and do not question the chess player. (But we could never be certain about this, and the chess pieces could be influencing the way the chess player thinks without the chess player being aware or conscious of it. Who knows, inanimate objects such as rocks or wood might have consciousness or emotions though this is not apparent. Animals such as dogs and cats probably have great ideas too, great ideas that they apparently are not able to share with human beings, i.e., animals and human beings have a communication problem. If inanimate objects do indeed have consciousness or emotions, it is not obvious or apparent to us at all. Could we then not say that inanimate objects and human beings have a terribly great communication problem?)

To the author, the theory of remote control of human beings by a higher Being or Beings is not a remote possibility. Outer space is apparently infinite, and, in all likelihood, it did not have a beginning in time, having apparently been in existence for an infinite span of time. In other words, outer space has apparently been around for far, far more than zillions of zillions of years, an uncountable number of years with no starting point. In such a "world" is it possible that the human species represents the most intelligent beings that could be found? Could there not be different forms of beings, and an infinite number of them too, which differ from each other in terms of appearance, personality, and intelligence, some more intelligent and some less intelligent than the human race, some even more intelligent and powerful than our God(s) and unimaginably so? This theory then would be the Theory of Everything and could explain the two often-mentioned but little understood things called "luck" and "fate". (However, "luck" is a controversial subject. There is a school of thought, which stipulates that "luck" is created by the self and not a gift from some higher Being or God. This school of thought states that people actively on the lookout for opportunities would get the right opportunities and would thus appear "lucky".) First of all, to be given the right opportunities, one's benefactor or benefactors have to have confidence in one's capabilities, trust one, or like one. Therefore, having obvious high intelligence, good personality, looks, and charm would help assure one gets the right opportunities. There are those fortunate or "lucky" ones who were apparently born with these good attributes, or, born into well-to-do or privileged families, who would naturally be somewhat "destined" for success. It would apparently be a "destined for failure" case for those born with undesirable attributes such as low intelligence, lackluster personality, lack of good looks, and lack of charm, or born into poor or underprivileged families. We have to remember that what attributes or genes and family backgrounds or environments we were born with were practically beyond our control. The question is whether such things have been determined for the person born into this world by a higher Being or Intelligence, a God. And, how could there be fairness when people have been apparently born with varying personal attributes and family backgrounds? Therefore, are not some people luckier or born luckier than others?) Einstein had said "God does not play dice". This statement implies that our laws of nature are the result of willful, intelligent design, by some higher, more intelligent Being(s).

However, we are still having uncertainties over our origin and existence and there is the possibility that the Being responsible for the creation of the human race and other facets of nature, if He exists at all, has wanted His design and handiwork to be shrouded in mystery and obscurity. As there is apparently an infinitude of universes, which could be termed the infinite multiverses, it implies that anything, both imaginable and unimaginable, is possible at some point or points in the infinite multiverses, e.g., incredible or absurd phenomena such as the dead coming back to life at some point in time, beings with more than one mind or consciousness (even with multi-minds or consciousness), beings who could instantaneously appear and disappear at will as though by magic, beings who do not age, or immortal beings or beings who do not die, et al.; in other words, there are infinite opportunities, possibilities and phenomena in the infinite multiverses – infinity implies limitlessness, endlessness, without beginning and ending, impossibility of measuring or counting, mind-boggling limitless expansiveness. That is to say, an infinite number of phenomena, conceivable and inconceivable, have infinite chances of taking place in the infinite multiverses – that is, we could consider that practically nothing is impossible in the infinite multiverses.

22 The Logic of the Act of Thinking

Is it possible to think without words, numbers, pictures, images, or symbols, or even emotions?

The author thinks it is not possible. Thinking means having ideas. Thoughts or ideas actually comprise words, numbers, pictures, images and/or symbols. Words, numbers, or symbols may represent pictures or images and vice versa.

As to emotions, when a person is conscious, he has feelings or emotions. In order to think, a person has to be conscious. Therefore, thought must necessarily be accompanied by emotions.

https://doi.org/10.1515/9783112214367-047

23 The Logic of Probability or Luck

The tossing of a coin 580 times has given the author some insight into probability. The author has found that the proportionate difference between the total number of heads and the total number of tails peters out when the number of throws increases, generally:

$$\int_{i=1}^{\infty} h_1 - t_1, h_2 - t_2, h_3 - t_3, \ldots\ldots\ldots\ldots = 0$$

This indicates that the chances of the coin turning up heads or tails are 50:50.

A number of things have to be assumed when framing the above-mentioned theory. Firstly, it has to be assumed that the coin is not "defective" or manufactured such that it would always tend to fall on one side. Secondly, it has to be assumed that the thrower would not consciously or subconsciously will and manipulate to have the coin falling on one side all the time. Thirdly, it has to be assumed that the ground on which the coin falls is not "biased" in favor of one side of the coin. Fourthly, it has to be assumed that the shape of the coin is such that the coin is not "off-balance" or "unfair".

It is possible in practice that biases and errors cancel out each other so that their influence is practically nil.

It is also possible that an objective estimate of the probability of some occurrence would not turn out to be correct at all. That things just turn out to be entirely different and surprise or shock everyone is not impossible. For example, a top scholar who has topped his class since he started schooling might be expected to be at the top of his class when the results of the final examinations are out. This is extrapolating or forecasting his examination results based on his past examination results. But, what really are his chances of repeating his scholastic achievements? Would it be 100%, or, absolute certainty? But there have been cases wherein top scholars as such have not achieved the expected top honor, in some cases even failing the examinations altogether, though this seems rare. It would be reasonable for the statistician to carry out some research on the various top scholars or valedictorians to find out how many of them invariably or always place first in competitive examinations; for example, if out of 20 valedictorians for the respective last examinations sat for 18 valedictorians have been found to have been consistently at the top of the cohort (for every examination each sat for since they started schooling), the statistician could, based on this record, reasonably rate the chances of the top scholar described above repeating his past successes as 90% or so, provided that the scholar did not fall sick or feel unwell during the examinations. This estimate of 90% or so chances of being top would be better if the examinations concerned were more of the objective type, e.g., examinations of a mathematical kind, rather than the more subjective, essay-writing type. As for the essay type examinations, some experiment conducted has proven that the same

https://doi.org/10.1515/9783112214367-048

answer script evaluated by different examiners could end up with a wide range of awarded marks varying from failed to distinction. Apparently, the psychological confidence or morale of the top scholar is a big plus factor in his favor.

However, surprises and shocks do occur when expectations or predictions fail. Is the law of probability reliable? The law of probability, if rigorously worked out, appears to be reliable in general, as is well manifested in the thriving insurance industry, which insures against the various risks. An undertaking might not appear risky as the calculated risks of failure are 20% only, for example, but the risk-taker should realize that he could be one of the 20% of "unlucky" ones. To estimate risks or make forecasts, one could just venture an intuitive guess, a guess not based on solid data but just gut feeling. However, if time permits, the person should go about it the more scientific way, by gathering statistical data and applying statistical techniques when computing or calculating risks or making forecasts.

Risk takers such as insurance companies have statisticians or mathematicians computing risks and probabilities and despite underwriting many risks faced by commercial enterprises and individuals, practically all of them pull through without being too badly hit with "bad luck". Many insurance companies are heavily backed by banks, which could absorb their financial losses. The banks themselves insure against the risks of giving out loans by obtaining collaterals from the loan takers, which could be easily sold for cash to recover the loans should the loan takers default on repayment. However, there have been banks, which have gone bust though this is apparently rare. On the whole, banks appear strongly favored to succeed and prosper, and play an important role in the economy.

Consider the case of a person, whom let us call Mr. Good Birth, who enjoys a favorable environment. He is, say, highly educated, rich, from good family background, with a great deal of important connections, in good health, happily married, with practically all the good things in life, and living a happy, ideal life. Could anyone else be any "luckier" than Mr. Good Birth? Does this mean that Mr. Good Birth could be successful in whatever endeavor he undertakes all the time, or, at least, most of the time. Logically, he should succeed at least most of the time. However, something untoward or unfortunate could crop up and cause problems, e.g., there could be a war or some other catastrophe, which could destroy his properties and life, he could meet with a bad accident, perhaps a fatal one, he could be harmed by somebody envious of him, or any other incident that is considered "bad luck", or untoward things that are normally beyond the person's control – surely, no person could have prevented or avoided things such as a war, an earthquake, a flash flood, and a back stabber. When considering a person's chances of or potential for success, we could look at the tangible factors such as the person's admirable qualities and his privileged background, but we should not forget the unforeseen obstacles, which might only be too ready to destroy their chances. The author recalls the great British economist, John Maynard Keynes, imploring everybody not to think too much about what was going to happen to the economy because (as he put it) "in the long run we are all dead". Probably, in

Mr. Good Birth's case, his chances of being successful in life would be very good. It would also depend on the external factors, some of which might be beyond his control; if he were living in tumultuous times, e.g., during a period of great anarchy or a war, he could have died before his time, before he could achieve any success, or even greatness, in life, such as, for instance, being dispatched to the front line during the war and being shot to death there, as had happened to many capable or brilliant youths during the two World Wars, including a number who had been primed for greatness. On the other hand, some of these people could have the "luck" to survive the tumults or the threats of death and live on to achieve great success in life. For instance, the great Irishman, Michael Collins, who had held high political office, and the charismatic American president, John F. Kennedy, who incidentally came from a wealthy, privileged family, had both been assassinated while holding their respective high offices, while in the case of the great Chinese leader, Premier Zhou En-Lai, he had once been imprisoned by the enemy Kuomintang forces and had almost been beheaded by his captors but had been helped to escape from imprisonment just before beheading and had lived on to become one of the most important leaders in China, having achieved the powerful position of prime minister – had he not escaped from prison and had he been beheaded he would never have had become the great Chinese leader and prime minister that he was, and, his name, Zhou En-Lai, might never have been known to posterity at all. Such are the vagaries of fortune or luck.

Some optimists might claim that we control our fate by creating our own fortune or luck. It is perhaps true that to some extent we determine our own fate. For example, if we consciously work hard toward achieving our goals in life we stand a good chance of achieving them. But is it not clear that there are many external factors beyond a person's control, which could upset his plans or goals?

The branch of mathematics, which deals with the computation of probability or chance, is known as statistics; there are tables and formulae, which could be utilized to compute probability and correlation between events. In our important decisions with regards to risks, statistics provides a relatively reliable way to deal with the matter. There are risks involved in many activities in life, e.g., business activities, which require proper management, and statistics helps us to properly manage these risks. Many people, e.g., business people, might rely on mere intuitive guess-work and past experiences while managing their risks. Though this might prove greatly successful in some cases, would it not be safer to combine intuitive guesswork with statistical techniques (which represent the more scientific method)? Of course, some would prefer to rely just on pure statistical techniques and debunk the human intuition, but the effectiveness of the human intuition, especially one that has been honed by many years of experiences, should not be underestimated. The risks attendant to any activity, as well as their extent, should be known and prevented, or minimized. In certain hazardous activities such as shipbuilding or repair and building construction, which are prone to often fatal accidents, there are usually safety specialists to look out for potential hazards, which, if spotted, would be rectified so that accidents would be prevented. It is actually quite pointless to know the risks and their probability of becoming a reality unless something is done to eliminate or at least minimize them; at least, then, we would have some control over our fate, fortune, or luck, whichever we might wish to call it.

Fortune or luck somehow implies that there is or are probably some higher, powerful entity or entities influencing or manipulating the courses of our lives, which explains why so many people pray to various deities asking for help, good fortune, or good luck. Though such things are evidently beyond logical explanation or proof at the moment, they are intuitively plausible. If all this were indeed true, a logical way to overcome life's obstacles would be to plead for help from the higher, powerful deity or deities influencing or controlling our lives, though some quarters who only believe in what could be concretely proven would probably not approve of such an idea. Though to the really rational, all this might not seem logical and might appear laughable, we have to admit that many aspects of life cannot be explained or controlled by rationality or logic, which seems to point to some higher, incomprehensible power or powers at work. As Blaise Pascal, a famous philosopher and mathematician, had posited, there is everything to gain and nothing to lose by praying to and asking for help from a higher, powerful Being, a God, which is actually common sense, especially if it gives one a feeling of calm, security, confidence, peace, and happiness, and does no harm to others. In fact, many have believed and claimed to have achieved success, peace of mind, and happiness by doing so. Such an apparently illogical act of seeking help from an unseen, unproven higher, powerful Being might be logical after all if it brings confidence, sense of security, and happiness to the person who also achieves what he sets out to achieve. The end justifies the means.

24 The Discrepancy in the Logic of Algebra

All of us are positive that $1 + 1 = 2$ or $2 + 3 = 5$. But, has anyone ever thought why this must always be so?

In the author's opinion, $1 + 1 = 2$ or $2 + 3 = 5$ are, at best, only approximations in the abstract. What exactly do these two equations mean? How can we interpret these two equations? We can interpret them (or misinterpret them) in at least several ways.

Firstly, we ask ourselves 1 (what?) + 1 (what?) = 2 (what?) and 2 (what?) + 3 (what?) = 5 (what?).

Secondly, we may interpret the above equations as 1 (pole) + 1 (pole) = 2 (poles) and 2 (poles) + 3 (poles) = 5 (poles), for example. Hence, we are saying $1 + 1 = 2$ and $2 + 3 = 5$.

Thirdly, we may construe that $1 + 1$ is not necessarily equal to 2 or that $2 + 3$ is not necessarily equal to 5. "How come"?, you may wonder. Look at the following illustrations:-

a) Can't 1 (2 ft. long pole) + 1 (1 ft. long pole) = 3 (1 ft. long poles), i.e., $1 + 1 = 3$, for example?
b) Can't 2 (2 ft. long poles) + 3 (1 ft. long poles) = 7 (1 ft. long poles), i.e., $2 + 3 = 7$, for example?

Fourthly, can 1 (apple) + 1 (pear) = 2 (apples), or, 2 (apples) + 3 (pears) = 5 (pears), for example? Apparently, these two equations will hardly hold water now.

We can "justify" the equations $1 + 1 = 2$ and $2 + 3 = 5$ by ensuring that the numerals such as 1 and 2 in equation $1 + 1 = 2$, and, 2, 3 and 5 in equation $2 + 3 = 5$, share a "common property". For example, 1 (gram) + 1 (gram) = 2 (grams), and hence $1 + 1 = 2$ ("gram" being the "common property"); 2 (2 ft. long poles) + 3 (2 ft. long poles) = 5 (2 ft. long poles), and hence, $2 + 3 = 5$ ("2 ft. long poles" being the "common property").

But, can we be absolutely certain, e.g., that 1 (apple) + 1 (apple) = 2 (apples), "apple" here being the "common property"? In other words, could 1 (small apple) + 1 (small apple) = 2 (big apples) ? This evidently could not be so.

Let us look at the following equation concerning apples:

$$1\,(\text{small apple}) + 1\,(\text{big apple}) = 2\,(\text{big apples})$$

Is the above equation valid? This equation can or cannot hold water, depending on whether some "common property" exists or not. Let us look at the following illustrations:-

a) If, e.g., 1 (small apple) + 1 (big apple) = 3 (grams of apple) and 2 big apples = 6 (grams of apple) then, of course, this equation will not be valid (for $3\text{ g} \neq 6\text{ g}$).
b) Similarly, e.g., if 1 (small apple) + 1 (big apple) = 7 (cubic centimeters of apple) and 2 (big apples) = 14 (cubic centimeters of apple), this equation will not be justifiable (for $7\text{ cm}^3 \neq 14\text{ cm}^3$).

https://doi.org/10.1515/9783112214367-049

This equation can be justified when the following conditions are present:

a) If, e.g., 1 (small apple) + 1 (big apple) = 4 (grams of apple) and 2 (big apples) = 4 (grams of apple), in which case, the “4 g of apple” is the “common property” belonging to each side of the equation, i.e., when we, in effect, have “4 g of apple = 4 g of apple” (here, common sense will tell us that the big apple on the left side of the equation is heavier than each of the two big apples on the other side of the equation).
b) Also, if, e.g., 1 (small apple) + 1 (big apple) = 7 (cubic centimeters of apple) and 2 (big apples) = 7 (cubic centimeters of apple), in which case, the “7 cm^3 of apple” is the “common property” belonging to each side of the equation, i.e., when we, in effect, have “7 cm^3 of apple = 7 cm^3 of apple” (here, common sense will again tell us that the big apple on the left side of the equation is of greater volume than each of the two big apples on the other side of the equation.

In (a) above, 1 (small apple) + 1 (big apple) = 2 (big apples) in terms of weight (grams), while in (b) above, they are equal in terms of volume (cubic centimeters).

Thus, “common property” here should be a form of “precise description”, e.g., precise measurements such as weight and length. For example, “4 g” and “7 cm^3” above are “precise properties”. Things can only be considered equal if they display similar measurable, even discernable, properties or qualities such as weight, volume, and length, and, perhaps, *physical features or appearance (*here, we should specify that things are equal because they are “equal” or similar in physical features or appearance – physical features or appearance tend to be subjective and difficult to judge, unlike the measurable properties such as weight and volume; physical features or appearance would give rise to sets, classes, types, or species, hence, the algebra of sets or classes (set theory), for example).

Imprecise descriptions, which cannot be precisely measured or discerned, will not do. For example, 1 (unhappy man) + 1 (unhappy man) = 2 (unhappy men) may not be justifiable, for we cannot measure happiness or ascertain the degree of happiness in each of the men.

So, will, e.g., one plus one always equal two, or, two plus three always equal five, et al.? It depends on how we interpret these algebraic equations. There should be more precision (of interpretation) to algebraic equations. Otherwise, algebraic equations may not really make much sense if we look deeply into them, as illustrated above.

25 The Logic of Infinity

What is infinity? Is it a number, a quantity? Perhaps, we could define infinity as an unquantifiable number, a rather self-contradictory definition. The lay person practically does not bother about infinity as he is apparently concerned about the now, and the not too distant future. After all, no one could live forever or to infinity. Scientists have even predicted that life on earth would end some time in the future, the result of some terrible catastrophe. So, what is the point of being so concerned about infinity?

Infinity is apparently an invention of the intellect, an abstraction. Everyone would have no problem understanding what infinity is all about and would realize that it is something that would always be physically and empirically unattainable; one would certainly be never able to prove infinity by any physical experiment, though one might think that one could prove infinity by logic as has been the case of mathematics, wherein mathematicians have been proving or trying to prove the infinitudes of, e.g., the primes, the twin primes (as in the twin primes conjecture), the even numbers greater than the number 2, which are each the sum of two prime numbers (as in the Goldbach conjecture), and the zeros of the Riemann zeta function (as in the Riemann hypothesis). We could of course conceive of an infinite outer space but could we be absolutely sure that outer space is infinite, a space that extends straight away from us (and not in a circular way) forever, or, to infinity? For all we know, outer space might be just a circle or circular entity, or, a Mobius Strip-like space, where we could go round forever in circles, as a number of scientists has conjectured, a concept that is more in line with our empirical experiences and which appears more acceptable or credible. However, as the saying goes, fact could be stranger than fiction. Immortality, or being able to live forever, e.g., is an imaginable concept and could exist in fiction, e.g., in a fairy tale, but everyone in his right mind knows that it would never be attainable; hence, nobody ever tries to prove that immortality is possible, achievable, or attainable. When mathematicians prove or try to prove that the prime numbers, twin primes, even numbers greater than 2 that are each the sum of two prime numbers, and the zeros of the Riemann zeta function, et al., are infinite, they are apparently implying that the numbers concerned, which are the objects of their study, are real and have a “life” of their own. This is apparently disputable, and all mathematicians apparently do not have a consensus on this. If the above-mentioned numbers, which are the objects of study of a number of mathematicians, might be just inventions of the mathematicians’ minds, then should they not be studied or considered only with a “pinch of salt”?

To prove that these numbers, the objects of study of a number of mathematicians, could go on forever, ad infinitum, it is apparently necessary to prove that the space (physical space) in which these numbers would occupy (e.g., when being listed, i.e., put on paper, such as a computer printout) is infinite (mental space could never be

https://doi.org/10.1515/9783112214367-050

infinite, as nobody could live forever and count forever). However, this space (which contains the listing or computer printout) would extend into outer space. Should we not therefore try to prove that this space, outer space, is infinite, i.e., this space goes on forever, which would allow the above-mentioned numbers, which are the objects of study of a number of mathematicians, to go on forever as well? But, our understanding of outer space appears limited. We apparently do not know what is actually out there in outer space and what it is actually like. We could of course conceive, or, imagine it to be infinite, but its infinity is neither really proven nor a certainty. As mentioned elsewhere in this tome, an infinite outer space would imply infinite possibilities, including things that are apparently impossible. Perhaps, at some point toward infinity in outer space, our physical laws and logic would not apply at all. Perhaps, at this point and beyond, life, consciousness, and logic could not and would not exist, and something, or many things, or an infinite number of things unimaginable, miraculous, or incredible, would exist or occur.

It is probably easier to prove the existence of God than the infinitudes of the above-mentioned mathematical objects, the various kinds of numbers. In fact, two reputable mathematicians in the past, viz., Rene Descartes and Kurt Godel, had produced proofs or reasoning that supported the existence of God. God as a sentient Being (man is also a sentient being) might be able to provide evidences of His existence, e.g., through the workings of physical phenomena, and might even be able to make His presence physically felt, e.g., someone might claim to have seen God or heard the voice of God. As a sentient Being, God could be expected to have His existence confirmed or proved in this manner. But the infinitudes of mathematical objects such as those described above belong to a different genre. An infinitude is not a sentient being that could be touched, heard or seen with the eyes, but an abstract noun, an abstract object, an object that might not be real at all but is just the result of the human imagination. For example, we might use our imagination to think of something absurd, e.g., we might imagine people living without food and drink, people living without breathing, people living forever, and things falling toward the sky on their own without outside force. Since we could conceive such absurdities, should we regard them as realities? Obviously, no. Such things have never happened in life and would apparently never happen in life. Similarly, no one has ever lived to, counted things to, or carried out activities to infinity, though everyone can conceive infinity. Therefore, infinitude or infinity should not be considered real, or taken so seriously (only imaginary or conceived, though neither "could be experienced", really knowable nor, thus, really provable, though practically all mathematicians appear to think it is provable).

The author recalls the responses of some editors of some of the mathematical journals to which he has submitted his article on the solutions for the twin primes conjecture. In the article, the author has shown several "constructions" or ways of producing an infinite list of twin primes. But these several editors commented that these "constructions" would not work at some point toward infinity, without giving any counterexample or other reason to prove their point. Of course, such a postulation based apparently on gut feeling and not concrete evidence or sound reasoning, such as a counterexample, is unfair and unacceptable. However, in accordance with the arguments presented above, there seem to be some grain of truth in what these editors said. It seems that mathematical problems such as the twin primes conjecture and the Goldbach conjecture are meaningless and really have no sensible solutions, i.e., problems involving infinity, such as the above-mentioned mathematical problems, are meaningless and not really solvable at all (and any so-called solutions found would probably be illusions only). This implies that practically everyone involved with such problems might have wasted their effort and time. In our working life, we deal with all kinds of numbers and quantities but there are surely some finite limits to how large these numbers or quantities would be, which would depend on the capacity of our computers. A number that is extremely large, e.g., a number with millions of digits, would be impossible to name and would be meaningless, and would thus have practically no utility – it would be just a curiosity. It would appear more reasonable, realistic, practical, and, useful to carry out search for the above-mentioned mathematical objects, e.g., the twin primes and the zeros of the Riemann zeta function, up to only a certain realistic but sufficiently large or gargantuan numerical limit, e.g., numbers with some millions, billions, trillions, or zillions of digits only, computing facilities permitting (which would of course result in a checkable proof, a proof that could be checked and confirmed with a computer, though many mathematicians of the older school might consider this kind of proof inelegant). There is no practical use in being concerned with what happens beyond this limit, this point, toward infinity, where at some point, our laws of nature and logic might not apply anymore. It would probably be better (not to mention easier too) to prove the existence of a Supreme Being; if He really exists, the human race could really implore Him to help or bless them. The human race certainly cannot expect any help from the infinite prime numbers, the infinite twin primes, the infinite even numbers that are each the sum of two prime numbers, the infinite zeros of the Riemann zeta function, et al., and, solving such mathematical problems would apparently contribute practically nothing at all to the well-being of the human race.

Just understanding the nature of infinity should be good enough. However, the great mathematician Georg Cantor had another interpretation for infinity, which has apparently been so bizarre that he had been greatly criticized by a number of his peers, though it is now generally accepted by mathematicians. According to Cantor, there are different levels or sizes of infinity, which could be denoted by the respective cardinal numbers. This seems to be paradoxical, as it appears to be against common sense. Could we then not utilize Cantor's interpretation of infinity to conjecture that there are different infinities of prime numbers, twin primes, even numbers that are each the sum of two prime numbers, zeros of the Riemann zeta function, etc? (Cantor's interpretation of infinity might be construed to imply this, which would complicate matters.) To Cantor, the infinite set of all real numbers (algebraic plus transcendental) has a higher cardinal number than the infinite set of positive integers, i.e., the infinity of the infinite set of real numbers is larger than the infinity of the infinite set of positive integers. What could be the rationale behind this? The author's interpretation of this rationale is as follows. Assume that x is a point at infinity for both the infinite set of real numbers and the infinite set of positive integers. Assume that a is the number of real numbers between 0 and x in the infinite set of real numbers. Assume that b is the number of positive integers between 0 and x in the infinite set of positive integers. Then, $a > b$, which means that the set of real numbers between 0 and x is larger than the set of positive integers between 0 and x. If $x = \infty$ (infinity), then the infinity of the infinite set of real numbers is larger than the infinity of the infinite set of positive integers. Is this reasoning really sound, though on the surface, it looks plausible? The author here ponders about a corollary. Consider a house, y (infinite set y). There are chairs c (real numbers) and tables t (positive integers). Does it mean that the house (y) with p number of chairs (c) is larger than the same house (y) with q number of tables (t) when $p > q$? It does not make sense anymore here. Similarly, Cantor's interpretation of infinity appears unsound.

As infinity should not be considered real (as explained above), mathematical problems having to do with it, such as the twin primes conjecture, the Goldbach conjecture, and the Riemann hypothesis, are not realistic and are hence meaningless. A suggestion here is to replace the "infinity" of such mathematical problems with a "realistic numerical limit approaching infinity", a realistic but sufficiently large or gargantuan numerical limit that is countable, quantifiable, or computable (and large enough to reasonably represent infinity). This "realistic numerical limit approaching infinity" could be regarded as a "point at infinity" or one of the elements, components or atoms of infinity, and could hence be regarded as representing infinity itself. There is actually no point in having a numerical limit or "point at infinity," which is so forbiddingly large that it is incomprehensible, meaningless, and useless. Numbers or magnitudes range from the infinitesimal or infinitely small to the infinite or infinitely large, i.e., from $-\infty$ to $+\infty$ (∞ being the symbol for infinity). This could be represented by a diagram showing the various magnitudes or graduations, e.g., infinitely small, very small, small, medium, large, very large, and infinitely large, as follows:-

infinitely small ($-\infty$)	very small	small	medium	large	very large	infinitely large ($+\infty$)

Our "realistic numerical limit approaching infinity" or "point at infinity" could be any of the numbers within the box designated "infinitely large", e.g., a number immediately after the boundary between "very large" and "infinitely large"; the "infinitely large" box could be regarded as representing infinity, or, infinity itself, and each and everyone of the numbers within the "infinitely large" box could be regarded as an element, component, or atom of infinity. This way of looking at infinity or treating infinity would make infinity a realistic and meaningful concept, and the mathematical problems associated with infinity would become more tractable. This actually represents another way of interpreting infinity and might appear self-delusive. Perhaps, it is better for mathematics to do away with the problems associated with infinity, problems such as the twin primes conjecture, the Goldbach conjecture, and the Riemann hypothesis, and modify such problems to problems dealing with the quantifiable or computable, problems that deal with the more concrete and more practical, problems that deal with the realistic, wherein finding an x number, a sufficiently large or gargantuan number, of twin primes, even numbers that are each the sum of two prime numbers, or zeros of the Riemann zeta function should be sufficient.

Finally, even if we could live to infinity, we could never reach infinity because infinity implies that it is inherently unreachable (without end or limit), i.e., if infinity is reachable, it is not infinity but a limit. All this implies that infinity is probably not real, and not realizable at all, if it is real. That is, if it is real, there would really be no way to prove or tell that there is indeed no end or limit to the infinite progression. To say that one is able to prove infinity is to say that one is able to achieve it, reach it, experience it, know it, indeed, or empirically exist or realize it, whence it would not really be infinity (infinity implies the opposite of all of these), a contradiction or paradox. Proving infinity by some other means, e.g., reasoning by "reductio ad absurdum" or contradiction, or, induction, would probably be just illusory, without any empirical or really solid basis, perhaps just a play of ideas (infinity, unlike other quantities, is not something that could be counted, measured, weighed, or easily visualized or conceived, but is a very highly abstract, even mind boggling, entity); both the proof by contradiction and the proof by induction, though they have been generally accepted mathematical proofs, are apparently not "foolproof" and without critics (e.g., those belonging to the intuitionist school of mathematics have raised objections to the proof by contradiction), having some apparent weaknesses (as explained elsewhere in this tome). This means that infinity is probably or likely to be just imaginary, an invention of the mind, a curiosity. This means that infinity should not be taken so seriously.

26 The Logic of Reality

So far, the many aspects of logic have been dealt with. But, does logic represent reality? What, in fact, is reality?

Reality has to do with existence and certainty – when something exists and its existence is certain (without doubt) to us, we could say that it is real. Certainty of existence implies consciousness – consciousness of the certainty of existence. A normal person relies on any (or, at least one) of his physical senses, viz., sight, hearing, taste, touch and smell, to ascertain whether a thing, e.g., a physical object, really exists. Without the physical senses, it would be virtually impossible to ascertain (or confirm) whether a thing really exists. The person could then at best only "feel", "guess" or "intuit" whether the thing really exists, but he could not be certain of whether that thing really exists; the physical senses which were needed to ascertain or confirm that thing's real existence could be unavailable to the person because he was unconscious, asleep, or even dead.

Reality could be regarded as the *consciousness* of the *certain existence* of an object or event. In other words, reality is the consciousness and the certainty of the existence of an object or event. As certainty (i.e., the consciousness of certainty) could normally only be achieved through the physical senses, viz., sight, hearing, taste, touch and smell (through the act of confirmation with any of these physical senses), consciousness and the physical senses must go hand-in-hand toward the creation of reality, which is actually the consciousness (or awareness) of the certain (or undoubted) existence of an object or event. We have to distinguish between consciousness and the physical senses. Are these two actually the same thing? The answer to this question boils down to perception or interpretation. However, they could be regarded as being the same thing. Sense implies consciousness. The physical senses imply specific consciousness or consciousness of some thing or things, or, event or events. In a person, the physical senses might give rise to the consciousness of the existence or presence of an object (the reality of the object, so to speak), i.e., the person physically senses (see, hear, touch, taste or smell) the object near him and regards it as real. But, when the person does not physically sense (viz., see hear, touch, taste or smell) the object and is thus not aware of the object's existence or reality (e.g., he could be blind and deaf), does it mean that the object is not real, i.e., does not exist? We could state that to this person the object in his presence is nonexistent and is not real. However, to another person nearby who could physically sense (i.e., see, hear, touch, taste or smell) the object, the object exists and is real. What would happen if the latter informs the former (presumably a blind and deaf person) that there is such an object near them? The former would then become conscious of the possible reality or existence of that object and could either choose to believe the latter (an act of faith and not a real consciousness and certainty of the existence of that object) or disbelieve him (whose reputation for trustworthiness would play a part).

https://doi.org/10.1515/9783112214367-051

It could thus be stated that reality is variable and subjective. That is, what is reality to one person might not be so to another person. Since this book deals with logic, we should be concerned with the question of whether logic per se is part of reality. However, what is logical needs not be real. For example, a fictitious train of events which is the work of the imagination might be technically logical but is certainly not real at all (only fiction or make-believe). On the other hand, what is real needs not be logical (and could be illogical), comprehensible or explainable, e.g., apparent miraculous or supranatural happenings, which defy explanation, comprehension, or, common sense, but which take place before our eyes and are thus undoubtedly real. So, what is the position of logic in this setting? Logic is but one aspect of reality. Both logical and illogical people could be found in the real world. We could state that reality comprises of the logical, illogical, and, unexplainable or incomprehensible, any such things which happen, take place or exist.

Reality depends on consciousness, on the interpretation of things as they are by consciousness. We have to have a clear understanding of what reality actually is. Reality is an abstract reality (something which is beyond our physical senses – we cannot physically see, hear, touch, taste or smell reality). We could only be conscious or aware of reality, which could be regarded as a thing of the mind, an abstraction. Reality is hence the consciousness and certainty that a thing, object or event exists. It might sound like a hazy or vague concept but it is an important concept nonetheless.

It would all boil down to consciousness. Without consciousness, no reality, and, of course, no logic could exist. Consciousness is paramount, most important, to human existence. Living is actually being conscious. A human vegetable, e.g., who has no consciousness, is technically speaking not living (though he might display other signs of living, e.g., breathing and involuntary movements of some parts of the body). Apparently, the consciousness and certainty of real existence (i.e., reality of an object) experienced by human beings concur in many or most instances. There would be greater rifts and troubles if they were to vary wildly among the human species, and, it would be very unfortunate. Since reality is intangible, i.e., it could not be directly sensed by the physical senses, it is in this respect not real, an apparent paradox. In other words, reality is unreal since we could not physically experience it. Yet, we apparently know it is there when we come across it. It is a mere thought, idea, concept, consciousness, feeling, abstraction. It is unlike other physical realities or physical objects which are characterized by tangibility, concreteness or solidity, which could be directly sensed by and physically affect some of or all our physical senses, thereby making our feeling of certainty about their existence stronger than otherwise would be the case.

We shall here consider a hypothetical case regarding reality. Now, consider a physical object such as a pen, torch-light or chair right in front of you, and, the reality of this physical object. How do you prove the reality of this physical object? In other words, how do you ascertain that this physical object is real? To determine whether a thing or object is real or not we need some criterion or criteria. One practical, certainty-boosting criterion is being capable of being sensed by the physical senses (sight, hearing, touch, taste and smell). Clearly, any physical object such as the pen, torch-light or chair would easily satisfy this practical, certainty-boosting criterion. In other words, the idea, or, abstraction, that the physical object is real is easily confirmed by this practical, certainty-boosting criterion, and, we know for certain that the physical object, e.g., the chair, exists, i.e., the chair is real.

We now have another abstraction. For example, assign a quality to the physical object; say, we state that the physical object is beautiful, useful or harmful (which is another idea or abstraction). In other words, we are stating that the beauty, usefulness or harmfulness of the physical object is real. But, how do we ascertain or prove that? Unlike the physical object (pen, torch-light, chair, etc.), beauty, usefulness or harmfulness is not physical or tangible, but intangible, abstract, and not sensible by the physical senses. These abstract entities are unphysical or invisible "objects" within the physical object itself. We could measure the physical dimensions of a physical object, e.g., a chair, weigh it or draw a picture or take a photograph of it. Could we do likewise for the unphysical or invisible "object" or "objects" within it, e.g., the physical object's beauty or usefulness? No, we could not and there lay the great difference and problem. The serious problem is how do we ascertain the reality of these invisible "object" or "objects" (beauty, usefulness or harmfulness, etc.)? What criterion or criteria could be used to ascertain its reality or existence? Certainly not the same as those used for physical objects as described above. We could even state that such invisible "objects" (or qualities) do not belong to (and are actually not found in) the physical object, but, resides or are found in our consciousness or mind. In other words, they only exist in our consciousness or our mind – they are created by our consciousness or our mind and are linked by our consciousness or mind to the physical object. That is, these invisible "objects" are actually outside the physical object itself and not part of it. Our consciousness or mind only assigns these invisible "objects", or, qualities, to it. If we, e.g., assign the label "beautiful" or "useful" to the physical object, the physical object would be beautiful or useful to us; if, on the other hand, we assign an "ugly" or "useless" label to the physical object it would be so to us. There are many cases of such physical objects being given different labels by different people, e.g., with some people assigning the label "beautiful", some others assigning the label "unsightly", some assigning the label "useful", some assigning the label "undesirable", some assigning the label "space-saving", some assigning the label "heavy", and so on. In simple terms, practically everyone would not doubt the reality or existence of a physical object, e.g., a table, as long as they could see the physical object, but they are less likely to agree fully about the qualities, e.g., beauty, usefulness, and so on of the physical object (which is subjectivity). Are such invisible "objects" or qualities real at all, or, just a creation or invention of the consciousness or mind?

Perhaps, we should divide reality into two types – the physical type and the theoretical, abstract or mental type. These invisible "objects" or qualities would belong to the theoretical, abstract or mental type of reality. The physical type of reality is objective, clear-cut and indubitable, while the theoretical, abstract or mental type of reality is subjective, not clear-cut and disputable. In ascertaining or proving the reality of any object, our physical senses are very important. For the physical objects such as chairs, tables, trees and cars, and so on, which are amenable to our physical senses, their existence or reality is indubitable as it could be physically verified by practically everyone. It is the objects of a nonphysical kind, e.g., ideas or qualities, whose reality is normally difficult to ascertain, prove or verify and doubtful (more like inventions or creations of the consciousness, mind or intellect), as they are beyond our physical capabilities (sight, hearing, touch, taste, and smell) to verify them. Take a physical object such as a chair with a modern design. We could see the chair, touch or feel it, sit on it, measure its dimensions, weigh it, lift it, etc., but we are unable to do the same with a nonphysical "object" or quality such as beauty, modernity or usefulness. For example, ten persons see the chair. Practically all the ten of them would be positive that the chair exists and is real. But, probably only five out of the ten of them are likely to agree that there is beauty in the chair, while the rest either disagree or are neutral. So, are nonphysical "objects" actually real?

We here probe further. Since reality is not a physical object, is it real? Is reality also one of the aforementioned creations or inventions of the consciousness, mind or intellect? Perhaps, we could state that there are realities (truths and logic) which could be easily ascertained, proved or verified and there are those which could not be so or could only be so with great difficulty. How do we overcome this faux pas? The author believes that this faux pas could be overcome as follows. As reality is an abstraction (an abstraction is a creation or invention of the intellect or mind), a consciousness, mind or intellect which has never heard of the term "reality" and the concept of reality would not know or understand reality; therefore, to such a consciousness, mind or intellect, e.g., a child or a pure, innocent mind such as that of an imbecile, reality does not exist, though the tangible, physical objects of the reality itself could be sensed by the physical senses and could be regarded by the physical senses as being seeable, audible, touchable, taste-able or smell-able, but these sensations would not be equated with the (unheard of) reality. Perhaps, we could describe reality simply as a thought. So, if we do not think of reality it does not exist, and, vice versa.

On the other hand, could we state that when we do not physically experience, i.e., see, hear, touch, taste, or smell, a physical object, though we had physically experienced it earlier, it does not exist? We could have a choice of interpretation here. We could either regard this physical object as nonexistent or unreal if this physical object were out of sight, hearing, touch, taste or smell, i.e., out of the range of our physical senses, or, alternatively, still in existence or real if we realize or know that this physical object has not been destroyed by some agency. Of course, if this real, existent physical object has never been seen, heard, touched, tasted or smelt by the person, i.e., sensed by the person's physical senses, it would be nonexistent (not a reality) to that person. Since, as stated earlier, reality or existence is a thought, we could say that for a real object, an object which really exists, when we think of such an object it exists (in our consciousness or mind) and when we stop thinking of this object it ceases to exist (in our consciousness or mind). All this is a matter of philosophical or personal interpretation, personal inclination or taste, and is thereby arbitrary. We could state that reality is variable and subjective, differing from person to person.

Now, there could be an instance described as follows. A physical object is "seen", "heard", "touched", "tasted", or "smelt" by a person who is certain of its existence or reality, but it is only an illusion and not really in existence, while all the other people around him could not see, hear, touch, taste, or smell the physical object at all, which means that to these people the physical object does not exist at all and is not a reality. Would an observer looking at this scene consider such a "physical object" real or existent? A philosophical observer could conclude that the "physical object" is a reality or existence to the person who is suffering from the illusion while to all those around him the "physical object" is not real and nonexistent, meaning there are two kinds of reality here. This further proves that reality is not fixed, but variable and subjective. (In fact, dreams sometimes seem so real as to be confused with, or, regarded as real happenings.)

Perhaps, we could state that the only true, indisputable reality is consciousness (or, life itself, as it were). Without consciousness (i.e., if we have no life) there could never be thought, ideas, logic, emotions, senses, feelings, actions, and, reality. According to Arthur Schopenhauer, reality is actually the will, which drives all human beings to act in ways which satisfy their desires. But, without consciousness (or life), the will could never exist. Right? Assuming that mankind does not exist anymore, and, hence, human consciousness is extinct, does the existence of physical objects such as the rocks and the trees then have any meaning, any reality, anymore? Physical objects per se would not matter anymore, as there would not be any human beings around to experience (see, hear, touch, taste, or smell) them. There would then not be any reality, since reality is only a conception of the mind and a result of consciousness. The most fundamental requirement for the existence or reality of all objects is therefore the presence or existence of consciousness. That is, consciousness must first exist before reality could follow suit – in other words, reality is dependent on the mind or consciousness. This is the author's interpretation of reality. There is another school of thought, which stipulates that reality is independent of the mind or consciousness. This means that if the whole human race (i.e., consciousness) were extinct, the reality of all the remaining physical objects in the world or universe would still remain. In this school of thought, the interpretation of the term "reality" must be different from that of the author's. Reality to the author is a thought, concept, abstraction or idea, and, would cease to exist with the extinction of consciousness. At the metaphysical level, we could perhaps state that as long as we are conscious of an object (as long as we think of an object), immaterial of whether the object is tangible or intangible, physical or abstract, the object exists (at least in our mind or consciousness), and vice versa.

Reality, being an abstract entity, might only be understood by the relatively intelligent mind, which has attained a certain level of intellectual sophistication. To the less intelligent or the uninitiated, e.g., an illiterate person, reality could be meaningless and, hence, nonexistent; this group of people might interpret what we understand to be reality in another manner, e.g., to them reality could be mind, inventiveness, imagination, desire, and emotion, not forgetting that to a renowned philosopher such as Arthur Schopenhauer it is the will. All this seems very much like a play of ideas or words. Perhaps, reality is simply a play of ideas. It could thus be concluded here that there are different types of reality, subject to the mind's capability at conceiving them, subject to the mind's way of interpretation. In fact, the level of thought or abstraction could be so great, as in the case of highly abstract mathematics, that new and bizarre reality could result. That is, new realities could be created or invented (some would prefer to call it "discovered", which is again a matter of interpretation) by the mind or consciousness, whose importance is undeniable. So, is reality (which could be equated with truth or fact) something which is there all the time which has to be discovered, or, something which is created or invented in the mind? It depends on individual interpretation. To the author it is both. Some realities are apparently more obvious than others; the physical objects, which are amenable to the physical senses (sight, hearing, touch, taste, and smell), belong to this category, and, no one who is in possession of these physical senses would be able to deny the existence or reality of a physical object which is right in front of him – this may not be the case with a nonphysical, intangible or abstract "object" or entity. Perhaps, only the physical objects are real, which could be another interpretation of reality. A strict criterion for judging or evaluating reality would be indisputable, absolute lack of doubt or indisputable, absolute certainty that something is real, and, the reality of the physical objects would doubtlessly meet this criterion. By this interpretation, therefore, only matter is real (the physical objects are matter, which is made up of atoms). We could extend this reasoning. Earlier the author has stated that perhaps consciousness is the only true, indisputable reality. But, consciousness or mind comes from the brain, which is matter (often referred to as "gray matter"). In other words, without the brain, or matter, consciousness would not exist. Hence, matter is the true reality.

There could be so many ways of interpreting reality that reality could probably be regarded as just an interpretation of nature. Since coming to terms with reality has been circuitous so far, this is perhaps the most clear-cut and best definition of reality. This definition implies that reality is different things to different people; it implies that there are different types of reality (which depend on individual interpretation). This apparently causes problems and conflicts among people. It is of course better to have only one type of reality, a reality which is indisputable. Apparently, only the reality of the physical objects would meet this criterion of being indubitable to practically everyone.

We elaborate further on this point. There are objects which are real and whose reality could be very easily ascertained or proved, e.g., physical objects, whose reality is obvious, and, objects which are apparently real but whose reality could not be ascertained or proved, or, could only be ascertained or proved with great difficulty. It appears that the more abstract the object is, i.e., the higher the level of abstraction is, the less real the object would seem, and, the more difficult it would seem to ascertain or prove the object's reality. The level of abstraction could be so high as to be totally out of touch with physical reality and common sense, e.g., the case of paradoxes and very complex or unconventional ideas. Some might regard this high level of abstraction, high level of thought or consciousness, as a higher reality or a different reality, e.g., the very abstract areas of mathematics which are beyond the grasp or comprehension of the uninitiated or nonspecialist. The important point about reality is the degree or strength of the feeling of certainty, the degree or strength of the intuition, that a thing or object is real, i.e., the reality should be convincing. The problem is that this feeling or intuition tends to vary from person to person in the case of abstract entities, e.g., ideas, thoughts or concepts, i.e., the reality of abstract entities tends to be disputable, controversial or uncertain. It is a truism that each of us lives in a mental world (or world of realities) of our own, and, this is a problem.

Reality is merely an idea, concept or abstract entity, and, a descriptive term or quality assigned to a thing or object, whether tangible such as a physical object, or, abstract and intangible such as a concept. It is an abstraction and a quite highly abstract entity. As mentioned above, it is also a kind of feeling or intuition. Earlier, it has been stated that there are different types of reality, subject to the mind's capability at conceiving them, subject to the mind's way of interpretation. This apparently leads to disputes and conflicts. If only there were just one standard, fixed, and objective reality. We would all then become the better for it. A simple, practical criterion for reality, whether of a physical or a nonphysical, abstract entity, could be indubitable certainty (i.e., indubitable obviousness) of the entity's reality, existence or truth felt by practically everyone who encounters the entity. Such indubitable realities would include the physical objects of course and nonphysical, abstract "objects", or, truths, such as, e.g., the universal knowledge that gravity affects all matter or no human being can live forever, and so on. Apparently, this indubitable certainty could best be achieved through the utility of the physical senses (sight, hearing, touch, taste, and smell), e.g., through experimentation and observation, which represents the more scientific way. This is probably the most practical and straightforward solution to the problem of reality.

The problems pertaining to reality imply the problems of logic, an abstract entity which is an aspect of reality itself.

27 Will the Laws or Logic of Physics Always Be Useful?

An expert physicist on the earth will probably be an ignoramus in a faraway universe. There is apparently an infinitude of universes out there, and there are bound to be infinitely many universes which can be expected to be completely different from ours. In other words, while we may be experts on the physics of the earth we are likely to be ignoramuses where the physics of extreme outer space is concerned. On the earth, we may be familiar with gravity, light, the common gases such as oxygen, nitrogen and hydrogen, atoms, and, consciousness. In a very faraway universe, all these may not be found, and, instead, different material forms may exist. We human beings cannot be expected to survive in such an environment. But there may be different forms of life, or, existence. Here, we have to be clear about what we mean by "life". Clearly, to be alive on the earth is to have consciousness, and for trees, plants and vegetation and other related living things to be able to consume some forms of energy and change accordingly. Life-forms, if any, in a faraway universe may be so different as to be unnoticeable, even invisible to the human eye and inaudible to the human ear. There is therefore no way that we can understand these vastly different environments with the knowledge of our very own physics. To understand them, we have to learn the physics of these new environments. Of course, humans may not survive in such environments. Beings fitter and able to adapt to such environments may survive. Human logic will probably not be useful, i.e., it will probably not work, as causes and effects can be expected to be very different from the earthly ones.

What can be expected to be found will probably be so incomprehensible that they are regarded as miracles?

If you are an Einstein or a scientific genius sent to carry out research in such a faraway universe, will you not be reduced to a scientific moron as there will probably be so little that you will understand with your great knowledge of physics? It will probably be like watching a magic show, a miracle happening, with plenty of surprises and shocks. And, what will happen if harmful or malignant life-forms are encountered. You may not even be aware of them, being so different. What happens if these hostile forms attack you and even try to attack the Earth? How will you fight an unknown, probably unseen, and inaudible enemy?

All this is mind-boggling and thinking seriously about it seems beyond our wildest imagination. In such an environment, do you not think that our knowledge of the laws of physics offers hardly any benefit and comfort? Do you not think that we will not be able to reason or think logically in such an environment but will instead be awed by what is happening around us due to lack of comprehension?

https://doi.org/10.1515/9783112214367-052

28 Practical Logic

Does being logical always pay? The author has been "experimenting" with logic since his younger days, his teens, especially in his relations with his fellow men, e.g., his peers, friends, associates, even strangers, all kinds of people, foolish, average, and smart people. In his forays or "adventures" in logic the author has received all kinds of reactions from other people – he has been appreciated, praised, scorned, rebuffed, envied, argued with, and faced all kinds of different reactions from all sorts of people. The trouble is, the author has found out, not everyone could appreciate logic, i.e., pure logic; many even had their own prejudiced logic. Worse, there seem to be many kinds of logic, many kinds of ideas, concepts, or opinions. Most human affairs are apparently characterized by ego, cultural bias, religious bias, motives, self-interests, emotions, prejudices, and preconceived ideas. It is apparently far from being a case of precision of thought (whereby, e.g., if you add one to one you would always get two). Unfortunately, most human affairs do not work out with this kind of arithmetical precision, though it would be what we apparently wish for.

The author has found out, often to his own disadvantage, that logic, even in a highly intellectual field of activities such as mathematics, does not always work fine. One might argue that the author had been having defective or faulty logic, which had resulted in rejection. But, assuredly, that is not always the case, most probably, often not the case, and, it would be wrong for the author to praise or regard his own logical capacity too highly. Life is indeed full of contradictions. For instance, if one were to be foolish, one would be scorned of course. So, one should be smart, to earn the esteem and respect of one's fellow men. But, when one is too smart, especially when one acts too smart, i.e., becomes a smart aleck, one could be envied, rebuffed, and scorned as well. So, act average? Perhaps, acting or being average might be the "safest". But, the author thinks that it is better to be natural. Ironically, a person's potential for advancement in life is based on his apparent intelligence or smartness. So, one has to display one's intelligence or smartness and get recognized for it in order to gain advancement or promotion. But this intelligence or smartness could just be a display, without any substance. Unfortunately, the powers-that-be, e.g., the bosses, apparently are often impressed, and duped, by such inconsequential displays of intelligence. Many people might not have an intrinsically smart or logical intelligence, but a low animal cunning. They are comparable to those low animals which, though apparently not as intelligent as humans, are able to outwit their human opponents, e.g., escape the traps laid out for them, raid their food stores, attack and eat them, etc. In this respect, there are criminals with apparent low IQs who are able to outwit the authorities. Perhaps, these animals or criminals are intelligent in some peculiar ways. Intelligence could here be equated with logical capacity.

The author had been naïve in his younger days in believing that being logical would help avoid or solve problems. This is true to some extent only. It is no point

https://doi.org/10.1515/9783112214367-053

reasoning with people who are either not capable of reasoning or not interested in reasoning or both. He had found to his chagrin that his so-called logic had frequently fallen on deaf ears, which had made him wonder what the use of logic was. Then, there are some people who are apparently intelligent but whose ideas are different from yours. If logic or intelligence were so useful and helpful, we could expect our top officials and leaders to be the most intelligent of the lot. But, often, this is apparently not the case. In a democracy, leaders are elected. If a person were perceived as capable and trustworthy, though he might not really be so, and if he were liked, he would probably be elected as a leader. In a dictatorial or harsh regime, a leader need not even be smart or intelligent. Those who rise to positions of power are likely to be courageous, risk-takers, and ruthless, without necessarily being smart or intelligent, though it would probably help.

The problem with human relations is not that there is not enough of logic being used. There is probably a lot of logic being used by many people. But the logic is used on many different premises, with many simply wrong, many simply prejudiced, many not true at all. The brain is actually a computer whereby if one inputs garbage the output is also garbage. The author has, e.g., come across apparently intelligent people, some of whom are staunchly religious, while some others are avowed atheists, and some rather neutral or "free-thinking". Is there any point in arguing with the staunchly religious that God does not exist, or, the reverse with the atheist? After all, such things belong more to the spiritual than the intellect. There are all kinds of people with all kinds of interests, likes, and dislikes, and their logic or thought would obviously very much revolve round these things. For example, it would be quite pointless to argue with a sports car enthusiast about the bane of fast driving or with the staunch Christian that attending church is a waste of time. Either party would probably have many reasons, many of which valid, to counter one another. Since such "problems with logic" are not easy to resolve, it would be necessary for the parties involved to respect each other's ideas, thoughts, or opinions and to compromise or give and take in order to achieve harmony.

Everyone should be concerned with using logic for results rather than just for the sake of being logical. Being logical or appearing logical might win the respect of one's peers, which is likely to give one pleasure or self-esteem. But logic should be employed to achieve more positive results, e.g., for solving problems, creative work, and betterment of society. The important point is that logic should help bring harmony and happiness to everyone in the community, and not conflicts, rifts, and problems, the latter appearing to be more so the case. This represents a "macro" view of the role of logic. Societal harmony and happiness are important, as all human beings are interdependent. For example, if only one out of the 100 people in the community is happy while the other 99 are sad and suffering, would the sadness and suffering not affect the only happy soul? Is it not better to have 100 winners than one winner and 99 losers, e.g., in our community or society? Unfortunately, in the world today, the gap between the haves and the have-nots seems to be widening. Would it not be better to have everyone in the world living together in relative comfort, harmony, understanding, cooperation, and happiness, than to have some very well-off, some not that well-off, and the others faring badly? Unfortunately, in the "dog-eat-dog" world of today, success seems to be often achieved at the expense of others. Of course, not every scholar, athlete, or business entity could be top or number one – if there were a top, there has to be the lower levels, e.g., the mid-ranking and the bottom. If there is a first, there has to be a last, unless positions and other status symbols are abolished. Human logic dictates that democracy and meritocracy are ideals in a free world, but how is it that there are still so much injustices and sufferings? A democracy might not be better off than a feudal state, if it were only a "guided" democracy. A meritocracy might discriminate against those highly intelligent people who dislike academic studies or are unable to perform well in academic studies due to some personal problems. Our logic should be applied wisely to the solution of such very real societal problems.

So far, we have been dealing with the various areas in human activities involving logic. Evidently, we need logic to get things done, i.e., we have to be able to convince others with logic, whether in sales or business where we have to convince the customers with business logic, in politics or the government where we have to convince others of what we could do to better their lives so that they would give us their votes, the boyfriend who has to convince the girlfriend to marry him, etc. Various kinds of logic might have to be involved. Of course, emotional appeals could also be used, which are beyond the scope of logic. Our concern here should be how to use logic for results rather than just for the sake of being logical. The first important step toward achieving results with logic is clearly to use logic on people who are really capable of logical reasoning. The problem is how are we to know whether a person has a logical mind? Perhaps, we could judge such a person by his credentials, e.g., his education and his intellectual achievements. Next would probably be his conversations and ideas. Then, his occupation, his manner and appearance, his family background or ancestry, his interests and hobbies, his apparent interest in things intellectual, the books he reads, and others. Unfortunately, even if the person were indeed logical, logic might not be able to convince him, for, as mentioned before, logic is variable and not something definite or "encased in concrete". Evidently, highly intelligent people, people with apparent great logical capabilities often disagree or differ in thought and reason, i.e., there are apparently many kinds of logic. We should of course not be logical just for the sake of being logical, though we ourselves have first to be logical if we want others to be logical, or, reasonable. It is pointless trying to reason or use logic with a moron, a human vegetable, or a brute lacking in logical prowess.

Here, the author would digress a little and touch on business logic, which is apparently an important area in human affairs. Many successful businessmen pride themselves on being smart and capable when they have achieved business success. Ironically, many business people are not academic star performers, being quite the reverse, giving the impression of being dull or lacking in intelligence, and hence, by association, lacking in logical capacity. On the other hand, many successful businessmen have been credited with good qualities such as foresight, shrewdness, intelligence, and, of course, luck. As compared to the academic or intellectual sphere, where a kind of more abstract intelligence (some call it intellect) is needed, businesses require a more practical kind of intelligence, basically an intelligence in understanding human psychology, motivation, and behavior, i.e., emotional intelligence (EQ), a relatively new terminology now being frequently used. It is apparent now that emotional intelligence is as important, if not more important, than academic intelligence, which is that intellectual capacity required for doing well in intellectual pursuits. What is business logic precisely? A successful businessman has to understand what the customers really want or need – price (low price, free gifts?), place (convenience of purchase?), promotion ("pull" and "push" tactics, e.g., advertising to create awareness and attract attention of customers, sales promotion to coerce customers to purchase), and product (good product features/quality?) – these are known as the four Ps in marketing terminology. At the same time he has to understand how his competitors and the other aspects of the environment, e.g., the government and the legal aspects, are affecting his business. He has to plan and conduct his business according to this understanding, which might be comparable to a general planning and conducting a military offensive. Actually, all this involves pure practical common sense, and abstract academic intelligence is not needed. Unfortunately, not everyone possesses this important practical common sense or logic, and, even apparently highly intelligent people with high academic credentials might be lacking in this practical intelligence or practical logical capacity. A brilliant scholar might fail as a businessman and a businessman might fail as a scholar, though some people are endowed with the capacity to succeed in both. Basically, business relies much on the understanding of human psychology, in this case, the psychology of the customers, who are only human. The element of risk or luck apparently figures more in business pursuits than the scholarly or academic pursuits. For human beings, in this case, customers are whimsical and might be hard to predict with a fair degree of accuracy. A businessman who thought he had understood what the customers wanted and acted accordingly might be in for a surprise later when he found that he had misunderstood his customers, or, his competitors had beaten him at his own game, or, his customers had changed suddenly, i.e., changed their requirements or expectations whimsically as it were. We should understand the logic or law of human nature. On the whole, human beings seem to be a whimsical and fickle-minded lot – they seem to be easily influenced by fashions, trends, hearsay, rumors, and politics, which are beyond the control of any business person. Therefore, the businessman, as long as he deals with human beings, especially his customers, has to face all these uncertainties, which are not found in the scholarly or academic pursuits, and, therefore, greater risks are involved. In this sense, because business involves uncertainties and risks, it is more difficult than scholarly, academic, or intellectual pursuits. No one could be certain of how the market (the customers) would behave at any point of time, but could only make an estimate or intelligent guess at best and plan and execute the business in accordance with what he feels (intuition) the customers would want. The businessman could well be wrong and could lose money as a result.

Not everyone, including the highly intelligent ones, could or want to live with uncertainties. Therefore, not everyone would want to be in business because they could not bear to face uncertainties and take risks. As customers have a tendency of being subject to whims and fancies, it is difficult to apply logic when dealing with them; the exception is perhaps in technical sales where technical matters requiring logical understanding are involved. In some business sectors, such as trading in shares, stocks, and currencies, where the "mind" of the market which comprises the fanciful, whimsical, rather unpredictable human beings is volatile, great uncertainties, and, thus, great risks, are involved, and, the market could crash overnight (when many people suddenly rush to sell their shares, for example), resulting in many bankruptcies. If one could be certain of convincing one's customers to buy onc's products or services, one would not have any hesitation about going into business. But, this is apparently never the case, and, business entails relatively great uncertainties and risks (even personal sacrifices), due to the general fancifulness, whimsicalness, and fickle-mindedness of the customer, who is only human and apparently tends to act more with his heart than his head, i.e., more with his emotions than his intelligence, more so with the "designer" consumer products, such as clothes and cars, where businesses tend to appeal to the customers' emotions through appropriate advertising and promotion. Though logic is involved to some extent, apparently the nonlogical aspects, such as emotional appeal and relationship, are more important in business. Therefore, can successful businessmen claim to be smart? Perhaps they are smart in the social (or emotional) sense – their emotional intelligence (a terminology which is fast catching on now) is high.

There are apparently many kinds of intelligence or logic. Emotional intelligence is one of them. Academic intelligence, where pure logic is involved, is another. And many others, e.g., athletic, artistic, and musical. All these various kinds of intelligence apparently combine to give general intelligence. Therefore, the more things one could do or the more capabilities one has, the more intelligent one should be. Being adept at abstract logical reasoning seems to be greatly honored. But, perhaps, we should value a more practical intelligence, a more practical logic, such as emotional intelligence, more. A successful businessman could be said to have a high practical intelligence, high practical logical capacity, or high emotional intelligence (EQ), as against high abstract intelligence or high intellect (IQ). Importantly, a businessman is someone who is not afraid of and is willing to take risks. Perhaps, he is not smart enough to realize the extent and possible repercussions of the risks involved, i.e., he is a bit stupid, in this respect, while the others are afraid of and tend to shun risks though they might be more intelligent.

The language we use, the way we use language, might also be the culprits, which give rise to problems. Some people apparently resort to big, bombastic words to impress, confuse, or confound their listeners. Lawyers, e.g., seem to thrive on such play of words. But this often leads to miscommunication, misunderstanding, vagueness, confusion, and, hence, problems, such as disputes, arguments, quarrels, or conflicts. Even a poor memory causes problems sometimes. For example, someone might have said something earlier or made a promise earlier, but subsequently denied it, because he had forgotten that he had said it or promised it, giving the impression to the other party, a wrong impression here, that he was a sneaky liar. It is therefore important to be able to use language or words properly when reasoning or persuading with logic. The choice of the correct, precise words, the correct expressions and connotations, the precision of ideas, clarity of thought, all these are important. As a rule of the thumb, we should say what we mean and mean what we say. In the first place, we have to be clear ourselves about what we want to express, i.e., not be confused or fuddy-duddy. Then, we have to make sure that what we are communicating by speech or writing is understood and clear to the other party. Many a times, it is not that ideas are too difficult to grasp or understand, but poorly communicated, e.g., using plenty of unfamiliar technical jargon, or even symbols. For example, lawyers apparently present their "legalese" regally, impressively but it is often opaque to the nonlegal mind. Ideas should never be communicated in this way. As far as possible, plain, simple, easy-to-understand language, words, and terms should be used. We should try to be as simple and plain as possible in our communication – use words, terms, or expressions that would certainly not lead to various interpretations (vague) but lead to one certain interpretation only.

How nice it would be if we could always put several ideas, especially complex ones, side by side, and be able to come to one fixed, undoubted, universal conclusion. However, this is often not possible. Perhaps, the day would come when each one of us carries a special laptop computer (which has perhaps some special software) whenever we have dealings with others – a laptop computer, which is actually a data bank or "encyclopedia", for easy reference and consultation, and which contains all kinds of information, including successful outcomes or results of cases of disputes, disagreements, conflicts, and other important areas in human affairs, so that whenever we encounter difficulties in negotiations or are embroiled in disputes we can obtain solutions from this data bank or "encyclopedia", i.e., it could "reason" or decide for us. Of course, the other alternative is to resort to arbitration when there are disagreements or conflicts but it is probably too time-consuming and expensive a process.

The big, important question is how logic should be used in our trouble-torn world where there are conflicts galore, disagreements, disputes, raising of tempers, violence, and wars. Just ask yourself this, "If I were to rebuild the human race, the world, or, the universe, from scratch, as a God, how would I make human logic, what characteristics would I create in human logic?" The author would like to create or make the following logic:-

1) Logic is easily accessible to everybody, regardless of training or education level.

2) The reasoning or logical deduction is clear-cut, indisputable, and evident to everyone, as indisputable and clear-cut as "1 + 1 = 2" or "green is green to all and sundry", i.e., no confusion, doubt, or subjectiveness.
3) Everyone would use logic objectively.
4) Logic would assuredly bring positive results.
5) Logic is universally understood as one concept or idea only and not different things to different people.
6) Everybody trusts logic, its use, has great faith in logic, and swears by it.
7) Logic is conflict-free, dispute-free, and problem-free, as people all think alike (one way only) where logic is concerned.
8) Logic would always reveal the facts or truths, and, there is no need for courts of law and arbitration centers.
9) Facts and information that support logic are easily available, e.g., in books, libraries, computers, and CDs.

Since there are apparently so many problems with logic, we should find a way out of this. To prevent these problems, we should perhaps abide by the following dos and don'ts:-

1) Recognize that logic is not perfect and is likely to bring problems.
2) Do not be overconfident about logic, and, adopt a flexible attitude.
3) Be objective and do not let your prejudices, preferences, and likes and dislikes affect your thinking.
4) Only seek out and use logic with those whom you are rather certain have a great respect for logic; never reason with a fool.
5) Be more prepared to listen than talk. Give others a chance to say out their ideas, listen to them, and consider them with care.
6) Do not expect perfection, do not expect others or yourself to be always right; we are all not perfect and make mistakes from time to time, but we should minimize the mistakes.
7) Do not expect that the truths or facts would always result from logical deduction, for there might be insufficient information, or, the wrong information or falsehood might have been used unwittingly.
8) Imagine that you are a God looking down from above at how humans think and behave; consider how you, as a God, would want yourself and your fellow human beings to think and behave, and try to think and act accordingly.
9) Expect to face dead-ends, contradictions, or, even paradoxes.
10) Do not expect easy answers or even an answer, all the time.
11) Recognize that certain ideas, e.g., very abstract ideas, might not be based on reality and might exist only in the imagination. Do not have too high expectations of such "unrealistic" ideas.

12) Be flexible, open, and tolerant of opposing viewpoints and note that some ideas might not be totally wrong, but partially correct, and, might even be correct under some other circumstances.
13) Be prepared to compromise or give and take.
14) Try to understand others, their viewpoints; try to put yourself in their shoes.
15) Avoid being emotional and egoistic.
16) Always try to get down to the basic, absolute truth.
17) Do not lie. Detect lies and expose them.
18) Use logic to achieve happiness, harmony, and positive results.
19) Do not use logic to put others down or make others feel stupid.
20) Avoid using logic to achieve a win-lose situation if possible. It is best to use logic to achieve a win-win situation.
21) There are more important things than logic in life, e.g., happiness, health, peace, and harmony.
22) Use logic to achieve happiness, peace, and harmony not only for yourself but others as well. Aim for a happy, peaceful, harmonious, conducive environment.
23) Be prepared to learn and improve your logical reasoning prowess.
24) Do not show off your logical reasoning prowess, but be down-to-earth and remember that there is always room for improvement. Know your limitations or weaknesses, as well as strengths.
25) Remember that there are always some others who are more capable, smarter, or cleverer than you, at least in some ways.
26) Remember that you are not perfect, can never be perfect, and will have some faults. Learn and try to overcome your faults and improve.
27) Strive to be a more logical and better person – there is unlimited potential for improvement. (It is said that we only utilize about ten per cent of our full mental capacity, and, therefore, we should try to use and develop as much of the other 90% mental capacity as possible.)
28) Live and let live.
29) Do not look down on others who are less capable or intelligent; do not be an intellectual snob.
30) Realize that there are bound to be some others who could think more logically and are smarter than you. Do not envy them but try to emulate them.
31) Do not overrate logic. There could be an alien race, which is so smart that they could arrive at the truths without the need for logical deduction (a waste of time) but by using perhaps some kind of strong intuition.
32) The greatest purpose of logic should be to bring happy results to oneself as well as others. (We are all interdependent. When others are happy we are happy, and when we are happy others are happy. We depend on each other for survival.)
33) There is a philosophical saying that it is better to be an unhappy Aristotle (representing a wise man) than a happy fool. Do you agree? Of course, one has to be smart enough to know what is happening around one to avoid harm's way. The

best is to be both smart and happy. But, given only one choice, it is perhaps better to be happy (so that life is worth living) than smart. On the other hand, for self-survival, it is probably better to be smart than happy, if given only one choice.

34) Always make an effort to write, speak, or communicate clearly and well by using the right words, expressions, and terms. Ensure you are understood, for otherwise you might not only not be understood, but misunderstood, which is worse. That is, have a good command of the language of communication. Be as simple as possible.
35) At the same time, ensure that you understand and not misunderstand what the other party is saying.
36) Exercise patience and care for the other party, i.e., be considerate.
37) Be sincere and trustworthy.
38) Encourage free exchange of ideas and thoughts.
39) When discussing the more controversial subjects, be cautious not to offend sensitivities and be prepared to seriously consider the differing viewpoints of others, even changing your own viewpoints or dropping them to accept the viewpoints of others if there were merit in them. Do not let ego or pride get in the way.
40) In general, be prepared to change or drop your own viewpoints or ideas in favor of the viewpoints or ideas of others if the latter have more merit. That is, do not be stubborn and inflexible. Be flexible and concerned with only the correct ideas and thoughts.
41) Be well-read and knowledgeable so that one's powers of comprehension might be increased.
42) Always think of the truths and always make an effort to seek the truths, and, avoid or get rid of falsehoods. (Be philosophical.)
43) Have intellectual honesty. Mean what you say and say what you mean.

The author has had his scrapes with the use of logic, especially in his younger days when he was still feeling his way around, due to lack of experience and exposure. After all the years of exposure and interactions with so many kinds of people, he has formed some conclusions of his own and has more or less developed his philosophy of life, and, as in this book, logic. He has learnt some important lessons, which he is happy to share.

To the point of not appearing to show off or boast about one's capacity for logical thinking, it is best to be seen as being a logical person, if one were indeed a logical person, whose logical views would be respected and heeded, while not being seen as a logical thinker, though one is truly a logical thinker, whose logical views are generally not respected or heeded, as it serves no useful purpose and is a waste. To be seen as being a logical person, while truly a moron, is the worst that could happen, for the wrong ideas of such a person not only do not benefit society but might detract from society. Ultimately, logic is useless and futile if it were to fall on deaf ears and, hence, have no chance of making a positive contribution to society.

The principles of logic should not be there just for show or to only look impressive. It is more important for them to be applied and put into practice, so that good, positive results could be gained. Hence, the importance of this chapter.

29 Conflicts Arising from Logic and Their Solutions

If only the conclusions of logic, in general, were as certain and indisputable as the simple conclusion that one plus one is equal to two. If the logic involves concepts or principles that are relatively simple, then conflicts would be minimal. It is when the concepts are more abstract that disputes and conflicts are more likely to be encountered. Take for example, the solutions to some of the long unsolved mathematical problems such as the twin primes conjecture and the Goldbach conjecture, with which the author have been involved. It is apparent that most mathematical journals would not be keen to publish the solutions to these problems, even if they were in all likelihood correct, if the solutions were the fruits of an unknown mathematician's creative labor or that of an amateur mathematician. Firstly, the solutions might be controversial and not what the professional mathematicians expect. Secondly, publishing the solutions of an unknown amateur could make the established professional mathematicians look bad – it could make them look incompetent or stupid, or both. Thirdly, it could make the problems themselves look unworthy of the effort of the professional mathematicians since they could be solved by amateurs. Fourthly, though mathematical logic is theoretically more rigorous than other forms of logic, its approval and acceptance are subject to peer review and approval; mathematics like any other endeavor, is subject to peer pressure and prejudices; there have been instances where if the author of a paper is known to the editors and referees of a journal, the paper is easily approved and published; there had even been an editor of a journal who approved and published his own paper on chaos in the same journal, of which he had been the editor. On the other hand, it is not necessary for an important mathematical idea to be published by a journal to gain respect and approval. For instance, the idea of self-similarity in chaos theory, which has been posited by Mitchell Feigenbaum in the 1970s has never been published in a journal, but Feigenbaum has conducted many seminars and given out many preprints of his paper on self-similarity, and he has made a name for himself as a result, and, self-similarity has ever since been accepted and approved by his peers. Fifthly, many published papers in mathematical journals have been reformulation of old or existing ideas; papers positing entirely novel ideas are comparatively few – such papers might even face problems getting published, especially if they are heretical or against conventional wisdom. Sixthly, solutions to long outstanding and famous problems such as the twin primes conjecture and the Goldbach conjecture are prone to attempts by supposed cranks, who might wish to achieve fame by getting the solutions to these famous problems; so, a solution offered by an unknown tends to be viewed with skepticism and suspicion; it is apparent that solutions to such problems are to have the wide approval and endorsement of the mathematical community before they are deemed fit for publication in a respectable journal. All these are well and good as many experts are there to vet the solutions to ensure that the arguments and logic of the solutions are faultless and per-

https://doi.org/10.1515/9783112214367-054

fect or near-perfect. Nevertheless, unless a person is a reputable mathematician, getting such endorsement for a mathematical solution would not be easy; being able to sell your ideas is thus important. Of course, mathematicians do not always concur on how they view mathematical logic, with there being at least three different schools of thought – the formalist, the logicist, and the intuitionist schools of thought, and, this is indeed a great problem. If the logic of such a relatively more exact science as mathematics is so fraught with the interplay of personalities and peer pressure, just imagine the degree of conflicts that is likely to be encountered in the other less exact sciences and areas of human endeavor, e.g., economics, psychology, and sociology.

Now, how do we resolve such conflicts? We may have the solution to a problem of logic but how do we know for sure that that is the correct, the best solution? The author would like to propose the following:

(1) We should try to determine what constitutes a good or acceptable solution, the criteria for acceptability of which should be free from bias and be based on objective principles. Otherwise, we may end up with differing opinions, with some claiming that the solution or proof is acceptable while the others rejecting it – this scenario does occur in mathematics, which is theoretically an exact, objective science. For example, what constitutes a good, acceptable proof for the Goldbach conjecture, which posits that every even number after the number 2 is the sum of two prime numbers? Computer searches completed in 2000 have found that this is the case for all even numbers up to 400 trillion. While, to the scientist, this may be acceptable as proof that the Goldbach conjecture is true, the mathematician will not accept this as a proof. So, how then should the proof of the Goldbach conjecture be, despite the evidence that 400 trillion even numbers fulfill the requirement stipulated by the Goldbach conjecture? What should be the criteria for an acceptable proof for the conjecture? Of course, proving the case for an infinite number of even numbers is an infinitely difficult task. First of all, infinity is an abstract, difficult, and impractical concept. It is practically impossible to live, count, calculate or compute to infinity. We might use a clever argument such as the proof by contradiction, whereby we make an assumption and then find a contradiction to this assumption – though many mathematicians accept this argument as a proof, those mathematicians belonging to the intuitionist school of thought do not – the proof by contradiction assumes two possible states (right or wrong), whereas to the intuitionist, there are three possible states (right, wrong, or undecidable). An argument or proof, e.g., could also make use of the concepts used in set theory, group theory, topology, algebraic geometry, complex analysis, or any other areas of mathematics to demonstrate a point. Such a proof might be acceptable to a mathematician familiar with any of these mathematical concepts but to the nonmathematician, who has little or no understanding of these concepts, such a proof might have little or no meaning. Shouldn't a mathematical proof be a universally accepted one, appealing and acceptable to all and sundry, mathematician or nonmathematician? Shouldn't the mathematical proof or logic

be as certain or indisputable as the logical conclusion that one plus one is equal to two? It is therefore apparent that mathematical logic is culturally biased, and this detracts from its appeal as an exact, objective science, though its striving for rigor is commendable. The author thinks that a mathematical proof should fulfill the requirement of Ockham's razor – be simple, short, and sharp. Specifically defining the requirements for an acceptable mathematical proof would certainly ensure that conflicts are nonexistent or minimal, but, this very act of defining the requirements itself might be subject to controversy and conflict. So, a flexible, realistic, open mindset is always a boon.

(2) A mathematical proof should have the consensus or majority approval of many mathematicians, not just the approval of a few mathematicians, as in the case of peer reviews. This is already the case for solutions to well-known mathematical problems, where publication in respected journals are normally endorsed by respected mathematicians or referees, and a waiting period of a number of years is called for, whereby if there are no counterexamples or defects found in the solutions and no objections are raised based on valid reasons from members of the mathematical community, then the solutions could be regarded as approved and accepted.

(3) A proof should be accepted as correct only, until proven wrong. With regards to this, a flexible, realistic, open mindset is still important. No one should insist that the proof is *absolutely* correct. Logic, mathematical or not, might not be simple or straightforward; its validity might even be undecidable or unprovable (as postulated by Godel).

(4) This is a very very important point here. The solution or proof should be capable of being certified, checked, or confirmed to be correct by some more concrete means, not just the mere approval or acceptance by most or all of those who matter. In the case of a scientific theory, its validity has to be confirmed by physical experiments or physical data. Though mathematics might be dealing with more abstract ideas, the validity of a mathematical proof or reasoning could still be checked and certified. For example, in the case of the twin primes conjecture, the methods proposed by the solution could be implemented to find out how capable they are of generating or creating new twin primes pairs – this generating or creating of new twin primes pairs is known as a "construction" in mathematical terms; this is akin to the physical experiments of the physical sciences.

Therefore, in all our endeavors with logic and problem-solving, we should have some common understanding of what constitutes an acceptable solution. Otherwise, instead of solving problems, we might just create more problems, more conflicts. The exercise of logical reasoning is to solve problems, not to create more problems. What is a proof? It is information or evidence that supports a hypothesis or theory and thus confirms its validity. This information should be real and not just the product of the imagination and should comprise the more concrete information, i.e., information that could be physically checked or confirmed, or, information that is tangible or truly observable. The author insists that if the proof or reasoning could not meet this requirement, this requirement mentioned in (4) above, then the proof or reasoning should only be regarded at most as being tentatively correct, though it might have the approval of all or most of the experts. So, logic, even the relatively more abstract aspects of logic, should fulfill the requirements mentioned above, especially the requirement mentioned in (4). For mathematics, at least it deals with numbers and geometrical shapes, which are checkable – numbers could be counted and checked and shapes could be observed and measured – so it would still be practicable to check mathematical reasoning involving numbers and geometrical shapes – that is why, mathematics is known as an exact science. But, how do we check the validity of the logic of, e.g., a conclusion in economic theory such as "The reduction in taxes would boost the economy as people would have more money to spend on goods"? This conclusion is not immediately obvious or "checkable" and might not necessarily be true, as people might hoard the savings from taxes rather than spend them (unless they could be forced to spend the savings). The validity of this conclusion could only be proved by the actual results of the reduction in taxes, i.e., when, after the reduction in taxes, people are found to spend more and the economy is found to be improving as a result. The validity of a mathematical theory or proof involving numbers and geometrical shapes that have no choice of their own and do not lie, if certifiable, is either true or untrue. But the validity of a theory or conclusion from one of the social sciences, such as, e.g., economics, involving human beings who have choices of their own and who may be fickle might not be certain, depending on the outcomes or results that can be obtained through a statistical survey; the validity of such a theory or conclusion, such as the above-mentioned conclusion pertaining to the reduction of taxes, should be regarded as "probably true" or, at most, "likely to be true" (if a number of past cases of reduction in taxes had been found to produce the said result). Thus, proving the validity of the logic of a theory or concept from one of the social sciences (the soft sciences) such as economics is relatively more difficult as compared to that for a mathematical idea (mathematics belongs to the hard sciences). It is apparent that dealing with people and conclusions about people is generally more difficult than dealing with numbers and geometrical shapes and conclusions about them. And yet, people tend to regard mathematics, especially higher mathematics, as more difficult than a social science such as economics. This is not surprising, as the higher mathematics uses plenty of symbols and notations that are as incomprehensible as Greek or Hieroglyphics is to the uninitiated; in all probability, a good explanation of what those mathematical symbols represent would make the mathematical ideas accessible to the non-mathematician. (As mentioned earlier, it is desirable for there to be a universal, commonly accepted language that facilitates logical reasoning, such as, e.g., mathematical reasoning, which would help to prevent miscommunication, misunderstanding, or lack of understanding.)

The trouble with most of us when disputing on logical principles is that we do not have a clear idea of how the validity of the logic could be certified, checked, or confirmed. We have to know the requirements that have to be met by a solution or reasoning before we can reasonably settle the question regarding its validity. Otherwise, there would be controversy or dispute pertaining to the validity of the solution or reasoning, which could be long-drawn. The author emphasizes that logic should as far as possible deal with real, observable, or tangible objects, and, be certifiable by real, observable, or tangible objects as well, in order to be real, in order to be worthy of consideration. Otherwise, we would get problems such as circularity of reasoning, paradoxes, and confusion. Our problems of social unrest and social disintegration are apparently a reflection of this failure of logic, which is characterized by circularity of reasoning, confusion, and paradoxes, and hence, disputes galore. To put it more directly, these problems are the apparent result of the lack of ability to think straight and think objectively.

The author laments that though the human race prides itself on the ability to reason and to use logic, the use of reasoning or logic is often biased, even wrong, due to the influences of different cultures, education systems (in the past, the education systems emphasized classical studies such as philosophy and literature, but now the emphasis is on technology), and even languages (different languages present different connotations and nuances). There is also at play what is called the "halo effect", whereby if a person is held in high esteem, e.g., a respected leader, a renowned mathematician, a famous scientist, or a powerful person, whatever this person espouses is likely to be viewed or taken more seriously, i.e., this person would be more influential and his views, his so-called logic, would carry more weight. Thus, the assessment of logic can be as subjective as the assessment of beauty in a beauty contest. Why can't there be absolutely, indubitably, best ways of doing things in all our enterprises? Why are there so many cliques, different philosophies, schools of thought, systems, religions, political ideologies, standards, and opinions, which often lead to disputes and conflicts? The ones with more power or influence would impose or effect their "logic" on others, who would have little or no choice but to follow this "logic". Logic should be so pure and objective that it transcends all the above-mentioned barriers, that, even if it is the result of the thought processes of a laborer, a gravedigger, or even a criminal, it would still be given due recognition. Though logic might not be perfect and might even prove elusive, man is also to be blamed for tending to use logic to his own selfish advantage without much regard for true objectivity (as compared to assumed or "faked" objectivity) and fairness (it could even be stated that logic is invented by man). Because of his failure to use logic objectively, realistically, and effectively, the so-called rational human being can often be found contradicting himself and others as well. There seems to be so many standards of logic (schools of thought) that logic looks as though it is artificial and unreal. It is indeed a sad day for logic; it is sad that logic often not only cannot help us settle our problems, but has helped to add to our problems. Don't we have enough of problems already? It is high time to really revamp the whole logical system to really seek a more foolproof logical system. At the same time, we have to modify or change our stance toward logic. We have to do away with dogmatism. We have to let go of our fixed ways. We have to be flexible, realistic, and open-minded. We have to be objective and fair-minded.

To further illustrate the problem with logic, the author would like to quote verbatim several paragraphs pertaining to peer review by scientific journals taken from a book about a scientific odyssey pertaining to a heretical speculation that the speed of light is variable, published in 2003. The author of this book is a theoretical physicist, who did his graduate work at and received his doctorate in Theoretical Physics from Cambridge University, followed by a very prestigious research fellowship at St. John's College, Cambridge, and a Royal Society research fellowship at Imperial College, London, where he now lectures in theoretical physics. He has made a revolutionary speculation that the speed of light is variable and not constant, the latter being what conventional wisdom holds to be true. The following paragraphs taken from the book are illuminating, especially when written by such a highly credentialed and respectable scholar:

> Another factor may well have been that John had rather complicated relations with certain scientific journals. (A notable exception was the "International Journal of Modern Physics".) I quote from an illuminating e-mail I received from John Barrow in early 1998: "Joao, I asked Janna Levin about Moffat as I remembered she spent time in Toronto . . . . She said that while he is very nice personally he seems to be perpetually engaged in major disputes with journals and their editors. She thought he [was] actually banned from having papers considered by some journals . . . . Sounds ideal! [The editor of PRD] will enjoy dealing with it."
>
> It must be said that in his hatred for journals and their idiosyncrasies, John Moffat is in good company. Indeed, many famous scientists fell out with one journal or another at some point. Einstein is perhaps an unexpected example, but consider this incident: In the late 1930s, Einstein and Rosen wrote a groundbreaking paper on gravitational waves and submitted it to *Physical Review*. Back came a four-page report rejecting their paper. According to Rosen, Einstein was so furious that he shredded the report into little pieces, threw the debris into a bin, and kicked the bin in a rage, shouting and swearing for the next half hour. He also swore that he would never again submit a paper to *Physical Review*, and apparently stuck to his word. (I was told this story by John Moffat, who heard it from Rosen. The funny thing is that, according to Rosen, the referee had a valid point.)
>
> Talking to John Moffat over beers, I came to share his views on scientific journals. A few years later, I would write a damning article titled "The Death of Scientific Journals", curiously enough as an invited contribution to a major publishers' conference. In this article, I began by describing how fraudulent scientific publishing had become. Referee reports are often empty of scientific content and reflect nothing but the authors' social standing, or their good or bad relations with the referee. The senior scientists adorning the authors' list more often than not contributed nothing to the paper, other than their illustrious identities – a process that smooths the refereeing process enormously. To cap it all, editors can be totally illiterate (to give due credit to the editor of *PRD*, I should add that Andy and I were truly lucky in this respect).
>
> I then explained why despite all this corruption, people still bothered submitting their papers to scientific journals: They have no choice. The establishment is set up so that one's official scientific record takes into account only publications in refereed journals, a totally artificial im-

> position. As a result, I myself publish all my papers in refereed journals, but regard the process cynically, as a chore not dissimilar to flushing the toilet or emptying the garbage bin. But it is also an unstable enterprise, and it contains the seeds of its own destruction. The most vocal anti-journal young people I know are now becoming quite senior themselves and have not changed their views on the matter. For this reason alone, things do not bode well for scientific publishing.
>
> But more importantly, the Web has changed everything because it has created a situation in which journals may be bypassed altogether. I have mentioned several times how physicists have started putting their papers on Web archives at the same time they submit them to journals. This has created a situation in which no one reads journals anymore because archives have replaced them. In 1992, I may have missed Moffat's paper on the Web, but nowadays, every morning before I start working, I read the new arrivals in the archives. When I need a reference, I find the relevant article in the archives, display it on a computer screen, and read it there and then. It's been ages since I looked in a journal, let alone went to the library to consult one. Journals are an anachronism, and they have already effectively been phased out.
>
> Some people think this is bad. They claim that Web archives have no quality control. That is true, but I would argue that the refereeing process associated with current journals provides no real quality control either. And in any case we don't need it; anyone should know which papers are worth reading without the need for prior filtering. Another argument is that Web archives will destroy our beloved concept of copyright. Again that may be true, but isn't the senior author in every paper already an insult to copyright? And on the few occasions someone has tried to plagiarize by using the archives it has ended badly, the person in question becoming the laughingstock of the community.
>
> In my article I went on to argue, controversially, that this may well spread into all forms of publishing. Perhaps one day all books will be Web-based, becoming organic and ever-evolving, more copy-able and part of an environment shared by all. This may sound utopian, and in its details it certainly is, but whatever happens in the future I don't believe the printed word in the form we know it will survive the computer revolution. We should come to grips with the reality that, one way or another, the Gutenberg galaxy is dead.

The above says much about scientific logic and, by implication, human logic in general. It is apparent here that getting a paper published in a research journal depends on who you are (your reputation) and your relationship with the editor or referee and, of course, luck. (The corollary is that success with logic depends on the human relation factor and luck as well.) The author himself has perhaps experienced more difficulties with research journals than the above-mentioned scholar. All this might be a blessing in disguise as it more or less indicates to the author that logic and its protagonists are generally beset with problems – if the supposedly exact and objective sciences are so beset with problems and "politics", what more of the other less exact human enterprises in general? The author wonders how the editors and referees of research journals such as scientific and mathematical research journals are selected. Here, the author would like to make a bold suggestion. The editors and referees of these research journals should be carefully selected, perhaps, elected from the rank and file, for their objectivity, rigor, and fair-mindedness. They should represent a correctly, properly trained group of intellectuals who know how to and would go about assessing intellectual merit with rigor, objectivity and, fairness. The author would say that knowing how to use logic effectively and how to assess logic correctly should go together and complement one another. Such training should have been introduced and emphasized in schools, colleges, and universities, as well as other professional institutions. For example, for the purpose of promoting rigorous, objective, and fair logical thinking, an association or associations, or, a professional body or professional bodies, could be formed, which conducts appropriate training, courses and exams, and assessment or certification tests on the use and assessment of logic; we could call this the Association for the Promotion of Logic and Its Proper Assessment, or, the Institute for Effective Logical Reasoning and Effective Assessment of Logic, and so on; these institutions could award credentials to those who successfully complete their training and/or pass their tests or exams, such as the Professional Diploma in Logic, Certified Logician, and Certified Professional in the Use and Assessment of Logic. Besides editors and referees of research journals, the above should also apply to all those who are directly or indirectly involved in the use and assessment of logic, such as magistrates, teachers, lecturers and professors, scientists, mathematicians, social scientists, managers, political leaders, and anyone who has anything to do with or is interested in the effective use and assessment of logic; in effect, this means practically everyone.

Unless more positive action like those described above is taken, logic would be likely to continue to wallow in problems and controversies.

30 The Serious Problem with Logic and Its Correction – Why Logic Goes Wrong and How to Prevent It from Going Wrong

It is often difficult to get logic to yield the necessary, correct results due to some reason, which would be explained here.

If logical deduction were to result in a conclusion that is fruitful, correct or true, it must be based on premises which are truths or facts, and, not premises which are falsehoods or partial truths. It appears that the greatest problem that people actually face when carrying out a reasoning process or an act of logical deduction is not that the reasoning or deduction itself that is wrong but that a premise or more, or, all the premises, upon which a conclusion is based, are wrong. The following simple example illustrates this point. Let us look at the following chain of reasoning:

Man is immortal.
John is a man.
Therefore, John is immortal.

The above deduction is valid, or, correct, but, the conclusion is untrue or wrong. Why? That is because the first premise, "Man is immortal", is untrue and wrong. It is an indisputable fact that man is mortal and nobody is immortal. In this instance, the conclusion, "Therefore, John is immortal", is valid, correct, but is untrue, wrong.

The above is only a simple example but there are many other more complex examples of deduction based on wrong or faulty premises. Therefore, the problem is not really the reasoning itself but that the person who executes the reasoning has got some or all of his facts or premises wrong in the first place. For example, two persons assessing the same situation might arrive at different conclusions. Why? Not surprisingly, you would probably find that they hold different views (facts, information, or premises) in the first instance, upon which their respective conclusions are based. This would result in disagreement, even conflict. So, how could such an impasse be overcome? The author would like to suggest that each of them carry out the following, instead of arguing or quarrelling:

1) Each party is to state the premises or facts upon which their respective conclusions are based.
2) Each party is to closely and carefully examine these so-called facts or premises and determine whether they are totally true, partially true, or, wrong.
3) It should hence be clear that the party harboring the wrong premise or premises would be having a conclusion that is *untrue*, though the deduction or reasoning might be valid, and, it is possible that *both* the parties' conclusions are *untrue*.

https://doi.org/10.1515/9783112214367-055

What happens then if the facts could not be obtained or if the information obtained is only probably true? When logical deduction is carried out with inconclusive evidence or information the conclusion should be a probabilistic one with "probable" or "perhaps" as a prefix. That is, the conclusion should only be a provisional one and accepted as probably true until proven untrue. It seems that, unfortunately, logic is often not handled that way.

To the author, everyone is capable of logical deduction or reasoning, to varying degrees of course. The problem really is not that people do not know how to reason or think but that facts or truths, upon which logical deduction must be based in order for a *true* conclusion to be possible, are hard to obtain, and, it is often difficult to ascertain whether a statement or a premise is a truth. The discoverers and finders of truths are our scientific workers and philosophers. That is why, when people become prejudiced, distort the facts, or accept partial truths or even falsehoods as truths, and utilize these in their reasoning, they get ineffective or wrong conclusions, which might result in harm to themselves and/or others. This is precisely the reason why the author has exhorted the reader to adopt an objective, realistic, flexible mind-set when using logic.

Therefore, in logical deduction, it is of paramount importance to have the actual facts or truths to work with in the first instance. The author would exhort the reader who is really keen to get his logical deduction correct and effective to ensure first of all that he has really got all the correct and necessary information and facts to work with, otherwise, it might just end up being a case of garbage in and garbage out. That is, fact-finding should precede logical deduction.

However, fact-finding or research might not be an easy task – it might require much effort and time – fact-finding might even prove to be harder than logical deduction. Education should have trained us in the arts of fact-finding and logical deduction besides equipping us with knowledge. Getting the right facts for one's decision-making or logical deduction should be efficiently carried out – it should be taken seriously and should be regarded as being as important as, if not more important than, the act of logical deduction itself.

Deduced logic would only be correct and useful when we have our facts correct in the first instance. Therefore, everybody should be as concerned with facts or truths as he is with logical soundness, and, for this to be possible a person should adopt an objective, unprejudiced, realistic, curious, and flexible mind-set.

31 The Simplification of Logic

The natural world is abound with symmetries and similarities. These represent the patterns of nature. When we recognize these patterns, e.g., the fractal pattern, we know what to expect, i.e., we could arrive at a conclusion. In actuality nature appears simple in all its patterns, beauty, and magnificence. We should thus value and appreciate simplicity in its variegated forms in nature, in the way we conduct ourselves, in our dealings with others, in the way we think, and so on. Simplicity appears to be one of the fundamental aspects of life.

Everything could in fact be simplified. Even the most complex theories could be reduced to a few simple statements, which are easily understood. There is no doubt about this. Philosophers have long subscribed to the concept of Ockham's razor, whereby it is held that given two reasoning – one convoluted and long, while the other is simple and brief – the simple and brief one is more likely to be the more correct one. In school, we have been taught to express ourselves clearly and simply, without any unnecessary adornment, and, we also tend to be impatient with somebody who gives an unnecessarily long-winded explanation when a simple and brief explanation is sufficient. Then, why do we often have to put up with texts or presentations which appear to have an over-exuberance of details? Worse, these details might be technical and difficult to comprehend, though they might seem profound and impressive. The effect is that "the trees we are looking for are obscured by the jungle". In this way, our simple life has been made unnecessarily complex and difficult.

For every complex subject, there is always a simple version, e.g., we have a wide range of "made simple" publications to enlighten the confused and the uninitiated. We would always prefer a simple solution to a complex one, as we naturally do not want to spend too much time on a problem, which is especially true in the commercial world where time is equated with money. Even professional mathematicians have been known to favor simple mathematical solutions, equating simplicity with elegance. Simplicity is in fact the efficiency, time-saving, productive, economy, and even elegance. A dancer who simply executes smooth, short movements is certainly more impressive than a dancer whose movements are ungainly and convoluted. The same should apply to the "dancer of the intellectual world". Certainly, expressing, e.g., ten points in two or three sentences requires more skill than expressing the same ten points in ten sentences. (Remember that we have had to pick up this skill in school through précis-writing.) It is therefore difficult to understand why intellectuals, apart from wanting to appear profound, resort to complexities in their theorizing.

In logical deduction, we could also rely on approximation, a qualitative sort of reasoning, whereby there is no need to be too precise. Often, a precise solution or conclusion is not possible, and, it might only be possible to produce an approximate, or, probabilistic, solution. We might use analogies. We might look for equivalences (in lieu of equality). We might compare and note the similarities. Then we conclude.

https://doi.org/10.1515/9783112214367-056

We could also rely on substitution to simplify the reasoning process. For example, we could simplify the following mathematical equation:

$$(\Delta_G E)^2 = \partial^2 (\text{In}\check{Z}) / \partial\beta^2 = \partial^2 (\text{In}\check{Z}) / \partial(\beta\mu)^2$$

by letting $\partial^2 (\text{In}\check{Z}) / \partial\beta^2 = S$

$$\therefore (\Delta_G E)^2 = S = S/\mu^2$$

Substitution is a common phenomenon in mathematics. As many mathematical equations have many arcane symbols and are thus complicated, simplifying them through the substitution of the more complex terms with simpler terms would make their manipulation more manageable. The symbols are the "short-hand" of mathematics and perhaps only the mathematician manipulating them knows precisely what they stand for. Two apparently very different equations could be equal as long as they each represent the same thing.

Any mathematician, or anyone for that matter, could create or invent his own symbols or notations. As long as he knows what his symbols or notations represent or stand for, he should be able to perform chains of reasoning with them, using certain mathematical rules. To support his reasoning, the mathematician could use axioms (which are obvious, assumed, and unproved truths), lemmas (a lemma is a subsidiary result proved in order to simplify the proof of a required theorem) and theorems (which are proven results).

In logic, as in life, simplicity should be the best policy. We should avoid obscurity and complexity as much as possible. After all, we need to be understood and appreciated. And we hate not being understood, or worse, misunderstood.

32 Being Logical

It would not be an overstatement to say that it is important to be logical. After all, being logical is supposed to be a distinguished characteristic of the human being, though at many times human beings have apparently failed to display this quality. This is not to say that other life-forms do not display this quality. Perhaps, an intelligent dog or horse has this quality too. Often, some human beings embrace a logic, which on close examination appears faulty or wrong. Wrong logic results in wrong decisions and wrong actions, which may cause harm or inconvenience to the self as well as to others.

The important question here is: How can we be logical? As stated elsewhere in this tome, it is important, first of all, to have the right attitudes: open-mindedness, broad-mindedness, objectivity, healthy skepticism, good command of a language, or a set of symbols for carrying out the act of logical reasoning, patience, willingness, and eagerness to learn and improve, humility, treating others as equals, willingness to listen to and consider other people's views, having an eye or head for details, etc. It is also good to know of or be familiar with some techniques or methods of logical reasoning. The problem is that often an apparently irrational or unreasonable person thinks that he is rational. People often make sweeping statements, which are not justifiable or generalizations that are not valid, apparently without being aware of their logical bungling.

How should we think or act in a logical manner? We should exercise pain, caution, and care when making statements, e.g., do our research first and get our facts or premises right, make a careful choice of words, terms, or expressions to be used, aim at clarity and at being understood, listen to, consider, and accept or adopt others' ideas or points of view if they are relevant, win the support of or acceptance by others for our logical propositions or ideas. A logical statement could be simple and short or it could be complex, detailed, and lengthy. The facts or premises contained in the statement should be true facts, facts whose truthfulness could be verified. It is important that the statement could be verified or proved to be true, e.g., confirmed by an experiment or experiments, or, some other kind of test. For those statements whose truths are not certain or verifiable, we could make them with some qualification or caveat – we could make such statements with a probabilistic, but reasoned approach, e.g., we could state that something is probably true, most probably true, unlikely to be true, has little likelihood of being true, in all probability true (or untrue or false), true under certain circumstances, or false under certain circumstances, depending on our intuition and how strongly we felt about the probability of its being true, or, false, though we were not certain of or able to verify its truth or falseness. It is of course important to have clarity and alertness of mind while making statements.

The following is a listing of the kinds of statement we may encounter from our fellow beings or counterparts:

https://doi.org/10.1515/9783112214367-057

1) The statement is entirely true. (The truth is verifiable.)
2) The statement is entirely false. (The falseness is verifiable.)
3) The statement is partially true and partially false. (The partial truth and partial falseness are verifiable.)
4) The statement is partially true and partially false, while the rest of the statement is not verifiable as true or false, but may be considered probably or most probably true or false. (The partial truth and partial falseness are verifiable.)
5) The statement is partially true, while the rest of the statement is not verifiable as true or false, but may be considered probably or most probably true or false. (The partial truth is verifiable.)
6) The statement is partially false, while the rest of the statement is not verifiable as true or false, but may be considered probably or, most probably true or false. (The partial falseness is verifiable.)
7) The statement is neither true nor false, but may be considered probably or most probably true or false. (The statement could not be verified to be true, or, false.)
8) The statement is entirely true under certain circumstances. (This is verifiable.)
9) The statement is entirely false under certain circumstances. (This is verifiable.)
10) The statement is partially true and partially false under certain circumstances. (These partial truth and partial falseness are verifiable.)
11) The statement is partially true and partially false under certain circumstances, while the rest of the statement is not verifiable as true or false, but may be considered probably or most probably true or false. (These partial truth and partial falseness are verifiable.)
12) The statement is partially true under certain circumstances, while the rest of the statement is not verifiable as true or false, but may be considered probably or most probably true or, false. (This partial truth is verifiable.)
13) The statement is partially false under certain circumstances, while the rest of the statement is not verifiable as true or false, but may be considered probably or most probably true or false. (This partial falseness is verifiable.)
14) The statement is neither true nor false under certain circumstances, but may be considered probably or most probably true or false. (This statement could not be verified to be true or false.)
15) The statement is neither true nor false under certain circumstances, and there is little or hardly any probability that it is true or false. (This statement could not be verified to be true or false.)
16) The statement is none of the above, i.e., it is neither true nor false under any circumstances, nor probably or most probably true or false – we cannot make anything or form any conclusion about the statement at all. (This statement could not be verified to be true or false.)

The above listing of 16 kinds of statements, which we may encounter, may not be an exhaustive listing. The listing may be further classified, e.g., there may be a statement that is three-quarters true and one-quarter false, a statement that is two-thirds true and one-third probably or most probably true or false, a statement that is one-third true, one-third false, and one-third probably or most probably true or, false, a statement that is one-fifth true, two-fifths false, and two-fifths probably or, most probably true or, false, etc. There could be lots of fine, subtle distinctions in the logical statements or propositions, which we should be keenly aware of. We should be alert to all the possible implications.

Recall that by Godel's incompleteness theorem there are true statements whose truth is not provable and false statements whose falseness is not provable. The question is if we could not prove or verify the truth or falseness of a statement how could we be certain or know that the statement is true or false? The most we could say about such a statement is that it is probably or most probably true or false. Godel appears to have had erred in this respect, for it is practically not possible to know whether a statement or proposition is true or false, if its truth or falseness is not verifiable or provable.

Thus, in handling, dealing with, or analyzing any logical statements, we should be alert to any subtleties or complexities, wary of anything wrong, inappropriate, or amiss and able to think clearly. There could be so many ways or angles of looking at things, so many possible interpretations. Our best recourse should be to have an open and broad mind and patiently consider as many views, angles, possibilities, and options as we could, so that we could have the best possible, if not perfect, conclusions or ideas. When working with logic we should really think carefully and hard before making any statements. Logical reasoning or being logical could indeed be hard work. But the hard work is worth it when there are good, tangible results.

How about the practical aspects of logical reasoning? Every person, whether he is a child or an adult, must be familiar with logical reasoning, an activity, which is normally regarded as thinking. Thinking or working with ideas and logic is in fact as natural as breathing and is carried out with a variety of effort; of course, simple things such as, e.g., interpersonal or social relations, driving a car and shopping for groceries, may not require strenuous mental exertion, but highly intellectual activities such as studying advanced calculus and solving problems in quantum mechanics may require not only strong mental exertion but a high intelligence as well. Generally, human beings carry out their logical reasoning in a "trial and error" sort of way, i.e., they do not normally follow any system or any fixed set of rules in doing so; how they reason would depend on their respective ability and resourcefulness, and, it is apparent that some people are simply smarter or cleverer than others. Unfortunately, the educational system nowadays apparently aims more at equipping students with the necessary knowledge for embarking on a professional career later on, rather than the skills required for logical reasoning and for coexisting harmoniously with their fellow beings. The author himself, having been both a student and an educator before, had never studied logic or ethics (except for bible knowledge) in his student career, and, it is apparent that our educational system hardly teaches these subjects, except for students of philosophy, and, mathematics where logic is one of the specialties or options. One may think that studying intellectually demanding subjects such as mathematics and engineering in school are sufficient training for logical reasoning. This may be true to a certain extent. But, it is important to note here that the mathematics we learn in our mathematics and engineering courses requires a type of reasoning that is quite different from other forms of reasoning. In mathematical reasoning we play around with and manipulate symbols and equations or formulas according to certain set rules and we are expected to arrive at the same answer and conclusion; many mathematical students have been known to practice their problem-solving so hard that they could remember the formulas and techniques so well that when they see similar problems appearing in their exams they could just solve them straightaway, instinctively, with apparent ease and hardly any mental effort. (In the olden days of the Wrangler system at Cambridge University, which has since been abolished, the Senior Wrangler in the mathematics tripos exams, who was the top mathematical scholar in the cohort, and the other high-ranking Wranglers such as the second, third, fourth, and fifth Wranglers, normally had to have experienced personal tutors, who had normally been Wranglers themselves and had been familiar with the tricks and twists of the "game", to coach them for the demanding mathematics tripos exams. This kind of mathematical training, though it might be good training for hard mental work and concentration, was found to have had an adverse effect on mathematical creativity.) In mathematics, if we use the correct formulas, adopt the right methods or steps, and follow the set rules we would all get the same answers and conclusions; there is precision of thought – in mathematics (which is called an exact science), we could achieve this as mathematical objects are normally quantifiable and measurable, e.g., force, area, volume, length, breadth, height, mass, perimeter, circumference, radius, diameter, gravitational force, coefficient of friction, torque, stress, strain, energy, centrifugal force, centripetal force, inductance, resistance, capacitance, frequency, electromagnetic force, displacement, acceleration, deceleration, luminance, rate of change, pressure, distance, velocity, and so on. However, in other forms of reasoning, the objects of reasoning are normally not amenable or that amenable to quantification, measurement, or verification, and this apparently gives rise to controversies and disputes. In economic science, for instance, we deal with demand, desire, supply, opportunity costs, propensity to spend, scales of preferences, satisfaction, anticipation of demand, and so on, which are abstract, intangible, unmeasurable, and unquantifiable concepts. In the other social sciences, and in our everyday human affairs, we are dealing with objects such as happiness, hatred, love, commitment, control, democracy, freedom, dictatorship, ruthlessness, evil, popularity, satisfaction, sorrow, calamities, terrorism, patriotism, greed, self-centeredness, commitment, cooperation, ambition, nobility, wisdom, religious-mindedness, kindness, dissatisfaction, fashion-consciousness, beauty, professionalism, efficiency, capability, feelings, emotions, hunger, war, weather, strength, harmony, cohesiveness, elitism, racism, democracy, socialism, friendliness, foolishness, unity, neglect, provocation, harm, anger, recklessness, faithfulness, devotion, aesthetics, punishment, advantages, disadvantages, hope, complexity, errors, support, interference, adaptability, hospitality, and so on, which are abstract and intangible objects. How are we to quantify and measure, e.g., kindness, support, emotions, elitism, and popularity? We could only make an estimate of their degree, make a rather subjective judgment. If only we could use a measuring instrument such as a ruler, a measuring tape, or a weighing scale to measure them. How could we mathematize all these objects? Even if we could invent some formulas to describe and reason with these objects, the results would still be unquantifiable, unmeasurable, and hence rather subjective. Reasoning with unquantifiable, unmeasurable objects such as the above-mentioned and ending up with unquantifiable, unmeasurable and subjective conclusions or results would apparently result in disputes and conflicts, unless we are prepared to be flexible and "give and take". So, in our everyday, nonmathematical, non-precise, thinking, whereby we do not deal with numbers or quantifiable, measurable objects, we may make statements, state our reasons, and jump to conclusions, but we may not know for sure whether our conclusions are indeed correct, unless we could verify them, which may not always be possible. Moreover, unlike mathematics, where the answers and conclusions are normally standard and precise, the answers and conclusions here are frequently open-ended, not so conclusive, i.e., there are frequently several possible conclusions.

We should not forget the part intuition plays in logical reasoning. Intuitively, all logical statements or propositions should make sense or appear sensible. Intuition perhaps plays a greater role in the nonmathematical, non-precise kind of reasoning such as that described above, whereby we are generally unsure whether our deliberations with logic have given us correct or valid conclusions, which are generally not that amenable to verification, and we have to rely strongly on our intuition and perhaps our past experience to sass out or get a strong feel of the correctness or validity of our conclusions; we may be so unsure of the conclusions that on receiving objections pertaining to their validity, regardless of whether the objections are justified or not, we become even more unsure of their validity and may even renounce them; on the other hand, if we receive more support or assurance from others with regards to their validity we become surer or more confident of their validity. To the uninitiated, mathematics and mathematical reasoning may seem difficult, even incomprehensible (which is perhaps also due to fear), though the mathematician or expert manipulating the equations or formulas may find it a breeze as he is familiar with the material. The nonmathematician and nonexpert may find the symbolism used, which may, e.g., comprise of Greek and fancy characters, arcane and daunting. But with some practice and familiarization he may not find the mathematical reasoning so difficult after all. Having patience and a head for details also help. The nonmathematical kind of reasoning described above however does not suffer from this problem as it is normally carried out in a common language which is familiar to practically everyone, unless the terms used are obscure, bombastic, inappropriate, or wrong. It does not mean that this kind of nonmathematical reasoning is any less rigorous than mathematical reasoning because it is apparently less precise than mathematical reasoning. Mathematical reasoning deals with precise, measurable objects, which lead to precise, definite conclusions; from beginning to conclusion, mathematics generally deals with numbers or quantifiable objects. The reverse is true of nonmathematical reasoning such as that described above. In this sense nonmathematical reasoning such as that described above is more difficult, reliant perhaps more on intuition, tends to be more subjective, the conclusions being somewhat uncertain, open-ended, indefinite, controversial, or disputable, which could be frustrating.

The uninitiated, who is not privy to the intricacies of mathematics and mathematical reasoning and finds them daunting is likely to feel like a mugwump, a fool, and even the mathematician who is not familiar with that particular field of work may feel that way – both would probably describe the experience as a headache.

The other difference between mathematical reasoning and the above-mentioned nonmathematical reasoning is that while in the former the symbols may be quite standard or may be understood to represent certain specific objects, words or terms from a common language such as English may not be that precise and may have several meanings leading to problems of interpretation or confusion.

It may be possible to convert many of the words or terms in nonmathematical reasoning to symbols but that would require knowledge of and skill with symbolism, which to the mathematician is a sort of shorthand and aid for logical reasoning. The nonmathematician may simply find the symbolization too much of a hassle and too daunting and may prefer to stick to common, everyday words and terms. For instance, if we were comfortable with symbolism, we could convert the following statements:

John is business-minded. Peter is clever. Both of them make good business partners.

to the following symbolism:

$$J^b + P^c \longrightarrow B^g$$

As could be seen, this symbolism is more compact, less cumbersome, and easier to manipulate than words, as it is brief and direct. But the nonmathematician may beg to differ.

The other question is: Can we not assign numerical values to properties such as honesty, patriotism, and kindness and manipulate these numbers in our logical reasoning as we do in mathematics? It is possible but not very meaningful, for assigning numerical values to such abstract, intangible, unquantifiable, unmeasurable properties is subjective, arbitrary, and not easy, as they are not physical properties, e.g., height and weight, which are amenable to precise measurement with some measuring device. Besides, a person could fake honesty, patriotism, or, kindness, i.e., act and appear honest, patriotic, or, kind, when at heart or basically he is not. Such properties could only be subjectively judged or estimated according to degrees such as "very" honest, patriotic, or kind, "quite" honest, patriotic, or kind, and "not so" honest, patriotic, or kind, which are not very precise, even arbitrary, and not very meaningful.

Mathematical reasoning is apparently a combination of both symbolism and common words or terms, with a tendency toward having more of the former than the latter. It is apparent that the total use of symbolism without any common words or terms, in mathematics, is very rare (except perhaps for the case of the massive three-volume *Principia Mathematica*, a tome co-authored by Bertrand Russell and Alfred North Whitehead to defend logicism, i.e., the view that mathematics could be reduced to logic, which was practically all symbolism and devoid of common words or terms, which very few people had read or understood) and is not very practical. Symbolism should be used where it does indeed aid thinking or logical reasoning. But, if it is found to be cumbersome and daunting, as is generally the case with the nonmathematician, then it is better to stick to the more familiar language, the more familiar words and terms, and do without the symbolism, or formalism, as it is commonly called in mathematical terms. If one were uncomfortable with English, one could communicate, e.g., in French, if one were well conversant with that language. However, though to the mathematician concerned, the symbolism, or, formalism, used might be shorthand and an aid to his reasoning, it might be obscure to others, e.g., the other mathematicians not specialized in his field of work and of course the nonmathematicians. Though his mathematical logic might be excellent, even brilliant, it serves no useful purpose if his fellow mathematicians and others do not understand his thoughts, and his thoughts are not recognized, adopted, or implemented. Therefore, he should make it a point to ensure that his ideas are readily understood and not obscure and incomprehensible. He could perhaps conduct a conference or a seminar where he could explain and clarify his ideas, so that they could gain acceptance and recognition, and could be adopted or implemented. Even if his ideas were understood by his peers a number of them might not consider them valid or correct and might not accept or recognize them. This is not surprising, even for an exact science such as mathematics, for there is apparently arbitrariness in mathematical logic as well. Take for example the cases of two mathematical geniuses, Georg Cantor and Evariste Galois. Cantor had been so severely criticized for his ideas on infinity that he suffered nervous breakdowns and had to be hospitalized; his ideas on infinity have become widely accepted and he became famous only posthumously. Galois had lived a tragic life and died of his injury sustained in a duel when he was 20; he was rightly the father of group theory, and group theory gained wide acceptance and he became famous only after his demise.

Despite the abstruse and seemingly impenetrable symbolism, the ideas behind the symbolism in mathematical reasoning may be simple, perhaps simpler than some of the ideas encountered in nonmathematical reasoning. It would be appropriate to clarify some points here with regards to mathematics and mathematical reasoning. It is apparently true that most people find mathematics difficult, including the mathematicians themselves. Of course, as stated above, the symbolism of mathematics may be unfamiliar, which impedes comprehension. Anyone not familiar with the words or terms of one of the languages, e.g., English, would encounter similar problems with comprehension. Mathematical symbols represent (symbolize) some objects, like the words or terms of a language such as English, French, German, or Spanish, i.e., mathematics itself is a language like the other languages – all of them convey ideas, all of them are capable of conveying abstractions, which are intangible, lacking in physical substance, and hence not as evident or obvious as the actual, physical objects of nature such as trees, rocks, and water – an abstraction is a qualitative kind of object, e.g., the patterns or symmetries and the connections or links between objects studied in mathematics may not be so evident and may require careful and close study to be realized or confirmed; otherwise, they would have been missed by the mathematician. This generally abstract nature of mathematical objects is one of the difficulties. The other difficulty is that mathematics generally has little to do with our everyday experiences, i.e., we normally find difficulty relating with mathematical objects, e.g., the square root of minus one (which may seem like nonsense to some), infinite series (no one has ever directly experienced infinity, which appears to have no reality at all besides being an idea or concept), and prime numbers (many do not know what they are but it does not matter as life would still go on as usual, though they are very important in mathematics as they are considered the atoms or building-blocks of the number system). In this respect, other subjects such as electronics, computer science, and physics, are similar; many of the concepts encountered in these subjects, e.g., the electrons and positrons we speak about in electronics, the codes we talk about in computer programming, and the superstrings in physics, are also beyond the scope of our normal, everyday experiences and appear strange, foreign; hence, the difficulty level for such subjects are normally rather high. As for the soft sciences or the social sciences, where the objects of study relate more to our everyday life and have some human interest, the difficulty level is generally not so high. Yet, some students of economics encounter difficulties with some of its concepts, e.g., concepts relating to the supply and demand curves, profit maximization, and equilibrium, which are rather abstract in nature. To the nonspecialist and the uninitiated, mathematics tends to be difficult as it is a rather specialized language, a language well understood and used only by the relatively few specialists, whereas a commonly used language such as English, which is an international language, a language with worldwide and widespread usage, generally does not face such a difficulty. Mathematics is to the mathematician what English is to the English-speaking person, i.e., the mathematician is capable of thinking only in mathematical symbols while the English-speaking person is capable of thinking only in English words or terms. Certainly, understanding mathematics is going to be easier than understanding a language with which we have no familiarity at all, e.g., Swahili or Tamil. Does this mean that languages such as Swahili and Tamil are more difficult than mathematics? It is hard to say but some of these languages may be more difficult to master than mathematics. But mathematics as a language, i.e., the symbolism and terms of mathematics, should be more or less similar in difficulty to those of the other languages. The difference between mathematics and the other languages is that the language or symbolism of mathematics is normally used to convey rather specialized, perhaps more abstract, ideas, whereas the other languages are normally used in everyday conversations to convey the more common or mundane, normally not so abstract, ideas. In other words, it is not really the symbolism behind mathematics which is causing the difficulty in understanding, it is the generally abstract nature of mathematical ideas that is the problem. Though doing away with the symbolism and conveying the mathematical ideas in simple, commonly used terms quite often helps in understanding, this is not possible all the time – the mathematical ideas, though expressed in simple, nontechnical terms, may still be difficult to grasp as they may be extremely abstract, complex, or have very little relation with reality or everyday experiences. Likewise, any abstract or complex ideas, or ideas which have little relation with our everyday experiences, when expressed purely in a simple, commonly used language, e.g., English, and devoid of technical terms or jargon, may also be difficult to grasp. Ideas that are abstract, complex, and difficult to grasp may require clear explanation and clarification, possibly including examples and analogies, in order to be readily understood. In any case, an idea may be incomprehensible because it has been poorly or not clearly explained.

Have you seen people arguing or quarreling over the multiplication table? Have you seen people disputing over their heights or weights? Have you seen people disputing over the built-up areas of their properties? No? Of course not, for such things could be easily verified, by counting, measuring, and weighing. If only we could count, measure, or weigh objects such as patriotism, racism, and honesty, how much more harmonious and peaceful our lives would be, for we could then easily verify our logical conclusions pertaining to these objects and do not have to feel unsure and dispute or quarrel over them. However, there could be some ways of verifying the results or conclusions of the reasoning with these unquantifiable, unmeasurable, intangible, and abstract objects. For instance, an economist or economic advisor through his logical deliberation might have an economic plan to revive the country's economy. He could convince his government to adopt and implement his economic plan. If after the economic plan had been implemented, the country's economy grew and thrived as predicted or forecasted, this could be regarded as verification that the economist's logical reasoning pertaining to the country's economy had been correct. Likewise, a marketer might apply his logical thinking and knowledge of the market to devise and implement a marketing plan, which includes the sales forecast for the year. If at the end of the year the forecasted sales results were achieved and all the stipulated marketing objectives were met, the marketer could be given a pat on the back and rewarded for his correct logical thinking and his proficiency. However, what if in each of these two examples only part of the targeted results were achieved? Then we could say that for each of these two cases the logical reasoning had been partially correct.

From all this, it is inevitable that it is highly important that logical thinking is result-oriented, for logical thinking with verifiable results would certainly obviate any disputes and conflicts, which would be redundant. Hence, the author's emphasis that logic should be result-oriented in this tome.

Furthermore, to strengthen our powers of logical reasoning, we should attempt to familiarize ourselves with the various types of reasoning, e.g., economic reasoning, the various modes of mathematical reasoning, scientific reasoning, legal disputation, technical analysis, social reasoning, religious and spiritual thinking, criminal investigation, and what have you. It would be interesting to find out how the various types of brain-workers think, apart from the knowledge gained. One should also have a strong interest in logic for its own sake. If everyone takes an active interest in logic and tries to find out and learns how others really think then there is likely to be greater understanding amongst all fellow beings and less controversies and conflicts. The author's very strong interest in logic has made him want to play an active part with logic, which includes producing this tome, and to read widely into the intellectually challenging subjects, such as the higher, advanced mathematics, quantum mechanics, and relativity. Though the author might not have a full understanding of these subjects, at least he has got a feel of how logical reasoning in these intellectually demanding, abstruse subjects is carried out, so that he could personally see or judge for himself whether this kind of reasoning is really any different, more effective, or better than our normal, everyday kind of reasoning. Of course, some of this kind of reasoning has been used in the attempt to achieve scientific breakthroughs such as a better understanding, interpretation, and explanation of nature and the invention of better and more utilities for mankind. The important question is if such a kind of reasoning is so good and effective, and would be so beneficial to society, why is society still so bogged down by sicknesses and diseases, natural calamities, political and social conflicts, including wars, terrorism and mass destruction, psychological problems, and social problems such as poverty, famines, high crime rates, violence, murders, and loss of jobs? It seems that we are now more in need of a great social thinker who could revamp and improve our current social system than a great mathematician or a great scientist.

The author would here like to caution readers against intellectual snobbery, which is itself a social problem. It is apparent that there are ivory tower intellectuals around, e.g., scientists, mathematicians, economists, sociologists, political leaders, and the like, who practice some form of intellectual elitism, some form of intellectual snobbery. They may have their own "old boys" network and may not give anyone with original and good ideas but not a member of this "old boys" network a chance, perhaps viewing them as upstarts; they may prefer to stick to the conventional wisdom, and the good, original ideas may never get off the ground but become a waste. We may become good, efficient thinkers through our effort or through some appropriate training or education, but we should not put ourselves in a more superior, snobbish position. As mentioned before, logic should be result-oriented and productive. By being intellectually snobbish and proud we would be inviting dislike, possible ostracization and social problems, which would be counterproductive, even foolish.

Being logical must include being sensible enough to know what is right and proper behavior, behavior that brings as much happiness and comfort to as much people as possible. This would be a sort of intellectual leadership. A person may impress others with his brilliant, logical mind and get their respect. This is brilliant as long as he could coerce others to do the right things and not feel intimidated or threatened by him. He should not use logic to belittle or humiliate others or make others feel inferior or foolish, i.e., be the irritating smart aleck, which is bound to cause resentment, unhappiness, and perhaps backbiting or revenge. He should be humble and should be able to lead those people who accept him as their leader. Otherwise, he should not force his leadership on others. He should only persuade and convince others to accept his ideas, and, should only lead when clearly given the approval to do so. As a logical person versed in the different modes of thinking, he should be able to contribute to society through his productive, result-oriented logic, which means bringing as much happiness and comfort to as much people as possible. This would be a logical ideal for a logical person to abide by. If more and more people become more logical, if more and more people abide by this ideal, the logical result would be a better and more harmonious society.

Let us hope that more and more people would learn to be more logical and learn to use logic properly so that it would benefit as much people as possible. This would be the noble utilization of logic and would be ideal.

33 A Short, Simple, and Effective Way of Dealing with Logical Problems and Disputes in Everyday Affairs

Conflicts, arguments, quarrels, and even fights are evidently everyday occurrences. Practically everyone would find all kinds of ideas and ways to defend their stand on a matter, mainly for egoistic, selfish, or "saving face" reasons. Often, lies, untruths, exaggerations, or distortions of facts, dishonest tactics, and even harmful methods such as threats, veiled, or open, are used by the parties to the dispute or conflict. Practically every disputant wants the other party to give in to him, acknowledging tacitly at least that he is right, though many times it might just be a case of the party who gives in not wanting to prolong a dispute which could bring about more difficulty or hardship. Frequently, neither party would give in and would press hard for their stand. The conflict or dispute drags on, and feelings and emotions get ruffled. The disputing parties might, of course, go for mediation, arbitration, or a court hearing to settle their disputes or differences. In the worst case, they fight each other, as when people come to blows with each other, or even go to war.

The author has encountered many, in fact, too many, of such "problems with logic". Such problems, the author feels, tend to be "unsolvable".

After many years of frustration, the author thinks that there is really one short, simple, and effective way to overcome such "failures in logical practices". It might be radical, but it should work. The author suggests the very important step of making lying and speaking, and acting dishonestly a crime and a sin, to be met by heavy punishment for the guilty. This simple measure should, all at once, get rid of many problems, such as slander, libel, political manipulations through lies and slanders, and disputes or conflicts. Of course, certain harmless "white" lies, e.g., telling a dying person that he would recover with proper medical treatment to prevent mental suffering, should be the exceptions. It is when people have to speak the truth, be honest, and not lie, exaggerate, or distort the truth under such serious circumstances that they would be intellectually honest and resolve disputes more objectively, more speedily, more efficiently, and with more success, even though some disputes might still be unresolved, whereby proper mediation or arbitration would be the next logical step. The problem now lies in the fact that practically everyone treats lying and dishonesty too cavalierly, as though they are expected. When honesty and objectivity are not to be expected, and, it is regarded as being alright, in fact, human, to lie and be dishonest, at least at times, we cannot expect fellow human beings to be able to jointly arrive at logical, objective, and sound conclusions and settlements on their own.

https://doi.org/10.1515/9783112214367-058

Once lying and dishonesty are outlawed, we could expect to see more peace and harmonious relations, and fewer wars or no wars. To the author, this would be a short, simple, effective, and surer way of handling logic where the affairs of man are concerned.

34 The Last Say on Logical Reasoning or the Use of Logic

Say, we are faced with several premises or statements. How shall we proceed from here to a truly logical conclusion? How do we proceed, hence, with the act of logical reasoning? How we wish logical reasoning could proceed so smoothly and surely as "1 + 1 = 2" or "1 + 1 + 1 = 3" or "A is taller than B. B is taller than C. Therefore, A is taller than C." or "A is taller than B. B is taller than C. D is shorter than B but taller than C. Therefore, A is taller than C and D." Notice that such "conclusions", which are the result of the act of logical reasoning, are measurable, quantifiable, or verifiable entities (in this case, quantity and height, respectively). But, unfortunately, in many cases, the conclusions derived from the act of logical reasoning is not measurable, quantifiable, or verifiable; the more abstract and less "practical" or "practicable" the conclusions, the less measurable, quantifiable, or verifiable they would be. (This would be, as stated earlier, a case of "quantitative" logic versus "qualitative" logic.) This is indeed the greatest problem with logical reasoning. Since a conclusion is not based on concrete, tangible, or verifiable facts or premises, there is no way to tangibly or physically to measure, check, evaluate, or verify the conclusion. At most, we could have a very strong intuitive feeling or "gut feeling" that the conclusion based on those premises is correct or logical.

In logic, "seeing is believing" must be everything. What we could not see, measure, or physically verify must be subject to some doubt. There is no point in insisting and swearing that one's conclusion is absolutely correct when there is no "physical basis" or "tangible, verifiable basis" for one's so-called logical conclusion. Where the conclusions are not measurable or verifiable, the best recourse should be to regard the conclusion as *tentatively* correct or logical; the best, most practical, and most logical stance is to wait and see, compromise, or give and take. Otherwise, we might end up with "unconvincing" arguments, circular reasoning, or confusion.

The author would like to propose the following ways of going about the act of logical reasoning:

1) List all the possible implications of the premises (or statements).
2) Find and list the logical links between these implications, if any.
3) Form a logical conclusion from these logical links.
4) Finally, and very importantly, attempt to *physically verify* the conclusion, if this is at all possible. (Just like the scientist carrying out a physical experiment to verify a hypothesis) At the very least, try to find out whether there was or were any parallel(s) in the past, i.e., verify whether any similar happenings or "conclusions" have taken place in the past as a result of similar circumstances or "premises". (Carry out thorough research work.)

https://doi.org/10.1515/9783112214367-059

Once again, the author stresses that logical reasoning should be, as much as possible, based on solid, tangible, measurable, quantifiable, verifiable facts in order to be "trouble-free", truly correct, or indubitable. If the logical reasoning is based on abstract premises alone, it is likely to be hard to follow or understand and to cause doubt. It is when logical reasoning is based on solid, tangible, and verifiable facts that it is easily comprehended, clear, and indubitable. When this happens, disputes would be minimal, and there would be comparative harmony and peace. In short, "quantitative" logic, whose results are measurable and verifiable is easier to handle than "qualitative" logic, whose results are not tangible or verifiable and might be subjective.

One additional bit on the use of logic here. Mathematicians, who are theoretically well-trained in the use of logical reasoning, have been attempting to apply logical reasoning to solve mathematical problems pertaining to infinity, e.g., Euclid's proof of the infinity of the primes, the twin primes conjecture pertaining to the infinity of the twin primes, the Goldbach conjecture pertaining to the infinity of the even numbers, each being the sum of two primes, and the Riemann hypothesis pertaining to the infinity of the solutions of the Riemann zeta function, which lie on a straight line which would imply the infinity of the primes and describe their distribution. It is apparent that, to the mathematicians, logical reasoning is a very powerful tool for solving very difficult problems such as those mentioned above. Now, could they or any logician not use the same kind of rigorous logical deduction to prove the existence of God (or, even the existence of the Devil or Satan), just as the mathematician, Descartes, had tried to use logical deduction to prove that God exists? In actuality, the concept of infinity is rather "mind blowing". No one could count, check, verify, or live to infinity. It is just an abstraction, a concept. We could conceptualize the idea of a Being that lives to infinity or eternity and that would be the idea of God. If a mathematician could prove the infinity of some mathematical objects, such as the infinity of the twin primes or the even numbers that are each the sum of two primes, they could, in theory, prove or disprove the possible existence of a Being, who lives for infinity or eternity, i.e., God, who could be regarded as the embodiment of infinity or Infinity itself personified. The mathematician, Georg Cantor, had proved that there are different orders or degrees of infinity (by using a "diagonal" method), an idea so bizarre that he had been viciously criticized by his contemporaries, which had led to his mental breakdown, though this concept of infinity is now embraced by mathematicians. In other words, some infinities are larger than others. All this is a question of interpretation of how one chooses to interpret infinity, an interpretation that is arbitrary and not subject to any definite and clear-cut rules. To the author's interpretation, an infinite sequence or progression is some object that goes on forever (with no ending), and there has never been a higher or greater, or, lower or lesser degree or magnitude of "forever". Mathematicians, in effect, tell us now that there are different degrees or magnitudes of "forever". Nobody has ever directly experienced infinity, something that goes on forever, and suddenly we are told that there are infinities that are larger than others. To practically all of us, there is only one "forever", one infinity. It is as though we are hearing someone tell us that one corpse is more dead than another, or, this 1 in is longer (or shorter) than the other 1 in, or, worse, some 1 in are longer (or shorter) than other 1 in. How ridiculous it all seems. (Here, it may be appropriate to elaborate a little further on this question of the different degrees of infinity as postulated by Cantor. Cantor had used a very clever technique, which could perhaps be viewed as a cunning "trick", if one may regard it as such. He had listed rows/columns of various numbers, the numbers representing members of the various sets. Then, by the "diagonal" method, he had created new sets, whose members were distinct from the members of the original list. Cantor had interpreted this phenomenon as the existence of different infinite sets with different degrees, magnitudes, or orders of infinity. But all these various sets of numbers could perhaps be viewed en masse as one set of numbers that are "infinite in all directions" – horizontally, vertically, diagonally, zigzag even, in fact any other direction, i.e., there is only one infinity, and not various infinities. If we, e.g., substitute the "quantity" infinity, here with the "quantity" 1 in, we could now consider ourselves having a one inch (which is actually equal to 25 mm) with different degrees or orders of magnitude. We could then consider ourselves having, e.g., a 1 in that is 7 mm, 1 in that is 8 mm, 1 in that is 9 mm, 1 in that is 10 mm, and so on, i.e., we have 1 in of various degrees of magnitude, which are the analogues of Cantor's infinities with different degrees or magnitudes. Is this not ridiculous?) Thus, it appears, logic is subject to different interpretations and rules, and hence, its resultant subjectivity and problems.

Earlier, it has already been stated that one's capacity for logical thought is a function of one's ability to use a certain medium (in this case, a language) for carrying out the act of logical thinking. Thoughts and ideas, logical thoughts and ideas, are conceived in words (that could be read), sounds (that could be heard) or symbols (that could be seen/heard and interpreted). Mathematicians are experts in the use of symbols, mathematical symbols to be precise. Others who make use of logic, such as, e.g., lawyers, make use of the medium of communication, e.g., the English language. Therefore, one's mastery of a medium of communication, a language in which to carry out logical deduction, is of very great importance. The English language, in all its complexities and nuances, is a very important medium for the performance of the act of logical deduction, despite its disadvantage of being able to lead to misinterpretation of meanings. The English language, it should be noted, is able to convey the subtlety of ideas or logical concepts, and is often subject to abusive usage by those who like to "twist and turn" with ideas or arguments. The author, however, is of the opinion that the English language, being an international language and a most popular and universal one too, is a most effective tool for logical deduction; its mastery is of paramount importance for logical deduction; even in mathematics, the English language is of fundamental importance. Mathematics may rely on a copious use of symbols (signs which have meanings), which are a form of "shorthand" in which to carry out mathematical reasoning; in other words, the mathematical symbols facilitate mathematical reasoning. But each mathematical symbol, which represents an idea or thought, would have to be interpreted by the human brain in a language with which it is familiar, a medium of communication, e.g., the English language, Greek, French, German, etc. And, the language itself, be it English, Greek, French, German, or any other, is the set of "symbols" that represent the objects that could be experienced by our senses, viz., sight, hearing, smell, taste, and touch. The mathematical symbols could be equated with the words of any language or medium of communication. An examination of any mathematical journals or texts would reveal that mathematical symbols alone are insufficient and are normally accompanied by words, the language of communication, e.g., English. It is apparent that most mathematics journals are published in English. Someone not familiar with the symbols in the mathematical text, though he might be an expert in the English language, would tend to find the text rather incomprehensible, which is not surprising. For example, the classic three-volume mathematical tome, *Principia Mathematica*, which had been co-authored by Bertrand Russell and Alfred N. Whitehead and which brought fame to its authors, are fully embellished with symbols and practically devoid of the normal language, and are by all accounts incomprehensible even to the keenest mathematicians, though the ideas in the tome, which are incomprehensible because of the arcane symbols used to convey them, are not that difficult. The author is of the opinion that the mastery of the language of communication, e.g., English, is more important than the mastery of the mathematical symbols (which could be translated into, or has to be understood in terms of, the language of communication, e.g., English). Even the mathematical symbols are normally created from the letters of a language of communication, e.g., English and Greek, something which is familiar. This has to be the case for, as stated earlier, ideas and thoughts cannot exist without the meaningful words or media of communication with which our minds are familiar. It is practically impossible to think without language or words, and, therefore, the better one's mastery of a language is, e.g., the English language, the more capable one is of logical deduction. We may make use of certain images or signs and may even communicate with signs, but our thoughts are in the terms of a language, or, words, with which we are familiar. All the seemingly abstruse symbols in mathematics can be translated into plain English words or the words of any other language, and, vice versa.

Many people fear mathematics and have problems understanding mathematical ideas, especially higher mathematics with its arcane symbols (which may also have the dual objective of keeping "outsiders" out or in the dark). The problem is due to the obscurity or ambiguity of the mathematical symbols themselves. It is because we are uncertain what the mathematical symbols represent and are unsure of how to interpret these symbols that it makes us want to throw up our hands in trying to follow or understand a chain of mathematical reasoning, especially so if the chain is a long and convoluted one (a good memory would be of great help in order to grasp with ease the long chain of reasoning, involving plenty of abstruse symbols and unfamiliar terms – hence, the importance of a good memory to logical deduction). However, these "frightful" mathematical symbols can be translated into plain, simple words of a language with which we are familiar, and every mathematical reasoning can be as clear as daylight. Therefore, in the very act of logical deduction, we should avoid obscurity and ambiguity and make our ideas or reasoning as clear, precise, and simple as possible. The author believes that no idea or logical reasoning is too difficult to be understood. Where the idea or logical reasoning seems too difficult to comprehend, it is most likely to be the fault of the act of exposition or the use of the wrong words or symbols. Take the following example; if the author were to use the following "abstruse" symbols to convey a message that is, in actual fact, very simple, how can anyone fathom what he is saying?:

$$\Xi \ \nabla \varpi \ \approx \zeta \phi \ \Theta \ \varnothing \ \lambda \not\subset \alpha$$

Perhaps, Sherlock Holmes could have deciphered this message.

In life, in politics, practically everyone would favor democracy. We want to have the right to vote, the right to express our views and be heard, the right to live freely, etc. Why not extend these democratic principles to our intellectual life as well? We preach equality when we espouse democracy. Whether rich or poor, Dutch or English, Christian or Buddhist, etc., we want to be treated equally. Then, we should treat everyone as our intellectual equals as well and give everyone a chance to express their views and thoughts and respect their views and thoughts, instead of castigating others for holding certain views and subjecting them to ridicule or showing disrespect or discourtesy to them for having those views. In the use of logic, we should be democratic enough to listen, respect, share, and compromise.

It is the habit of experts or professionals in various specializations to use jargon, terms, or expressions peculiar to their specializations when communicating. For example, lawyers are fond of legal terms or jargon, what we call "legalese", engineers are prone to using their own technical jargon or technical terms and mathematicians or logicians have their own special symbols and terms. It is supposed that this is to add a touch of "professionalism" or "image of special knowledge" for these various specialists, and the uninitiated who is not privy to their specializations is not supposed to understand what they are saying and so would be kept out – this could be regarded as a form of "knowledge protectionism". This could also be made out in such a way that these specialists would appear more profound than what they actually are. In fact, some philosophers have been known to have been accused of couching their ideas, which might indeed be very simple, in high-sounding or profound-sounding terms so as to be impressive. Is all this bad? All this is likely to lead to lack of understanding (here, the ideas serve no purpose – if nobody understands them, they are as good as not having been said – in fact, it is a waste of the target audience's time to utter these ideas if the philosopher knows that these ideas are going to be incomprehensible – the philosopher may just want to appear profound and impress, but this is selfish and should be discouraged), misunderstanding (which is worse than lack of understanding as it may cause problems), confusion, or even conflict, on the part of the uninitiated. It is as though the specialists are keeping out the uninitiated.

Therefore, in the exercise or use of logic, it is proper and wise to avoid incomprehensible or difficult to comprehend jargon or specialized terms, and utilize clear and simple terms in order to have the logical connections easily and clearly understood, and for there to be no confusion. The important aim of logic should be to convince or convert and not to confuse. Unfortunately, this is often not the case. Here, the author emphasizes that logic should rely on simple, clear, concise, and precise terms and expressions as far as possible. The English language, e.g., is capable of meeting this requirement but, unfortunately, its usage is often abused, with untoward consequences.

To the author, logic could be regarded as the beauty of ideas or concepts, which is not unlike physical beauty. Physical beauty is subjective and qualitative, and not quantitative, but at least it describes a tangible person or object. On the other hand, the beauty of ideas, though also subjective and qualitative, describes something intangible (not physical or concrete) or abstract. As logic, like physical beauty, is thus subjective, all the more, it should always be held with suspicion or even skepticism. We should never force it on others, as others might have their own standards of logic, their own standards of beauty for ideas. (As we normally do not quarrel about or dispute over physical beauty, claiming that it is only "in the eyes of the beholder", so, similarly, should we regard logic, or, the beauty of ideas, as such.) Logic could be regarded as an invention of the human mind, which cannot exist independently of the human mind, unlike say, the existence of tangible objects such as rocks, trees, or water. It could be regarded as merely a play of words or language (as the great philosopher, Wittgenstein, had postulated, for, also, as mentioned earlier, it is not possible to conceive ideas, logic, without words or symbols; can a blank but conscious mind think logic, can one carry out the act of logical deduction without words or symbols?) Logic could even be a tautology, describing an object in different terms but actually conveying the same thing, e.g.:

It has long hair and wears a skirt and high heels.
It is not a man.
Therefore, it is a woman.

In any case, even without the above logical deduction, we all know that if a person is not a man, he is, he must be, a woman (though there are very rare cases of hermaphrodites or persons with both male and female features), i.e., the negative of man is woman. In other words, "woman" has the same meaning (synonym) as "not a man" – a tautology so to speak, using different expressions but meaning the same thing.

The author would like to bring up some examples of logic being subjective and merely some rules invented by the human mind for arriving at some logical conclusions. These examples are taken from mathematics, where logic is supposed to figure prominently, where its practitioners take pride in the rigor of its logic. As the reader is probably aware, mathematical reasoning relies much on certain axioms or assumptions, which are obvious truths that do not require proofs (the problem here is what is obviously true to one person, might not be obviously true to another), lemmas, and theorems (the problem of lemmas and theorems, which are truths that have proofs, is somewhat similar to that for axioms, in that what are considered as the proofs of the truths' validity might not be acceptable or obvious to some others). Without them, no mathematical reasoning might have been possible. Of course, if they were wrong, then the results of the mathematical reasoning would have been wrong and meaningless. This seems unique only to mathematics, for no other type of reasoning, e.g., scientific, legal, and economic reasoning, work in such a manner, i.e., rely on axioms, lemmas, and theorems, though they might make assumptions (which could be considered the equivalent of axioms). Since mathematical logic is reputed for its rigor, should not other types of reasoning, like the three mentioned above, proceed ahead with their respective axioms, lemmas, and theorems, if this were at all possible and practical? On the other hand, why couldn't mathematical reasoning be more mundane and more like our normal, usual way of reasoning? Is our normal, everyday mode of reasoning, though it might have helped produce great results and has brought great problems as well, be really inferior to mathematical reasoning? The author has had some doubts about some of the so-called mathematical proofs or reasoning, which seem rather artificial, arbitrary, and lacking in rigor.

Consider the following mathematical theorem:

If α and β are cardinal numbers (finite or infinite):

$$\alpha \leq \beta, \beta \leq \alpha \rightarrow \alpha = \beta$$

Stated in everyday language, if alpha is smaller than or equal to beta, and, beta is smaller than or equal to alpha, then alpha is equal to beta. But, there seems to be a contradiction here – on the one hand, it is stated that alpha is smaller than (or equal to) beta, and, somewhat contrarily, it is stated that beta is smaller than (or equal to) alpha, which casts some doubt on the credibility of the logic of the conclusion that alpha is equal to beta, there being not quite a logical link between the premises, "alpha is smaller than or equal to beta", and "beta is smaller than or equal to alpha", and instead, there appears to be a contradiction between them – "alpha is smaller than beta" contradicts "beta is smaller than alpha". There appears to be some other way of interpreting, explaining, or proving this theorem, but the author has no knowledge of it – this theorem has been proven independently by E. Schroder (1896) and F. Bernstein (1898), and is known as the Schroder-Bernstein theorem. To the author, we should not definitely conclude from the two statements or premises, alpha is smaller than or equal to beta, and, beta is smaller than or equal to alpha ($\alpha \leq \beta$, &, $\beta \leq \alpha$)that alpha is equal to beta ($\alpha = \beta$); indeed, three possibilities could be deduced from these two statements or premises: either alpha is smaller than beta ($\alpha < \beta$), or, beta is smaller than alpha ($\beta < \alpha$), or, alpha is equal to beta ($\alpha = \beta$) – the probability of each of these three possibilities being realized being 33.33%. On the other hand, if alpha is an integer or whole number smaller than 2, and, beta is also an integer smaller than 2, then we could certainly conclude that alpha is equal to beta, that is: $\alpha < 2$, $\beta < 2 \rightarrow \alpha = \beta$ – in this case, alpha is equal to 1 and beta is also equal to 1, wherein alpha is obviously equal to beta, that is: $\alpha = 1 = \beta$.

Let us look at the following proof or mathematical reasoning:

Proposition: If A, B, C are sets with $A \subseteq B$ and $B \subseteq C$, then $A \subseteq C$.
Stated in everyday language, this proposition stipulates that if A, B, and C are sets, with A being a subset of B and B being a subset of C, i.e., every element of A is an element of B and every element of B is an element of C, then A is a subset (or element) of C.

Proof. Every element of A is an element of B and every element of B is an element of C. Therefore, every element of A is an element of C, so $A \subseteq C$.

It appears that the above proof is merely a restatement, or, repetition of the proposition, and seems hardly a proof, though it is regarded as one. ($A \subseteq C$ could be read as or interpreted as "A is a subset of C, wherein every element of A is an element of C".)

The following mathematical reasoning might appear questionable:

Let $P_{(a)}, \ldots\ldots\ldots\ldots\ldots, P_{(n)}$ be a sequence.

Then, by induction, this sequence implies the presence of other sequences, as shown below:

$$P_{(a)}, \ldots\ldots\ldots\ldots\ldots, P_{(n)} \rightarrow P_{(b)}, \ldots\ldots\ldots\ldots\ldots, P_{(n+1)}$$

$$P_{(b)}, \ldots\ldots\ldots\ldots\ldots, P_{(n+1)} \rightarrow P_{(c)}, \ldots\ldots\ldots\ldots\ldots, P_{(n+2)}$$

$$\ldots\ldots\ldots\ldots\ldots\ldots\ldots\ldots$$

The arrow ($\rightarrow$) represents "leads to". If we, e.g., let $_a$ above equal 1 ($_b$ is one step after $_a$ and $_c$ is one step after $_b$, as in the alphabet) and $_n$ above equal 10, then the above mathematical statements could be translated into the following:

$$1, \ldots\ldots\ldots\ldots\ldots, 10 \rightarrow 2, \ldots\ldots\ldots\ldots\ldots, 11$$

$$2, \ldots\ldots\ldots\ldots\ldots, 11 \rightarrow 3, \ldots\ldots\ldots\ldots\ldots, 12$$

$$\ldots\ldots\ldots\ldots\ldots\ldots$$

In other words, induction tells us that if we see 1 ($P_{(a)}$), we could expect to see 2 ($P_{(b)}$) next, and if we see 2 ($P_{(b)}$), we could expect to see 3 ($P_{(c)}$) next, and so on. Induction is commonly used in mathematical proofs or mathematical reasoning. The question is whether its results are justifiable or rigorous. If, for instance, we see seagulls appearing at the beach every summer, we naturally conclude that seagulls would be at the beach every summer. This is induction, whereby a pattern or trend is observed. Is induction foolproof? Definitely not. In fact, some mathematicians frown on induction as a method of mathematical reasoning. There is no solid reason why seagulls must appear at the beach every summer and it might not be logical, and be too much to expect seagulls to appear at the beach in the future summers. How could we be certain that the seagulls would turn up at the beach again in the future summers; is it not possible for them to find a better beach to frequent by then?

In mathematical reasoning, axioms or obvious truths that are not proven are frequently used. The author has some doubt about the use of axioms in mathematical reasoning. Mathematicians require "air tight" logic or rigorous proofs in solutions to mathematical problems and yet could accept the use of axioms. To the author, this is a contradiction. What is an axiom or obvious truth to one person might not appear so to another. Earlier too, the author mentioned some well-known mathematicians who needed 100 pages or so to prove such an obvious truth as: $1 + 1 = 2$, which could have been simply treated as an axiom. Is this not too much? It is as though one is trying to split hair, which is quite impossible and absurd. The author understands that axioms often have to be used before mathematical reasoning can proceed, for they act as the starting points or bases. But, at least, they should be used with discretion and qualification, e.g., we could state that the axioms might be subject to doubt by some quarters but they could, for feasibility, be regarded as true, or, they could be regarded as provisionally true till proven otherwise. The imposition of this condition makes the reasoning all the more reasonable, all the more logical.

Also, mathematicians frequently resort to the "reductio ad absurdum" proof or proof by contradiction. It has been mentioned earlier and would be emphasized here that this kind of proof has a weakness. Here, the mathematician makes an assumption that a conjecture is wrong, or, correct, and then goes on to find evidence that contradicts this assumption, proving otherwise. For example, if an assumption that a conjecture is wrong is proven incorrect:

1) It might not necessarily mean that the conjecture is correct. (It could be partially correct, or, correct under certain circumstances and wrong under other circumstances. It could be half-correct and half-wrong. Some things can neither be classified as correct or wrong, e.g., an act might not be based on a moral or correct, or, immoral or wrong principle – it might be amoral. It could be just "average" or "in-between correct and wrong".)
2) Its correctness or wrongness might be undecidable, i.e., it is neither possible to prove its correctness nor wrongness. (Reference can be made to Godel's incompleteness theorems.)

As many mathematical proofs are proofs by contradiction, there is the implication that there is a flaw or defect in much of mathematical logic. Godel's Incompleteness Theorems testify this – there is a serious defect in the foundation of mathematics. Hence, mathematical logic appears flawed in many aspects.

Though mathematics might be a comparatively difficult subject to master, it nevertheless plays a very important role in our lives. We need mathematics for calculations and computations, e.g., in engineering and computer work. Mathematics uses equations and formulae, which are comprised of symbols and letters, as well as the operations of addition, subtraction, multiplication, and division. It deals with abstract concepts such as equality, sizes, degrees, elements, sets, groups, convergence, divergence, and infinity, and, this is perhaps where the difficulty really lies, especially when the abstraction of mathematical concepts is carried out to such an extent that the ideas appear to be detached from reality.

Pure mathematics, like chess, might be an intellectually demanding art, with sets of rules to follow that have little utility or practical value. Even a mathematician and a real genius with incredible powers of comprehension such as John von Neumann had difficulty with mathematics. Here, we quote him, "One never understands mathematics, one simply gets used to it". In pure mathematics, there are not only many rules one has to be familiar with but one has to get used to the many symbols and terms as well. Learning higher mathematics is not unlike learning a difficult foreign language. But the author thinks that learning a truly difficult foreign language is more difficult. What is the point of learning mathematics then, if it is difficult and has little utility or practical value? The renowned British mathematician, G. H. Hardy, had lamented that as a pure mathematician by profession, his contribution to society had been practically nil. The only utility perhaps is that pure mathematics and chess, like the beautiful paintings of an artist, may bring joy and happiness to those who practice them or deal with them. However, some areas of pure mathematics, such as group theory and topology, e.g., have found practical uses in such fields as physics and biology. As to logic itself, the author would like to subscribe to the "good salesman" concept. Though logic may bring joy to those who appreciate its "beauty", it is in effect no better off than beautiful pieces of art, if it does not enable us to solve problems and progress. Good and effective logic is comparable to a good and effective salesman. What are the criteria for being a good and effective salesman? Is a good salesman a salesman who just carries himself well and is capable of making superb sales presentations? If you were an employer who has to choose between firing one of your two salesmen, one of whom is very impressive and capable of superb sales presentations but produces poor sales results, while the other salesman is not impressive at all and not so capable of making good sales presentations but produces excellent sales results, who would you fire? Wouldn't it be the former? The same concept should apply to logic as well. Logic should not just appear nice and impressive and bring happiness or pleasure. It should be capable of bringing concrete results, it should be capable of being "creative" or productive, it should be result-oriented. Like the beauty of art, where they say that "beauty is in the eyes of the beholder", the validity or truth of logic is "in the mind of the thinker". As the saying goes, "One man's meat is another man's poison." Perhaps, we could state, "One man's logic is another man's nonsense".

We know that logic is imperfect and creates problems for practically everyone. The author has all along been trying very hard indeed to find a way out of this rut. Since logic often cannot convince us as to what is right and what is wrong, we need a third party who is neutral and impartial to decide for us what the right paths are, a third party, whether human or artificial intelligence, who is sanctioned by the law, the authorities, to guide us through the right and logical paths. We need a highly intelligent computer that is capable of storing vast quantities of data and using quick and efficient computation to assist us, to guide us, a powerful thinking or logic-analyzing machine that would indeed be impartial and neutral; all we have to carry out is to input our requests to this supercomputer and the answers or advice would spew forth expeditiously, efficiently, and impartially from it. This is a distinct possibility and an apparently ideal solution pertaining to our perhaps overwhelming problems with logic. When this happens, if it does happen, there could be lasting harmony and peace. And, we could all sit back and sigh, "Peace at last!"

But, before we have the luxury of such a thinking, logic-analyzing machine, we have to have an efficient and proper way of handling logic. We pride ourselves in our intelligence, our ability to understand and to reason (i.e., use logic). For example, the engineer uses logic (including mathematical reasoning) to analyze and solve engineering problems, the scientist relies heavily on reasoning or logic in analyzing natural phenomena, the politician has to understand the needs of his constituents and has to find solutions to their needs, which requires logical reasoning; the chef has to figure out what food ingredients to use and in what proportions to arrive at dishes that stimulate the taste buds and, to some extent, this requires logical reasoning; even the house-wife has to figure out how to spend on consumables within a given budget, an act that requires some degree of logical reasoning, etc. We see that logic plays a vital role in our lives. Earlier, the author has stated that the human brain is hardwired for logic (which means that logic is a function of the human brain). This implies that if the human brain had been hardwired differently, human logic could have been different too. In logical deduction, the human mind has to strain itself over premises, statements, or hypotheses to arrive at the correct connections or conclusions. Earlier too, the author has stated that there is the possibility of a mind, perhaps one belonging to an extra-terrestrial (whose brain might be hardwired differently, if he does possess one), to know, without having the need for logical deduction. This would be a super-smart or super-intuitive mind. Our minds are intuitive and to some extent, we are capable of sensing and knowing, without the need for logical deduction. If only we could have instant understanding and knowledge all the time without having to perform the rigmarole of the act of logical deduction. Our minds would have to be much more intuitive than they are – be super intuitive. Imagine a mathematician or logician not having the need to use axioms, lemmas, theorems, and elaborate logical steps to form valid conclusions. Imagine that through a fantastic leap of intuition, the mathematician or logician has instant understanding and knowledge, just like a seer or fortune-teller, who could forecast future events before they took place. If our intuition had been much greater, if our brain had been much better constructed, we would perhaps encounter less problems with logic, and we would probably have created different rules or laws of logic – we would probably have a different kind of logic, perhaps a logic that is much more effective; we might even have a much more hassle-free way of understanding and knowing, e.g., instant understanding and knowledge. In other words, we would have to be much smarter than what we are, to be able to reduce or overcome our problems with logic.

How reliable is human logic? Human logic does not seem so reliable after all. The author is of the opinion that the human being should rely more on his physical senses (sight, hearing, touch, taste, and smell) than the mere use of logical reasoning. In other words, he should try to find out, to figure out, or know by relying more on his physical senses. Logical deduction might tell us that a certain phenomenon should be the outcome. But, we should verify the outcome with our physical senses, i.e., physically observe whether the outcome would occur indeed if such and such conditions were present; in other words, observe the effect of the cause. Empirical evidence or physical evidence, which is the essence of existence, should rule over logic, which is abstract evidence. Would you prefer to believe something that could be physically observed or something that could not be physically observed? The answer must be obvious. Indeed, seeing is believing! This is precisely the reason why scientists carry out physical experiments (to verify or confirm hypotheses obtained through logical deduction), why manufacturers participate in exhibitions (where their products could be displayed, and, physically studied by potential customers), why judges occasionally visit the scenes of crimes (to physically ascertain for themselves) or have acts of crimes acted out in the court (to physically ascertain certain possibilities), etc. It would indeed be foolhardy to rely entirely on pure logical deduction without carrying out physical verification, especially now that we are now aware of the shortcomings of logic.

Therefore, we should always regard logic with an open mind and a healthy skepticism. If only we are so smart that we can understand and know without logical deduction, without cracking our brains. It would not be surprising that such an intelligent race (or races) already exist(s) in another universe (or universes). Our problems with logic hence appear to be the outcome of our faulty mental equipment or brains. For example, due to our imperfect memory, we are easily lost in a long and complex chain of logical reasoning, as by the time we arrive at a later stage of the long chain of reasoning, we would normally have forgotten about the earlier stages, resulting in us failing to see the logical link in the long chain of reasoning; by the time we reach the later stages of the long chain of reasoning, we would normally have to refer back to the earlier stages in order to see the logical link in the whole chain of reasoning; understanding a long and complex chain of logical reasoning, such as that normally encountered in mathematics, is thus an arduous and time-consuming process requiring a lot of patience; of course, a better and more efficient memory would have made the task an easier one; of course, it would have helped in understanding if the long and complex chain of logical reasoning were presented in a clear, simple, precise, and easy style. If only human beings were much more intelligent, much smarter. If only our intuitive minds were much more intuitive. If only our memories were much better. Then, we would certainly not have so much problems with logic. Human beings should be humble enough to acknowledge the shortcomings of their mental faculties. When this happens, disputes pertaining to logic would be reduced.

It appears man is generally affected by intellectual pride. He not only thinks he is the smartest animal on earth; many a person thinks that he is smarter than others; many also try to be smart to outdo others in the smartness game. But alas, little do these smart alecks realize that logic is imperfect, and "tricky", and that they are just making "much ado about nothing". To these people, the author would like to say, "Take it easy. Do not take intelligence and logic so seriously". It has been acknowledged that there are many types of intelligence and that our I.Q. or intelligence tests measure only some of them, which means that a person with a higher I.Q. score is not necessarily more intelligent.

After so much talk and discussion about logic, how important is logic really? If the author might offer his view, it would be, "It does not appear to figure prominently in the behavior of human beings". Human beings can be seen behaving no differently, or, worse, than animals, e.g., butchering each other and blowing each other to pieces. All these are perhaps manifestations of different logics or conflicting logics at work. This implies that human logic is defective and "problematic". Thus, despite our "high-level" logic, we are basically brutes, animals.

There appears to be a great barrier to logical thinking or logical behavior, i.e., emotions or passions. It seems that human beings are ruled more by emotions and passions than logic, and, often allow their emotions or passions to color their logic, i.e., the emotions and passions seem to have a greater influence on us than logic. Our logic is often based on prejudiced, biased, or faulty premises. The result is confusion, self-contradictions, wrong conclusions, and, of course, problems. When emotions and passions interfere, a person might find himself unable to think straight, think clearly. Though it is inherent in man's nature to be emotional and passionate, he should not allow emotions and passions to interfere with his thinking, i.e., he should be as objective as possible, which means that he should be disciplined and should be able to control or rein in his emotions and passions. Many a time, a person, in a moment of great emotion or passion, does something that he later regrets. Therefore, through proper education, a person should have been trained, not only in logical thinking, but in self-discipline, in being capable of controlling the emotions and passions. Human beings generally have whims, fancies, and urges, and might carry out an action in a fit of fancy or strong emotion, against their better judgment, against logic. So, the emotions and passions seem more important than logic, and being able to deal effectively with them seems very important. Thus, besides learning how to think logically, we should learn how to live balanced, happy, and contented lives. What is the point of being intelligent and being able to think logically if we feel unhappy and discontented all the time? Then, is it not better to be happy and contented, though one might not be intelligent or capable of effective logical thinking? Is it not better to be a "happy fool" than an "unhappy Aristotle"? Some might prefer to be "happy fools", while others might prefer to be "unhappy Aristotles", as they value intellect over everything else. The ideal of course is to be smart, happy, and contented. It might be argued that if a person is smart, he would be able to apply his intelligence to find happiness and contentment, find the objects that bring him happiness and contentment. Yet, many apparently intelligent people are clearly unhappy, even depressed. In actuality, a person's happiness has much to do with his environment than anything else. How could a wise and intelligent man, surrounded by fools and evil people, be really happy? He might attempt to change them for the better with his intellect, but there is no guarantee of success. He might even be unhappy simply because he is smarter and cleverer than his peers, who might find him "queer" and avoid or ostracize him. He might try to make himself more ordinary, less "queer", to be accepted by his more inferior peers – if the acceptance of his peers were so important to him, which is a question of personal taste. It is perhaps only the superiority of a person's intellect that enables him to realize the seriousness of a situation and makes him sad or depressed, while the person of lesser intellect might not even be aware of the seriousness of the situation, and hence would be able to sleep over it – remember the adage: Ignorance is bliss.

But, intelligence and logic are important for the success of any human enterprise and the survival and progress of the human race. And, intelligence, aka logic, could be used for good as well as for evil. Let us hope that there would never be any evil genius out there to wreak terror and havoc in society. Let us hope that society would be capable enough to prevent such an untoward situation from arising and good enough to see to it that intelligence would always be used for good and peaceful purposes. It would boil down to the question of good triumphing over evil, intellectually and physically.

Cheers to logic for the good and betterment of mankind, a logic that is noble, pure, objective, and indisputable; nay to twisted, crooked, or pseudo logic.

A few blunt remarks on logic: What is the damned use of logic if, instead of helping us solve our most pressing problems, especially problems pertaining to interhuman relationships, it helps to create rifts in our society, with the result that there are conflicts galore? This would probably be logic at its most counterproductive and lowest ebb. It might then be better off to be without logic. And, logic might not even be a precondition or prerequisite for survival. For instance, a tree can survive for hundreds of years or more without any apparent need for logic, e.g., through adaptation to its environment. (Being able to adapt to one's environment is a sign of intelligence in a human being. So, can a tree not be regarded as intelligent, since it could adapt to its environment?) Can man with his logic ever hope to emulate this? Is it better to be not too reliant on logic? Is it not better to have a free and open mind on logic and regard it with a healthy skepticism? At this point here, we should all know that logic is imperfect and is not to be taken too seriously. To be ignorant of this is to be so at our own peril. Perhaps, we would be really better off if all our brains had been hardwired to see logic in only one way (instead of so many different ways, as it is apparently so now), just as our physical senses make us all, e.g., see the color green as green or red as red. (Imagine the chaos if our physical senses make us see colors differently, e.g., red might be red to some, blue to some others, green to yet others, and so on. How ridiculous it would then be! How disruptive!) Enough is said. The lesson should have been learnt.

In lieu of the human brains being hardwired to see logic in only one way, to think alike, they could be coerced to do so. Our society functions with hierarchy, a social structure. At the very top, we have the leaders and governments, who lay down the rules or laws that have to be adhered to by everyone down the social ladder. (Hierarchy and rules or laws ensure some form of conformity and uniformity, and, preserve some necessary order and prevent chaos.) These leaders and governments can of course lay down the rules or laws pertaining to all matters related to logic or logical reasoning so that disputation does not exist, or is minimal. Alternatively, we can have a carefully and democratically elected intellectual elite who are, theoretically, trained and experts in logic and the art of logical deduction to draw up codes of conduct and rules pertaining to the use of logic, which are to be adhered to by everyone. In other words, we should attempt to get everyone to "think alike". Does this not mean that originality would be bad? Originality, if it "rocks the boat" and causes disputes and problems would of course be undesirable. All this might be easier said than done. But, to rid society of the various schisms and rifts that threaten to break it up something, to the effect described here, should be attempted. As they say, if there is no venture, there would not be any gain. If there are no laws to guide and control the actions of people, everyone would do as they like without any concern for the general well-being of everyone in the group, in society, and, the result would be utter chaos, pandemonium. In short, if people were allowed to let their thoughts, their logic, run wild, there would be terrific chaos, terrible disharmony.

The author would like to say something more about the above-mentioned carefully and democratically elected intellectual elite. This specially elected group of intellectuals should be capable of producing the logic tables and logic guiders, a kind of regulations or laws for everyone to follow, as well as ensuring that they are effectively implemented. As society needs laws and regulations in order to function in an orderly manner, it also needs laws and rules for regulating intellectual activities or the activities of brainworkers. With rules and guidelines to abide by or refer to, there should be no necessity to engage in disputes or arguments. Arbitration of disputes and court cases might now be the norm for unsettled disputes. But, with the effective implementation of these logic tables and logic guides, arbitration and court hearings should more or less be a thing of the past. These logic rules and regulations should have the objective of being as fair as possible to all the parties concerned, as well as being able to bring as much good as possible and cause as little harm as possible to the society at large. In practically all other aspects of human activities, there are rules and regulations to act as guide, e.g., we have rules and regulations in the game of chess, soccer, badminton, and golf. But, why are rules and regulations lacking in the more serious activities such as intellectual activities, or activities involving the use of logic? This appears to be some sort of discrepancy. We might have the various systems of logic created by the logicians and mathematicians, but unfortunately, these systems appear arbitrary and difficult to put into practice, though on paper, they might appear fine. Hence, the logic tables and logic guides should be convincing, acceptable, and easy to put into practice. They should be legislated in order to have an effective, beneficial impact on the society at large. They should have an orderly effect on the way intellectual activities are carried out. Original ideas, or even revolutionary ones, should still be welcomed, but should not be allowed to be forced upon others; they should only be adopted by others when the majority approves them and wishes to have them adopted. So, in the intellectual sphere, in the realm of logic, the principles of democracy could still be put into practice. And, democracy assumes that people are mature and wise enough to know what is best for themselves, and hence are able to elect or choose the best or most appropriate option for themselves. But this need not be the case. What happens if the people who are to choose, elect, or vote are really lacking in wisdom and maturity? In that case, we might have to be sorry for the society, for the way activities, including intellectual activities and activities involving logic, would be carried out. There is a theory that states that the group is generally wiser and more capable of making better decisions than the individual. Let us hope that this theory would hold true, or, there could be problems in a democracy, when people are not wise or mature enough to exercise their voting rights in the most appropriate manner. What happens if they elect the wrong leaders to run society? It seems dicey. Perhaps, the people who elect the leaders should be qualified first, i.e., meet certain standards or criteria, e.g., educational level, intelligence, and moral character. The people who elect the leaders, who run the society, should not be just any Tom, Dick, or Harry. They should either be carefully selected or go through a proper election process. If everyone, including morons, immoral people, and criminals, can just cast their votes and elect their leaders, might it not happen that inappropriate or wrong leaders could end up running the show, causing more problems than solving them? This appears to be a chink in the armor of democracy. Let us stop here now and not digress further.

Logic has also been beset by paradoxes, which indeed show up their weakness. As mentioned earlier, logical reasoning should be carried out with true statements if the conclusion or result were to be true. Deduction, based on untrue statements, would of course end up with a conclusion that is also untrue. Unfortunately, it is not always possible to differentiate truths from untruths, and many a time, the premises for a logical deduction have to be assumed to be true, if it were not possible to ascertain their trueness, if the process of logical deduction were to proceed at all. A typical example is mathematical reasoning, where axioms (obvious, assumed, but unproven, true statements), and lemmas and theorems (which are statements that have been proven to be true) are used as premises.

The author is of the view that paradoxes are the result of reasoning with untrue premises or premises not based on reality or real facts. Take for example, the statement, "This statement is false". This statement is self-contradictory. If this statement is true, it is not false, and, if it is false, it is true. In reality, in real life, would anyone ever pass such a remark, except in jest? There is another statement, which has tickled the author. It is, "I am humble and am proud of my humility". In this case, am I proud or humble? I seem to be contradicting myself when I make this remark – on the one hand, I say that I am "humble", then I contradict myself by saying that I am "proud" (of my humility). There is yet another interpretation: I am humble about other things beside my humility, but I am proud of my humility (only). The following statement is found in Chapter 1:

> Man is so intelligent that he could make his stupidity look like intelligence.

This statement is self-contradictory. So, is man really intelligent, or, stupid? Thus, logic has a great deal to do with the meanings of words and their interpretation, which implies that if logic were to be problem-free, or paradox-free, the words, terms, and premises used in the process of logical deduction should be carefully and precisely chosen – for precision of thought, the correct, proper words or terms are a necessity – that is, the words or terms used should be precise in meaning, and not have subtle or multiple interpretations. The author would like to emphasize again that the mastery of a proper language, or, a proper set of symbols (especially for mathematicians), is necessary for effective logical reasoning.

Another aspect of logic involves generalization, or, forming conclusions. For example, mathematicians are known to generalize things with their mathematical logic. Unfortunately, in the real world, things are not that simplistic, things are much more complex than that. A mathematician might create a mathematical model of how the economy behaves; there is this branch of mathematics known as econometrics, which uses mathematics to carry out forecasts of the economy. But, such a model is at best only a simplistic model, which might have little or no relevance to how the economy would actually behave. In reality, the economy is probably much too complex for an accurate forecast to be possible. In any case, to have a forecast, a plan, even an unlikely to be accurate one, is better than to have none at all. This is one of the shortcomings of logic. It should be stated that though logic works on the basis of "cause and effect", it appears to fail miserably where there is chaos. Chaos is the situation, where a slight perturbation, interference, causes an "out-of-portion" or overwhelming result or effect, and, chaos theory is that relatively new branch of mathematics, where mathematicians are trying to come to terms with the fact that in chaos, it is practically impossible to make any forecasts or predictions at all. The thoughts and ideas of human beings appear at times to be characterized by chaos, and, that is when, logic would not be of use. Hence, logic would seem to work only when ideas and thoughts conform to certain accepted rules or conditions, certain conventional wisdom. Thus, a person whose ideas and thoughts are too original and do not conform to conventional wisdom is likely to find himself not understood, and, worse, he might be misunderstood. It appears to be a case like "we are logical and right in our ideas and thoughts, and, therefore, anyone whose ideas and thoughts are different are wrong and not logical". Unfortunately, logic is not as simple as this, and, what is logical or sensible to one sane person might not be so to another sane person, i.e., logic is arbitrary. And, how do we judge or prove sanity or soundness of mind? This is yet another hurdle. As logic is a function of the sanity or soundness of mind of the person, does it mean that there are at least two kinds of logic: the logic of the person of sane mind and the logic of the person of insane mind? Here, we probably have a situation wherein to the person of sane mind, the person of insane mind is illogical and to the person of insane mind, the person of sane mind is illogical. If a third party of sane mind were to arbitrate a logical dispute between the two, what would be the result? On the other hand, if a third party of insane mind were to arbitrate a logical dispute between the two, what would be the outcome? Indeed, it could be a situation wherein to the sane person, the insane person is insane, and, to the insane person, the sane person is insane. What happens if the psychiatrist were of insane mind? Would he treat a person of sane mind as a person suffering from insanity? Indeed, how do we prove sanity or insanity? How do we distinguish between the logic of the sane and the logic of the insane; what are the differences, if any, between the two? Perhaps it is all relative; logic is relative.

Logical reasoning is one of the ways for us to have knowledge. Other ways of knowing are, of course, by using our physical senses, e.g., seeing and hearing. We can see, hear, or witness, something, and we would then know that it is true. However, if there were no opportunity to see, hear, or witness an incident, we could still arrive at the conclusion that the incident had occurred by the process of logical deduction or logical reasoning. But then, we normally still have to "physically confirm" our conclusion that has been arrived at by the process of logical reasoning (as in the case of the scientist at work). That is why, the author has earlier stated that one should rely more on one's physical senses than on mere logic. In other words, it is better and easier to deal with the more concrete things of the real, actual world than the abstract entities of logic.

When we consider the logic of ideas or thoughts, we are in effect thinking of what these ideas or thoughts imply, what other ideas or thoughts these ideas or thoughts are linked to or connected with. For example, we think of A. Then, we realize that A implies B or is linked to B, B implies C or is linked to C, and so on and so forth. The result here would be a chain of reasoning. To be familiar with effective and result-oriented reasoning or deduction, the reader should peruse the Sherlock Holmes stories created by Arthur Conan Doyle. The author himself has read practically all these stories and has thereby definitely gained much insight of logical deduction. Though Sherlock Holmes is a fictitious character, his creator, Arthur Conan Doyle, who was himself a medical doctor, had modeled him after a real-life medical doctor, who had acute powers of observation and deduction, by the name of Joseph Bell, a person that Doyle had admired.

The author has indeed been interested in many subjects where logic plays an important role, e.g., mathematics and its so-called mathematical logic, which could appear deep, profound, complex, abstruse, and incomprehensible, but is generally impractical; the physical sciences such as physics, where the logic appears to be of a more practical or applied nature (in fact, there have been instances of mathematicians who found pure mathematics (as opposed to applied mathematics) "useless" and having little practical applications, if any at all, switching over to the sciences such as physics where their mathematical knowledge could be put to good, practical use), the social sciences such as economics and psychology, where the logic has a more practical, mundane, and "human interest" slant, and, philosophy (the author's favorite), which is concerned with the truth of everything, which is to be arrived at through analysis and the use of logic. The author has been especially keen on mathematics, whose symbolism is often arcane and impenetrable, but which has a claim to rigor or a high standard of logical precision. Being able to solve a very difficult mathematical problem might give one the "high", like being able to play excellent chess and being able to beat a chess grandmaster. Mathematics and chess might provide some good training for thinking and bring some self-satisfaction and pleasure, but they cannot directly benefit society unless they can be "applied" to solve or tackle our practical problems, our real problems, e.g., in the case of applied mathematics such as statistics and operations research. For example, consider the author's solutions to the two very difficult problems in pure mathematics, viz., the twin primes conjecture (Is there an infinitude of twin primes?) and the Goldbach conjecture (Is there an infinitude of even numbers that are each the sum of two prime numbers?). The author has in his solutions proved that there is an infinitude of twin primes and an infinitude of even numbers that are each the sum of two primes. The question is, "What does it matter at all?" The author might be happy for a while that he has the solutions, one of which has been published in a respected mathematics journal. Some others who had perused his solutions might appreciate them and be pleased for a while. These solutions might have a little "entertainment value". Of course, without them, we can all certainly find other things to please us, things which are even likely to please us more, things that have higher "entertainment value". Without the prime numbers, without the even numbers, even without mathematics, life would still be able to go on. There are certainly many people with practically zero mathematical knowledge who have lived successful, happy lives, lives that are much more successful and happier than those of the top mathematical geniuses. On the contrary, many mathematical geniuses and famous ones too, e.g., Galois, Cantor and Riemann, had lived tragic lives. The author is here not crying "sour grapes" because he is not a mathematical genius. He is just stating a plain, simple fact.

Logic per se is in fact a waste, of no importance, if it could not be applied to the affairs of the society at large, if it could not be put to practical use. It would then just be another beautiful “work of art”, which is more or less on par with a beautiful piece of painting or sculpture, or a beautiful musical piece. On the other hand, it seems that logic has often been wrongly applied, i.e., it has often been twisted, manipulated, exaggerated, and forced upon others, resulting in problems galore (whereby, one man's logic could be another man's nonsense). So far, in this tome, the author has been painting the picture of logic as a troublesome, imperfect creature, which needs to be handled with care, an open mind, and a healthy skepticism. Here, the author is suggesting that we should be more concerned with applied logic than pure logic, i.e., logic should be result-oriented.

Concluding here, we could say that logic has an important role to play in our search for the truth. Of course, the truth could be directly witnessed through the physical senses (sight, hearing, touch, taste, and smell; especially seeing, whereby seeing is believing). Where the truth could not be directly found through the physical senses, logical deduction might be able to help us to arrive at the truth. Thus, the predominance of the physical senses vis-à-vis the faculty of logical reasoning.

35 The Future of Logic

It has been said that a person, in general, uses only about ten per cent of the capability of his brains. This means that about 90% of the capability of his brains is unutilized, and, that is a lot of waste.

If man could learn to fully utilize the capability of his brains, or even uses up to say, 60–70%, of the capability of his brains, what wonders of brainwork would result, what feats of logic and creativity would arise?

It is thus important that our education systems and our educational psychologists are assiduous in tapping this vastly untapped potential of the human brain. If this is the direction our society is moving in the future, then we could expect logic to play a more effective and successful role in our lives in the future. Indeed, logic and logical thinking should be made compulsory subjects, together with ethics, in our education systems. To cram our students' brains with professional knowledge for their future working life is not a big deal if they cannot or do not know how to think logically. They are then no better than parrots that remember and repeat words.

The question that constantly irks the author is why logic is not more fool-proof, why it is not as indisputable as, e.g., the statement that one plus one equals two, and why it is not generally more systematic or more "formalizable"? Leibniz, the co-inventor of calculus (with Newton) and one of the pioneers who had attempted to systematize all aspects of logical deduction, had failed to put logical deduction on a firmer footing, not for want of trying. Another pioneer, Boole, had tried to formalize the higher aspects, more abstruse parts, of logic but found this impossible. Would someone be able to succeed thus in future whereby logical deduction would be as straightforward as solving problems in algebra? With the great advances in technology, especially computer technology, whereby society is going to depend more and more on it, instead of arguing or disputing over logical points people might just turn to their computers for help. For example, as mentioned before, with simulation software for decision-making, or with the aid of large data-banks of factual or relevant information, disputants might access the information stored in their computers freely in order to arrive at a correct conclusion, or simply refer their dispute to the computer for "arbitration" and let the "judicial" computer settle their dispute for them – this would be as simple as using a calculator to carry out numerical computations – this would mean the end of arguments, quarrels, and possibly, wars – moreover, the computer is cold, unemotional, precise, constant, and unpartisan, and is thus trustworthy and dependable, as it is now. This scenario for the future of logic and logical deduction is not an absurd one but, from the author's point of view, a great possibility. Computers and robots with artificial intelligence have taken over many of our jobs and are clearly here to stay. Would society still need leaders and managers to lead and make decisions when computers with great artificial intelligence, an intelligence that is more powerful, more efficient, and more effective than human intelli-

https://doi.org/10.1515/9783112214367-060

gence, are available? Could the reader not envision a future society guided by an extremely intelligent and competent computer? This might not belong to the stuff of science fiction but could very well turn out to be true, though frighteningly so. Perhaps the powers that be would never allow such smart and powerful computers to usurp their powers, their decision-making powers. But, this is indeed a possibility. This is when human logic would become subservient to computer logic. This would mean that human intelligence is now ranked below computer intelligence, or, artificial intelligence. Perhaps, with the amount of human conflicts that the world is embroiled in now, which is due in no small part to the "failure of human logic", this would be a desirable end. The author recalls the words of a computer scientist who wrote computer programs, which might seem frightening to some. This computer scientist stated that he was surprised when he found out that the computer could perform some tasks that it had not been programmed to do. It is as if the computer had a life of its own.

The author feels that there is much that the human brain is capable of, but this potential is as yet unexplored or under-explored. A deeper study of how the human brain works and developing methods for enhancing the powers of the human brain are thus the right directions to be adopted. Today, there are mental techniques such as the Silva mind control technique and the Buzan mind-mapping method, which have their own following; other techniques involving influence on the mind are hypnosis and mind-altering drugs. As the capability of the human brain in abstract thinking has been found to depend on the number of glial cells found in it, artificially developing more glial cells in the human brain would enhance its capability for abstract thought. There are also a number of other techniques to boost the capability of the human brain, e.g., deep breathing techniques, food that nourishes the brain, hanging upside down, and allowing more blood to flow to the brains, though the efficacy of some of these techniques might be in doubt. A futurist who forecasted future trends in human affairs even predicted that, at some point in the future, increasing one's intelligence could be achieved by swallowing a pill, a so-called "high IQ" pill. Yet, logicians might simply develop a more advanced, more effective system of logic, though, as mentioned earlier, this is not going to be an easy task. But, as mentioned before, with the correct education and the development of personal traits such as tolerance of opposing viewpoints, objectivity, and flexibility of thought, the exercise and use of logic could become more of a "cooperative" exercise than a "combative" exercise.

There is indeed much potential for the development of logic, with a better utilization of the capacity of the human brain. A look at what idiot savants, talented individuals with well below average intelligence, could achieve – for example, being capable of computing extremely large numbers in their head and being able to draw very complex figures – testify to the vast, perhaps unlimited, potential of the human brain. Then there is another example in the Indian mathematical genius, Ramanujan, who had been the protégé of the renowned British mathematician, G. H. Hardy, who had little or no formal training in mathematics, but yet had been able to conceive beautiful mathematical theorems – Ramanujan himself could not figure out how he achieved this ingenious feat, apart from attributing it to the divine – perhaps the concepts of beautiful mathematical theorems had been "hard-wired" in his brains, which were perhaps unusual. Other geniuses such as Henri Poincare, e.g., had attributed their creative genius to the powers of the unconscious mind. Thus, it would not be surprising to find out that the brains of a person of average or above-average intelligence are, in fact, capable of much more feats than they have been accustomed to, feats which could perhaps be regarded as "superhuman" even. It is really up to us to realize this and train our minds to achieve precision of thought and logical consistency, and this is very important for our ideological and societal harmony.

So, in parallel with the act of examining logic closely and attempting to develop more fool-proof systems of logic, it is appropriate to make a concerted effort to understand the human brain more fully and attempt to fully utilize its vast potential. There is one aspect of the human brain that seems very important, especially to logic, and, that is the power of intuition. There have been books on how to develop one's powers of intuition. There are at least two ways of arriving at valid conclusions. One method involves strict logical deduction, while the other involves just plain intuition. Both methods could bring the same conclusions. The author would not be surprised if, in the end, or, long run, intuition were found to lead to more correct conclusions, predictions, or forecasts than strict logical deduction.

Therefore, no one should ever underestimate the powers of the human brain, and everyone should attempt to fully utilize his mental capacity to achieve precision of thought and logic.

36 Final Conclusion: What Logic Really Is After All

Logic is the perceived link or connection between objects, be they tangible things or abstract ideas – it results in the ascertaining of causes and effects, generalizations, and conclusions about these things or ideas by the process of selection and elimination. The perception of this link or connection is however a matter of interpretation, hence the subjective nature of logic (though diehard logicians might want to dispute this). If logic is subjective, then it is not very reliable. In fact, there are plenty of instances of the lack of reliability of logic, or, the problems caused by logic, especially if logic is treated as cast in stone.

We are all the time exposed to ideas, statements and facts, and all kinds of data. This may set us thinking. For example, we may think about what some ideas, statements, or facts imply. In other words, we may analyze these ideas, statements, or facts and may come up with more related ideas, statements, or facts. With all these data we may jump to a conclusion or conclusions, which represent the perceived logical link or links between all the ideas, statements, or facts, i.e., with the totality of all these ideas, statements or facts we may end up with a logical deduction or deductions.

Logic is a function of the senses, viz., sight, hearing, touch, taste, and smell, as only the senses could input the data for logical reasoning. The senses being imperfect and differing from person to person to varying degrees therefore input data of varying degrees to different people. As these sensual data upon which logical reasoning is based vary from person to person, logical reasoning must hence differ to varying degrees from person to person. So, how could anyone ever claim that human logic is foolproof? Logic is subject not only to the data input of the physical senses, but also to the perhaps inherently subjective interpretation of these data by the mind.

Logic, being subject to the interpretation by the mind, is somehow judgmental. Anything that depends on judgment could never be 100% objective. Therefore, the problems of logic. Why does the author say that logic is subject to interpretation? Well, logic depends on how one looks at things or interprets things. For example, "brightness is similar to darkness" is an apparently illogical, contradictory remark. But, if interpreted in a certain manner, this remark could make sense or appear logical. For example, the brightness of an object with its light switched on, e.g., a light bulb, could be as scary or frightful as its darkness when its light was not switched on. So, in this sense, could it not be *interpreted* that the brightness is similar to the darkness, in that they similarly bring about a scare or frightfulness?

As the author sees it, logic presents plenty of problems, as all human minds are more than capable of interpreting things in many different ways. To ensure that logic works smoothly and conflict-free for everybody, we have to ensure that we interpret or observe things in the same way.

https://doi.org/10.1515/9783112214367-061

Let us digress a little here. The distinguished logician, Godel, proved that any collection of axioms, e.g., those underlying Euclidean geometry or even arithmetic, would lead to a mathematical system containing true statements, which could be neither proved nor disproved on the basis of those axioms. Thus, a theorem could be undecidable. Adding more axioms would not necessarily help, as some theorems would still slip through the inevitable cracks. Godel's work further implies that in some cases mathematicians would not even be able to decide whether a theorem could or could not be proved. In other words, there is uncertainty in mathematics. There is also inconsistence. The following is an example of an inconsistent statement:

> This statement is false.

If this statement is false, it is true. If this statement is true, it is false. Whichever way one looks at it, there is a contradiction. Hence, there is inconsistence. For a statement to be logically consistent, it has to be either true or false and not both true and false (contradictory), like the above-mentioned statement. It has been established that such paradoxes are the result of a statement referring to itself.

Godel had resorted to amending the above-mentioned statement by replacing the word "false" with "not provable", in making his point, as follows:

> This statement is not provable.

This statement implies that it is not provable, whether it is true or false. This conclusion is the result of a very subtle reasoning, which is as follows:

If this statement, "This statement is not provable", is true, then this statement is true, but not provable. That is, this statement, "This statement is not provable", is not provable. On the other hand, if this statement, "This statement is not provable", is false, then this statement is provable. That is, this statement, "This statement is not provable", is provable. Hence, this statement, "This statement is not provable", is *not provable, all the time,* regardless of whether it is true or false, whether its non-provability is provable or not provable. All this might sound very confusing and self-contradictory.

To make the conclusion clearer and more obvious, we have a rephrasing here. If this statement, "This statement is not provable", is *true,* there is *no proof* that this statement, "This statement is not provable", is *not provable.* If this statement, "This statement is not provable", is *false,* this statement, "This statement is not provable", is provable, which is the same as saying that there is *proof* that this statement, "This statement is not provable", is *not provable.*

So, either, there is *no proof* that this statement (This statement is not provable.) is *not provable,* or there is *proof* that this statement (This statement is not provable.) is *not provable.* And, this statement (This statement is not provable.) would always be *not provable,* regardless of whether *there is no proof of its non-provability* or *there is proof of its non-provability.* (We might regard Godel's proof or incompleteness theorem itself as an example of a *proof of the non-provability of some statements (in mathematics).* Hopefully, the logic of the reasoning is now obvious. It is by dint of such a very subtle mode of reasoning that Godel proved that there are statements (in mathematics, and, by extension, logic), which could neither be proved to be true or false, i.e., there are statements (in mathematics) that are not provable.

What had Godel done to achieve his earth-shaking results? He introduced an arithmetical system (using the operations of mathematics) of coding all axioms leading to a theorem, including the theorem itself. He used the code numbers, 1, 2, 3, 4, 5, 6, 7, 8, 9, 10, 11, 12, 13, 14, and 15, to respectively represent the 15 symbols, 0, s, +, x, = , (,), ., x, 1, ¬ , &, ∃, ∀, and →. With these code numbers he introduced a numbering system making use of the list of consecutive prime numbers (2, 3, 5, 7, 11, 13, 17, 19, 23, 29, 31, 37, 41, 43, 47, 53, 59, and so on). Consider the formula: x + s x = x + x + x. As the code number for x is 9, the code number for + is 3, the code number for s is 2, and the code number for = is 5, the formula, x + s x = x + x + x, is hence translated into: 9, 3, 2, 9, 5, 9, 3, 9, 3, 9. These 10 code numbers (i.e., 9, 3, 2, 9, 5, 9, 3, 9, 3, 9) would be the powers of the first 10 prime numbers in the list of consecutive prime numbers (i.e., 2, 3, 5, 7, 11, 13, 17, 19, 23, 29). The Godel number for the formula, x + s x = x + x + x, is obtained by multiplying together these 10 consecutive prime numbers, with the 10 code numbers as their powers, as follows:

$$2^9 \text{ x } 3^3 \text{ x } 5^2 \text{ x } 7^9 \text{ x } 11^5 \text{ x } 13^9 \text{ x } 17^3 \text{ x } 19^9 \text{ x } 23^3 \text{ x } 29^9$$

The resulting Godel number is huge but finite. In this manner, a Godel number could have the whole mathematical proof encoded within it.

The Godel numbers are of course all natural numbers, and, as such form part of the standard arithmetic. This means that Godel could construct statements about the Godel numbers, just as he could for the ordinary integers. In other words, within the standard arithmetic, he could frame predicates, which took the Godel numbers as their subject matter, e.g., the predicate: x is the Godel number for the proof X of a one-variable formula Y with Godel number y and with the integer z substituted into it, which could be expressed in the following simple symbolic form: Proof (x, y, z).

However, in the process of constructing these statements about the Godel numbers, he discovered anomalies and paradoxes, which led him to conclude that the standard arithmetic was inconsistent. The only way out of such a predicament was to assume that there were statements, which were true but were not provable.

All this shows that mathematics (and logic) is inconsistent, harboring contradictory theorems, or incomplete, harboring unprovable theorems. Importantly, this implies that that we should handle logic with caution, and, even some skepticism. However, the author feels that it is better not to be too concerned about this problem, for it has no serious practical impact on our lives, except perhaps to make some logicians or mathematicians who take logic seriously feel insecure. The real test for the usefulness of anything, including logic, is whether there is any practical application. Many paradoxes could be found in life, but that does not mean we have to have sleepless nights over them. All we should do is to maintain an open mind and a healthy skepticism where logic is concerned.

As human beings living together in a fast-paced internet-linked world, we should all be more concerned about "social logic" than mathematical logic and what other logic have you. All this hype about the paradoxes in logic discovered by the likes of Godel such as the aforementioned liar paradox (pertaining to the statement, "This statement is false".) might be a case of overdoing things or "making a mountain out of a mole-hill". The author could prove to the reader that he would not be able to catch up with him (the author) if the author were moving ten steps ahead of him (the reader), by citing the earlier described Zeno's paradox concerning the race between Achilles and the tortoise as proof. Should the reader accept this unrealistic, paradoxical, ridiculous, but apparently solid reasoning of the author though? Reverting to the statement, "This statement is false", there are a few subtle points to note here. If this statement refers to itself and states "This statement is false", then the liar paradox would apply. Imagine the statement itself (personified) pointing a finger at itself and saying "This statement is false", It is similar to a person telling another person "I am lying". But, if an outside party were to refer to or point at the statement, "This statement is false", and remark that "This statement is false", the liar paradox would not apply. That is, when some outside party makes the statement, "This statement, "This statement is false", is false", the liar paradox would not be valid. Also, someone could claim that a (different) statement, e.g., the statement, "John is a liar" or "David is a fool", etc., is false – he just proclaims "This statement ("John is a liar" or "David is a fool") is false", by which he means John is not a liar, or, David is not a fool. We have to distinguish the subtle differences. When the statement (personified) refers to itself and proclaims "This statement is false", it is in fact the same as the statement (personified) saying "I am lying". In other words falsehood here implies lying. And the liar paradox would pop up. But when an outside party refers to the statement, "This statement is false", and says "This statement is false", he is calling the statement, "This statement is false", a false statement. Here, "false" takes on quite a different meaning and it has nothing to do with the statement (personified) lying – it means "untrue", i.e., the statement, "This statement is false", is untrue, against reality, against logic, against common sense, unacceptable, and nonsense. Yet another subtle difference would present itself when an outside party refers to another statement he disagrees with or does not accept as true, e.g., the statement, "The sun is black in color," et al., through the remark "This statement is false". For example, when someone refers to the statement, "The sun is black in color", and proclaims "This statement is false", he is actually stating that the sun is not black in color. So, there could be three different interpretations to the statement, "This statement is false". It is only when the statement is self-referential that the liar paradox rears its head.

The author is uncertain whether Godel had been aware of these subtle differences when he considered the statement, "This statement is false" and the liar paradox. Why make a fuss over the unrealistic, paradoxical, inconsistent interpretation (liar paradox interpretation) of the statement, when there are two other interpretations, which show that the statement makes sense and is consistent? Why fuss over an interpretation, which seems to be the machination of an overactive imagination and is not based on reality? In reality, has any statement ever called itself false or lying? (Could a statement speak?) Has anyone ever heard any serious and sane person (let alone a statement) proclaim "I am lying"? Only a madman or a prankster would ever say "I am lying". This is a statement that has truly no possibility or basis in the real, sane world. Has anyone ever lost a race to the tortoise (as Achilles had in Zeno's paradox)? There are also other paradoxes. For example, the corpse says, "I am alive", the living man says, "I am dead", the blind man says, "I can see", and the man with sight says, "I am blind". What else? Many more. Just use the imagination. Godel had even devised a proof for the existence of God through logical deduction.

The author is more concerned about the social implications of the effect of logic rather than all this nitty-gritty about logic. Whether logic is right or wrong, though important, is secondary to the whole human society being able to use logic amicably. If thoughts and logic differentiate people so much that there are serious rifts and conflicts in society, then some appropriate "laws of thought and logic", if available and practicable, should be enforced, and these have better be universal throughout the world.

The other (very important) aspect of logic concerns the logic of morals. Why is the world so full of evil? Are there a lot of evil people around? Is Satan really at work bringing forth these evils? And, what is evil? There could be many interpretations about what is evil? Evil, to the author, is the state of having the intention to cause harm or bring trouble. Any evil act is, strictly speaking, not evil. It is the person carrying out the evil act, who intends to cause harm by doing so, who is evil. But according to the Italian statesman and political opportunist, Nicolo Machiavelli, renowned for his treatise, *The Prince*, which shows how political leaders and rulers should govern, evil might be good after all, he had brought up examples of good rulers who had brought tragedy to their subjects in the end and evil and ruthless manipulators and politicos who had done well for their subjects in the end. Machiavelli held the belief that people cared not whether their rulers were good or evil but whether in the end they could give them what they wanted. Thus, according to Machiavelli, using evil means to achieve a good end whereby everyone in society benefits is justifiable. This should be preferred over having a good and kind ruler who had everyone suffering and in tatters in the end. In other words, the end justifies the means. But, of course, the ideal situation is to have a good and kind ruler who brings peace, harmony and prosperity to everyone all the time, but Machiavelli, who had been a shrewd student of history, seems to have thought that this ideal was not possible. Machiavelli seems to have had been a realist. The simplest logical statement concerning ethics or morals appears to be "Do not do unto others what you do not want others to do unto you". As mentioned earlier, there are many standards of morals. What is evil or unacceptable in one society might be acceptable in another. Are we human beings not aware that for every live stock we consume, some living thing, which is able to experience pain and suffering has to suffer and pay with its life? Yet, we continue to have the intention to consume live stocks, being aware of the suffering and loss we would cause these living things. In other words, we continue to have such "evil" intentions. So, could we not say that all meat-eating, carnivorous human beings are "evil"? For this reason, some amongst us turn vegetarian, and even vegetables and plants are some kind of life-forms, which might have feelings and might be able to experience suffering, as some scientists have postulated. But, if we do not eat anything, we cannot live, we would die. Some scientists have suggested that we go for synthetic meat, which appears a possibility now, if we were to overcome our hang-up, our guilt over the consumption of live stocks. Thus, even the logic of morals is not without contradictions or paradoxes. Perhaps we need a Godel to settle this problem.

There is another point that should be brought out. Often people carry out apparently evil deeds without any intention to harm others. To others unaware of this these people appear evil. But due to their ignorance or stupidity or both, they might not have had prior awareness of the harmful consequences of their actions. Since they had never had the intention of causing harm we should not classify them as evil despite their apparently evil deeds. Simply put, they could have made a wrong decision or a mistake by carrying out the apparently evil deeds. The evil deeds might imply that the evil doer had evil or harmful intentions, but this needs not be the case. If the evil doer, who might have committed murder, is truly insane even the law would exonerate him. If the evil doer has carried out the evil acts out of sheer stupidity or ignorance or both without having been aware that they were evil acts and without any evil intention, could he be really regarded as evil? By the above-mentioned definition of evil, this evil doer is not really evil. We should not tag any person or action as evil indiscriminately without realizing the state of his mind and whether he had any evil intentions. Stupid or insane people might not only harm others, they might also harm themselves, inadvertently of course.

Evil and good are elusive concepts. How could we read a person's mind and really know whether he has evil or good intentions? It seems very difficult to find a saint or extremely good person or an extremely evil person. Most people are sometimes good and sometimes bad or evil and it would be difficult to classify most people as either good or evil. In other words, most people are good and bad or evil at different times, making them good and evil on the whole, neither good nor evil on the whole, good on the whole, or, bad or evil on the whole, depending on how one interprets their actions and behaviors. Good and evil are subjective, qualitative attributes, which could not be measured or judged with precision like the height or weight of a person. How about the person who harbors the intention to carry out a harmful deed to achieve good? How about the person who intends to carry out or fake a good deed in order to mislead and harm somebody? What are the criteria for good and evil? Evil, as has been mentioned earlier, is the state of having the intention to cause harm. Then, good, being the reverse, is the state of having the intention to bring something desirable, pleasure or happiness. What if the person has had the intention to do good but through his actions had inadvertently brought harm instead? To some people, to those not aware of his good intentions, outwardly at least he might appear bad or evil. How about the evil-intentioned person who has achieved a good or desirable outcome instead? Should he be regarded as a good person? Should a person be judged good or evil based only on his good or evil intentions or based solely on whether his actions brought good or harmful results? These are difficult questions to answer. While some might favor results over intentions, others might favor intentions over results. The best is to have both good intentions and good results, which might not be achievable all the time. Could one ever trust an opportunist who has no hesitation in carrying out an evil deed to achieve a selfish aim, e.g., more power and riches for himself? Then how about an unselfish opportunist who has no hesitation in executing an evil deed in order to achieve the general good, the most good for the most people? These are complex matters for pondering.

To all this confusion, the author has this to say. Society should welcome and support any intention or action that is concerned with the well-being and/or progress of society as a whole. In the spirit of democracy, everyone should wish for and act for or strive for the greatest good for the greatest number of people. Therefore, any motive and action, which support this principle of the most good for the most people, are "good" and any motive and action that detract from this principle are "bad" or "evil".

Mathematics is reputedly the science of reasoning with numbers, the integers, fractions, transcendental numbers, abstract symbols, etc. Numbers, at least, could be counted. But when mathematicians grapple with mathematical concepts involving infinity, e.g., the twin primes conjecture, the Goldbach conjecture, and the Riemann hypothesis, they tacitly assume that infinity exists though they could not attach a number to infinity or quantify infinity. Take the twin primes conjecture which postulates that there is an infinite number of twin primes, for example. If the number of twin primes is infinite, theoretically it is possible to list out (on paper) the twin primes forever, as far as the assumed infinite space (including the outer space beyond our universe) around us allows. Now, we could not be certain that the outer space beyond our universe would not end or terminate at some point, though we could assume that this outer space is infinite. There is no apparent evidence that this would never be the case. There is a subtle point of contention here. Of course, it is practically not feasible to list out (on paper) the twin primes forever, for no one lives forever, and, we could not be certain whether human civilization would not become extinct some time in the future, perhaps millions of years from now, though this does not seem a remote possibility. Should the day come when scientists discover that outer space has a termination point and is not infinite, then it could perhaps imply that infinity does not exist. If infinity does not exist, then the list of twin primes could not be infinite obviously, i.e., one could never list out (on paper) the twin primes forever (there is a limit in physical space for doing so). In this sense, there would never be infinite mathematical objects. However, conceptually, we could still reach beyond this newly discovered termination point in outer space and go on forever, with our imagination and our consciousness. In this sense, there is still infinity, which is just an idea or abstraction, though it has now been proven in reality not the case. Here, the question of infinity is a philosophical one. The point is that infinity might not exist at all, except in the human consciousness.

Next, let us dwell a little on the controversial and difficult logic of the existence of God. We might wish to compare God to a robot-maker. A robot-maker makes robots that are amenable to his control and his influence. Certainly he could not and would not make a robot that is able to question him, argue or quarrel with him, or, rebel against him. If God had created man, it is strange that there are people who do not believe in Him, question Him, or even abuse Him, people who we might call atheists. In fact, a world-renowned scientist who is an authority on black holes has expressed this sentiment. Another renowned scientist, a physicist, who was a Nobel laureate and an atheist, had held the belief that God is there to make people behave. He had felt that the belief in a God is necessary for people to behave morally. In other words, God is there to instill fear so that people would behave morally. Was man really created by God? How do we reconcile Charles Darwin's theory of the evolution of all living things including man, which is now accepted as a scientific theory and which contradicts the point that man was created by the Almighty? Does God really exist? Well, Godel had a proof that God exists. Let us come straight to the point here. Let us consider the following hypothetical case. On the one hand, we have, e.g., a person who claims he has seen God and who has even taken a photograph of Him (and the photograph, though it might be suspected to be a fake, was, e.g., confirmed to be not so). On the other hand, there is, e.g., a distinguished logician, who has the stature of Godel, who has a rigorous proof, a logical proof, that God does not exist. Would you believe the person who has seen God and has the evident proof or photographic proof of God or the logician with a solid reputation who has the logical proof that God does not exist? If you yourself have really seen or witnessed God with your own eyes (and perhaps other physical senses), would you now believe this reputable logician? If the above-mentioned physical observer, yourself, and more and more people, have seen and witnessed the presence of God, would not the evidence that God exists be more and more convincing, even overwhelming? On the other hand, what if more and more reputable logicians refute the existence of God with their logical reasoning? Should we rely on or trust logic, or, the capacity for reasoning, more, or, the physical senses, especially the sense of sight, more? Have we not often heard the cliché, "Seeing is believing."?

One of the axioms of mathematics is that a statement could only be either true or false (and, not both true and false, which is the state of being inconsistent). However, Godel had found a flaw in this axiom, when he proved the incompleteness theorem. Could we not have an axiom, which stipulates that a statement could be true when viewed from a certain angle or aspect and false when viewed from a certain angle or aspect? Such an axiom seems more attuned to reality. After all how the mind interprets data or evidence plays an important part.

Logic appears to be a really slippery subject, giving rise to many differing theories, ideologies, and schools of thought, many at conflict with each other. It is easy for the nimble-minded, the cunning fellow, and the crooked person to twist their logic in order to persuade others to accept their ideas. This is the crooked, dishonest use of logic. This would create problems. This is not good for the well-being of society. This should be prevented, controlled, and stopped. There should be intellectual honesty. There should be the overriding concern for the truth and nothing but the truth, to the exclusion of the self, the ego, and all selfish motives. When people are intellectually honest and selflessly truth-seeking we could expect there to be social harmony and peace. (In this respect, any competition in society is not really good, as it coerces people into acting selfishly or self-centeredly, even to the extent of twisting logic or misusing logic in order to achieve their selfish ends. Competition might be the root cause of our social problems.)

Logic should have some desired practical result. It should see that objectives are achieved. Logic should also be used to persuade or influence other people, assuming that others are logical and are thus capable of appreciating or accepting a chain of logical reasoning. It is bad if a really logical person is not able to persuade or convince somebody with his true logic, whereas a real crank is able to convince or persuade somebody with his brand of "logic". The fact is that this does happen in life. This means that logic does stand on precarious ground.

For the sake of justice and the well-being of our society, everyone has to learn, to be taught, to be coerced, to use logic properly. To be able to use logic properly requires a host of desirable qualities besides the possession of a logical mind, qualities such as intellectual honesty, lack of ego, emotional stability or high emotional I.Q., friendliness, and objectivity.

Logic, to be useful (and trouble-free), should have practical applications – logical deduction without any practical purpose is like shadow-boxing (whereby we could, e.g., imagine ourselves boxing with the world heavyweight boxing champion, knocking him out in the process and emerging the new world heavyweight boxing champion, which is an unrealistic dream at most, and a waste of time); for logic to be trouble-free and smooth-sailing, the words or symbols used should be precise, concise, and clear in meaning (at least to be so as far as possible though this might be difficult at times), and we should interpret or view things or ideas from as many angles or aspects as possible (review as many possibilities as one could). However, in discussions, to avoid or minimize disputes or conflicts, the parties concerned should agree with each other to carry out their interpretation of the details from the same or common angle or aspect. At least, they should make known to each other from which angle or aspect they are looking at things. Otherwise, the differences might be great or even irreconcilable. They should also agree with each other not to get their egos involved and to be intellectually honest.

It could be said that logic, any kind of logic, consists of a set of rules for arriving at conclusions, which are considered valid, and these rules could be rather arbitrarily made, unmade, and revised. This difficulty is of course compounded by the fact that the terms, expressions, and ideas used in the reasoning process are subject to varying interpretations – their nuances, their meanings, might be subject to different interpretations amongst the reasoners. (It is hence important to have as precise, clear, and unequivocal a language (or terms or expressions) as possible for reasoning, though this might not be attainable all the time.) Therefore, there is no absolute certainty as far as logic is concerned. Logic might be valid if the interpretation of the premises or statements is done in a certain way and be invalid if the interpretation is something else. This is the subjectivity of logic, and this has spawned many disputes and conflicts. In this respect, logic is just another aspect, one of the aspects, of human activities, which are subjective – subject to the prejudices and cultural influences, and, whims and fancies, of the human psyche. Logical deduction, e.g., in mathematics, physics, economics, finance, et al., is in a sense no different from a game of chess, soccer or golf in that there are prescribed rules or principles to abide by. These rules are created or invented and could be changed, unchanged, modified, as and when considered appropriate or necessary. They could be arbitrary. There lies the serious problem with logic. To avoid or prevent unnecessary problems, the rules of logic, the principles governing thought, should be as carefully and thoughtfully conceived as possible, and avoid or be devoid of as much arbitrariness or controversy as possible. Unfortunately, it has not been found possible to have a foolproof set of rules for logic, a foolproof system of logic, or a foolproof method for carrying out logical reasoning, not for want of trying though. Logicians since Leibniz had not been able to produce a sound system or method of logical reasoning, a "calculus for logic" so to speak. The only sound system of logical reasoning, which has no apparent practical problem, appears to be arithmetic, which is in effect the science of logical reasoning with numbers involving the four operations, addition, subtraction, division and multiplication. In any case, numbers are more or less child's play and any six- or seven-year-old child would have little problem understanding and playing with them. Unfortunately, other objects of logical deduction are generally not as simple or straightforward as the numbers. As we all know, many objects of logical deduction, which are objects or ideas, are very complex, abstruse, difficult to grasp and explain, difficult to express in simple words or symbols, even confusing and bewildering. Hence, the great difficulty of logical deduction involving such difficult objects or ideas. This difficulty is in fact manifested clearly in our legal cases, legal wrangles. It is apparent that in practically every legal dispute there is always an opposing angle or more, which a defending attorney or a prosecuting attorney could exploit. Attorneys, who rely on logic to argue their cases, could achieve very good results if they were really good at their jobs, good at using logic. It is here clear that logic is hardly ever seen in "black" and "white" and more often in "gray". Had logic been "perfect" and were available in "black" and "white" attorneys who argue legal cases would have been redundant, and, life would have been much simpler and more harmonious for everyone. Unfortunately, this appears to be far from the case.

Logical deduction is therefore often not as simple and smooth-sailing as we want it to be. We should regard or treat logic with an open, flexible mindset, and not with a "one-size-fits-all" mindset. It is important to have an open mind that is receptive to all views from all angles. When the mind is closed, unreceptive, and prejudiced, there would be problems. And, it seems that the human mind tends more to be prejudiced, more favorably disposed toward some pet peeves, than to be open and unprejudiced. No amount of logical training would help if the mind is closed and prejudiced. It is thus greatly important that the individual is trained or educated to have an open, unprejudiced, and unencumbered mind. If this is cultivated, we could expect trouble-free, smooth-sailing logic or rules of thought. After all, many aspects of logic are "open-ended", controversial, and without clear-cut or obvious answers or solutions. It is indeed foolish to argue or quarrel over something that is "open-ended", something that has no obvious, clear-cut answer or solution. In such a situation, the correct, most logical option should be the "give-and-take" attitude, the "subject to the benefit of a doubt" attitude, the "correct until proven wrong" attitude. Unfortunately, in our dealings with our fellow beings, more of the hard-headed, stubborn, or dogmatic variety are likely to be encountered. When hard-headed and stubborn meets hard-headed and stubborn the result would be a clash, conflict. For the well-being of society as a whole, the intellectual "give and take", the intellectual fluidity, are all-important. As they say, it takes two hands to clap, two to tango.

In this respect, a course on negotiating techniques might be more important than a course on logic. The greatest problem encountered is that human logic is not definite but subjective much of the time. If we do not realize this, if we are not keenly aware of this, if we treat logic like it is written in stone, and if we are not capable of the intellectual "give and take", then there would be problems, conflicts.

It is also important that we do not use logic for egotistical reasons or personal glory, but to seriously and honestly utilize it to seek only the truths and for noble and ethical objectives. The author has emphasized again and again that logic be related to the practical things. Why? Because it is only through the practical, the tests and the experiments, that the facts could be confirmed and the real truths could be obtained, thus preventing unnecessary disputes and conflicts. In any case, it is the real, practical world that matters most. Logic, ideas, theories, and abstractions, while they might bring happiness, pleasure, or pride to their perpetrators, are nevertheless useless if they have no practical applications and could not help bring useful practical results. Logic, ideas, concepts, or theories, if they could not be tested and confirmed, are not only disputable, they could be the source of serious friction and conflicts, especially when the reasoning or argument becomes circular and protracted (and there is no practical way (test or experiment) to settle it except by "give and take", or arbitration).

The aim of logical deduction is to arrive at correct or valid conclusions. For this to be possible there have to be facts or truths. Logical deduction based on lies, or even half-truths, would result in nonsense. Now, in our working life, whether we are a scientist, a mathematician, an engineer, a doctor, a lawyer, an accountant, in fact, in any profession or trade, we might depend much on our own experiences, but much of our knowledge or skills are derived from text-books, reference books, professional or trade journals, or books, in fact any text, which purports to equip us with the required knowledge. Has anyone ever wondered how reliable or accurate are the so-called facts presented in these media or texts? For example, some of the facts presented by an oriental medical text might be in opposition with some of those presented by a Western medical text, and some of the historians who had written our historical texts might revise, or even retract, some of the so-called historical facts. Practically all the time, we take all such so-called facts for granted, regarding them as "gospel truths". We generally do not question the authenticity of these so-called facts and tacitly regard them as truths. This accepting, unquestioning attitude has been ingrained in us by the educational system since our school-days, it seems. Of course, we do not expect students to question their teachers about the authenticity of the so-called facts presented in the textbooks, the teachers whose job is to instruct the students and explain the materials presented in the text-books. If this ever happens there would be chaos in the classrooms and little would have been learnt by the students. And yet it is expected of intelligent students to express their doubts and ask questions. Some of these students might rewrite the textbooks when they become researchers later.

Here probably lies the root problem with logic. The educational system has fed us knowledge, which is tacitly assumed to be true. The fact is that some of this knowledge might be untrue, though this would probably be miniscule. The point here is that in order for a logical conclusion to be a truth or a fact, and not nonsense, the premises for the logical conclusion or act of deduction have to be truths or facts. Therefore, we should be more concerned with the facts first and then the logic of the conclusion. The problem is that there is generally no way to tell whether the facts presented in the textbooks or reference books are indeed facts, unless we carry out a research, e.g., by perusing the archives, conducting experiments, talking to the experts, et al., and there is no way or time for research when we are immature learners or schoolboys. It is hence important for everyone to scrutinize the premises of a chain of reasoning carefully and ascertain their truth before arriving at a logical conclusion. In other words, get the facts right first, before reasoning. Therefore, knowledge workers should ascertain the facts or be very sure of the facts, before proceeding; they should not take things for granted for this could lead to serious errors.

On this cautionary note, the author here wraps up all the talk on logic and hopes to walk the talk. We should not just be logical, we should be practical as well, as stated above, for logic is of no real use to everyone if it is not able to help bring about beneficial, good practical results. Worse, logic has often brought about problems and rifts to society. The author hopes that this tome would successfully address this problem.

Appendices

Appendix 1: Is Our Physics the Only Reality?

Introduction

A number of points are brought up below for pondering and possible discussion. We should not regard our physics as the one and only reality. What is described below may seem like science fiction, but should not be brushed off as inconsequential.

Life and Existence

We take many things for granted. For example, we take it that all living things need oxygen and water to survive. We take it that planets are affected by gravity and must revolve round the Sun. We take it that all living things have consciousness and intelligence. We take it that all living things propagate themselves and increase their population. We take it that all matter is comprised of atoms. And so on. All this has become the standard reality in our minds; all this is life, existence.

Can anyone imagine a universe where all the above do not happen, that what we regard as life here in our planet have no semblance at all to any living thing that possibly exists in this universe? For example, there may be life without mind or consciousness (whereby life takes on a different meaning), life without the need to eat or respire whereby life may have an altogether different characteristic, life may even be just a vegetable-like existence (though possibly with consciousness), it may even be immortal. In other words, life or living beings may exist in this strange, vastly different universe but may be unrecognizable, even unseeable by our human sight or invisible. The life-forms there may be so different from ours that in our human minds they may not be regarded as life-forms at all. Surely, in a universe or environment where, for example, essential life-giving elements such as oxygen and water are non-existent, any life-forms, if they exist, must be vastly different from the life-forms here in our planet, so much so that they may not be regarded as life-forms by us mortals. In short, what we may see as barren in this vastly different universe may in fact be teeming with unrecognizable, perhaps "ghost-like", life-forms which may be invisible to our human eyes. This should not be a surprise. Even in this earthly existence of ours, some people and even animals, for example, dogs, are apparently capable of seeing "spiritual entities" or "ghosts" while the rest of their brotherhood live in this world in apparent blissful ignorance of the existence of such "frightful entities". It should be noted that the forces of nature in another universe may be vastly or even totally different from ours here, which probably account for the different life-forms that may exist there, that is, the physics there may be different.

https://doi.org/10.1515/9783112214367-062

Intelligence and Reasoning Power

Man has always prided himself on his high intelligence and reasoning power. We think we are so smart. But we still need to reason to arrive at the truths. That is, we need to have premises, facts, theorems, to reason with, we need to rack our brains. This reasoning process is often a time-consuming, even time-wasting, process. There are probably intelligent beings in other universes who have some sort of telepathic or intuitive mind which is capable of knowing or arriving at the truths straight away without any use of our kind of reasoning process, thus saving plenty of time and mental energy. Can we then say that the reasoning human being is damn smart (as compared to these extra-terrestrials with the possibly telepathic, intuitive, lightning-quick mind)? These extra-terrestrials may know or understand straight away without any need of a reasoning process (unlike us mortals), as if by magic, a seeming miracle. All this represents a different reality.

Possible Danger Posed by Extra-Terrestrials

A few scientists have postulated that there is the possibility of the Earth being attacked by extra-terrestrials or beings from other universes some day, and, our astronauts or space-travelers may also be attacked by them in outer space. Is the Earth prepared to defend itself against these hostile aliens?

Conclusion

Is humanity prepared for the above-described possible new physics, new reality? Does anyone feel threatened by all this?

Appendix 2: The Logic of Space-Time and Time Travel

Introduction

Time is actually an abstract entity which became part of special relativity. There has in fact been postulation that time is unreal but just an invention of the intellect. Whether time is an invention or actually real, it cannot be denied that it plays a very important role in our lives, e.g., without a watch to tell us the time practically all of us would be lost in time. This chapter takes a look at the philosophical ramifications and difficulties of time, as well as time travel, and clarifies things.

Space-Time Continuum

When Einstein first propounded the concept of the space-time continuum (which implies the possibility of time travel), in general relativity there were critics who scoffed at this revolutionary idea, which posits the existence of four dimensions in the universe: the three physical dimensions of length, breadth, and height, plus the fourth dimension of time; they could not conceive time as the fourth physical reality, and could not understand the relativity of time.

What is the actual problem with time? The problem is that while the three dimensions of length, breadth, and height are physical and visualizable, the fourth dimension of time is abstract, intangible, and not possible to visualize. Any object could easily be physically observed or seen moving or traveling along any of the three physical dimensions of length, breadth, and height, and even along all three of these physical dimensions simultaneously (which could be described by a vector diagram, utilizing vector analysis), but it is not possible to physically observe or see an object move or travel in the time dimension, i.e., one could use a ruler to physically measure or gauge the length, breadth, and height dimensions but it is not possible to use a ruler to physically gauge the time dimension. However, the time dimension is directly related to the length, breadth, and height dimensions – moving or traveling along any or simultaneously all of the three dimensions of length, breadth, and height from one point to another point in space requires a certain amount of time which depends on the velocity or speed of movement or travel. This amount of time could only be measured or gauged by a clock or watch. As the three dimensions of length, breadth, and height are physical, tangible, visible, and concrete, while the time dimension is not physical at all and is strictly intangible and abstract (time is strictly an abstraction), we should not regard or treat the time dimension as similar to the dimensions of length, breadth, and height. It is apparent that the confusion caused is due to us treating the non-physical time dimension as similar to the physical dimensions of length,

https://doi.org/10.1515/9783112214367-063

breadth, and height, i.e., treating non-physical time as physical (giving rise to the absurd speculation about (physical) time travel as well). This is actually a philosophical difficulty and should first be tackled. However, movement precedes time, the latter being the measure of the former.

The concept of Einstein's space-time continuum seems to go against common sense. If time is the fourth dimension, can we move or travel in the time dimension as we can do so along the three physical dimensions of length, breadth, and height – many appear to think we can? As per the reasons stated just above, this is clearly not possible. Can there be more than four dimensions in the universe? Kaluza, e.g., had modeled a universe with 21 dimensions, reducible to 11 dimensions, a universe that could be described physically in 5 dimensions. Einstein found his idea elegant and encouraged its publication.

Einstein's concept of time as the fourth dimension of the universe may be "interpreted" in a way which makes more sense (which implies subjectivity) although some people cannot accept it as sensible because time to them is abstract, intangible, and lies outside our physical senses. If a person, e.g., were to enter a completely dark room, whereby he could not see anything at all, and hence, any movement, we can say that he could not see time (passing) and would lose his sense of time. But when there is light, and he could see movement (even the movement of sunlight, moonlight, and day and night), e.g., he could see the hands of a clock moving, he could see the points in space, he could see people and objects moving from point to point in space, he could "see" and is in fact "seeing" time (passing), "seeing" here is "sensing" rather than physically seeing, bearing in mind that time is actually intangible and cannot be physically seen or handled as though it were a solid, a physical, a tangible, object. (Here it invokes the question of how time travel is ever physically possible, though mentally or consciously it is possible – to travel to the past just think of the past and to travel to the future just think of the future.) This gives rise to another philosophical point, viz., the presence of light, which results in visibility and the physically observing or seeing of movements, is necessary for the existence of time (the sense, awareness, consciousness, or idea of the passing of time), i.e., light is necessary for sight (we cannot see in the dark), which in turn is necessary for the existence of time (sensation, awareness, consciousness, or idea of the passing of time due to the observing or seeing of a physical object moving from point to point or location to location in space). This is an important point to note.

Movement, any movement from point to point in space, e.g., a person's or object's movements from place to place, the Earth's orbit around the Sun, and the orbits of the other planets, is analogous or equivalent to the fourth dimension of time (each of these movements takes a certain amount of time to complete, time which could be measured or gauged only with a clock or watch) – movement is therefore the equivalent of Einstein's fourth dimension (i.e., time) – movement could be translated, or transformed into time. Hence, Einstein's continuum of the four dimensions of length, breadth, height, and time, i.e., his space-time continuum, is equivalent to the length, breadth, height, and movement continuum which is visualizable, as movement is tangible, observable, seeable, unlike time.

Time could in a (subjective) way be looked upon as a physical reality and not as a higher reality existing outside our senses, especially our sense of sight, not as such an abstract entity as is deemed, i.e., it is a matter of interpretation. For instance, when we are looking at the physical objects that are making their movements along points in space, we cannot fail to be aware or conscious of time passing. (A moving object is deemed as taking a certain amount of time to move from a certain point to a certain point (distance) in a certain direction, whose actual movement could be observed and gauged in terms of time, distance, velocity, force, and direction.) If we touch these objects, we could feel the force of their movement, or on the other hand, the force of our movement (along the objects), and this would also give us a sense of time.

To put it another way, time could be interpreted (implying subjectivity) as seeable by the naked eye and sensable by physical touch (in other words, it is a matter of interpretation). But *time* is actually psychological, not physical – it is the *sense, consciousness*, or *idea* (of time passing), which is associated with the movements of physical objects in any, or, simultaneously all, of the three physical dimensions (of length, breadth, and height, which are also evidently the result of (subjective) interpretation, an idea derived from renowned French mathematician and philosopher Rene Descartes' invention of the co-ordinate system, a system of describing a physical object's location in space by using three co-ordinates or axes, commonly called the *X*, *Y*, and *Z* axes, which are in fact the *length, breadth*, and *height* axes; the other possible (subjective) interpretation of the dimensions of space is to regard space as a Hilbert space, as is the case with the scientific treatment of the micro-world of the quantum particles – a Hilbert space is a space with an infinite number of dimensions, which would make the interpretation of the dimensions of space more complicated and more intractable). In particular, time is always associated with the movements of the second, minute, and hour needles of the clock or watch, which measure or gauge the precise passing of time, without which our sense of time or of time passing would be imprecise and vague.

To circumvent the controversy pertaining to whether time is an invention or actually real, time being actually an abstraction which is evidently derived from the movements of objects that could be physically observed or seen (a physical object is deemed, understood, as taking time to move or travel from one location to another location in space, time which could be gauged by a watch or clock) and which is apparently causing some misunderstanding or lack of comprehension, we could describe space as a space-movement continuum or space-vector continuum (which should be visualizable, more evident, more comprehensible, and less controversial) instead of space-time continuum, which are in fact all equivalent. Space may also be described as a space-consciousness continuum instead of space-time continuum as time is directly related to both movement and consciousness, being a psychological entity as is described above, with consciousness being also an intangible, abstract entity like time, but this interpretation would be more abstract, and more problematic, than the space-time continuum (at least, time can be gauged or measured with a clock or watch, whereas there is no way to gauge or measure consciousness). The velocity or speed of movement of an object is derived by gauging the time the object takes to move or travel from one location to another location in space (distance) in terms of units of measurement for distance such as feet, meters or miles per unit of measurement of time such as second, minute or hour. It is evident here that for time (and distance) to be gauged there has first to be movement. We can actually physically observe or see an object move in the length, breadth, height dimensions, axes or coordinates, but we can never physically observe or see the object move in the time dimension (though we can mentally or consciously sense it, the movement of the object from location to location in space is known or understood to take time (an abstract, intangible entity) to do so – time being psychological while the dimensions of length, breadth and height are physical). In any case, no one could say movement is not real but an invention of the intellect, unlike the case of time, as movement is physically observable, seeable, and thus indisputable (i.e., visualizable), unlike time. We should also bear in mind that movement can be described by, besides time, distance, velocity (units of distance per unit of time), force (amount of mass per unit of acceleration), and direction, which could all be physically measured or gauged, as is stated above. (In vector analysis, movements of objects with directions are depicted with vector diagrams, a vector being a term that describes an object or force moving or traveling in a certain direction.)

Now, back to the important question: Is time real or is it just an invention of the intellect? Time is like beauty and intelligence; they are all very abstract, intangible, and seem unreal. However, we somehow know they are there when we encounter objects or beings endowed with them. But, unfortunately, we cannot physically count or quantify them as we physically count our coins, and we, e.g., cannot present a bottle of time to a person, we cannot present a bottle of beauty to him, and we cannot offer him a bottle of intelligence. How nice if we can do so! On the other hand, we can offer a TV set which physically has length, breadth, and height to a bored person to provide him entertainment, we can offer a bottle of soft drink to a thirsty person to quench his thirst, and we can offer a hungry person a packet of food to satisfy his hunger. The purpose of bringing up these examples, and explaining the philosophical logic above, is to really show that we should not confuse the intangible entities such as time, beauty, and intelligence with the tangible entities such as a TV set with length, breadth, and height, a bottle of soft drink and a packet of food, and treat them all the same, which would create logical problems. Therefore, to avoid such logical problems, the time dimension, which is abstract and intangible, of the space-time continuum should not be confused with and treated the same as the tangible, physical dimensions of length, breadth, and height. Is there some other better way to model space? How about a space-gravity or space-graviton continuum, for example? So, is time real? The intellect can consider it real or consider it an invention of the intellect, i.e., not real. It is a matter of interpretation. Both can be acceptable. All this sounds like Einstein explaining the theory of special relativity, isn't it? Bearing in mind that time, which is intangible, is a sort of consciousness or awareness or conception – it is the consciousness or awareness or idea of the period covered by the distance a moving object moved or traveled (i.e., the consciousness or awareness or idea of the time a moving object took to move or travel from one point to another point in space), if consciousness is real, time is also real. This is yet another interpretation of time, which the author tends to favor.

To really physically travel in the time dimension, we have to be able to walk, run, move, travel on time, as we do on the road. Can time be the road for us to physically travel on? All this is absurd. Thus, physical time travel is absurd. It seems difficult to convince the time travel die-hards that time travel is an impossibility. However, it appears that through ambiguities many have been misled into believing that time travel is possible. We should bear in mind that Einstein himself had thought that time travel is not possible. More on time travel below. The author had in the past been confused by all the mumbo-jumbo about time and time travel and has here gone to some length to explicate the philosophy of time in order to convince the readers, as well as himself, that physical time travel is absurd and impossible.

Time Travel

How nice it is to time travel back to the happy childhood days, the good times, the memorable events! All time travel enthusiasts are probably aware of the "grandfather" paradox, i.e., travel back in time to the time of one's grandfather and then kill him so that he will not be able to father one's father, or mother, and, therefore, oneself will not exist. (Such paradoxes may be construed as *reductio ad absurdum* evidence that physical time travel is an impossibility.)

When Einstein published the theory of general relativity 10 years after his theory of special relativity, it brought out the possibility of traveling into the past. Taking the mouths of two wormholes, which are tunnels through space made possible by general relativity, and putting one mouth on a spaceship and flying the spaceship off at close to the velocity of light would introduce a time difference between them due to *time dilation, i.e., the clock slowing down, and the aging process slowing down in the mouth on the spaceship. The person who jumps into the "future" mouth of the wormhole would emerge from the other mouth in the past. (*Time dilation, which is predicted by Einstein's theory of special relativity, states that a person flying off on a spaceship at close to the velocity of light would return home a short number of years later to find that many years had passed back on the Earth. In other words, the space traveler had jumped into the future.)

However, the technical difficulties of implementing the above scheme are considerable, though some physicists have speculated that it may become possible in the centuries to come.

Is the above phenomenon really time travel? Let us look at an analogous situation involving, e.g., two 60-year-old men, applying the same "aging", time dilation principle; for instance, one of them ages more slowly due to a better genetic make-up, e.g., he still has a head of black hair and looks 20 years younger, while the other man's hair is all white and he looks his age. Can we say that the younger looking man on meeting the older looking man has jumped into the future? Can we say that the older looking man on meeting the younger looking man has jumped into the past?

Another situation of "time travel" here. The Earth rotates clockwise from east to west, at say, x miles per y second. If John were to travel in the same clockwise direction as Earth's rotation toward the west but at a speed greater than x miles per y second, would he be traveling into the future? If John were to travel in the opposite, anti-clockwise direction toward the east at any speed, would he be traveling into the past?

Are all these time travels not metaphors, or illusions? All this is apparently evidence of the inventiveness or creativeness of the contemplating intellect which has been at work (and is apparently subjective).

Conclusion

Hopefully, all the apparent confusion, wooliness, and misunderstanding about time and time travel are straightened out.

Appendix 3: The Puzzling Logic of Gravity

Introduction

Newton accidentally discovered gravity after observing an apple falling from a tree. What exactly is gravity? This chapter takes a look at gravity, which evidently remains a subject of mystery.

Interpretation of Gravity

Einstein had attempted to unify the four forces of nature, i.e., gravity, weak nuclear force, strong nuclear force, and electromagnetism, but had failed. As in the past, he was unable to derive the electromagnetic field equations, even for the weak-field approximation. He was to live to the end of his life without any success with the unified field theory.

To arrive at this grand unified theory, a better understanding of gravity would be necessary, e.g., its roles in the micro-world of the quantum particles and in our macro-world. It is thought that quantum particles are free from the effect of gravity, which seems only to have a negligible effect on them, unlike in the macro-world. Nobel laureate Richard Feynman had posited that gravity could be totally unlike what it had been thought to be, thus possibly making unification with the other three forces impossible. Feynman had suggested that anti-particles are like ordinary particles moving backward in time, which implies that anti-particles should have anti-gravity. In fact, there is the belief that there is an anti-gravitational force for every gravitational force, just as there is an anti-particle for every particle.

In the theory of special relativity published in 1905, Einstein introduced the gravitational force which was considered responsible for the orbits of planets, as was described by Newtonian gravity. In Newton's law, gravity was an instantaneous force that propagated through space infinitely fast. This was evidently at odds with Einstein's theory of special relativity, which postulates that nothing can exceed the velocity of light.

In 1915, Einstein published the general theory of relativity, which introduced a new theory of gravity that was compatible with the special theory. In this theory the space-time continuum was introduced wherein empty space was likened to a flat rubber sheet which was flexible – a massive object creates an indentation in this empty space, or "rubber sheet" – this indentation is hence interpreted as a gravitational effect, a curved space-time, a geometry. In fact, in general relativity massive objects affect the way space-time curves. The link between an object's mass and the space-time curvature can be worked out, which is encapsulated in Einstein's all-important "field equations". In this way, Einstein was finally able to bring gravity in line with relativity.

https://doi.org/10.1515/9783112214367-064

Gravitation has always been thought of as a pulling or attracting force, just like the force of attraction between two magnets. Gravitation and magnetism may be different manifestations of the same thing. And, gravity may be a pushing force instead, a force that presses down on all objects in the direction of the center of the Earth. In fact, a push is equivalent to a pull, the former originates at the back of an object while the latter originates at the front. So far, gravitational forces are seen as forces of attraction only, while magnetic and electric forces are forces of attraction and repulsion. There may be a gravitational force of repulsion.

Gravity, which is crucial in the formulation of a unified field theory, can be described by the following formula:

$$F = G\,M_1M_2/R^2$$

where F is the gravitational force of attraction, G is the gravitational constant, M_1 and M_2 are the masses of two objects, and R^2 is the distance between masses.

Before Newton discovered gravity, nobody had known it existed or had thought that there was such an attractive force. However, Einstein in his general theory of relativity interpreted it as a curvature of the space-time continuum, a geometrical form, as is described above. Can gravity be the fifth dimension, in addition to the four dimensions of general relativity comprising the three physical dimensions and the time dimension, as Theodor Kaluza had suggested, which impressed Einstein greatly?

According to Einstein's theory of gravity, the hypothetical quantum of gravity, the graviton, which is a spin-2 boson, interacts extremely weakly with other matter, far more weakly than neutrinos; it is so weak that no instruments so far have been able to detect it. In the supergravity extension of this theory of gravity, the graviton finds a superpartner, the gravitino, which is a spin-3/2 fermion. Under local supersymmetric transformations, these two particles transform one into the other. When quantum calculations were carried out using supergravity theory, it was discovered that the infinities which plagued the earlier gravity theory with only the graviton were now being canceled by equal and opposite infinities produced by the gravitino. This is evidently the result of the deeper consequence of the presence of supersymmetry. Though it is not certain whether the supergravity theory is completely renormalizable, this "softening of the infinities" appears to be a step toward a viable theory of quantum gravity. As simple supergravity theory includes only the graviton and the gravitino, this hardly corresponds to the real world with its many particles. Most of those who have worked on supergravity feel that some crucial idea is still missing. Without this crucial idea, the theories simply do not describe the real world.

How do we make supergravity theory realistic? If we can solve this problem, we can have supergravity theory as a completely unified field theory. It has been shown that the principle of local supersymmetry is so restrictive that only eight possible supergravity theories exist, which are each labeled by an integer N = 1, 2, . . ., 8. Supergravity theory shares the same features with its progenitor, the theory of general relativity, namely, conceptual power and mathematical complexity. Perhaps, by postulating the existence of a single master supersymmetry we can have a unified field theory that accounts for the whole universe.

The following is Einstein's equation for general relativity:

$$G_{im} = -\mathbf{K}(T_{im} - 1/2g_{im}T)$$

This beautiful equation expresses the curvature of space-time. The left-hand side refers to a set of terms that characterize the geometry of space, while the right-hand side refers to a set of terms that describe the distribution of energy and momentum, i.e., the left is the geometry side, while the right is the matter side. Reading from left to right is space-time telling mass how to move, while reading from right to left is mass telling space-time how to curve. In general relativity, there is neither absolute time nor space and gravitation is not a force, or pull between one object and another but a property of space and time. All this represents a great conceptual leap by the theory's creator, Einstein. As for the coordinate system of Einstein's general theory of relativity, it has no basis in reality and is only a mental construct used to describe the space-time continuum of the general theory of relativity.

One may wonder when a unified field equation will be discovered. However, without a good understanding of gravity, this unified field equation will not come by easily. Once the experimental confirmation of the existence of the graviton, the hypothetical quantum of gravity, is achieved, we should be sure of obtaining a unified field equation that accounts for the whole universe, which may be supported by the existence of a single master supersymmetry. Superstring theory now appears to pave the way toward achieving this difficult goal.

The search for gravitational waves is still ongoing. The existence of gravitational waves so far is inconclusive – there is only indirect evidence of their existence. In the gravitational effect known as frame dragging, objects occupying the space near to a rotating object get swept around with it; rotating objects drag space around with them rather like a spoon in treacle (syrup). In 2004, scientists analyzing data from two Earth-orbiting satellites apparently found evidence of frame dragging – they claimed to have detected the minute frame dragging effect of our planet. There is the possibility of extracting useful energy from the rotation of our planet. There is speculation that this could serve as a power source for an advanced civilization.

Is Gravity Really What It Is Thought to Be?

Is all the space occupied by us and around us really one immensely large, flexible sheet of curved space-time as is posited by the general theory of relativity? Or, is there an infinitude of such "flexible sheets"? That is, is gravity just one flexible sheet of curved space-time or more, perhaps infinitely more? The Earth is a sphere but all of us are evidently "stuck to this sphere" and do not fall off it. What is the force, if it exists, which causes this? Will this force be an attractive or pulling force, or a pushing force? Can gravity be a kind of fluid, whether air, gas, or liquid? (We can descend, glide, or fly up through the air like how a bird or an airplane does. We can dive into and descend in, surface in, float in, and swim in, a pool of liquid. All these may be interpreted as manifestations of "gravitational effect".) Can gravity be electrical, electronic, quantum, or nuclear? Or, is gravity none of these but something else, as Richard Feynman had speculated?

It could be any intelligent person's guess.

Appendix 4: The Logic Behind a Scientific Discovery (Maxwell Equations)

Maxwell had discovered that electricity and magnetism join together to become light. To understand this theory it is necessary to understand what we call the Maxwell equations. The following are the four Maxwell equations, written using a branch of mathematics called vector calculus, for the behavior of electricity and magnetism in matter:

$$\nabla \cdot \mathbf{E} = \mathrm{p}/\varepsilon_0$$

$$\nabla \cdot \mathrm{B} = 0$$

$$\nabla \times \mathbf{E} = -\dot{\mathbf{B}}$$

$$\nabla \times \mathrm{B} = \mu_0 \mathrm{j} + \mu_0 \varepsilon_0 \dot{\mathrm{E}}$$

A vector that is written in bold-face type is any quantity that has both a magnitude and a direction. For example, 30 miles an hour is not a vector but 30 miles an hour due north on Express Highway is. **E** and **B** in the equations represent the electric and magnetic fields. The triangle is called a nabla (due to its resemblance to a certain ancient Middle Eastern harp) and expresses how the electric or magnetic fields vary in three-dimensional space, while the "dot product" and the "cross product" after the nablas are statements of two different kinds of spatial variation.

E and **B** represent the time variation (the rate of change of the electric and magnetic fields). **j** represents the electric current. The lowercase Greek letter p (rho) stands for the density of electrical charges. ϵ_0 (pronounced "epsilon zero") and μ_0 (pronounced "mu zero") are not variables, and are properties of the substances **E** and **B**, which are measured and determined by experiment. In a vacuum, ϵ_0 and μ_0 would be constants of nature.

It is striking how simple these equations are considering how many different quantities are being brought together in them. (They could have gone on for pages, but they do not.)

The first of the above four equations tells how an electric field due to electrical charges (for example, electrons) varies with distance – it becomes weaker the further away we go. However, the greater the charge density is (the more electrons, e.g., in a given space), the stronger the field would be.

The second equation says that there is no comparable statement in magnetism, as Mesmer's magnetic "charges" (magnetic "monopoles") do not exist. For example, if we saw a magnet in half we would not be having an isolated "north" pole and an isolated "south" pole – each piece would now have its own "north" and "south" poles.

The third equation describes how a changing magnetic field induces an electric field.

https://doi.org/10.1515/9783112214367-065

The fourth equation describes how a changing electric field (or an electrical current) induces a magnetic field, the converse.

These four equations are basically distillations of generations of laboratory experiments, mainly by British and French scientists. Though what have been described here might have been done so vaguely and qualitatively, the Maxwell equations describe exactly and quantitatively.

Maxwell then asked the following strange question: What would the equations look like in empty space, in a vacuum, where there are no electrical charges and no electrical currents? We might think there would be no electric and no magnetic fields in a vacuum. However, he suggested instead the following right form of the Maxwell equations for the behavior of electricity and magnetism in empty space:

$$\nabla \cdot \mathbf{E} = 0$$

$$\nabla \cdot \mathbf{B} = 0$$

$$\nabla \times \mathbf{E} = -\dot{\mathbf{B}}$$

$$\nabla \times \mathbf{B} = \mu_0 \varepsilon_0 \dot{\mathrm{E}}$$

He had set p equal to zero to indicate that there are no electrical charges and **j** equal to zero to indicate that there are no electrical currents. He did not discard the last term, $\mu_0\epsilon_0\mathbf{E}$, the feeble displacement current in insulators, in the fourth equation because he wanted to preserve the symmetry between the magnetic and electric fields. He thought that even in a vacuum, in the total absence of electricity, or even matter, a changing magnetic field elicits an electric field and vice versa. The equations were to represent Nature, which, Maxwell believed, is beautiful and elegant. There was a more technical reason for preserving the displacement current in a vacuum.

The four Maxwell equations for a vacuum briefly state the following:

1) Electrical charges do not exist in a vacuum.
2) Magnetic monopoles do not exist in a vacuum.
3) A changing magnetic field generates an electrical field.
4) A changing electrical field generates a magnetic field.

By presenting the equations in the above manner, Maxwell was able to demonstrate that **E** and **B** propagated through empty space as though they were *waves*. He could even calculate the speed of the wave, which was 1 divided by the square root of ϵ_0 times μ_0. (ϵ_0 and μ_0 had been measured in the laboratory). When the numbers are plugged in, it could be found that the electric and magnetic fields in a vacuum propagate, surprisingly, at the same speed as light. To Maxwell, this agreement was too close to be accidental. All of a sudden, electricity and magnetism were found to be deeply implicated in the nature of light.

As light now appeared to behave as waves and to derive from the electric and magnetic fields, Maxwell called it “electromagnetic”. These results had baffled Maxwell. The vacuum could be "electrically polarized", and it seemed to act like a dielectric. As it was then the mechanical age, Maxwell felt compelled to offer some sort of mechanical model for the propagation of an electromagnetic wave through a perfect vacuum. Thus, he imagined space filled with a mysterious substance which he called the aether, which supported and contained the time-varying electric and magnetic fields, a kind of throbbing and invisible Jell-O that permeates the universe. The quivering of the aether caused light to propagate through it – the same way as water waves propagate through water and sound waves through air.

What are the qualities of the aether? Theoretically, it is very thin, ghostly, almost incorporeal. The Sun, the Moon, the planets, and the stars had to pass through the aether without being slowed down, and, yet it had to be stiff enough to support all these waves traveling at very great speeds. In fact, in the early days of radio when they said "On the air", they were actually thinking of the aether. Of course, radio waves travel readily through a vacuum, which was one of Maxwell's main results. Radio waves do not need air to propagate. Actually, the presence of air is an impediment to the radio waves. The idea of light and matter moving through the aether had led in another 40 years to Einstein's special theory of relativity, $E = mc^2$, and much else. However, relativity, and experiments leading up to it, had shown conclusively that there was no aether supporting the propagation of electromagnetic waves. Many physicists were greatly troubled by the demise of the aether for they had needed some mechanical model to make the notion of the propagation of light in a vacuum reasonable and plausible. Here common sense does not serve anymore. Are electric and magnetic "fields" magically moving objects at a distance a reality or mere mathematical abstractions?

The equations of Maxwell show that a rapidly varying electric field (making **E** large) should generate electromagnetic waves. The German physicist, Heinrich Hertz, carried out the experiment and found that he had generated a new kind of radiation, radio waves, in 1888. British scientists in Cambridge transmitted radio signals over a distance of 1 km 7 years later. By 1901, Marconi was using radio waves to communicate across the Atlantic Ocean. Our modern telecommunications marvels such as radar, television, microwave relays, broadcast towers, and communication satellites are indebted to the abstract ideas of Maxwell. We are always surrounded by the very tiny wiggles of the electric and magnetic fields which are actually waves or signals, e.g., radio waves or signals; for instance, waves that originated billions of light years away, from galaxies in the remotest parts of the universe, could be detected by very large antennas.

Appendix 5: A Reply to the Editor of a Prestigious Mathematical Journal Raising Some Important Points

Dear Sir,

Thank you very much for your time in replying to my e-mail and providing the interesting information.

There is an interesting book by the late Professor Imre Lakatos, *Proofs and Refutations*, which I understand has been popular among mathematicians. He posited that all mathematical proofs, even if approved as correct, should be treated as tentative as like any approved and accepted scientific theories or concepts they are capable of being refuted and improved on in the future. Professor Lakatos had been a mathematician but he had an undogmatic and open mind about mathematical logic. I understand that he had been idolized by many mathematicians.

Yes, I am aware that Hadamard proved the Prime Number Theorem (PNT) using complex analysis. But, I think, his proof involved the Riemann hypothesis, which involves a complex number with a real part (1/2), the imaginary number (i), and prime numbers (to mathematicians, the distribution of the so-called zeros shows the distribution of the primes) – in this case, since the PNT proof involves a complex number (via the Riemann hypothesis), complex analysis would have been needed.

On hind sight, and to be frank, problems involving infinity such as the infinity of the primes and twin primes or the infinity of the zeros in the Riemann hypothesis are quite meaningless as infinity is unquantifiable, cannot be experienced, and it is impossible, impractical to count, compute, or live to infinity; in any case, infinity has no practical importance or bearing in our lives. A mathematician (I cannot recall who: perhaps Keith Devlin, Ian Stewart, or John Barrow, one of these popularizers of mathematics) in a book he published when he explained Euclid's proof of the infinity of the primes has stated to the effect that "perhaps at an extremely distant point toward infinity Euclid's proof of the infinity of the primes would not apply anymore"; in other words, we could never be absolutely certain that the number of primes is infinite. Even Euclid's proof (by contradiction) of the infinity of the primes could be doubtful. By Godel's incomplete theorems, a theorem could be true but not provable, i.e., a theorem could be undecidable, which means that the contradiction of an assumption does not necessarily mean that the contrary is true – if it is not right it does not necessarily mean it is wrong, it could be undecidable, unprovable; not tall does not necessarily mean short, how about being of average height? This is precisely what Brouwer's intuitionist school holds to be true – a proposition is either right, wrong, or undecidable, which would invalidate any proofs by contradiction, e.g., Euclid's. But

https://doi.org/10.1515/9783112214367-066

the undeniable strength of Euclid's algorithm/proof of the infinity of the primes is that larger primes could be generated all the time – a new prime or prime factor(s) could always be found or generated which would be larger than P, the assumed largest existing prime. (In this respect, Euclid's proof is a constructive proof, which could be regarded as excellent.) Incidentally it was also Brouwer who had expressed doubts about transcendental numbers – he had stated that there is a possibility that pi's transcendence would end at some point toward infinity. Even for complex analysis, some mathematicians lament the use of the imaginary number, i, in complex numbers, for it is logically absurd to propose a square root for minus 1. I bring up all these to show the imperfection and controversial nature of mathematical proofs, which makes mathematics look not as objective and "fool-proof" as it is supposed to be – in other words, as Morris Kline in his book, *Mathematics: The Loss of Certainty*, lamented, the image of mathematics has been tarnished. All this goes to show the elusiveness, even artificiality, of human logic, all kinds of logic – mathematics, the sciences, economics, psychology, politics, and so on. This is a serious problem with human logic which has resulted in disputes and conflicts in our society. All right-thinking people should be undogmatic about logic and try to address this problem in logic. Professor Lakatos and his admirers or followers would have had in all probability agreed with all this.

In any case, my effort at the twin primes conjecture is for me a study of logic, especially mathematical logic. I think that I have now improved on the solution by a great deal and have added several interesting ways of generating new twin primes (my proof is a constructive proof, which is supposedly the ideal mathematical proof), though convincing some mathematicians about its correctness and completeness would probably be a great obstacle. I might just be regarded as another crank. I do not discount the possibility that I have tried to solve a problem that is impossible or meaningless to solve. Just a few simple questions here. Could anyone give a real, actual example of infinity? Is not infinity the result of abstraction or even the result of the human imagination? How could anyone prove something that could not be experienced, that could not be quantified or meaningfully described, that has no tangible or observable form, that might not even be real, such as infinity? Philosophically, I think that we should first prove that infinity exists and is real, which might not be possible. At the very least, there should be some attempt to put infinity on a firmer, surer footing. (Cantor seems to have tried this; he had tried to show that there are different degrees of infinity, a concept that is now generally accepted, though in the past his ideas had been scoffed at.) Otherwise proving mathematical or any other problems involving infinity would appear like putting the cart before the horse, or like proving something that has no certainty of being real, something whose existence is doubtful, a not really worthy task – an inconsequential play of ideas or working of the imagination, though this might involve plenty of hard intellectual labor and might be a good exercise in logical deduction. I might be making it look as if mathematicians do not know what they are doing; I think that I am guilty of this myself by attempting the twin primes problem; I should not take my attempt or solution so seriously but should regard the solution as "tentative". I hope that there is a way to convey all these sentiments to the community of mathematicians and logicians.

I very much appreciate your feedback and your time.

With thanks and best regards.

Yours sincerely,
Author

Appendix 6: Notes – Some Logical Concepts of Calculus (Mathematics)

1) $(1+x)^{1/2} = 1+1/2x - 1/8x^2 + 1/16x^3 + \cdots$ and so on ad infinitum.
2) $v = \partial s/\partial t$, as $\partial t \rightarrow 0$ (i.e., reaches its limit at zero); v reaches the limit which is its exact speed at the given moment.
3) $y = x^n$

the derivative $\partial y/\partial x = nx^{n-1}$ (this mathematical operation is known as differentiation).

If $n = 2$, then $y = x^2$.

The rate of change at any point, $\partial y/\partial x = 2x^{2-1} = 2x$.

In other words, the gradient of the tangent at any point on the curve will always be 2x.

4) Velocity of a point (v) can be expressed in terms of the infinitely small distance ∂s traveled in the brief moment ∂t.

Thus: $\mathbf{\partial s = v\ \partial t}$.

5) Measurable distance, s, which the point travels between t_1 and t_2 is found by continuously summing the changes in this interval (which is known as integration).

Thus, $\mathbf{s} = \int_{t_2}^{t_1} \mathbf{v}\ \partial \mathbf{t}$ ($\int$ is the sign for integration)

For example, for $y = x^n$, instead of $\partial y/\partial x = nx^{n-1}$ (differentiation), for integration we get $\int x^n\ \partial x = x^{n+1}/n+1$.

(This immensely useful technique could be used for such problems as finding the area of any kind of shape described by a formulaic curve (and rotating about the axis produced by a volume)). It could also be used for any problem where continuous summing of infinitesimal changes was required.

https://doi.org/10.1515/9783112214367-067

Appendix 7: Notes – Some Logical Concepts of Quantum Theory (Physics)

1) The first law of thermodynamics:

$$\Delta W = \text{Work}$$

$$\Delta H = \text{Heat}$$

$$\Delta E = \Delta W + \Delta H$$

$$(\Delta E = \text{Energy})$$

2) Entropy is given by

$$S = k \operatorname{Log} W \ \ldots$$

where k is the constant (now called Boltzmann's constant) and W is the probability that a particular arrangement of atoms will occur.

3)

$$E = C_1 f^5 / \exp(-c_2 f/T) - 1$$

where E is the energy, f is frequency, exp is the natural exponent, and T is the temperature.

(The constants C_1 and C_2 are numbers chosen by Planck to make the equation fit the experiments.)

4) S = k log W (Boltzmann's version of the second law of thermodynamics)

5) Total energy E = Ne,

where N is an integer and e an arbitrarily small amount of energy (e would eventually become infinitesimally small as the chunks of energy (Boltzmann's method of dividing the energy of the oscillators into arbitrarily small but finite chunks) became infinite in number, consistent with the mathematical procedure).

6) Energy units have to be chosen which are proportional to the oscillator frequencies, namely, e = hf, in order to obtain the correct form for the total energy (f is the frequency and h is a constant which would eventually decrease to zero).

7) *But* Planck's experimental radiation law only applies if *the energy is discontinuous.*

Planck was thus forced to postulate that the quantity e = hf must be a finite amount, and h is not zero.

(Thus, if this is correct, it must be concluded that it is not possible for an oscillator to absorb and emit energy in a continuous range. It must gain and lose energy discontinuously, in small indivisible units of e = hf, which Planck called "energy quanta".)

https://doi.org/10.1515/9783112214367-068

8) In early 1901, the constant – today called Planck's constant – appeared in print for the first time. The number is small – h = 0.000000000000000000000000006626 – but it is not zero! If it were, we would never be able to sit in front of a fire. In fact, the whole universe would be different. Be thankful for the little things in life.

9)
$$E = \alpha f^3 \exp(-\beta f/T)$$

where α (alpha) is a constant, f is the frequency, β (beta) is the constant, and T is the temperature.

10) $E = n k \beta f$ (The result for the radiation is very similar to that of the compressed gas.)

Einstein's hypothesis: With the validity of Wien's law (i.e., at high frequency), radiation behaves thermodynamically as if it consists of mutually independent energy quanta of magnitude k β f. In other words, like light particles.

11)
$$\text{Kinetic energy} = \begin{matrix} \text{hf(Energy of the incoming photon)} \\ \text{Less P(Work to get out of the metal)} \end{matrix}$$

12) If the plate must be changed to a voltage V_o in order to offset the kinetic energy and reduce the electron current to zero (i.e., stopping even the most energetic ejected electrons), it follows that:

$$q V_o = hf - p$$

where q denotes the electron charge.

13) Johann Jakob Balmer's formula:

$$f = R\left(1/n_f^{\,2} - 1/n_i^{\,2}\right)$$

where R is the Rydberg constant.

Using this equation, Balmer could predict the frequencies of the four hydrogen lines if n_f (final) was chosen to be 2; n_i (initial) = 3, 4, 5, and 6, and, R was given the value 3.29163 × 10^{15} cycles/s. This gave the best fit to the measurements.

14) Linear momentum = Product of the mass of the body and its speed.
p = mv (linear momentum)

15) L = mvr (angular momentum)
where m is the mass and v is the speed around the orbit.
constant angular momentum (no torque)

16) Bohr's first quantum postulate:
$L = mvr = n\,(h/2\pi)$ (This is the quantum orbital condition.)

17) The angular momentum L cannot take on any arbitrary value, as is the usual case in classical physics, but only certain values:

$L = 1\,(h/2\pi)$ in the first orbit
$L = 2\,(h/2\pi)$ in the second orbit
$L = 3\,(h/2\pi)$ in the third orbit . . . and so on.

Only orbits in which L is a whole multiple of the quantized unit $h/2\pi$ are allowed.

(This integer, n, is called the principal quantum number.)

18)

$$r = \left(h^2/4\pi^2 mq^2\right) n^2$$

where r is the radius of electron orbit, h is Planck's constant, m is the mass of electron, q is the electronic charge, and n is the principal quantum number.

(Bohr based his derivation on Newton's planetary model of the solar system to obtain a formula for the radii of the electron orbits.)

The smallest radius is for n = 1, when its value is 5.3×10^{-9} m or 5.3 nm.

19) A sudden transition of the electron between two stationary states will produce an emission or absorption of radiation with a frequency given by the Planck/Einstein relation:

$$hf = E_i - E_f$$

where E_i and E_f are the energies of the atom in the initial and final stationary states.

20) Bohr's derivative of Balmer's formula – he mixed classical and quantum physics together to obtain:

$$f = 2\pi^2 mq^4/h^3\left(1/nf^2 - ni^2\right)$$

This was exactly the same formula that Balmer had obtained for the frequencies of hydrogen, if the constant term R in Balmer's equation (called the Rydberg constant) could be shown to be equal to: $R = 2\pi^2 mq^4/h^3$.

Using values for q, m, and h available in 1914, Bohr calculated:
$R = 3.26 \times 10^{15}$ cycles/s, within a few per cent of Balmer's value.

21) The angular momentum of the spinning electron turned out to be the only one-half of the normal value $h/2\pi$ of atomic orbits, so-called spin 1/2:

$$S = +1/2\ (\text{up}) \quad S = -1/2\ (\text{down})$$

Two different orbits. Distinct energy levels.

22)

$$v = f\lambda,$$

where v is the speed, λ is the wavelength (length of the wave), and f is the frequency of a wave (no. of waves per second).

23) Louis de Broglie's derivation, as follows:
He started with Einstein's famous $E = mc^2 = (mc)\ (c)$
Now watch de Broglie's series of substitutions . . .
Since mc is just mass times speed, the momentum, p, of a photon . . .

$$E = (p)\ (c) = (p)\ (f\lambda)$$

using c (speed) = f (frequency) × λ (wavelength) for waves.
Equating $E = hf$ from the Planck/Einstein relation to the expression above, we obtain:

$$(h)\ (f) = (p)\ (f\lambda)$$

and some simple algebra gives . . .

$$h/p = \lambda\ \text{(photons)}$$

(When the wavelength of light is decreased, the momentum of the individual light photons is increased.)

By direct analogy, de Broglie proposed that his relation was true not just for photons, but for electrons . . . and all particles.

$$\lambda = h/p$$

or (wavelength) = (Planck's constant divided by momentum)

For electrons, the MOMENTUM P = (M) (V) = (MASS) (VELOCITY) can easily be determined in an experimental situation, and the wavelength could thus be predicted from de Broglie's equation.

To most physicists, the concept seemed preposterous. The electron was a PARTICLE known by classical physicists since J. J. Thomson's discovery in 1897!

24) Electron "standing waves" in action. Only certain wavelengths will fit around a circle . . .

$$2\pi r = n\lambda \quad 2\pi r \neq n\lambda$$

Watch, a little algebra . . .
$n\lambda = 2\pi r$ (standing waves)
$n\,(h/mv) = 2\pi r$ (using de Broglie's equation)
$n\,(h/28) = mvr$ (quantum orbital postulate)

Bohr's quantum condition is no longer a postulate – it is a reality.

25) Schrödinger's equation: Schrödinger had found an equation which can be applied to any physical system in which the mathematical form of the energy is known:

$$\partial^2\Psi/\partial x^2 + 8\pi^2 m/h^2(E - V)\Psi = 0$$

where ∂^2 is the second derivative with respect to x, Ψ is Schrodinger's wave function, x is the position, E is the energy, and V is the potential energy.

26) Fourier wave analysis of periodic functions:
Any periodic function f(t) is equal to the sum of simple harmonic functions.

Therefore, the solution of Schrödinger's equation – the wave function for the system – was replaced by an infinite series – the wave functions of the individual states – which are natural harmonics of each other. That is, their frequencies are related in the ratio of whole numbers, or integers. Ψ is the PROBABILITY AMPLITUDE for an electron in the state n to scatter into the direction m. It is, in a sense, its own intensity wave. When it is squared and the absolute value is taken, it turns out to be a PHYSICAL PROBABILITY of the associated particle's presence. One month later, Born stated that the probability of the existence of a state is given by the *square* of the normalized amplitude of the individual wave function (i.e., Ψ^2). This was another new concept – the probability that a *certain quantum state exists*.

No more exact answers, said Born. In atomic theory, all we get are probabilities.

The wave Ψ determines the likelihood that the electron will be in a particular position. Unlike the electromagnetic field, Ψ has no physical reality.

27) Dirac was at first puzzled by the appearance of non-commuting quantities (where the product of two quantities depended on their order, so that A × B does not equal B × A) and realized that this was the essence of the new approach. He quickly found a link to classical physics and used the new fundamental idea of non-commutability to develop his own version of quantum mechanics.

Wave mechanics of Erwin Schrödinger ≡ Heisenberg's original matrix mechanics: special cases of Dirac's more general formulation: in other words, THEY ARE ALL EQUIVALENT.

According to Dirac, light could be treated as WAVES or PARTICLES.

28) Non-commutability of the two variables – POSITION and MOMENTUM

$$(pq - qp = h/2\pi i) \cdot \cdot \cdot \cdot$$

29) According to Heisenberg, the minimum IMPRECISION IN THE MOMENTUM MEASUREMENT is approximately equal to the momentum imparted to the electron by a SINGLE PHOTON used to illuminate the particle, the smallest disturbance possible.

From the de Broglie/Einstein relation, $\Delta p \sim h/\lambda$, Heisenberg obtained the imprecision in the momentum. Multiplying the two inaccuracies together, Heisenberg showed that the product, $\Delta x\, \Delta p$, will always be greater than or equal to (≥) a certain amount . . .

$$(\Delta x)\,(\Delta p) \geq (\lambda)\,(h/\lambda) \geq h \text{ or} \ldots \;\; \Delta x\, \Delta p \geq h$$

where $(\Delta p) \geq (h/\lambda)$ is the de Broglie relation and $(\Delta x) \geq (\lambda)$ is from the diffraction limit.

This is Heisenberg's uncertainty principle (HUP), which states that:

The uncertainty in a simultaneous measurement of momentum and position is always greater than a fixed amount, approximately equal to Planck's constant, h.

Though we do not notice HUP in our everyday experience with the gross macroscopic world, the wave/particle duality defeats the atomic experimentalist who seeks perfection. Many believe there are serious PHILOSOPHICAL CONSEQUENCES for us all in this idea.

30) According to Bohr, although the wave and particle behavior of an object mutually exclude each other, both are necessary for a full understanding of the object's properties (complementarity).
31) Newton's formula (model for quantum theory):

$$F = G\,M_1\,M_2/R^2$$

where F is the gravitational force of attraction, G is the gravitational constant ($G = 6.67 \times 10^{-11}$ N-m^2/kg^2), M_1 and M_2 are masses of two objects, and R is the distance between masses.

Appendix 8: Economic Reasoning at Work (with Commentary)

Commentary

The author senses the absurdity, and apparent helplessness, of the world's economic problems and argues with economic logic that the dire world economic situation could be and should be rectified.

Economics has to do with how buyers and sellers behave in the marketplace, and since most countries have free economies, there is generally little control over how buyers and sellers in the marketplace behave. We could hope to influence them and change their behaviors through advertisements, incentives, tax policies, and so on, but this is no guarantee that the desired change for the better would be forthcoming. Commercial logic might just fail to convince or change their minds. The three articles here have some important suggestions on improving the world economic situation.

Economics is effectually the study of how people behave in the business environment. Mathematical models might be used to describe the interaction of buyers and sellers in the marketplace, and forecasting techniques might be utilized to foretell how buyers and sellers would interact under varying conditions. But, the fact is that all human beings tend to be whimsical and their behaviors in commerce and in every other thing would thus tend to be difficult, if not impossible, to predict. How many times economic forecasts are inaccurate and economic policies fail? Game theory, which had been pioneered by John von Neumann, a mathematician, and for which John Nash, the mathematical genius of "Beautiful Mind" fame, won the Nobel Prize for economics, has been used as a model of competitive behavior in business. But, it is based on the "cooperative" concept, whereby groups of businesses might team together or group together for mutual benefits to compete with the others. This appears simplistic, as in the real situation businesses might not cooperate at all but would instead be all out to cut each other's throats.

https://doi.org/10.1515/9783112214367-069

1 Neo-economics: We Should Control the Economy and Not Vice Versa (Edited Version of This Article Had Been Published in an International Economics Journal)

Abstract

In this article, the author proposes some solutions to our modern seemingly intractable economic problems. Keynesian theories, the bugbear of modern economics, seem to hold little water nowadays. Modern economies appear to be in need of more radical solutions.

Proposed Economic Solutions

We are now in the throes of an economic recession. We have become the victim of the economic recession. We hope, pray, and wait helplessly for the economy to recover. Governments are trying to do something that may help the economy to recover fast, but there is no guarantee that they can help bring about a reasonably quick recovery. They may introduce more public works to create jobs, implement fiscal and monetary policies, retrain retrenched workers, and carry out a myriad of other actions. But the economy may take its own time to pick up. The economy seems to live a life of its own, a life that does not seem easy to control at all. In fact, the fate of the economy is the fate of all human beings. Difficulty in controlling the economy simply translates to difficulty in controlling the human destiny.

Looking around us, human miseries are all too obvious. The jobless are living in fear and anxiety. Many have even turned to crimes (hence, the higher crime rates now), such as robbery and looting. For those who are less well off, they now have to tighten their belts, live on a smaller budget. A number have sought the assistance of the authorities. It is a helpless situation.

From the study of the history of past economic trends, everyone is more or less telling himself that when the economy swings downward it can be expected to swing upward again sometime in the future (perhaps, in a few years' time) and vice versa. That is, we assume that "history repeats itself". What, if in the future, history does not repeat itself? This is not an impossible scenario. It is indeed dicey.

This chaotic state of affairs which is not new and which brings about much uncertainty and misery seems to imply something. It seems to tell us that there is something wrong with our social system. It seems to imply at least some of the following, though they are perhaps not obvious to many as they are very complex indeed:

1) There may be a problem in the distribution of wealth. The gap between the haves and the have-nots may have become very wide.

2) People may have been "psyched" by adverse publicity into hoarding rather than spending as usual, causing a fall in demand for goods and services all round, hence, reduction in production, retrenchment, and loss of jobs.
3) Too much competition and undercutting result in loss of profitability, hence loss of incentive to produce or increase production, and hence, business closure, retrenchment, and loss of jobs. (Competition may not be really good, though consumers tend to deem it desirable.)
4) The free enterprise system, or the capitalistic system, without much governmental control may not be that efficient and good after all as it implies freedom for the economic system to be well beyond proper control which enables it to move along the correct, desirable path, ending up in the wrong, undesirable direction instead.
5) The wealthy has too much power and control over the workers and the have-nots. (In a recession, the workers and the have-nots will obviously suffer more than the wealthy, who may not hesitate to displace them from their jobs to save their own skins.)
6) The bosses exploit the workers, bringing misery to the latter one way or another. Though the labor legislations protect the workers and ensure that their minimal entitlements are fulfilled, bosses can still find subtle means to exploit their workers. (This is perhaps one of the greatest flaws of the capitalist system – exploitation, greed, or profit-maximization at the expense of the workers.)
7) The bosses may also exercise some form of subtle control over the government, e.g., by sponsoring politicians (who run the government) or threatening to relocate to another country, thus possibly causing job loss and a poor "employment situation" (a political headache for an elected government).
8) It may be a situation that even those who are desperate for a job and are prepared to accept a low salary may have hardly any opportunity of employment. The government cannot force companies to hire anyone they do not wish to hire. (In this respect, a socialist economy has an advantage in that its government-controlled industries can provide plenty of job opportunities for the populace.)
9) Instead of being the master of technology, we may have become its slave or victim. Technology can be changing so rapidly that it is very difficult or impossible to keep up with it.
10) The economy of one nation may have been adversely affected by the economies of the others and vice versa.

We, especially our governments, should manage the economy instead of allowing the economy to manage us. We have to rethink and revamp our economic policies on a global scale, with nations consulting one another. It appears that capitalism is far from perfect as an economic or social system as it encourages human greed and exploitation. It may be better to combine the desirable elements of both the capitalist and socialist systems. Otherwise, we may have to reformulate a new social/economic system, which is radically different from both the capitalist and socialist systems.

How about a world association to manage and coordinate the economies and economic activities of all the nations in the world, this association being represented by members from all the nations in the world? This association can fix or set quotas for imports and exports amongst all the world's nations. For example, the association can declare that Country A exports Item X to Country B and Item Y to Country C and imports Item S from Country B and Item T from Country C, Country D exports Item J to Countries E, F, G, and H and imports Item K from Country E, Item L from Country F, Item M from Country G, and Item N from Country H, with the quantities imported and exported amongst them stipulated, subject to periodic changes in the future. In this way all countries in the world will have to produce sufficient goods and services to fulfill their own consumption needs as well as meet the quotas set by the association. Thus, there should be enough jobs and incomes for everyone.

Contrary to general opinion, competition is really not good for business, especially in the long run. As mentioned in (3) above, too much competition and undercutting may lead to business closure, retrenchment, and loss of jobs, which is indeed not good for everybody. Fair enough, competition ensures that the consumer does not have to pay a high price for his good or service. Also, it is felt that competition will encourage businesses to be more productive (e.g., to be more competitive or better businesses will strive to be more efficient and more cost-conscious, but they may be so cost-conscious that the welfare and safety of their workers are compromised). Cooperation and teamwork amongst competitive businesses are better, for there are economies of scale (resulting in cost savings and thus, theoretically at least, lower prices for the consumer) and more heads to do the thinking and thrash ideas. However, to protect the consumer from overpriced goods or services, the government should have the right to step in and regulate prices.

Finally, a social or economic system should consider the general welfare and economic well-being or safety of every human being in the world. What we are having now is social and economic imbalance, with some people and nations having great wealth, some not so great wealth, and some having abject poverty (with problems such as famine, undernourishment, and disease). Economic problems tend to lead to social problems such as crimes, riots, looting, or even rebellion. The economic system should take into consideration the welfare of every human being and should not have allowed some to wallow in vast wealth while some others are down in the economic dumps. Our social or economic system should ensure a more even distribution of wealth. A more even distribution of wealth will lead to greater social equanimity.

References

[1] The Wealth of Nations, Adam Smith, The Modern Library

[2] The General Theory of Employment, Interest, And Money, John Maynard Keynes, Harcourt, Inc.

[3] The Making of Modern Economics, Mark Skousen, M. E. Sharpe

2 A New Monetary System for Combating Our Economic Ills (Edited Version of This Article Had Been Published in an International Economics Journal)

Abstract

Economic problems that plague us seem intractable and seem to require some radical countermeasures. One of the root causes of a recession is that people are not spending sufficiently. To overcome this problem a radical change in our monetary system could be effected.

Proposed Monetary System

The world economy is now generally gloomy and depressed.

The economy works through the market forces of supply and demand. The level of demand ultimately determines the quantity of goods and services that will be made available in the market, though sometimes the quantity of goods and services produced in the market are so produced in anticipation of demand.

The crux of the matter is that money held by people has to be spent or change hands in sufficiently large volumes for the economy to be buoyant. If people lose their jobs, have little or no money to spend, and curtail or stop spending the economy will be in trouble. Equally bad is the situation where employment rate is high, people have money to spend but are hardly spending. To rectify the situation of low employment governments can help to create jobs. But after the jobs have been created and people have jobs and money but are still not spending sufficiently, what will become of the economy? This is an interesting point to ponder. The author's view is that companies or producers of goods and services will flounder when their goods and services are not in demand as people have little inclination to spend the money they possess (unless the government now performs the charitable act of financially supporting them to ensure that their employees do not lose their jobs). When these companies or business units flounder, they will retrench or wind up, that is, those with jobs will now find themselves without jobs and income, and spending will become less. So, it is more or less back to "square one". Of course in the situation whereby people have money but are not spending enough but saving the bulk of their income, the government may well be urged to resort to suitable fiscal policies to rectify the situation such as reducing bank rates and reducing taxes. Companies and business enterprises may well be advised to advertise and market their goods and services more aggressively to create or increase the demand for them. Will these be certain to change people's attitude toward spending? The certainty could hardly be expected.

The point is that governmental policies and aggressive advertising and marketing can only help to coerce and persuade people to spend more but the final choice of whether to spend or not still belongs to the people. In other words, the "customer is king," the market belongs to the buyer. Put bluntly, business, the economy (even the government), is at the mercy of the buyer or customer. Can we force the buyer or customer to spend? No? The author's answer is yes. But, a radical change has to be implemented.

All these suggest that the economy is indeed difficult, even impossible, to control. We should not allow the whims and fancies of the buyer or market to create problems in the economy, problems for businesses who need a profit to carry on, problems for people who need jobs, and problems for the government whose responsibility is to see that their people have jobs and incomes. As mentioned earlier, creating jobs and incomes for the people may be fine and good, but when those with incomes are not spending sufficiently, it is not good for the economy, it is not good for everybody. We have to find a way to force everyone to spend sufficiently so that economically everyone will survive. The question here is how? The author would like to make a radical suggestion. The suggestion is that two types of currencies be introduced to ensure that there is sufficient spending. One type of currency has an expiry date while the second type of currency has no expiry date like our current ones. The issuance and administration of such currencies, especially the first type which has an expiry date, are to be the responsibility of a governmental monetary authority or statutory board. For example, the government can pass a law to the effect that all salaries have to paid in the following format: 70% in currencies which expire say 3 months from the date of payment and the balance 30% in currencies with no expiry date like our present currencies. The seller who receives the currencies with expiry dates from the buyer could then exchange them at the governmental monetary authority for the following currencies: 70% in currencies which expire in 3 months' time and 30% in currencies with no expiry date. In this way everyone, bosses and employees, and buyers and sellers alike, have to spend at least 70% of their earnings within the allotted time frame of 3 months. Here we are only talking about salaries and purchases within the same country.

At the international level, where business between countries is concerned, the same principle should also apply, except that the currencies with expiry dates and the currencies without expiry dates should now be issued and administered by an international authority, for example, a newly created division of the United Nations.

The question is whether this arrangement is practical. Except for some teething problems on initial implementation, the author does not think implementation is going to be difficult. Though it entails some administrative costs, the price is worth paying. For sure, implementing this new monetary system will result in new jobs being created, which is good for the economy. And, most important of all, this new monetary system will help to eradicate recessions, unemployment, and other economic and related social problems.

There is no point grumbling about the poor state of the economy and how miserable life is because one has no job and no income, no business, and no earnings. Everyone, especially the powers that be, should seriously attempt to analyze what causes the economy to fail and people to lose their livelihood and try hard to do something about it.

Hence, the authorities should seriously consider the author's reasoned proposal here.

References

[1] The General Theory of Employment, Interest, And Money, John Maynard Keynes, Harcourt, Inc.
[2] The Making of Modern Economics, Mark Skousen, M. E. Sharpe
[3] Principles of Economics, Paul Samuelson, McGraw Hill

3 The Economic Logic of Job-Hopping

There have often been complaints about job-hopping – it is often ascribed to the irresponsible attitude of workers. How much truth is there in this?

Is not job-hopping somewhat inevitable since our labor market is a free one, where the forces of supply and demand have free play?

Also, labor mobility is in some ways necessary to any economy, although it must be subject to reasonable restraint.

Entrepreneurs would not be keen on investing in a country or in expanding operations there if there was not a remote possibility of their attracting experienced workers from other corporations.

The companies which lose workers may complain but those who are hunting for workers may bask in the satisfaction of meeting their labor requirements.

Another point to bear in mind is that labor mobility helps to ensure a better spread of available talent in the country.

If favored and important industries face serious labor shortages, the government may even encourage workers from other sectors to join such companies.

This may be called redeployment of labor but when workers move on their own accord, it is called job-hopping.

So, labor mobility is a relative thing that may be viewed differently as circumstances change.

On principle, employees should have freedom of choice just as employers have the right to hire and fire their employees.

While employees should adopt responsible work attitudes, employers should also play their part by adopting an enlightened approach to personnel management.

It is after all the employer who determines the conditions and pace of work. They can either motivate or demotivate workers by correctly assessing skills and abilities and by reading intelligently the psychology of workers.

They should also be aware of poor human relations at the work place which seems to be often overlooked.

Workers leave their jobs sometimes because they cannot get along with their colleagues or their bosses or both. Not everyone likes getting involved in office politics. Those who do not get into it, get out.

We should thus take a deep at the causes of job-hopping before we begin to blame job-hoppers for their irresponsible attitude, as has often been the case.

But there are errant employers who contribute to job-hopping. Some employers may have bad attitudes too and are not incapable of mismanagement.

To help resolve this problem of job-hopping the employers' federations and the unions could draw up codes of conduct or ethics for both employers and workers who should adhere closely to them.

After all, job-hopping is not always bad. Sometimes it is desirable, for the reasons given above.

Appendix 9: Scientific Reasoning at Work (with Commentary)

Commentary

The six articles here deal with ideas which are hypotheses, ideas which are yet to be proven by physical data or experiments. The subject presented falls under the category known as theoretical physics. It is the task of the theoretical physicist to explore feasible ideas and produce hypotheses without spending time in the laboratory. On the other hand, the experimental physicist derives his hypotheses from physical experiments – he works in the laboratory.

The theoretical physicist uses mathematical models to describe nature, and since he has not had his ideas proven in the laboratory, they might be regarded with incredulity, especially if they were counterintuitive or seemed to be against common sense, e.g., concepts such as time travel and the existence of particles known as tachyons which travel faster than light (though it has been claimed that nothing could travel faster than light). Einstein had tried but had failed to produce a unified field theory which would explain why the four forces of nature, gravitational forces, weak nuclear forces, strong nuclear forces, and electromagnetism work together. The articles in this appendix have been ambitious attempts at a better understanding of nature.

The theories, forecasts, or speculations of theoretical physicists, which could be modeled with elegant mathematical equations that might be regarded as proofs of their validity, are still in need of physical confirmation. It is the experimental physicists who perform the experiments to confirm these theories and should the experimental evidences prove otherwise these theories would be discarded. No matter how sound the logic of these theories seemed to be, as detailed, e.g., in some elegant mathematical equations, they would be useless once they were contradicted by the experimental evidences.

Could the scientific community still insist on hanging on to the theories now and say that the experiments had been wrong and had thereby ended up with erroneous results, while insisting that the logic of the theories was so elegant that it could not be wrong? In other words, could they choose logic over empirical evidence? The answer is clearly "no". If they had so much faith in the logic, the experimental physicists would not be required. It would be contradictory, silly in getting the experimental physicists to obtain the empirical evidences and when they are found to contradict the theories, reject them. When it comes to the crunch, which is more important – elegant logic or contradictory empirical evidence? Of course, the empirical evidence should take precedence, as the physical senses are apparently "fool-proof" as compared to the logical faculty. The fact that experimental physicists are needed is a tell-

https://doi.org/10.1515/9783112214367-070

ing sign for logic; in this case, scientific logic. That is, logic is not that trustworthy or reliable, and empirical fact is more reliable and important than logic.

But, when and where the empirical facts are not obtainable, logic would come in handy, i.e., we would have to resort to logical deduction to arrive at the truths. Arriving at the truths through physical experiment would apparently be the preferred way, provided it is feasible.

Is this not telling: for example, we tend to argue about the ethics (right or wrong) of things, but, hardly do we doubt or dispute what we or others see with our eyes, for instance, colors such as red, blue, and green?

Hence, the importance of empirical evidence in science, as well as other areas of human activities, an importance which could be regarded as greater than that of logic.

1 On the Theory of Everything

Introduction

A theory of everything, or Grand Unified Theory (which Einstein had been working on without success, with superstring theory now being a good candidate), is one which unites all the forces of nature, viz., gravity, electromagnetism, the strong nuclear force, and the weak nuclear force. Important as this theory might be, it is lacking in one important fundamental aspect, viz., the role of consciousness, which could in fact be considered the most fundamental aspect of physics. This article explains that a theory of consciousness is more important than a theory of everything or Grand Unified Theory and should be the theory of everything instead, or at least, a part of the theory of everything.

An Alternative Theory of Everything?

It is apparent that consciousness is somehow connected with phenomena in the quantum realm. Consider Bell's double-slit experiment. In this experiment, are the particles and waves just sensitive to screens and other equipment or are they picking up messages from the physicist's brain? It could be said that particles of matter and particles of mind might come into being together, but in any case our self-awareness and what seems to be some kind of consciousness at the quantum level appear to be in deep communication. David Bohm had in his classic work, *Wholeness and the Implicate Order*, developed a theory of quantum physics which treats the totality of existence, including matter and consciousness, as an unbroken whole. There is the implication that at the sub-quantum level the observing device used to measure the quantum particles must have connections with all parts of the system, including the link with our consciousness, and through these a "signal" might be transmitted to the molecule that a certain observable was eventually going to be measured.

Consciousness is nonlocal; in other words, no one could say where the mind is or how far the effects of thought could reach. As a matter of fact, within the brain, all the forces are active, including gravity, which is the force that holds the entire universe together.

Though fundamentally the material brain and the other matters are composed of the same thing, viz., atoms, reductionists might wonder why the material brain as compared to other matters is so special. Is this due to the special composition of the atoms in the brain? If we take a number of atoms and arrange their composition so that they would be similar to that of the material brain, could we produce a brain and consciousness?

The universe might be in some sense a Great Mind and a theory of everything might have to include a theory of consciousness. Superstrings, which are a strong candidate for a theory of everything, might be thought particles with a life of their own. Many physicists are making attempts at deriving a Grand Unified Theory of the universe on the basis of particle physics. This effort might be incomplete as particles might be just a reflection of the information-processing foundations of the universe (but it is certainly not a waste of time as this research might help us to figure out how the information-processing system works). In the last analysis, we might not be able to completely understand the universe, if it is ever possible to do so, until it is examined as a self-evolving and organizing information-processing machine, one which produces intelligent minds to examine itself with. Hence, a theory of consciousness might be consolidated with the theory of physics (such as the superstring theory or the membrane theory) into a Grand Information Theory (GIT). This could be considered the theory of everything.

In quantum mechanics, the behavior of particles, which are regarded as waves, could be predicted, as it were, and they are thus known as probability waves or Dirac wave particles. Here, there is a wave/particle duality. When the particle is not observed (when consciousness is not present), it remains a wave (a probability wave), but on being observed (when consciousness is present) it becomes a particle. (Note that according to the uncertainty principle, for which Werner Heisenberg won a Nobel Prize, the very act of observing a quantum particle affects its behavior, i.e., consciousness affects a quantum particle.) Pauli, who was a Nobelist, was fascinated by subatomic particles and consciousness, collaborating for some time with psychologist Carl Jung, whose patient he was for a time. The mathematician, John von Neumann, the biologist, George Wald, and the physicists, David Bohm and Arthur Eddington, had declared that the universe is mind-stuff. The mathematical physicist, Sir Roger Penrose, sometime colleague of Stephen Hawking, considers that there is "definite possibility" that consciousness is connected with phenomena in the quantum realm. The Anthropic Principles (both strong and weak) stipulate that in man there are intellectual capacities which are there for a reason that somehow human beings with their minds are obliged to help the universe through the next stage; this is indeed manifested by the fact that scientists such as Einstein had been using their brains/consciousness to understand nature and many had been attempting to formulate a theory of everything or unified field theory. The Gaia hypothesis of biologist, James Lovelock, paints a picture of the Earth functioning as one large organism, which implies will and consciousness being at work.

A number of scientists had postulated that there has to be a "cosmic consciousness" pervading the universe; objects spring into existence when measurements are made, measurements which are made by conscious beings, which implies that there must be cosmic consciousness that pervades the universe determining which state we are in. Some scientists, e.g., Nobel laureate Eugene Wigner, had argued that this is evidence of the existence of God or some cosmic consciousness. Wigner had remarked that it was not possible to formulate the laws of quantum theory in a fully consistent way without reference to consciousness.

Classical philosophers such as Berkeley and Hume had in fact questioned whether the existence of any object was independent of the existence of the mind or consciousness: If I had never seen (never been aware of) an object, does that object exist?

Thus the great importance of the role of consciousness in nature. In fact, there appears to be an intricate link between nature and consciousness, the latter being apparently the common denominator in the workings of nature, especially at the quantum level, as is described above. However, far-fetched as it might seem, consciousness could be some sort of particles, not unlike quantum particles, which could explain why they are able to interact with one another, as is aforedescribed. This could also possibly explain phenomena such as intuition, mind-reading, and telepathy. For example, Rupert Sheldrake, a well-known biologist, in his "morphogenesis" theory, stated that all our minds or consciousness are linked or interconnected so that how people think and behave in one geographical area affects how people living in another distant geographical area act and think without any communication whatsoever between them, a phenomenon which applies to animals as well. Consciousness could therefore be regarded as a "force" of nature.

We present an important poser here: What is life, the theory of everything or the Grand Unified Theory without consciousness? This is in fact an irrelevant question as it is a tautology, for life and consciousness are synonymous, and without life or consciousness to contemplate a theory of everything or Grand Unified Theory, the latter is an impossibility and is redundant.

It is foremost, most important, to have consciousness, and, hence a theory of consciousness, which should precede a theory of everything or Grand Unified Theory, for to contemplate a theory of everything or Grand Unified Theory without taking into account the evident role of consciousness is like riding the horse cart without the horse. Isn't it more important to better understand ourselves, our consciousness, our nature, first, an understanding we are still evidently lacking, before we try to probe further the nature external to us? Isn't it evident that it is mind which controls external matter and not vice versa, for otherwise life would be overwhelmed by natural phenomena? Therefore, shouldn't a theory of consciousness really be the theory of everything, e.g., as a GIT, which is described above, or at least, a part of the theory of everything, a very important, fundamental part, consciousness and matter having some fundamental link as is described above? Importantly, this theory of consciousness giving us a more comprehensive understanding of our mind would be a boon or aid to our affairs.

2 Linking the Forces of Nature

Introduction

The article proposes a practical way to arrive at a unified field theory and theory of everything, based on the author's experiences with computer simulation. Having only a beautiful theory and beautiful equations to model nature are insufficient for affirming the characteristic of the universe. The affirmation has to be ultimately carried out by physical experiment. However, in lieu of physical experimentation, computer simulation is proposed, simulation with powerful software having been effectively utilized in areas such as aeronautical, electronic, mechanical, and marine designs, wherein it now becomes unnecessary to produce costly prototypes. A few possible ways of unification are also presented.

Forces of Nature and Unification

Our views of nature may be modified in the future. For instance, superstring theories, which had been neglected in the past, are now the "in" thing, being regarded by many scientists as beautiful, or, elegant, and a possible theory of everything.

Einstein had attempted to unify the four forces of nature, i.e., gravity, weak nuclear force, strong nuclear force, and electromagnetism, but had failed. As in the past, he was unable to derive the electromagnetic field equations, even for the weak-field approximation. He was to live to the end of his life without any success with the unified field theory.

Einstein had thought that David Bohm would be the first scientist to solve the unification problem. But ironically the latter regarded the concern with the unified field theory problem as merely an unnecessary fuss, dismissing it as an "illusion of parts" and simply relegating the problem to the logical constraints of topology; he resorted to the metaphysical way of interpreting the universe, calling it the "looking-glass" universe.

However, gravitation and electromagnetism can be linked, and the result can be antigravitational force and/or torsion in space-time. There is also the belief that there is an antigravitational force for every gravitational force, just as there is an antiparticle for every particle. Nobel Laureate Richard Feynman had suggested that antiparticles are like ordinary particles moving backward in time, which implies that antiparticles should have antigravity.

Gravitation has always been thought of as a pulling or attracting force, just like the force of attraction between two magnets. Gravitation and magnetism may be different manifestations of the same thing. And, gravity may be a pushing force instead, a force that presses down on all objects in the direction of the center of the Earth. In fact, a push is equivalent to a pull; the former originates at the back of an object while the latter originates at the front. So far, gravitational forces are seen as forces of attraction only, while magnetic and electric forces are forces of attraction and repulsion. There may be a gravitational force of repulsion. All this will affect our approach toward the unified field theory.

There is evidently a "looking-glass" characteristic in the universe. David Bohm, the British scientist, whom Einstein had once hailed as being the person who would one day solve the unified field theory problem, had postulated that our universe is a vast fluid nothingness or no-thingness in which everything is. He had spawned a surprising relationship between maps and terrains. According to him, in the "looking-glass" universe, our mapmaking changes the very terrain, and the terrain in turn changes our map; maps, mapmakers, and terrains intertwine to form an integral whole. In other words, different things are really one thing. Heisenberg's uncertainty principle postulates that the experimenter affects the experiment and vice versa, and the experimenter is also part of the experiment. Accordingly, all things are relative, how things are really depends on how or from what angles we look at them; there is a "yin" to every "yang", and the "yin" and the "yang" combine to make the "yin-yang" which is an integral whole – so good cannot be distinguished from bad for they belong to the same whole, the "yin-yang". Regarding Einstein's unified field theory, Bohm maintained that it is based on the illusion of parts (gravity, strong nuclear force, weak nuclear force, and electromagnetic force) and that it is a futile problem. The observer is the observed, the part is the whole – all this seems more metaphysics than physics, but nevertheless this has evidently been the state of physics. Bohm believed that unification could be expressed by using the logical relations of topology, as is stated above.

Failure

The failure in deriving the unified field theory is the failure to derive an equation which will link our visible macro-world with the invisible micro-world of the quantum particles, which will link the gravitational force with the weak nuclear force, strong nuclear force, and electromagnetism, an equation which should encapsulate the totality of information about the universe. If Bohm's interpretation were correct, all these four forces are actually one and the same force, being different manifestations of the same force. Can any of these four forces exist without the others? They are all evidently essential parts of our universe, making up the whole.

Gravity

Gravity, which is crucial in the formulation of a unified field theory, can be described by the following formula:

$$F = G\ M_1M_2/R^2$$

where F is the gravitational force of attraction, G is the gravitational constant, M_1 and M_2 are the masses of two objects, and R^2 is the distance between masses.

Before Newton discovered gravity, nobody had known it existed or had thought that there was such an attractive force. Einstein in his general theory of relativity interpreted it as a curvature of the space-time continuum, a geometrical form. Can gravity be the fifth dimension, in addition to the four dimensions of general relativity comprising the three physical dimensions and the time dimension, as Theodor Kaluza had suggested, which impressed Einstein greatly?

Hilbert Space

Can the physical world be considered infinite-dimensional, i.e., a Hilbert space, instead of four-dimensional, a commonly held view? It depends on how we look upon the physical world, on our mental inventiveness, and is thus subjective. Quantum particles in the micro-world, unlike the objects in the macro-world, are comparatively unpredictable where their actions or movements are concerned and can only be predicted if at all in a probabilistic fashion. We will never be able to know for certain where a quantum particle will turn up next. Moreover quantum particles are capable of being at two different places at the same time and also capable of instantaneous travel or teleportation, which is "spooky" and incomprehensible, and evidently in defiance of gravity. We can have a quantum field equation involving infinite dimensions. According to modern quantum mechanics, all possible physical states of a system correspond to space vectors in a Hilbert space. An infinite-dimensional Hilbert space will also fit in with the theory of the existence of an infinite number of parallel universes which are connected with each other through wormholes.

Supersymmetry

According to Einstein's theory of gravity, the hypothetical quantum of gravity, the graviton, which is a spin-2 boson, interacts extremely weakly with other matter, far more weakly than neutrinos; it is so weak that no instruments so far have been able to detect it. In the supergravity extension of this theory of gravity, the graviton finds a superpartner, the gravitino, which is a spin-3/2 fermion. Under local supersymmetric transformations these two particles transform one into the other. When quantum calculations were carried out using supergravity theory, it was discovered that the infinities which plagued the earlier gravity theory with only the graviton were now being cancelled by equal and opposite infinities produced by the gravitino. This is evidently the result of the deeper consequence of the presence of supersymmetry. Though it is not certain whether the supergravity theory is completely renormalizable, this "softening of the infinities" appears to be a step toward a viable theory of quantum gravity. As simple supergravity theory includes only the graviton and the gravitino, this hardly corresponds to the real world with its many particles. Most of those who have worked on supergravity feel that some crucial idea is still missing. Without this crucial idea the theories simply do not describe the real world.

How do we make supergravity theory realistic? If we can solve this problem, we can have supergravity theory as a completely unified field theory. It has been shown that the principle of local supersymmetry is so restrictive that only eight possible supergravity theories exist, which are each labeled by an integer $N = 1, 2, \ldots, 8$. Supergravity theory shares the same features with its progenitor, the theory of general relativity, namely, conceptual power and mathematical complexity. Perhaps, by postulating the existence of a single master supersymmetry we can have a unified field theory that accounts for the whole universe.

General Relativity and Curved Space-Time

The following is Einstein's equation for general relativity:

$$G_{im} = -K(T_{im} - 1/2g_{im}T)$$

This beautiful equation expresses the curvature of space-time. The left-hand side refers to a set of terms which characterize the geometry of space, while the right-hand side refers to a set of terms which describe the distribution of energy and momentum, i.e., the left is the geometry side, while the right is the matter side. Reading from left to right is space-time telling mass how to move, while reading from right to left is mass telling space-time how to curve. In general relativity, there is neither absolute time nor space and gravitation is not a force or pull between one object and another but a property of space and time. All this represents a great conceptual leap by the theory's creator Einstein. As for the coordinate system of Einstein's general theory of relativity it has no basis in reality and is only a mental construct used to describe the space-time continuum of the general theory of relativity.

A suggestion is to change the left-hand side of the equation, the geometry of space, which is here a four-dimensional space-time continuum, into an infinite-dimensional space, a Hilbert space, which is as follows:

$$G_{imH} = -K(T_{im} - 1/2g_{im}T)$$

What kind of geometrical form can represent this infinite-dimensional space? One geometrical form of this nature can be an infinite number of Mobius strips which are intricately intertwined and linked with each other (with each strip being cut lengthwise into several narrower strips that are connected together at narrow points, which represent parallel universes). It is thought that since we are only able to move around in the three large, observable spatial dimensions comprising length, breadth, and height, and one of time, all other dimensions must be very small and thus invisible to us, being curled up in a multidimensional space (which may be construed as representing the invisible micro-world of the quantum particles). This is in keeping with the concept of the unified field theory which Einstein had attempted to formulate by combining general relativity and quantum theory.

Success

One may wonder when a unified field equation will be discovered. However, without a good understanding of gravity, this unified field equation will not come by easily. This unified field equation will of course link both the macro-world and the micro-world (through gravity, which will be the common denominator for both). Once the experimental confirmation of the existence of the graviton, the hypothetical quantum of gravity, is achieved, we should be surer of obtaining a unified field equation that accounts for the whole universe, which may be supported by the existence of a single master supersymmetry. Superstring theory now appears to pave the way toward achieving this difficult goal.

Universal Laws

Can there be some yet to be discovered universal laws which govern everything that exists in the universe? It is difficult to tell but it will be very useful to know these laws. This then will really be the theory of everything.

To many, including scientists, even Einstein, the laws of nature had been created by a Supreme Being, a God (whom Einstein believed does not play with dice). Alternatively, can all of nature be a computer simulation carried out by a very advanced race of beings? In 2001, a philosopher called Nick Bostrom had begun circulating a paper titled "Are You Living in a Computer Simulation?" This has been considered a good possibility.

Practical Affirmation

It appears a good idea to get some computer game designers and/or computer programmers to collaborate with the scientists to produce a simulation of life and existence in the universe. This is probably a mammoth if not impossible task. These programming experts and scientists have to sort of play God. What will be the parameters involved, the coordinate system to be utilized (e.g., three-dimensional, three-and-a-half-dimensional, and four-dimensional), the algorithms or mathematical formulas to be used for governing the movements of virtual objects (which will be important as they may be equivalent to the field equations of the unified field theory), and so on? With such a simulation, we may be able to understand better how the forces of nature, e.g., the mysterious but all-important gravitational force, behave. In this manner, which can be considered a kind of reverse engineering, success with the unified field theory and the theory of everything may be achieved.

Conclusion

The missing link in this unification process is gravity, which is still not a fully understood phenomenon. We like to think that gravity affects our position and movement in space. But gravity may not be what we think it is. It may not even exist, which means that our search for gravitational waves and gravitons may be a futile exercise. The search for gravitational waves is still ongoing. The existence of gravitational waves so far is inconclusive – there is only indirect evidence of their existence. In the gravitational effect known as frame dragging, objects occupying the space near to a rotating object get swept around with it; rotating objects drag space around with them rather like a spoon in treacle (syrup). In 2004, scientists analyzing data from two Earth-orbiting satellites apparently found evidence of frame dragging – they claimed to have detected the minute frame dragging effect of our planet. There is the possibility of extracting useful energy from the rotation of our planet. There is speculation that this could serve as a power source for an advanced civilization.

Is all the space occupied by us and around us really one immensely large, flexible sheet of curved space-time as is posited by the general theory of relativity? Or is there an infinitude of such “flexible sheets”? That is, is gravity just one flexible sheet of curved space-time or more, perhaps infinitely more? The Earth is a sphere but all of us are evidently “stuck to this sphere” and do not fall off it. What is the force, if it exists, which causes this? Will this force be an attractive or pulling force or a pushing force? Can gravity be a kind of fluid, whether air, gas, or liquid? (We can descend, glide, or fly up through the air like how a bird or an airplane does. We can dive into and descend in, surface in, float in, and swim in a pool of liquid. All these may be interpreted as manifestations of “gravitational effect”.) Can gravity be electrical, electronic, quantum, or nuclear? Or, is gravity none of these but something else, as Richard Feynman had speculated?

What we regard or interpret as the attractive force known as gravity may be just a facet of electromagnetism, which comprises an electric field and a magnetic field, wherein one mass attracts or repulses another mass and pulls or pushes the latter toward or away from itself (a pull/attraction and a push/repulsion may be viewed as equivalent, the former being a force exerted in front of a mass to move it while the latter is a force exerted behind the same mass to move it in the same direction). If this were the case, the weak nuclear force, strong nuclear force, and electromagnetism would be sufficient to account for the phenomena of both our macro-world, the universe, and the micro-world of the quantum particles. The gravitational wave and the quantum of gravity, the graviton, may be nonexistent and may never be found, though evidence of their existence is still being sought after; a good reason why they are not directly found (the gravitational wave is only indirectly deduced) can be that they are nonexistent, and when physical experiments fail to detect them directly (though it is thought that they are very weak and instruments are not sensitive enough to detect them) after a long period of time we should be prepared to make a turnaround instead of continuing banging the head against the wall.

There is another possibility. All space wherein both the macro-world of the universe and the micro-world of the quantum particles occupy may be a sea of invisible, yet undetected, material, or fabric, much as all space is occupied by fluid such as air, liquid, or gas. This material may be called "ether" or any other terms. This is the substance where all matter in the macro-world and quantum particles in the micro-world are theoretically submerged in (this submerging effect being manifested possibly as the gravitational effect evident in the universe), much as objects exist and perambulate in fluids such as air, liquid, or gas. This is theoretically the most basic constituent of space such that a vacuum in space is theoretically nonexistent. The ether, which has been contemplated heretofore, is apparently comparable to this material or fabric, which may be regarded as the fabric of the cosmos. This will theoretically become the link between all the forces of nature in both the macro-world and the micro-world, without which nothing can exist, similar to the case of the very basic life-giving oxygen without whose existence in the air or in liquid no animals, insects, or marine life can exist. This will probably be the spark-of-life, the life-giving, the activity-causing, the mobility-causing, the motion-causing, the behavior-causing, the action-causing, agent which causes activity, motion, change in all particles, and matter, e.g., compositional and positional/directional/gravitational changes, including possibly consciousness, which may be regarded as the essence, soul, or spirit of existence or life. Quantum particles in the micro-world apparently have a "life" of their own, e.g., light particles or photons are always in motion, like the animate objects in the macro-world. What is the cause of this apparent life in the macro-world of the universe and the micro-world of the quantum particles if not for this life-giving, action-causing essence which for want of a better term may be called "ether", "life-giver", "life-force", "life-fluid", "change agent", "activator", "essence", "energizer", or any other suitable term? The cells which make up our bodies, the bodies of animals, and the bodies of plants are apparently living ones using up oxygen to generate energy while quantum particles also generate energy. This possible agent, which we may, e.g., call "life-fluid", which probably causes all this activity in the cells and the quantum particles can be regarded as the common link between these two things found respectively in the macro-world and the micro-world. The forces in both the macro-world and the micro-world may thus be unified in this manner. However, detecting and confirming the existence of this theoretical life-giving, action-causing agent common to both the macro-world, and the micro-world is likely to be a hurdle. We can now only define it abstractly as an agent which causes activity, motion, change in particles, and matter including possibly consciousness. This theoretical agent may also be regarded as the controller of nature, the director which guides all activities in nature, like how the DNA regulates bodily functions, and the computer software regulates computer functions. Figuring out what will constitute this important agent, e.g., whether it is a form of energy, whether it is made up of some particular atoms, whether it is a sort of fluid, and so on, is a difficulty, which has to be resolved before experiments can be carried out to detect it.

3 The Nature of Light

Introduction

This article touches on the subject of whether light really needs a medium, which is termed the luminiferous ether, for its transmission as is in the case of sound which requires a fluid such as air or liquid as the medium for its transmission, and the subject of the velocity of light.

Position of Luminiferous Ether

Einstein in his relativity theory had done away with the luminiferous ether being the medium for the transmission of light. But a number of scientists today are apparently trying to resurrect the role of the luminiferous ether. Is the luminiferous ether indeed necessary for the transmission of light?

Light should not be expected to act the same as sound as they are both intrinsically different. Sound is the result of vibrations in the air (a medium of transmission for sound), vibrations which reach our eardrums at varying frequencies, i.e., sound is actually moving, vibrating air which affects our eardrums. Light has been described as a "wave/particle" object – it could be viewed as both a wave and a particle. Does it make sense to expect light too to travel through a medium like sound does? The quantum particle of light is the photon, which is rather similar to the other quantum particles such as the positron, electron, proton, and neutron, whereas sound has no quantum particle within it, being just vibrations or physical movements of the air (which affect the sensitive eardrums). Quantum particles evidently do not need a medium for their functionality. Why should quantum light need a medium such as the luminiferous ether for its transmission?

The Michelson-Morley experiment and others did not find any evidence of this medium, the luminiferous ether. And, apparently from this "null" result of the Michelson-Morley experiment it had been concluded, evidently without much justification, that the velocity of light is invariable, and that nothing could travel faster than the velocity of light, which has apparently just recently been disproved by CERN. Moreover, quantum particles have been found to be capable of teleportation, i.e., transport to another location in space instantaneously, to display "weirdness" (i.e., appear strange and incomprehensible). From all this, it could be concluded that unlike sound, which requires a medium such as air or liquid for its transmission, for light such a medium is not a necessity.

On the other hand, if the luminiferous ether exists (as the medium for the transmission of light), doesn't it have to be composed of atoms or quantum particles as well (the luminiferous ether here appears comparable, e.g., to the carrier signal which carries the picture and sound signals to our TV set all the way from the transmitter at the broadcasting station which could be many miles away)? For those who have been toying with the idea of the luminiferous ether or who are convinced that it exists, how would they describe this medium, e.g., what they are made of and whether they are composed of atoms?

Tests

The only way to determine that the luminiferous ether does indeed exist is to physically detect it through experiments, which had evidently so far not been successful, instead of theorization, and since the experiments so far had not detected it, it could be concluded that the luminiferous ether does not exist (provided that the experiments had not been faulty).

Scrutiny

However, the following important questions should be lavished with some consideration: How is light able to travel very long distances at the very great velocity of 186,000 miles per second (mi/s) in a vacuum? Is some very strong force, a yet unknown, undetected, force, perhaps, a very strong, very high frequency, carrier wave (similar to the above-mentioned carrier signal), which might be interpreted by some as the medium of transmission, carrying light along? Or, is the movement of light entirely self-propagating, i.e., without the aid of any external force? Shouldn't the velocity of light be at, above and below 186,000 mi/s at various points in time, i.e., be variable, as clocks and watches, in fact, all mechanisms, natural and artificial, do go faster or slower to varying degrees at various times, which explains why all equipment, including precision equipment, have to be calibrated from time to time in order that accuracy at the accepted level is maintained?

Superluminal Motion

A person could be deceived about the time by a not perfect clock, and the distance by also a not perfect measuring rod. All this implies that time is subjective and not absolute, or as Einstein had put it, relative, depending on the situation.

There should be a sufficient reason to explain why the clock and the brain and bodily functions of the person slow down, and the length of the measuring rod contracts in the direction of motion, on approaching the velocity of light, while at the velocity of light the mechanism of the clock and time are at a standstill and the length of the measuring rod is zero, which is important. Though the intense gravitational field caused by travel at almost the velocity of light might account for the slowing down of the clock (for which experimental evidence had been obtained) and therefore time, as well as the brain and bodily functions of a person, it apparently hardly suffices as an explanation for the contraction of the length of the measuring rod in the direction of travel at almost the velocity of light (for which experimental evidence has yet to be found). Though the constancy of the velocity of light as gauged from the Earth is apparently a well-proven phenomenon, no one has yet been able to travel at almost the velocity of light and gauge the velocity of a light beam by traveling besides it in the same direction though it has been postulated that in this instance the velocity of the light beam would remain the same. Despite the experimental findings that at high velocities, though very much less than the velocity of light, clocks slow down, the contraction of measuring rods in the direction of motion at almost the velocity of light is apparently only an inference, with no experimental basis.

The following equation describes how the velocity of light (v) is derived:

$$v = d/t$$

where d represents the distance traveled by the light beam and t represents the time taken by the light beam to travel the distance d.

Since time is relative (and not absolute) and depends on the mechanism of the clock, which slows down on approaching the velocity of light, it could be arbitrary. The clock which is used to gauge the time t taken by the light beam to travel the distance d might not slow down uniformly (at the same rate) on approaching the velocity of light (under normal, earthly conditions time varies from clock to clock by minutes or more – there is apparently some uncertainty in the mechanism of clocks). Besides, the measuring rod used to gauge the distance d traveled by the light beam in time t might not contract in length uniformly (at the same rate) on approaching the velocity of light. If the clock does not slow down uniformly (at the same rate) and the measuring rod does not contract in length uniformly (at the same rate) on approaching the velocity of light there is the probability that the velocity of light (v) as represented by d/t would be variable, higher than 186,000 mi/s at times, below 186,000 mi/s at other times, or equal to 186,000 mi/s at yet other times. Moreover, in accordance with the special theory of relativity, for the velocity of light to really remain constant, on approaching the velocity of light, the clock must slow down to the same degree as the contraction in the length of the measuring rod. We describe these possible outcomes as follows:

i) $S^{\%} > C^{\%} \rightarrow I^{1}$
ii) $S^{\%} < C^{\%} \rightarrow D^{1}$
iii) $S^{\%} = C^{\%} \rightarrow S^{1}$

where $S^{\%}$ represents percentage of slowing down of the clock, $C^{\%}$ represents percentage of contraction in the length of the measuring rod, I^{1} represents increase in the velocity of light, i.e., exceed 186,000 mi/s, D^{1} represents decrease in the velocity of light, i.e., go below 186,000 mi/s, and S^{1} represents velocity of light, i.e., 186,000 mi/s.

Since light particles (photons) do not have mass or inertia, which prevents an object possessing it from accelerating beyond the velocity of light, viz., 186,000 mi/s, theoretically there is nothing to prevent light particles (photons) or other objects without mass or inertia from traveling at a velocity greater than 186,000 mi/s.

The following equation shows that no moving object could travel faster than the velocity of light, which is in accordance with the special theory of relativity:

$$v = (a + b) \div (1 + ab/c^2)$$

If we let a = velocity of moving train, b = velocity of light beam (which is sent from the back of the moving train to the front of the moving train) with respect to the moving train, which is the moving frame (i.e., the velocity of the light beam (which is sent from the back of the moving train to the front of the moving train) is gauged from the moving train, which is the moving frame), v = velocity of light beam (which is sent from the back of the moving train to the front of the moving train) with respect to the ground level, which is the stationary frame (i.e., the velocity of the light beam (which is sent from the back of the moving train to the front of the moving train) is gauged from the ground level, which is the stationary frame), c = velocity of light = 186,000 mi/s, and also let $a = b = c$, then

$$v = (c + c) \div (1 + c.c/c^2) = 2c/2 = c! \text{ (And not } 2c!)$$

Though theoretically no object could travel faster than the velocity of light because at the velocity of light the object's mass is infinitely great and therefore it is unable to accelerate, an object without mass, possibly, a quantum particle which is somewhat similar to a photon (a photon is a quantum particle without mass always in motion) might be capable of traveling faster than the velocity of light. Such an object or objects might be waiting to be discovered. As it is, a "theoretical" particle which travels faster than the velocity of light, which is termed "tachyon", has been thought to exist.

There have been a number of speculations pertaining to the variable speed of light (VSL), e.g., one theory states that the velocity of light varies with the various stages of the evolution of the universe, exceeding 186,000 mi/s at certain points of time.

Conclusion

The importance of the above points about light cannot be denied, in view of the fact that an important tenet of the special theory of relativity, namely that no object could travel faster than the velocity of light, has just recently been contravened, due to the recent discovery of the superluminal motion of neutrinos at CERN. (This was later confirmed an error, i.e., there was no superluminal motion of neutrinos.)

4 Nature and Consciousness

Nature

First of all, to understand all the intricacies of nature would require one fundamental factor. Namely, that there be consciousness. The latter is evidently the common denominator in the mechanics of nature. Next, there are two things that could be distinguished with one's consciousness. On the one hand, there is the "truly" real (or intrinsic reality), which must exist independently of consciousness and which the consciousness is only aware of when its attention is directed to it by purpose or by chance. On the other hand, there is the "artificially" real, which necessarily exists as a result of the existence of consciousness – it is created by the conscious mind (and is actually an abstraction or concept, whichever term one might choose to describe it).

All things which are tangible could be regarded as intrinsically real as they are known by the consciousness as objects existing outside its confines. For instance, a conscious mind could see a rock, a tree, water, or fire outside itself when its eyes are opened, but does this mean that when this conscious being closes his eyes these tangible objects cease to exist? The answer is simple – ask someone to look at them and tell whether they are there or not and their existence independently of consciousness would be obvious. There is no doubt to the conscious mind that they are for real. They have objective reality.

On the other hand, intangible entities such as force, attraction, time, and gravity, for example, are not exactly real or intrinsically real but are the "creation" or "interpretation" of the conscious mind – when we see an object moving away we might infer there is a force causing this, when we see two objects moving toward one another we might infer that there is an attractive force acting on them, when we observe movement in space in general we might infer the simultaneous passage of time (as signified by movement in the four-dimensional space-time continuum of general relativity), and when we see an object falling straight down from a high perch we might infer that the object is being subject to the influence of gravity (as Newton had postulated). In other words, their existence is dependent on how the conscious mind conceives them. They could be regarded as the embodiment of subjective reality, their existence being a variable function of the conscious mind. Much of the physics now and in the past appear to fall into this category, which is manifested by the disputation and disagreement between the proponents of the different schools of physical thought.

So, reality could be classified into (at least) two types. One, the intrinsic or objective reality that is being aware of by the conscious mind as existing independently of it. Two, the nonintrinsic or subjective reality which is the object of "interpretation" by the conscious mind (an "interpretation" which could be controversial or arbitrary). Perhaps, bearing this point in mind, the difficulties and paradoxes in physics would not appear so baffling.

A Theory of Consciousness

As consciousness evidently plays an important part in nature, we should formulate a theory of consciousness. Nature could be regarded as having two basic components, namely, the nonmaterial which is consciousness and the material which consists of physical things including our physical bodies. We could do with a few simple equations to describe nature, which are as follows:

Let η represent nature, ξ represent consciousness, λ represent the nonmaterial, and μ represent the material. Then

$$\eta = \xi + \mu = \lambda + \mu \quad (1)$$

If we let φ represent internal space which is actually consciousness and γ represent external space which consists of physical things and physical phenomena, then also

$$\eta = \varphi + \gamma \quad (2)$$

The existence of μ or γ without ξ is meaningless, that is

$$\eta \neq 0 + \mu \neq 0 + \gamma \quad (3)$$

But the existence of ξ without μ or γ is evidently plausible and ξ could possibly respond to itself (ξ itself) or other internal stimuli – ξ could possibly at least think of, contemplate, influence itself whether μ or γ exists or not, possibly as some intangible, invisible or spiritual (nonmaterial) form. (Though the part of μ which ξ evidently owes its existence to, namely, the brain, and the supporting body, normally has to exist in order that this is possible. Could consciousness really exist independently of the brain, the body? Nobody seems to know though this is evidently plausible.) So it is evident that η could possibly take any of the following forms:

$$\eta = \xi + \mu = \lambda + \mu = \varphi + \gamma \quad (4)$$

$$\eta = \xi + \xi \quad (5)$$

Thus, for η to be a reality it is imperative for ξ to exist. The reality of nature comprises consciousness and the object of its (consciousness') attention together making up the whole, like the yin-yang pair, whereby consciousness is the object and the object is consciousness, consciousness and the object being one and the same thing; nature without consciousness is meaningless; if the mind were blank, unconscious there would be void or uniform, dark emptiness, or nothing; we certainly could not equate nature with nothing. Evidently, reality or existence boils down to consciousness. The permanent absence of consciousness signifies nonexistence, nonbeing, or death, in which case nature would be redundant. What would become of nature in the universe if the human race were to become extinct? Nature, the laws of nature, are but an interpretation. Interpretation is impossible without consciousness. As is implied by (5) above, consciousness is the equivalent of nature, it is existence, life. It is little wonder that scientists, for example, Sir Roger Penrose, David Bohm, and Wolfgang Pauli, have had great interest in consciousness. (David Bohm's classic book, *Wholeness and the Implicate Order*, describes succinctly the link between matter, nature, and consciousness – David Bohm had been acknowledged as the possible successor to Einstein – the person who was considered likely to succeed with the unified field theory while Einstein had failed. There are also many other good books describing this link, including up-to-date literature from the world-wide-web.)

Lest there is still some lingering doubt about the ultimate role or importance of consciousness, the reader is here entreated to ponder how statistical mechanics or quantum mechanics (including the famous Schrodinger's Cat experiment wherein there is a 50/50 chance that the cat either lives or dies when the cover is opened) could come into play without consciousness, for are not the so-called probability amplitudes of quantum particles (and the 50/50 chance the Schrodinger Cat either lives or dies) a function of consciousness? What really is probability or likelihood? Evidently, only a conscious mind or person could ever conceive such a thing as probability, which is thought, intuitive/subjective feeling – consciousness itself. However, if we were to negate the role played by consciousness, like the good skeptic, practically all the concepts in quantum mechanics, for example, Schrodinger's wave mechanics and Heisenberg's uncertainty principle, would become untenable. We should not lose sight of the fact that all the laws of nature are but the interpretations of the consciousness, minds, of some human beings. The ultimate law of nature should therefore be the law of consciousness.

All this might seem a tad-like philosophy. Actually, there is nothing wrong with that. Philosophy is a subject which is concerned with arriving at the truths, all kinds of truths including scientific truths, while science is a subject which is concerned with arriving at the truths with regards to nature, a specific kind of truths – science has in fact been described as "natural philosophy". Many brilliant scientists, for example, those mentioned above, have had an avid interest in consciousness, believing in its important role in nature. The strange thing is that consciousness is so close (inside us) and yet so far away, for it is still very much an enigma, a mystery, evidently much more so than the subjects of their contemplation, for example, the natural phenomena. There is a very important and pertinent question: Would the mystery of consciousness ever be solved and if so how soon? This is one of the greatest challenges facing mankind, if not the greatest. There is indeed a whole "internal" universe waiting for us to explore while everyone is so busy with the "external" universe; however, these "internal" universe and "external" universe are evidently intricately linked, presenting an integral whole.

5 The Constancy of the Speed of Light

Introduction

The special theory of relativity postulates that the speed of light always remains constant at 186,000 mi/s at all inertial frames. This article describes an anomaly in the standard computation pertaining to the constancy of the speed of light.

The special theory of relativity posits that a person on a moving vehicle, e.g., a very fast moving train (moving frame), traveling at almost the speed of a beam of light (moving frame) in the same direction would find the speed of the beam of light (moving frame) to be unchanged at 186,000 mi/s, instead of the difference between the speed of the very fast moving train (moving frame) and the speed of the beam of light (moving frame), which would normally be the case. This is because, according to the Special Theory, on the very fast moving train (moving frame) approaching the speed of light the clock therein used to gauge the time traveled by the beam of light (moving frame) has slowed down by the same degree (say X %) as the ruler or measuring device (stated as meter stick or measuring rod in some texts) therein used to gauge the distance traveled by the beam of light (moving frame) has contracted in length in the direction of the very fast moving train's motion (also X%), the greater the very fast moving train's traveling speed the more the clock slows down and the greater the length contraction of the ruler or measuring device. This is expressed in the following equation (the speed of the beam of light (moving frame) being the distance it traveled divided by the time it took to travel this distance), which is in accordance with the special theory of relativity:

$$(186,000 \text{ miles} - \text{X\% of } 186,000 \text{ miles}) \div (1\text{ s} - \text{X\% of } 1\text{ s}) = 186,000 \text{ miles per second}$$

In other words, there has to be a same percentage decrease in the time gauged and the distance gauged due to the respective slowing down of the clock and contracting in length of the ruler or measuring device therein the very fast moving train (moving frame), in order for the speed of the beam of light (moving frame) to remain constant, which is consistent with mathematical logic – this condition must indeed apply in order for the speed of the beam of light (moving frame) to remain constant.

There is however something not quite usual related to the above concept. According to the special theory of relativity, the person on the very fast moving train traveling at almost the speed of light (moving frame) gauging the speed of the beam of light traveling in the same direction (moving frame) would not notice that the clock on his very fast moving train (moving frame) has slowed down and the ruler or measuring device therein has contracted in length in the direction of motion. In other words, everything would appear normal to him despite the fact that his clock has actually slowed down and his ruler or measuring device has actually contracted in length in the direction of motion, as is postulated by the special theory of relativity. But, according to the special theory of relativity, when he (moving frame) compares himself to a person on the ground who is not moving (stationary frame), he could even consider himself stationary (stationary frame) while thinking that the person on the ground (who is not moving) is actually moving (moving frame), i.e., all movements are relative. He (moving frame) would notice that the clock on the ground (stationary frame) is slower and the ruler or measuring device on the ground (stationary frame) is shorter. The person on the ground who is not moving (stationary frame) would also notice that the clock on the very fast moving train (moving frame) is slower, the ruler or measuring device therein is shorter, and the length of the very fast moving train (moving frame) is shorter. In other words, both the train-traveler (moving frame) and the person on the ground who is not moving (stationary frame) would notice that the other's clock is slower and the other's ruler or measuring device is shorter, and according to the special theory of relativity, the slowing down of clocks and the shortening of rulers or measuring devices would appear to be by the same degree (X%) for both.

But, it is actually the clock on the very fast moving train traveling at almost the speed of the beam of light in the same direction (moving frame) which has slowed down and the ruler or measuring device therein which has contracted in length (in the direction of motion) as is postulated by the special theory of relativity and not those on the ground (stationary frame). To the traveler on the very fast moving train (moving frame) who is gauging the speed of the beam of light traveling in the same direction (moving frame), the beam of light (moving frame) appears to take less time (time dilation) to travel a shorter distance (length contraction), which, according to the special theory of relativity and in accordance with the following equation, explains the constancy of the speed of light at all inertial frames:

$$(186,000 \text{ miles} - \text{X\% of } 186,000 \text{ miles}) \div (1 \text{ s} - \text{X\% of } 1 \text{ s}) = 186,000 \text{ miles per second}$$

The speed of the beam of light (moving frame) is obtained by dividing the distance traveled by the beam of light (as gauged by the ruler or measuring device on the very fast moving train traveling at almost the speed of light – moving frame) by the time it took to travel that distance by the beam of light (as gauged by the clock therein the very fast moving train – moving frame), the gauging being carried out by the traveler on the very fast moving train (moving frame). In the above example, the clock on the very fast moving train traveling at almost the speed of light (moving frame) slows down and gauges a slower time, X% slower. The ruler or measuring device therein also contracts in length in the direction of motion by X%; however because of this it should gauge any object as "longer" due to a change in the scale of the ruler or measuring device on contracting in length in the direction of motion (see Appendix 1), and this evidently gives rise to an anomaly when computing the speed of the beam of light (moving frame). This anomaly in the computation of the speed of light is described in the example below.

The person on the very fast moving train (moving frame) traveling at almost the speed of the beam of light (moving frame) in the same direction could be considered one inertial frame, and the ground level (stationary frame) could be considered another inertial frame. If, traveling at almost the speed of light, both his clock has slowed down and his ruler or measuring device has contracted in length in the direction of motion by the same degree or percentage, say 50%, the beam of light (moving frame) whose speed he is gauging, according to the special theory of relativity, would also appear to him to have traveled a distance (from one designated point to another designated point – reference, stationary frame) which is shortened by 50%, i.e., the distance between the two designated points (reference, stationary frame) appears to him to have shortened by 50%.

Let us look at the following self-explanatory diagrammatic example, which contains the abovesaid anomaly:

Gauging of the Speed of a Beam of Light at Almost the Speed of Light when, for example, the Clock Has Slowed Down by 50% and the Ruler Has Contracted in Length in the Direction of Motion by 50%

How do we gauge the speed of a beam of light (moving frame) while traveling at almost the speed of light on a very fast moving train (moving frame) in the same direction? We do so as follows:

a) The beam of light (moving frame) travels from one point to another point, say, from Point A to Point B (the distance between Point A and Point B being a reference, stationary frame next to the beam of light, say, on the embankment next to the railway tracks).

b) Let us say that the distance from Point A to Point B (reference, stationary frame) is 1 m and the beam of light (moving frame) takes x second to travel this 1 m from Point A to Point B (reference, stationary frame), i.e., the speed of the beam of light (moving frame) is 1 m per x second. In other words, as gauged from the ground level (stationary frame) the speed of the beam of light (moving frame) traveling from Point A to Point B (reference, stationary frame) is 1 m per x second, which is depicted as follows:

beam of light travels 1 m from Point A to Point B (speed is 1 m
--> per x second)
I---I x second (before clock
<-------------------------- 1 m --------------------------> slows down 50%)
(length of ruler before 50% length contraction)

1-m ruler	(ruler before 50% length contraction)

c) Next, the beam of light (moving frame) is gauged from a very fast moving train (moving frame) traveling in the same direction as the beam of light (moving frame), at almost the speed of light, when, say, the clock used to time the distance traveled by the beam of light (moving frame) from Point A to Point B (reference, stationary frame) on the embankment besides the railway tracks has slowed down by 50%, the 1-m-long ruler used to measure the distance traveled has contracted in length in the direction of motion by 50%, and the distance traveled by the beam of light (moving frame) from Point A to Point B (reference, stationary frame) has also shortened by the same degree, 50% (in the eyes of the train-traveler (moving frame) gauging the speed of the beam of light (moving frame), which is in accordance with the special theory of relativity) – i.e., the distance between Point A and Point B (reference, stationary frame) on the embankment besides the railway tracks has shortened from 1 m by 50% to 0.5 m as seen by the train-traveler (moving frame) who is gauging the speed of the beam of light (moving frame), which is in accordance with the special theory of relativity. Note that, in accordance with the special theory of relativity, the train-traveler (moving frame) does not notice that his clock has slowed down by 50% and his 1 m-long ruler has contracted in length in the direction of motion by 50% while his train is moving at almost the speed of light. All this is depicted as follows:

beam of light (contracts in length by 50% in the eyes of the train-traveler – speed is now 0.5 m per **1/8** x second or 4 m per x second)
-------------->
I--I ½ x second (after
<-------------------------------- 1 m ------------------------------> clock slows down
(length of ruler before 50% length contraction) by 50%, but train-
traveler does not
notice it,
= x second ÷ 2)

1-m ruler (ruler after 50% length contraction
which train-traveler does not notice)

I----------------------------------I ¼ x second (after clock slows down by 50%,
<----- reads "1 meter" -----> but train-traveler does not notice it, = ½ x
(length of ruler second ÷ 2)
after 50% length contraction = 50% of original length)
I-----------I **1/8** x second (after clock slows down by 50%, but train-
<-0.5 m-> traveler does not notice it, = ¼ x second ÷ 2)

(train-traveler sees this 0.5 m as the distance from Point A to Point B traveled by the beam of light after the distance has shortened by 50% in accordance with the special theory of relativity).

The train-traveler (moving frame) obtains the speed of the beam of light (moving frame) by dividing the distance between Point A and Point B (reference, stationary frame, which in his eyes, in accordance with the special theory of relativity, has contracted by 50% from 1 to 0.5 m) by the time the beam of light (moving frame) takes to travel this distance (0.5 m).

As is evident from the above diagrams, the train-traveler's clock, which has slowed down by 50%, would time the distance (0.5 m now) traveled by the beam of light (moving frame) 50% slower of course; it would gauge the time the beam of light (moving frame) takes to travel the distance of 0.5 m from Point A to Point B (reference, stationary frame) as **1/8** x second. This translates into a speed of 4 m per x second for the beam of light (moving frame), i.e., four times the speed of light, instead of just the speed of light which Special Relativity postulates. (The above description might be difficult to comprehend. For an alternative description, see Appendix 2.)

This is an anomaly and it contradicts an important postulate of the special theory of relativity, namely, the constancy of the speed of light at all inertial frames. All this requires attention and correction.

In order for the speed of light to remain constant in this instance, i.e., remain at 1 m per x second, one of the following has to happen:

1) When the clock slows down by 50%, the ruler should increase in length by 100%.
2) When the ruler decreases in length by 50%, the clock should quicken by 100%.
3) When the clock slows down by 50% and the ruler decreases in length by 50%, the beam of light (moving frame) should slow down by 400%.

Appendix 1

For example, we have an atlas of size X feet in length by Y feet in breadth with a scale of 1 is to 100,000 (1 in on the atlas represents 100,000 in on the actual ground) which shows only a portion of our globe. To have the whole globe represented in this atlas X feet in length by Y feet in breadth, it is evident that we have to change its scale, e.g., change its scale to 1 is to 1,000,000 (1 in on the atlas represents 1,000,000 in on the actual ground). The above-mentioned ruler or measuring device which has contracted in length is analogous to an atlas whose scale has been changed to allow it to represent a larger area (i.e., the whole globe), e.g., from 1 is to 100,000 to 1 is to 1,000,000 – with the new scale of 1 is to 1,000,000, 1 in on the atlas now represents a length of 1,000,000 in instead of 100,000 in on the actual ground. If we change the scale of the above-mentioned atlas which shows only a portion of our globe to 1 is to 1,000,000 from 1 is to 100,000, we would have a much smaller atlas showing the same portion of our globe, whose dimensions would be 1/10 X feet in length by 1/10 Y feet in breadth, a contracted atlas which is analogous to the above-mentioned ruler or measuring device which has contracted in length. A 1-m-long ruler which has contracted in length by 50%, e.g., would now gauge the length of 1 m as 2 m (and not 0.5 m in accordance with the special theory of relativity), and this is evidently the cause of the abovedescribed anomaly relating to the computation of the speed of light. In fact, for the 1-m-long ruler to gauge the length of 1 m as 0.5 m it would have to increase in length by 100%. A shortened ruler would gauge an object as "longer" while a lengthened ruler would gauge an object as "shorter".

Let us look at a simple example here. Say, the 1-m-long ruler used to gauge distance has contracted in length in the direction of motion by 20%. The clock used to gauge time, which has also slowed down by 20%, according to the special theory of relativity, would now gauge the time taken, say X, to travel the distance between two designated points (reference, stationary frame), say Y, as having decreased by 20% to become 0.8 X. Though the 1-m-long ruler, which has contracted in length by 20%, still reads "1 meter" in length, it is in effect shorter by 20% (actually only 0.8 m in length). Therefore, when it gauges the distance traveled, Y, above, this distance Y would now be gauged as 1.25 Y, and not 0.8 Y in accordance with the special theory of relativity. As stated above, the special theory of relativity theorizes that for a beam of light (moving frame) to remain constant in speed the beam of light (moving frame) has to take less time (time dilation) to travel a shorter distance (length contraction) – in effect, X % less time to travel a distance shorter by X%, in accordance with the following equation, which implies that the speed of the beam of light (moving frame) would remain constant, e.g., 0.8 X (time) to travel 0.8 Y (distance) after "time dilation" and "length contraction":

$$(186,000 \text{ miles} - \text{X\% of } 186,000 \text{ miles}) \div (1 \text{ s} - \text{X\% of } 1 \text{ s}) = 186,000 \text{ miles per second}$$

But, as explained above, this would not be the case; the beam of light (moving frame) would have been gauged as having taken 0.8 X (time) to travel 1.25 Y (distance). This is an anomaly in the special theory of relativity.

Appendix 2

Alternative Description – Gauging of the Speed of a Beam of Light at Almost the Speed of Light when, for example, the Clock has Slowed Down by 50% and the Ruler Has Contracted in Length in the Direction of Motion by 50%

The description below is based on the following conditions ((i), (ii), (iii), (iv), and (v)) which are related to the special theory of relativity:

i) Clocks slow down and rulers contract in length in the direction of motion, at almost the speed of light, but these would not be noticed by the person traveling at almost the speed of light (moving frame) who possess such clocks and rulers. These do not apply to the clock and ruler of the person on the ground who is not moving (stationary frame).
ii) The train-traveler traveling at almost the speed of light (moving frame) and the person on the ground who is not moving (stationary frame) would notice one another's clock and ruler as respectively slower and shorter than his own. Both of them would each notice the other's clock slowing down and ruler length shorten-

ing to the same degree (X% for the clock and ruler of the train-traveler traveling at almost the speed of light (moving frame) and also X% for the clock and ruler of the person on the ground who is not moving (stationary frame)). Though the train-traveler's clock has slowed down and his ruler has contracted in length in the direction of motion while he is traveling at almost the speed of light, as is described in (i) above, he would not notice these happenings. All this does not apply to the clock and ruler (which are not affected by (i) above) of the person on the ground who is not moving (stationary frame).

iii) All movements are relative. When the train-traveler traveling at almost the speed of light (moving frame) compares himself to a person on the ground who is not moving (stationary frame), he could consider himself stationary (stationary frame) while thinking that the person on the ground (who is not moving) is actually moving (moving frame) – both parties could each regard themselves as stationary (stationary frame) and consider the other party in motion (moving frame).

iv) The train-traveler traveling at almost the speed of light (moving frame) would see the distance from Point A to Point B at the embankment besides the railway tracks (reference, stationary frame – you might substitute this distance with a ruler) as having shortened. This shortening (by X%) of the distance from Point A to Point B (reference, stationary frame) would be by the same degree as his own clock has slowed down (X%) and his own ruler has contracted in length in the direction of motion (X%), both of which he does not notice. This point is the equivalent of Point (ii) above.

v) The speed of the beam of light (moving frame) is obtained by dividing the distance traveled by the beam of light (as gauged by the ruler on the very fast moving train traveling at almost the speed of light – moving frame) by the time it took to travel that distance by the beam of light (as gauged by the clock therein the very fast moving train – moving frame), the gauging being carried out by the traveler on the very fast moving train (moving frame), the greater the very fast moving train's traveling speed the more the clock slows down and the greater the length contraction of the ruler. For the speed of the beam of light (moving frame) to be measured as constant, the clock in the train traveling at almost the speed of light (moving frame) used to gauge the time traveled by the beam of light (moving frame) has to gauge a time which has slowed down by the same degree (X%) as the distance traveled by the beam of light (moving frame) which has shortened (also X%) as is gauged by the ruler therein.

1) Before the distance from Point A to Point B (reference, stationary frame) has shortened by 50% (as is gauged at the ground level – stationary frame):
   a) Distance from Point A to Point B (reference, stationary frame) = 1 m
   b) Time taken by the beam of light (moving frame) to travel this 1 m = x second, i.e., speed of the beam of light (moving frame) = *1 m per x second*

The above distance of 1 m in (a) is gauged at the ground level (stationary frame) with a ruler which has not contracted in length (by 50%, as yet). The above time of x second in (b) is gauged at the ground level (stationary frame) with a clock which has not slowed down (by 50%, as yet).

2) a) After the ruler on the moving train (moving frame) has contracted in length in the direction of motion by 50%, it would gauge the above-mentioned 1 m (distance from Point A to Point B (reference, stationary frame), before shortening) as *2 m*. (A shortened ruler gauges an object as "longer" while a lengthened ruler gauges an object as "shorter", due to the different scales, as is explained in Appendix 1.)
   b) After the clock on the moving train (moving frame) has slowed down by 50%, at the same time that the ruler has contracted in length by 50%, it would gauge the above-mentioned x second taken to travel this measured distance of 2 m as *1/2 x second.*
   c) That is, the beam of light (moving frame) is now gauged from the moving train (moving frame) as requiring 1/2 x second to travel the distance of 2 m (before the shortening of the distance between Point A and Point B (we assume here that the distance between Point A and Point B has not shortened) – in accordance with the special theory of relativity this distance would also be shortened (also by 50%), as is described below); this translates into a speed of 4 m per x second or 4 X speed of light.
3) After the distance from Point A to Point B has shortened by 50%, as is seen by the train-traveler traveling at almost the speed of light (moving frame), in accordance with the special theory of relativity:
   a) The 1 m ruler which has contracted in length in the direction of motion by 50% would gauge the above-mentioned 50% shortened distance from Point A to Point B as 50% shorter than before (i.e., 0.5 m instead of 1 m) and would *read "0.5 meter"* (instead of "1 meter" – full length of the 1 m ruler) on its 50% shortened length, which is in accordance with the special theory of relativity. (The ruler now has a different scale as compared to its scale before length contraction. See Appendix 1 for explanation.)
   b) When the scale of the ruler has changed due to the ruler's length contraction (by 50%), and the clock has slowed down (by 50%), the beam of light (moving frame) being gauged as requiring 1/2 x second to travel the distance of 2 m, as is described in (2) above, the time taken by the beam of light (moving frame) to travel 0.5 m now = (0.5 m ÷ 2 m) x 1/2 x second = 1/8 x second, i.e., the speed of the beam of light (moving frame) is now 0.5 m per 1/8 x second, or *4 m per x second/4 X speed of light*, which is an anomaly (the speed of the beam of light should have remained 1 m per x second/1 X speed of light, as is postulated by the special theory of relativity).

The above computations have been carried out in accordance with the conditions which are related to the special theory of relativity. The said anomaly is evident due to the change of the scale of the ruler which has contracted in length in the direction of motion by 50%.

As for faster than light travel, it should be noted that quantum particles are capable of teleportation or travel to another location instantaneously.

6 The Special Theory of Relativity: A Special View

Introduction

According to Einstein's special theory of relativity, the speed of light always remains constant at 186,000 mi/s regardless of whether it is gauged from a stationary reference point, a moving reference point or any other reference points, no object could travel faster than 186,000 mi/s (the speed of light itself) because the mass of the object would then be so great (infinitely great) that it could not accelerate anymore, on approaching the speed of light a moving object contracts in length in the direction of motion while a clock gauging the time slows down, at the speed of light the length of the moving object contracts to zero while the clock (and time) becomes at a standstill, and importantly, the mass of an object multiplied by the square of the speed of light gives energy ($E = MC^2$), i.e., mass could be converted to energy and vice versa; on approaching the speed of light the brain and bodily functions of a person slow down; observers do not agree on the simultaneity of events – two simultaneous events for one observer might not be simultaneous for another; in the Special Theory time travel (in the space-time continuum) is an apparent possibility. A deeper look at the special theory of relativity is presented in this article, employing some strong, subtle, and important mathematical reasoning in the process.

Relativity and the Speed of Light

The special theory of relativity posits that on approaching the speed of light clocks slow down, moving objects contract in length in the direction of motion, and a person's brain and bodily functions slow down. For example, a person on a moving vehicle (moving frame), which travels besides a beam of light (moving frame) in the same direction and almost as fast as the beam of light (moving frame) itself, gauges the speed of the beam of light (moving frame). The Special Theory postulates that this person on the moving vehicle (moving frame) traveling at almost the speed of the beam of light (moving frame) would find the speed of the beam of light (moving frame) to be unchanged at 186,000 mi/s, instead of the difference between the speed of the moving vehicle (moving frame) and the speed of the beam of light (moving frame), which would normally be the case; this is because, according to the Special Theory, on the moving vehicle (moving frame) approaching the speed of light the clock therein used to gauge the time traveled by the beam of light (moving frame) has slowed down by the same degree (say X%) as the ruler or measuring device therein used to gauge the distance traveled by the beam of light (moving frame) has contracted in length in the direction of the vehicle's motion (also X%), the greater the moving vehicle's traveling speed the more the clock slows down and the greater the length contraction of the ruler or measuring device. This is expressed in the following equation, which is in accordance with the special theory of relativity:

$$(186,000 \text{ miles} - X\% \text{ of } 186,000 \text{ miles}) \div (1 \text{ s} - X\% \text{ of } 1 \text{ s}) = 186,000 \text{ miles per second}$$

A person traveling at almost the speed of light (as in the case of the person in the above-mentioned vehicle traveling at almost the speed of light (moving frame) besides a beam of light (moving frame)) would experience the slowing down of time and age more slowly (without being aware of it), as postulated by the special theory of relativity. This "time dilation" effect is described by the following equation:

$$t = t' / \sqrt{1 - v^2/c^2}$$

where t is the time gauged by a clock at the ground level (stationary frame), t' the time gauged by a clock on the above-mentioned vehicle traveling at almost the speed of light (moving frame), c the speed of light = 186,000 mi/s; v is the speed of the above-mentioned vehicle traveling at almost the speed of light (moving frame), e.g., 0.9c; t' would be a fraction of t, i.e.,

$t' < t$, e.g., if $t' = 24$ years, $v = 0.8c$, then $t = 40$ years.

But (and this is a very important "but"), if the moving vehicle's clock had not slowed down and its ruler or measuring device had not contracted in length (i.e., under normal conditions), the speed of the light beam (moving frame), as gauged from the vehicle traveling besides it at almost the speed of light (moving frame), would have had been the difference between the speed of the light beam (moving frame) and the speed of the moving vehicle (moving frame), e.g., 186,000 mi/s (speed of the light beam) minus 185,990 mi/s (speed of the moving vehicle), which is equal to 10 mi/s.

Thus, the speed of the light beam (moving frame), i.e., 186,000 mi/s, as gauged from the vehicle traveling besides it at almost the same speed (moving frame) in the same direction is evidently an illusion, which is the result of the slowing down of the clock, and the brain and bodily functions of the person, on the moving vehicle (moving frame), and the contraction in the length of the ruler or measuring device (in the direction of motion) therein, as postulated by the special theory of relativity. This is somewhat similar to the situation, whereby a driver in a car which is actually cruising at 60 mi/h believes that his car is traveling at 30 mi/h because the car's speedometer, which happens to be faulty, gives a reading of the car's cruising speed as 30 mi/h instead of 60 mi/h, i.e., the driver is misled by the car's faulty speedometer.

In the above-mentioned case, the person on the vehicle traveling at almost the speed of light (moving frame), say a spaceship, would not notice that his clock is ticking more slowly, time is passing more slowly for him, he is aging more slowly, and the length of the ruler or measuring device on his spaceship (moving frame) has contracted in the direction of motion (because there is nothing to compare with). According to the special theory of relativity, when this traveler on the spaceship traveling at almost the speed of light (moving frame) looks at a clock on the Earth (stationary frame) he would perceive that the clock has slowed down and when he looks at a ruler or measuring device on the Earth (stationary frame) he would perceive that it is shorter. But, this is evidently only an illusion and not true, and the clock ticking away on the Earth (stationary frame) is actually ticking more quickly (which implies that time is passing more quickly) as compared to the traveler's clock on the spaceship traveling at almost the speed of light (moving frame), and the ruler or measuring device on the Earth (stationary frame) is actually longer as compared to the traveler's ruler or measuring device on the spaceship traveling at almost the speed of light (moving frame) – according to the special theory of relativity, the clock on the spaceship traveling at almost the speed of light (moving frame) has slowed down and both the length of the spaceship traveling at almost the speed of light (moving frame) and the length of the ruler or measuring device on this spaceship (moving frame) have contracted in length in the direction of motion, whilst the clock on the Earth (stationary frame) has not slowed down and the ruler or measuring device on the Earth (stationary frame) has not contracted in length.

According to the special theory of relativity, the person on the Earth (stationary frame) who takes a peep at the clock on the spaceship traveling at almost the speed of light (moving frame) would perceive that the clock ticks more slowly than his clock on the Earth (stationary frame). He would also perceive that the ruler or measuring device on the spaceship traveling at almost the speed of light (moving frame) is shorter than his ruler or measuring device on the Earth (stationary frame), and that the spaceship is foreshortened in the direction of its motion. The person on the Earth (stationary frame), according to the special theory, would see the clock on the spaceship traveling at almost the speed of light (moving frame) slowed down to the same degree as the traveler on the spaceship traveling at almost the speed of light (moving frame) sees the clock on the Earth (stationary frame) slowed down and see the length of the ruler or measuring device on the spaceship traveling at almost the speed of light (moving frame), as well as the spaceship's length, shortened to the same degree as the traveler on the spaceship traveling at almost the speed of light (moving frame) sees the ruler or measuring device on the Earth (stationary frame) shortened. That is, each of them, the person on the Earth (stationary frame), and the traveler on the spaceship traveling at almost the speed of light (moving frame), according to the Special Theory, would measure the above-mentioned differences in the other to the same degree, and the other thing each of them would agree on is the constancy of the speed of light (the speed of light would remain constant at 186,000 mi/s).

But, actually, on the Earth (stationary frame), the clock has not slowed down and time passes more quickly, and the length of the ruler or measuring device has not contracted, whereas on the spaceship traveling at almost the speed of light (moving frame), the reverse, as described earlier, is true. The dilemma is which time is to be regarded as the actual or correct time – the time on the Earth (stationary frame) or the time on the spaceship traveling at almost the speed of light (moving frame) – and, which length is to be regarded as the actual or correct length – the length of the ruler or measuring device on the Earth (stationary frame) or the length of the ruler or measuring device on the spaceship traveling at almost the speed of light (moving frame)?

According to Einstein, time is relative, i.e., there is no such thing as absolute or actual time, which is a Newtonian concept; Einstein had done away with Newton's concept of absolute time and space. Time depends on the speed of the clock, and the brain and bodily functions – the nearer the speed of light is approached the slower would be the clock and the passage of time as well as the brain and bodily functions. What does all this imply? Without a clock and consciousness it would not be possible to tell the time – time would not exist, i.e., the reality of time is not independent of the clock and consciousness. In fact, without the clock or watch everyone would be disoriented where the time is concerned.

However, like the case of the car driver misled by his car's faulty speedometer which is described above, the traveler on the spaceship traveling at almost the speed of light (moving frame) described above has also been misled into thinking that his clock is ticking normally and his time is passing normally, and his ruler or measuring device is normal and gauging lengths or distances normally. It is only after taking a peep at the clock ticking away and the ruler or measuring device on the Earth (stationary frame) that he would realize this might not be so. After having taken a peep at the clock ticking away and the ruler or measuring device on the Earth (stationary frame) and perceiving that the clock on the Earth (stationary frame) is ticking more slowly and the ruler or measuring device there is shorter (which is in accordance with the special theory of relativity), he should suspect that either his clock on his spaceship traveling at almost the speed of light (moving frame) or the clock on the Earth (stationary frame), and either his ruler or measuring device on his spaceship (moving frame) or the ruler or measuring device on the Earth (stationary frame) are not quite right. If he finds out that in fact his clock on his spaceship traveling at almost the speed of light (moving frame) has been ticking more slowly, time has been passing more slowly for him, he has been aging more slowly and the length of the ruler or measuring device on his spaceship traveling at almost the speed of light (moving frame) has contracted in the direction of his spaceship's motion, due to some change in the physical environment, viz., the presence of an intense gravitational field which is created through travel at almost the speed of light, he would realize that the accuracy of the time presented by his clock on his spaceship traveling at almost the speed of light (moving frame) as well as the length or distance measurements made by the ruler or measuring device on his spaceship (moving frame) are out. He would then think that perhaps the time presented by the clock ticking away on the Earth (stationary frame) and the length or distance measurements made by the ruler or measuring device there are the actual or correct time and length or distance measurements. But, he might not be able to find out that his clock has been ticking more slowly, time has been passing more slowly for him, he has been aging more slowly and the length of the ruler or measuring device on his spaceship traveling at almost the speed of light (moving frame) has contracted, due to some change in the physical environment, viz., the presence of an intense gravitational field, and, even if he found out that the clock on the Earth (stationary frame) is actually ticking more quickly than his and the ruler or measuring device there has actually not contracted in length and is actually longer than his (which is in accordance with the special theory of relativity) he might not understand why and might be puzzled as to whether his clock and ruler or measuring device on his spaceship traveling at almost the speed of light (moving frame) or the clock and ruler or measuring device on the Earth (stationary frame) are the clock and ruler or measuring device which are faulty (assuming he has no knowledge of the special theory of relativity).

If neither of the two parties (on the Earth (stationary frame) and on the spaceship traveling at almost the speed of light (moving frame)) had been aware that their respective clocks had been ticking away at different speeds, and the lengths of their respective rulers or measuring devices had been different, then each of them would have regarded the times shown by their respective clocks as the actual time and the lengths displayed by their respective rulers or measuring devices as the actual length, in which case there would be two sets of actual time and actual length, i.e., two sets of reality, a quite absurd situation.

As the third party looking on at the two cases described above and being aware of the circumstances, we could regard the time presented by the clock on the Earth (stationary frame) as the actual or correct time and the time presented by the clock on the spaceship traveling at almost the speed of light (moving frame) as the distorted time. Also, as the third party who is all too familiar with the special theory of relativity we could regard the length of the ruler or measuring device on the Earth (stationary frame) as the actuality and the length of the ruler or measuring device on the spaceship traveling at almost the speed of light (moving frame), which has contracted in the direction of the spaceship's motion, as the distorted length.

We here consider the example of two spaceships traveling almost next to one another in the same direction, one (we call it X, which is a moving frame) traveling at almost the speed of light, say, 185,000 mi/s (as gauged from the Earth, a stationary frame) and the other (we call it Y, which is another moving frame) traveling also at almost the speed of light, say, 185,500 mi/s (as gauged from the Earth, a stationary frame). (Theoretically, no spaceship could travel at the speed of light – the special theory of relativity postulates that at the speed of light everything would be at a standstill – the mass of the spaceship would be infinite and the spaceship would not be able to accelerate anymore, the spaceship's length would have shrunk to zero and any clock within the spaceship would have stopped beating, registering zero time.) The speed of Y (moving frame) as gauged from X (moving frame) or vice versa is computed by using the following formula, as is postulated by the special theory of relativity:

$$v = (b - a) \div (1 - ba/c^2)$$

where c = speed of light = 186,000 mi/s, b = speed of Y = 185,500 mi/s (= 0.9973118c), and a = speed of X = 185,000 mi/s (= 0.9946236c)

$$\therefore\ v = 0.9973118c - 0.9946236c/1 - 0.9973118c \times 0.9946236c/c^2$$

$$= 0.0026882c/1 - 0.9919498c^2/c^2$$

$$= 0.0026882c/0.0080502$$

$$= 0.3339295c\ (33.39295\%\ \text{of speed of light})$$

$$= 0.3339295 \times 186{,}000\ \text{miles per second}$$

$$= 62{,}110.887\ \text{miles per second}$$

∴ speed of Y as gauged from X = plus 62,110.887 mi/s (and not plus 500 mi/s (185,500 mi/s minus 185,000 mi/s), which should normally be the case – Y (moving frame) should appear to the traveler in X (moving frame) to be moving away from X (moving frame) in the same direction.

∴ speed of X as gauged from Y = minus 62,110.887 mi/s (and not minus 500 mi/s (minus (185,500 mi/s minus 185,000 mi/s)), which should normally be the case – X (moving frame) should appear to the traveler in Y (moving frame) to be moving away from Y (moving frame) in the opposite direction.

What would be X's and Y's respective speeds then (when gauged from the other), when X (moving frame) and Y (moving frame) travel in opposite directions (instead of the same direction)? The speed of Y (moving frame) as gauged from X (moving frame) and the speed of X (moving frame) as gauged from Y (moving frame) should each not exceed 186,000 mi/s, the speed of light, which represents the ultimate limit, the maximum possible speed any accelerating object could attain, as is postulated by the special theory of relativity (and not respectively 370,500 mi/s (185,000 mi/s plus 185,500 mi/s), which should normally be the case), and they are computed by using the following formula (which is described further on), which is in accordance with the special theory of relativity:

$$v = (a + b) \div (1 + ab/c^2)$$

where c = speed of light = 186,000 mi/s, a = speed of X = 185,000 mi/s (= 0.9946236c), and b = speed of Y = 185,500 mi/s (= 0.9973118c)

$$\therefore \text{v} = 0.9946236c + 0.9973118c/1 + 0.9946236c \times 0.9973118c/c^2$$
$$= 1.9919354c/1 + 0.9919498c^2/c^2$$
$$= 1.9919354c/1.9919498$$
$$= 0.9999927c \text{ (99.99927\% of speed of light)}$$
$$= 0.9999927 \times 186,000 \text{ miles per second}$$
$$= 185,998.64 \text{ miles per second}$$

∴ speed of Y as gauged from X = speed of X as gauged from Y = 185,998.64 mi/s (Y (moving frame) should appear to the traveler in X (moving frame) to be moving toward X (moving frame) in the opposite direction, and X (moving frame) should appear to the traveler in Y (moving frame) to be moving toward Y (moving frame) in the opposite direction).

How strange and counterintuitive it is to find the speeds of X (moving frame) and Y (moving frame) to be minus 62,110.887 mi/s and plus 62,110.887 mi/s, respectively, as gauged from Y (moving frame) and X (moving frame), respectively (and not minus 500 mi/s and plus 500 mi/s, respectively, which should normally be the case) in the first case above, and to be each only 185,998.64 mi/s (less than the speed of light (186,000 mi/s) and not, respectively, 370,500 mi/s (185,000 mi/s plus 185,500 mi/s), which should normally be the case) in the second case above, one may think. Evidently, the respective clocks in X (moving frame) and Y (moving frame) were slowing down at different speeds and the respective rulers or measuring devices in X (moving frame) and Y (moving frame) were contracting in length to different extents since the respective speeds of X (moving frame) and Y (moving frame) are different, viz., 185,000 mi/s and 185,500 mi/s, respectively (the higher the speed of the spaceship the more its clock would slow down and the more the length of its ruler or measuring device would contract). As postulated by the special theory of relativity, both the travelers in X (moving frame) and Y (moving frame) would each see the other's clock as being slower to the same degree and the other's ruler or measuring device as being shorter to the same degree. The dilemma here is to decide whether the clock on X (moving frame) or the clock on Y (moving frame) is giving the correct reading in time and whether the ruler or measuring device on X (moving frame) or the ruler or measuring device on Y (moving frame) is providing the correct measurement in the distance traveled/measured. It is evidently very difficult to decide thus. The travelers in X (moving frame) and Y (moving frame) would each naturally think that everything is fine and consider their respective gauging of the other's speed as correct (assuming that they have no knowledge at all about the special theory of relativity). However, if the travelers in X (moving frame) and Y (moving frame) noticed that the other's clock had been slower and the other's ruler or measuring devices had been shorter, they might each be puzzled and might each wonder whether whose clock and ruler or measuring device are accurate (assuming that they have no knowledge of the special theory of relativity). Of course, if they had known the principles behind the special theory of relativity they would have realized that this phenomenon had been the result of "distortion" due to the creation of an intense gravitational field through travel at almost the speed of light, i.e., the slowing down of their respective clocks and the contraction in the lengths of their respective rulers or measuring devices are transient (they are not permanent – X's and Y's respective clocks would beat at the normal rate and the lengths of their respective rulers or measuring devices would return to their original length once the speeds of X (moving frame) and Y (moving frame) have returned from almost the speed of light (185,000 mi/s and 185,500 mi/s, respectively) to the normal speeds, according to the special theory of relativity.

The important question is if the spaceships', X's and Y's, times and length or distance measurements are "distorted" or not real, what should be the real time and real length or distance measurement? However, to both the travelers in spaceships X (moving frame) and Y (moving frame), without being able to compare or look at one another's clock and ruler or measuring device and without any knowledge of the special theory of relativity, the time given by their respective clock and the length or distance measurement given by their respective ruler or measuring device would be the real time and real length or distance measurement. But, to us, the third party looking on at these two scenarios, who have knowledge of the special theory of relativity, both X's and Y's "real" times and "real" length or distance measurements are illusions and are indeed not real, and the real time and real length or distance measurement would be those read off a clock and a ruler or measuring device on the Earth (stationary frame), where the clock and the ruler or measuring device are free from the "distortional" effect of the intense gravitational field created through travel at almost the speed of light. Since to the traveler on spaceship X (moving frame), the traveler on spaceship Y (moving frame) and to us on the Earth (stationary frame), our times and length or distance measurements are "real" to each of us, it is implied that time and length or distance measurement are relative, i.e., time and length or distance measurement depend on environmental or situational factors.

All this appears to be a case of how we choose to interpret these three scenarios. For example, if we put ourselves in the X traveler's shoes, have no knowledge that Y (moving frame) exists or if we know that Y (moving frame) exists we have no knowledge that Y's clock is running at a different pace and that Y's ruler or measuring device is gauging length or distance differently from our ruler or measuring device and have no knowledge that our clock has slowed down and that our ruler or measuring device has shrunk in length, i.e., we have no knowledge of the special theory of relativity, then we would just think that our time and length or distance measurement in X (moving frame) are the real things. The same applies if we put ourselves in the Y traveler's shoes. On the other hand, if the person concerned were wised up to the special theory of relativity he could choose to regard the real time and real length or distance measurement as those only made by any clock and any ruler or measuring device on Mother Earth (stationary frame). This is only one possible interpretation. A philosophical-minded person could choose the other interpretation, viz., X's time and length or distance measurement are real to the traveler in X (moving frame), Y's time and length or distance measurement are real to the traveler in Y (moving frame), and Earth's time and length or distance measurement are real to the resident on the Earth (stationary frame), i.e., there are different realities. Should there be only one reality or should more than one reality be allowed?

Therefore, consciousness or knowledge of certain facts, e.g., the fact that there is another clock under a different set of circumstances ticking away at a different speed, as described above, or simply some other clock to compare the time with would affect our sense of time, time being the fourth dimension in Einstein's Relativity theory.

Also, a person could be deceived about the time by a faulty clock, a case which is similar to the above-mentioned case of the car driver being misled by the faulty speedometer of his car.

All this implies that time is subjective, or as Einstein had put it, relative (not absolute), depending on the situation, and consciousness has an important role to play.

There should be a sufficient reason to explain why the clock, the brain, and bodily functions of the person slow down, and the length of the ruler or measuring device contracts, on approaching the speed of light, while at the speed of light the mechanism of the clock and time are at a standstill and the length of the ruler or measuring device is zero, which is important. Though the intense gravitational field caused by travel at almost the speed of light might account for the slowing down of the clock (for which experimental evidence had been obtained) and therefore time, as well as the brain and bodily functions of a person, it evidently hardly suffices as an explanation for the contraction of the length of the ruler or measuring device in the direction of travel at almost the speed of light (for which experimental evidence has yet to be found, and, which seems like a "fudge on the figure" by the inventor of the theory to "ensure the constancy of the speed of light"). Though the constancy of the speed of light as gauged from the Earth is evidently a well-proven phenomenon, no one has yet been able to travel at almost the speed of light and gauge the speed of a light beam by traveling besides it in the same direction, as described above – despite the experimental findings that at high speeds, though very much less than the speed of light, clocks slow down, the contraction of rulers, or measuring devices in the direction of motion at almost the speed of light is evidently only an inference, with no experimental basis.

The following equation describes how the speed of light (v) is derived:

$$v = d/t,$$

where d represents the distance traveled by the light beam and t represents the time taken by the light beam to travel the distance d.

Since time is relative (and not absolute) and depends on the mechanism of the clock, as well as consciousness, which slow down on approaching the speed of light, it could be arbitrary. The clock which is used to gauge the time t taken by the light beam to travel the distance d might not slow down uniformly (at the same rate) on approaching the speed of light (under normal, earthly conditions time varies from clock to clock by minutes or more – there is evidently some uncertainty in the mechanism of clocks). Besides, the ruler or measuring device used to gauge the distance d traveled by the light beam in time t might not contract in length uniformly (at the same rate) on approaching the speed of light. If the clock does not slow down uniformly (at the same rate) and the ruler or measuring device does not contract in length uniformly (at the same rate) on approaching the speed of light there is all probability that the speed of light (v) as represented by d/t would be variable, higher than 186,000 mi/s at times, below 186,000 mi/s at other times, or equal to 186,000 mi/s at yet other times. Moreover, in accordance with the special theory of relativity, as described above, for the speed of light to really remain constant, on approaching the speed of light the clock must slow down to the same degree as the contraction in the length of the ruler or measuring device. (This explanation presented by the special theory of relativity is incorrect, as is described in the author's work, "An Aspect of the Special Theory of Relativity".) We describe these possible outcomes as follows (only for argument's sake here, we assume that the above statement "for the speed of light to really remain constant, on approaching the speed of light the clock must slow down to the same degree as the contraction in the length of the ruler or measuring device" is correct and would lead to the following possible outcomes; the examples which follow, for the sake of argument, would also be based on this assumption):

i) $S^{\%} > C^{\%} \rightarrow I^{1}$
ii) $S^{\%} < C^{\%} \rightarrow D^{1}$
iii) $S^{\%} = C^{\%} \rightarrow S^{1}$

where $S^{\%}$ represents percentage of slowing down of the clock, $C^{\%}$ represents percentage of contraction in the length of the ruler or measuring device, I^{1} represents increase in the speed of light, i.e., exceed 186,000 mi/s, D^{1} represents decrease in the speed of light, i.e., go below 186,000 mi/s, and S^{1} represents speed of light, i.e., 186,000 mi/s.

We ponder this point more deeply by reconsidering the two examples pertaining to spaceships X (moving frame) and Y (moving frame) which have been described above.

Let us look at the first case pertaining to spaceships X (moving frame) and Y (moving frame) traveling at speeds of 185,000 mi/s (as gauged from the Earth, a stationary frame) and 185,500 mi/s (as gauged from the Earth, a stationary frame), respectively, in the same direction almost next to one another. The speed of Y (moving frame) as gauged from X (moving frame) should normally be plus 500 mi/s (185,500 mi/s minus 185,000 mi/s) and the speed of X (moving frame) as gauged from Y (moving frame) should normally be minus 500 mi/s (minus (185,500 mi/s minus 185,000 mi/s)) as explained above. But, the speed of Y (moving frame) as gauged from X (moving frame) and the speed of X (moving frame) as gauged from Y (moving frame) should be plus

62,110.887 mi/s and minus 62,110.887 mi/s, respectively, as computed by using the formula below, which is in accordance with the special theory of relativity:

$$v = (b - a) \div (1 - ba/c^2)$$

where c = speed of light = 186,000 mi/s, b = speed of Y = 185,500 mi/s (= 0.9973118c), and a = speed of X = 185,000 mi/s (= 0.9946236c)

$$= 62,110.887 \text{ mi/s}$$

Speed of Y as gauged from X = plus 62,110.887 mi/s (and not plus 500 mi/s (185,500 mi/s minus 185,000 mi/s), which should normally be the case – Y (moving frame) should appear to the traveler in X (moving frame) to be moving away from X (moving frame) in the same direction).

Speed of X as gauged from Y = minus 62,110.887 mi/s (and not minus 500 mi/s (minus (185,500 mi/s minus 185,000 mi/s)), which should normally be the case – X (moving frame) should appear to the traveler in Y (moving frame) to be moving away from Y (moving frame) in the opposite direction).

Evidently, the clocks and rulers or measuring devices on X (moving frame) and Y (moving frame) have, respectively, slowed down and contracted in length in the direction of motion to different degrees while X (moving frame) and Y (moving frame) have been traveling at 185,000 mi/s and 185,500 mi/s, respectively, almost the speed of light. Whilst the respective speeds of Y (moving frame) and X (moving frame) as gauged from the other should be plus 500 mi/s and minus 500 mi/s, respectively, the slowing down of their respective clocks and the contraction in the lengths of their respective rulers or measuring devices to different degrees have resulted in the above-mentioned plus/minus 500 mi/s being gauged differently, as shown in the following examples.

For example, if the length of the ruler or measuring device on the spaceship contracts by 20% while the spaceship travels at almost the speed of light, the relative speed of plus/minus 500 mi/s of each of the spaceships, X (moving frame) and Y (moving frame), should be recomputed/gauged as follows to produce the "distorted" speed of plus/minus 62,110.887 mi/s, which is in accordance with the special theory of relativity:

$$\begin{aligned} 62,110.887 \text{ miles} \div 1 \text{ s} = {} & 400 \text{ miles (0.8 of 500 miles – due to ruler length} \\ & \text{contraction of 20\%, 500 miles are gauged by} \\ & \text{space – ship traveler as 400 miles)} \div 0.00644 \\ & \text{s (due to clock on space – ship slowing} \\ & \text{down by 15528\%, 1 s is gauged by} \\ & \text{space – ship traveler as 0.00644 s)} \end{aligned}$$

For example, if the length of the ruler or measuring device on the spaceship contracts by 40% while the spaceship travels at almost the speed of light, the relative speed of plus/minus 500 mi/s of each of the spaceships, X (moving frame) and Y (moving frame), should be recomputed/gauged as follows to produce the "distorted" speed of plus/minus 62,110.887 mi/s, which is in accordance with the special theory of relativity:

$$\frac{62,110.887 \text{ miles}}{\div 1 \text{ s}} = \begin{array}{l} 300 \text{ miles (0.6 of 500 miles} - \text{due to ruler length} \\ \text{contraction of 40\%, 500 miles are gauged by} \\ \text{spaceship traveler as 300 miles)} \div 0.00483 \\ \text{second (due to clock on space} - \text{ship slowing} \\ \text{down by 20703.93\%, 1 s is gauged by} \\ \text{spaceship traveler as 0.00483 s)} \end{array}$$

For example, if the length of the ruler or measuring device on the spaceship contracts by 60% while the spaceship travels at almost the speed of light, the relative speed of plus/minus 500 mi/s of each of the spaceships, X (moving frame) and Y (moving frame), should be recomputed/gauged as follows to produce the "distorted" speed of plus/minus 62,110.887 mi/s, which is in accordance with the special theory of relativity:

$$\frac{62,110.887 \text{ miles}}{\div 1 \text{ s}} = \begin{array}{l} 200 \text{ miles (0.4 of 500 miles} - \text{due to ruler length} \\ \text{contraction of 60\%, 500 miles are gauged by} \\ \text{spaceship traveler as 200 miles)} \div 0.00322 \\ \text{s (due to clock on space} - \text{ship slowing} \\ \text{down by 31055.9\%, 1 second is gauged by} \\ \text{spaceship traveler as 0.00322 s)} \end{array}$$

However, if the length of the ruler or measuring device on the spaceship contracts by, e.g., 20%, and the clock on the spaceship slows down by the same percentage, i.e., 20%, while the spaceship travels at almost the speed of light, the relative speed of plus/minus 500 mi/s of each of the spaceships, X (moving frame) and Y (moving frame) would remain the same, i.e., plus/minus 500 mi/s, which is in accordance with the special theory of relativity:

$$\frac{500 \text{ miles}}{\div 1 \text{ s}} = \begin{array}{l} 400 \text{ miles (0.8 of 500 miles} - \text{due to ruler length} \\ \text{contraction of 20\%, 500 miles are gauged by space} \\ \text{ship traveler as 400 miles)} \div 0.8 \text{ s (due to} \\ \text{clock on spaceship slowing down by 20\%, 1 s} \\ \text{is gauged by spaceship traveler as 0.8 s)} \end{array}$$

We now look at the second case pertaining to spaceships X (moving frame) and Y (moving frame) traveling at speeds of 185,000 mi/s (as gauged from the Earth, a stationary frame) and 185,500 mi/s (as gauged from the Earth, a stationary frame), respectively, in opposite directions. The speed of Y (moving frame) as gauged from X (moving frame) and the speed of X (moving frame) as gauged from Y (moving frame) should each normally be 370,500 mi/s (185,000 mi/s plus 185,500 mi/s), as explained above. However, the speed of Y (moving frame) as gauged from X (moving frame) and the speed of X (moving frame) as gauged from Y (moving frame) should each not exceed 186,000 mi/s, the speed of light, which represents the ultimate limit, the maximum possible speed any accelerating object could attain, as postulated by the special theory of relativity (and not, respectively, 370,500 mi/s (185,000 mi/s plus 185,500 mi/s)), which should normally be the case, and they are computed by using the following formula, which is in accordance with the special theory of relativity:

$$v = (a + b) \div (1 + ab/c^2)$$

where c = speed of light = 186,000 mi/s, a = speed of X = 185,000 mi/s (= 0.9946236c), b = speed of Y = 185,500 mi/s (= 0.9973118c)

$$= 185,998.64 \text{ miles per second}$$

Speed of Y as gauged from X = speed of X as gauged from Y = 185,998.64 mi/s (Y (moving frame) should appear to the traveler in X (moving frame) to be moving toward X (moving frame) in the opposite direction, and X (moving frame) should appear to the traveler in Y (moving frame) to be moving toward Y (moving frame) in the opposite direction).

Let us look at the following examples:

For example, if the length of the ruler or measuring device on the spaceship contracts by 60% while the spaceship travels at almost the speed of light, the relative speed of 370,500 mi/s of each of the spaceships, X (moving frame) and Y (moving frame), should be recomputed/gauged as follows to produce the "distorted" speed of 185,998.64 mi/s, which is in accordance with the special theory of relativity:

$$\begin{array}{ll} 185,998.64 \text{ miles} & = 148,200 \text{ miles } (0.4 \text{ of } 370,500 \text{ miles} - \text{due to} \\ \div 1 \text{ s} & \quad \text{ruler length contraction of } 60\%,\ 370,500 \text{ miles} \\ & \quad \text{are gauged by spaceship traveler as } 148,200 \\ & \quad \text{miles}) \div 0.79678 \text{ s (due to clock on} \\ & \quad \text{spaceship slowing down by } 125.51\%,\ 1 \\ & \quad \text{s is gauged by spaceship traveler as} \\ & \quad 0.79678 \text{ s)} \end{array}$$

For example, if the length of the ruler or measuring device on the spaceship contracts by 70% while the spaceship travels at almost the speed of light, the relative speed of 370,500 mi/s of each of the spaceships, X (moving frame) and Y (moving frame), should be recomputed/gauged as follows to produce the "distorted" speed of 185,998.64 mi/s, which is in accordance with the special theory of relativity:

$$\frac{185{,}998.64 \text{ miles}}{\div 1 \text{ s}} = 111{,}150 \text{ miles (0.3 of 370,500 miles – due to ruler length contraction of 70\%, 370,500 miles are gauged by spaceship traveler as 111,150 miles)} \div 0.597585 \text{ s (due to clock on spaceship slowing down by 167.34\%, 1 s is gauged by spaceship traveler as 0.597585 s)}$$

For example, if the length of the ruler or measuring device on the spaceship contracts by 80% while the spaceship travels at almost the speed of light, the relative speed of 370,500 mi/s of each of the spaceships, X (moving frame) and Y (moving frame), should be recomputed/gauged as follows to produce the "distorted" speed of 185,998.64 mi/s, which is in accordance with the special theory of relativity:

$$\frac{185{,}998.64 \text{ miles}}{\div 1 \text{ s}} = 74{,}100 \text{ miles (0.2 of 370,500 miles – due to ruler length contraction of 80\%, 370,500 miles are gauged by spaceship traveler as 74,100 miles)} \div 0.39839 \text{ s (due to clock on spaceship slowing down by 251.01\%, 1 s is gauged byspaceship traveler as 0.39839 s)}$$

However, if the length of the ruler or measuring device on the spaceship contracts by, e.g., 60%, and the clock on the spaceship slows down by the same percentage, i.e., 60%, while the spaceship travels at almost the speed of light, the relative speed of 370,500 mi/s of each of the spaceships, X (moving frame) and Y (moving frame), would remain the same, i.e., 370,500 mi/s, which is in accordance with the special theory of relativity:

$$\frac{370{,}500 \text{ miles}}{\div 1 \text{ s}} = 148{,}200 \text{ miles (0.4 of 370,500 miles – due to ruler length contraction of 60\%, 370,500 miles are gauged by spaceship traveler as 148,200 miles)} \div 0.4 \text{ s (due to clock on spaceship slowing down by 60\%, 1 s is gauged by spaceship traveler as 0.4 s)}$$

Thus, as is evident from the above examples, which are in accordance with the special theory of relativity, to arrive at the two speeds, i.e., 62,110.887 mi/s and 185,998.64 mi/s, as well as other speeds, obtained by using the formulas postulated by the special theory of relativity, $v = (b - a) \div (1 - ba/c^2)$ and $v = (a + b) \div (1 + ab/c^2)$, the clocks and the rulers or measuring devices on the spaceships traveling at almost the speed of light would have to each, respectively, slow down and contract in length at different rates (and definitely not at the same rate). The only exception is evidently the case of the constancy of the speed of light, whereby the clock and the ruler or measuring device have to each, respectively, slow down and contract in length at the same rate, giving the same percentage decrease in the time gauged and the distance gauged, as follows, as postulated by the special theory of relativity:

$$(186,000\ \text{miles} - X\%\ \text{of}\ 186,000\ \text{miles}) \div (1\ \text{s} - X\%\ \text{of}\ 1\ \text{s}) = 186,000\ \text{mi/s}$$

Why is the constancy of the speed of light the exception? Was it an adjustment or modification of the mathematics to "ensure" the constancy of the speed of light? Could the speed of light not be variable, below, at and above 186,000 mi/s at various times, as some have suggested?

Let us look again at the case of the constancy of the speed of light. We have stated that according to the special theory of relativity the speed of a beam of light (c = 186,000 mi/s) would always be found to be constant or unchanged when gauged by a person traveling close to the beam of light in the same direction at almost the same speed as the beam of light (say 0.9c = 167,400 mi/s) because the clock and ruler or measuring device used by the person traveling in the same direction at almost the speed of light in gauging the speed of the beam of light would have respectively slowed down and contracted in length at the same rate (say X%).

We have to remember that the speed of the beam of light is obtained by dividing the distance traveled by the beam of light (as gauged by the ruler or measuring device on the spaceship traveling at almost the speed of light – moving frame, which has contracted in length) by the time it took to travel that distance by the beam of light (as gauged by the clock therein the spaceship – moving frame, which has slowed down), the gauging being carried out by the traveler on the spaceship (moving frame). Let us, using a simple example, say that the 1-m-long ruler used to gauge distance has contracted in length in the direction of motion by 20%. The clock used to gauge time, which has also slowed down by 20%, according to the special theory of relativity, would now gauge the time taken, say X, to travel the distance between two designated points (reference, stationary frame), say Y, as having decreased by 20% to become 0.8 X. Though the 1-m-long ruler, which has contracted in length by 20%, still reads "1 meter" in length, it is in effect shorter by 20% (actually only 0.8 m in length). Therefore, when it gauges the distance traveled, Y, above, this distance Y would now be gauged as 1.25 Y, and not 0.8 Y in accordance with the special theory of relativity. As stated above, special relativity theorizes that for the speed of a beam of light to remain constant the beam of light would have to take less time (time dilation) to travel a shorter distance (length contraction) – in effect, X% less time to travel a distance shorter by X%, in accordance with the above-mentioned equation, which implies that the speed of the beam of light would remain constant, e.g., 0.8 X (time) to travel 0.8 Y (distance) after "time dilation" and "length contraction". But, as explained above, this would not be the case; the beam of light would have been gauged as having taken 0.8 X (time) to travel 1.25 Y (distance). This is an anomaly and it shows that there is something not right with the special theory of relativity. The author's other work, "An Aspect of the Special Theory of Relativity", describes this anomaly in detail.

Since light particles (photons) do not have mass or inertia, which prevents an object possessing it from accelerating beyond the speed of light, viz., 186,000 mi/s, theoretically there is nothing to prevent light particles (photons) or other objects without mass or inertia from traveling at a speed greater than 186,000 mi/s.

Faster than Light Travel

What would be the speed of the beams of light from the head lights of a vehicle traveling at, e.g., 0.0083 mile per second or 30 mi per hour? If gauged from outside the moving vehicle (stationary frame) it should normally be 186,000.0083 mi/s (speed of the moving vehicle (0.0083 mi/s or 30 mi/h) plus speed of each beam of light (186,000 mi/s)), but, according to the special theory of relativity, this is not the case and the speed of the beams of light is not 186,000.0083 mi/s but still 186,000 mi/s (constant). How come? This is because light travels on its own independently of or unaffected by its source; in the above case, the source of the beams of light being the head lights of the vehicle traveling at 0.0083 mile per second or 30 mi/h.

In the above case, the beams of light are so much faster (186,000 mi/s) than the speed of the moving vehicle (0.0083 mile per second or 30 mi/h) that they would be continually moving way ahead of the vehicle after they leave the head lights. Logically, if the vehicle had traveled at a speed exceeding the speed of light (which according to the special theory of relativity is impossible – no object could travel faster than the speed of light – it is only an assumption here for the sake of argument) and continually "overtaken" the beams of light that its head lights had emitted, the beams of light (whose speed remains the same at 186,000 mi/s – constant) should not be in front of the head lights. If the vehicle were traveling much in excess of the speed of light, the beams of light emitted from the vehicle's head lights should tag behind the vehicle. This is counter to our normal experiences with light beams from the head lights of moving vehicles, which always appear in front of the vehicles.

The following equation shows that no moving object could travel faster than the speed of light, which is in accordance with the special theory of relativity:

$$v = (a + b) \div \left(1 + ab/c^2\right)$$

If we let a = velocity of moving train, b = velocity of light beam (which is sent from the back of the moving train to the front of the moving train) with respect to the moving train, which is the moving frame (i.e., the velocity of the light beam (which is sent from the back of the moving train to the front of the moving train) is gauged from the moving train, which is the moving frame), v = velocity of light beam (which is sent from the back of the moving train to the front of the moving train) with respect to the ground level, which is the stationary frame (i.e., the velocity of the light beam (which is sent from the back of the moving train to the front of the moving train) is gauged from the ground level, which is the stationary frame), c = velocity of light = 186,000 mi/s, and also let a = b = c, then

$$v = (c + c) \div (1 + c.c/c^2) = 2c/2 = c! \text{ (And not 2c!)}$$

Though theoretically no object could travel faster than the speed of light because at the speed of light the object's mass is infinitely great and therefore it is unable to accelerate, an object without mass, possibly, a quantum particle which is somewhat similar to a photon (a photon is a quantum particle without mass always in motion) might be capable of traveling faster than the speed of light. Such an object or objects might be waiting to be discovered. As it is, a "theoretical" particle which travels faster than the speed of light, which is termed "tachyon", has been thought to exist.

There have been a number of speculations pertaining to the variable speed of light (VSL), e.g., one theory states that the speed of light varies with the various stages of the evolution of the universe, exceeding 186,000 mi/s at certain points of time.

Time Travel

In the special theory of relativity it is implied that an object traveling at more than the speed of light would go backward in time, which is bizarre, but has evidently been taken seriously by quite a number. If time travel were indeed possible, a person could go back to the time before he was born and murder his grandfather so that his father, and hence, himself, would not have been conceived, which is against causality, paradoxical, and absurd, thus implying that time travel is not really possible.

After an object had contracted to zero length at the speed of light, it would actually cease to exist. Could an object really exceed the speed of light, contract in length further to "minus something" length from zero length or nonexistence and travel backward in time, as have been postulated by a number of people who hold the view that time travel is possible? Negative quantities, e.g., negative lengths, as such, are evidently abstractions devoid of any real meaning or existence. For instance, a string which is minus one foot in length or a boat which is minus 10 feet in length is meaningless. The idea of time travel suggested by the special theory of relativity should be only regarded as a metaphor or curiosity; it should be regarded as something which might be theoretically possible but is not practicable. It is somehow comparable to one of those well-known paradoxes, e.g., Zeno's paradox, whereby it is shown that Achilles would never be able to overtake a tortoise which was given a head start in a race (which is absurd).

Stephen Hawking has stated that if time travel were possible we would have had visitations from tourists from the future (which we have evidently none so far), which, according to him, shows that time travel is not possible. But, what about the reports and books about alleged sightings of UFOs and encounters with or abductions by aliens or extra-terrestrials? Are these alleged aliens or extra-terrestrials not possible tourists from the future, beings with seemingly advanced technologies, technologies which seem more advanced than ours, e.g., flying saucers, instantaneous appearance, and disappearance which are suggestive of teleportation? It is also possible, and seemingly much likely, that such beings, if they indeed exist, are from a civilization or civilizations from another part of the universe or another universe, which are more advanced than ours. There have been much stories and speculations pertaining to time travel or teleportation. It has been discovered that quantum particles could be teleported, i.e., made to appear instantaneously at another location. If only there is a technology to teleport all the atoms in the body of a human being so that the human being could appear instantaneously at another place without physical and mental harm.

The postulation of the impossibility of exceeding the speed of light by the special theory of relativity means that time travel is not possible, and Einstein had stated that time travel is impossible. Though time travel in the physical dimension is evidently impossible, time travel in the domain of the consciousness or mind is possible.

Conclusion

The special theory of relativity postulates that on approaching the speed of light a person's brain, as well as bodily, functions slows down, which partly explains why the speed of light would remain unchanged to him under all circumstances; in other words, he experiences the slowing down of time on approaching the speed of light. It implies that the person's mind could travel backward and forward in the space-time continuum. The mind is evidently closely linked to the "time" dimension of the space-time continuum, while the three coordinates of length, breadth, and height make up the "space" dimension of the space-time continuum. We could equate the mind and time as follows:

$$C = S^t$$

where C represents consciousness and S^t represents sense of the passing of time.

Thus, there ought to be a theory of consciousness allied to the special theory of relativity, which would contribute to an important understanding of the workings of nature.

Finally, a number of important points raised in this article should be looked into.

Appendix 10: Mathematical Reasoning at Work (with Commentary)

Commentary

The possible solutions to the three long outstanding mathematical problems here are more or less based on new combinations of old ideas. The author might not have fully agreed with some of these old ideas, e.g., the argument by contradiction and the concepts relating to infinity, for instance, the different orders of infinity, but since they are now "conventional wisdom" which appears acceptable to practically all mathematicians, the author just adopted them as "common grounds for agreement" to get the job done. The author would therefore regard the possible solutions as being rather subjective as there was the element of the desire to include ideas which he surmised that the referees of any mathematical journals considering the articles for publication would find acceptable; to be too original and to include totally new ideas might only invite suspicion and rejection.

Are the referees of the mathematical journals really objective and fair-minded enough to give the solutions a fair consideration? Though the author might be wrong and risks giving the impression of crying "sour grapes", he could not help but feel that most journal referees are overly cautious of cranks and somewhat biased – they appear to generally favor a certain style of mathematical presentation and complexity of argument – they seem to foster the impression that mathematics is difficult and must appear difficult – the editor of one of these mathematical journals even had the affront to reply (even before the author had submitted the article to him) that if the author were not a professional mathematician, he could not be expected to have the correct solution. Is this how a professional mathematician should conduct himself? We should all be sorry for mathematics if many professional mathematicians are like that. There are much more pressing and important problems to be solved than all the very difficult, outstanding mathematical problems. So what if the twin primes conjecture or the Goldbach conjecture is right or wrong? Does it make any iota of difference at all to our lives or our well-being? Its only contribution perhaps is that it gives the author and some like-minded people a sense of satisfaction, or sense of achievement, which is indeed nothing of great importance.

However, mathematics has been effectively used to model nature, and it has been said that nature is mathematical. Mathematics is not that different from a game, such as chess, or any other game, in that there are rules to be followed. In presenting the possible solutions to the three problems, the author had to make an attempt to stick to some rules in order not to be "disqualified", rules which might appear arbitrary. It is as though the author was playing a game, e.g., chess.

Besides, the author always had the fear that the journal referees would not understand his ideas, and he had to attempt to present his ideas in as simple and least

https://doi.org/10.1515/9783112214367-071

complicated a manner as possible in order that they be understood and approved by the referees, whoever they might be. Of course, if the author's ideas were not understood, his articles would never stand a chance of publication by the journals.

In the five papers on the twin primes conjecture and the Goldbach conjecture, the author has shown a number of ways of creating, producing or generating respectively twin primes, and even numbers which are each the sum of two primes, all the way to infinity. This represents a constructive proof. On top of this, the author has used the "reductio ad absurdum" proof or proof by contradiction (which, as stated before, is not acceptable to mathematicians who abide by the Intuitionist school of thought and favor the constructive proof). Various other arguments have also been presented. If these were not acceptable as evidence which supports the conjectures, then what kind or kinds of evidence are acceptable? Strangely and apparently illogically, the managing editor of one mathematical journal had replied the author that the methods of producing or generating pairs of twin primes presented in his article would fail at infinity. If the managing editor had proffered a good reason, or proof, why the methods would fail at infinity, then the author could have been forced to change his stand, but none had been offered. What then constitutes an acceptable mathematical proof or evidence? This appears to be the central problem of mathematics – no mathematician seems to know, and it seems more like a matter of opinion, subjective.

In mathematics, even if trillions of twin primes, or even numbers which are each the sum of two primes, could be produced or generated with the help of very powerful computers, this is still not regarded as good enough evidence which supports either conjecture, though to the scientist such overwhelming evidence could be regarded as proof that a hypothesis is correct. The twin primes conjecture and the Goldbach conjecture involve infinity, and, since we could not quantify infinity and might not even have a good, decent definition or understanding of infinity, it would apparently be very difficult to prove the infinity of the twin primes or the even numbers which are each the sum of two primes.

A proof is information that supports a proposition or statement. It is something which certifies that a proposition or statement is correct. For example, how do we prove or certify that Sudan is hot? Could we do it by mere logical deduction? For instance, logical deduction such as the following:

1) Everyone I spoke to who has been to Sudan complains that it is hot.
2) The geography books I consulted state that Sudan is hot.
3) Therefore, Sudan is indeed hot.

There is certainly a much better way, in fact a scientific way, of proving or certifying that Sudan is hot. We go to Sudan, station ourselves at a representative place there, and monitor the temperatures there with thermometers at certain regular intervals for a specific period. Then we total all the temperature readings taken over this period and average them up. We have first of all to decide on and specify what temperature range is considered "hot climate". We compare the average temperature computed against this temperature range, and, if it falls within this range, we could conclude, or certify, that Sudan is indeed hot.

Next, how do we prove that a person is a thief, or murderer, for example? The chain of reasoning could be as follows:

1) The accused was not at home, or at work, at the time the crime was committed.
2) When questioned, he claimed he was drinking at a pub at the time the crime was committed.
3) A check with the pub manager, and the CCTV recordings showed he was never at the pub, which implies he had lied.
4) He had no other reason to lie except for covering up his crime.
5) We therefore conclude that he was the one who committed the crime.

On the other hand, what if there were other evidences, e.g., a CCTV record or film of the accused committing the act of crime, and his fingerprints or DNA (if he had injured himself and bled) at the scene of the crime? Wouldn't these be incontrovertible, incriminating evidence or proof of his guilt, as compared to the proof of his guilt by mere logical deduction?

As for the twin primes conjecture and the Goldbach conjecture, the only truly incontrovertible proof of the infinitude of the twin primes and the even numbers which are each the sum of two primes is to be at infinity (if only this were possible) and check or confirm whether the twin primes and such even numbers would still be found. To arrive at a conclusion for these two conjectures by mere logical deduction is at best only a second-guess, the next best method of proof in lieu of actual, experimental proof. Physical or experimental evidence or proof would be more assuring than proof by mere logical reasoning. In other words, physical and concrete evidence or proof should take precedence over abstract proof. However, for the case of the twin primes conjecture and the Goldbach conjecture, since infinity is something not physically achievable at all but is just an abstract (we might call it "invented") entity, there is no other way of proving these conjectures one way or another apart from logical deduction.

The author feels that trying to prove the infinity of anything might be just trying to do the impossible, which is absurd. Having said all this, does it not imply that the author has been a fool in attempting the solutions for the two conjectures? In a sense, the author has been a fool. He could not resist taking up a great intellectual challenge. He could not resist the pleasure of applying clever ideas or ploys in solving these two challenging problems. However, the author thinks that his solutions are reasonable, even clever, or cunning. And, he would suggest that they be only regarded, at best, as correct, until proven wrong. The author would be very happy if any of the solutions is proved wrong. He would then be happy because of the efficacy of logic, he would be happy for logic.

1 (a) On the Twin Primes Conjecture (This Paper Is a Much Expanded Version of a Paper Which Had Been Published in an International Mathematics Journal in 2003)

Abstract

The author has been contemplating the twin primes problem for many years and had published a paper on the solutions for the twin primes conjecture in an international mathematics journal in 2003. He has never stopped thinking about a better solution for the conjecture. This paper presents various approaches to the twin primes problem and consists of seven parts which are each self-contained.
MSC: 11-XX (Number Theory)

Introduction

In 1919, Viggo Brun (1885–1978) proved that the sum of the reciprocals of the twin primes converges to Brun's constant:

$$1/3 + 1/5 + 1/7 + 1/11 + 1/13 + 1/17 + 1/19 + \cdots = 1.9021605 \ \ldots$$

It is evident that the twin primes thin out as infinity is approached. The problem of whether there is an infinitude of twin primes is an inherently difficult one to solve, as infinity (normally symbolized by: ∞) is a difficult concept and is against common sense. It is impossible to count, calculate or live to infinity, perhaps with the exception of God. Infinity is a nebulous idea and appears to be only an abstraction devoid of any actual practical meaning. How do we quantify infinity? How big is infinity? We could either attempt to prove that the twin primes are finite, or infinite. If the twin primes were finite, how could we prove that a particular pair of twin primes is the largest existing pair of twin primes, and, if they were infinite, how could we prove that there are always larger and larger pairs of them? It is evidently difficult to prove either, with the former appearing more difficult to prove as the odds seem against it. This paper provides proof of the latter, i.e., the infinitude of the twin primes.

Part 1

Keywords: indivisible, impossibility, new primes/twin primes

Theorem. The twin primes are infinite.

Proof.

Let 3, 5, 7, 11, 13, 17, 19, . . ., n – 2, n be the list of consecutive primes, wherein n and n – 2 are assumed to be the largest existing twin prime pair, within the infinite list of the primes.

$$\text{Let } 3 \times 5 \times 7 \times 11 \times 13 \times 17 \times 19 = a.$$

Lemma. (a × · · · × n – 2 x n) – 2, and, (a × · · · × n – 2 × n) – 4 will never be divisible by any of the consecutive primes in the list: 3, 5, 7, 11, 13, 17, 19, . . ., n – 2, n, whether they are prime or composite (see Appendix 1).

This implies that:
If (a × · · · × n – 2 × n) – 2 &/V (a × · · · × n – 2 × n) – 4 are prime, then:

$$(a \times \cdots \times n-2 \times n) - 2 > (a \times \cdots \times n-2 \times n) - 4 > n > n-2$$

If (a × · · · × n – 2 × n) – 2 &/V (a ×· · · × n – 2 × n) – 4 are non-prime/composite, then:

(a) each prime factor, e.g., y below, of (a × · · · × n – 2 × n) – 2 > n > n – 2
(b) each prime factor, e.g., z below, of (a × · · · × n – 2 × n) – 4 > n > n – 2

$$(a \times \cdots \times n-2 \times n) - 2 = \text{prime V composite} \tag{1}$$

$$(a \times \cdots \times n-2 \times n) - 4 = \text{prime V composite} \tag{2}$$

(1) and (2) = twin primes, if both (1) and (2) are prime
(1) and (2) > n and n – 2

Let Y represent the prime factors of (a × · · · × n – 2 × n) – 2 if (a × · · · × n – 2 × n) – 2 is not prime (i.e., it is composite), each prime factor may pair up with another prime which differs from it by 2 to form twin primes. Let y = prime factor in Y.

y and y +/– 2 = twin primes, if y +/– 2 is prime

$$y \text{ and } y +/-2 > n \text{ and } n-2$$

Let Z represent the prime factors of (a × · · · × n – 2 × n) – 4 if (a × · · · × n – 2 × n) – 4 is not prime (i.e., it is composite), each prime factor may pair up with another prime which differs from it by 2 to form twin primes. Let z = prime factor in Z.

z and z +/– 2 = twin primes, if z +/– 2 is prime

$$z \text{ and } z +/-2 > n \text{ and } n-2$$

Therefore: (a × · · · × n – 2 x n) – 2 > (a × · · · × n – 2 × n) – 4 > y V y +/– 2 V z V z +/– 2 > n > n – 2

By the above, the following, which implies that n and n – 2 are the largest existing twin prime pair, is an impossibility:

$$n > n-2 > (a\text{x} \ldots \text{x}n - 2\text{x}n) - 2 > (a\text{x} \ldots \text{x}n - 2\text{x}n) - 4 > y\text{V}y +/-2\text{V}z\text{V}z +/-2$$

It is hence clear that no n and n – 2 in any list of consecutive primes can ever possibly be the largest existing twin prime pair and larger twin primes than them can always be found by applying the same mathematical logic (as is described in Appendix 1), e.g., by utilizing the evidently effective Algorithm 1 or Algorithm 2 described in Part 7. That is, a largest existing twin prime pair is an impossibility, which implies that the twin primes are infinite. It is possible to find larger twin primes than n and n – 2 no matter how large n and n – 2 are, with the following formulae involving the list of consecutive primes: (a × · · · × n) – 2 and (a × · · · × n) – 4, which by the nature of their composition are capable of generating new primes/twin primes which will always be larger than n and n – 2 (see Appendix 1); this operation is part of Algorithm 1 described in Part 7. This is an indirect proof or proof by contradiction (*reductio ad absurdum*) of the infinity of the twin primes, for our assumption of n and n – 2 as the largest existing twin prime pair will be contradicted by the discovery of larger twin primes, implying the infinity of the twin primes. Again, by applying the same mathematical logic (described in Appendix 1), by way of this evidently effective Algorithm 1 in Part 7, and going one step further, we can find that many twin odd integers found between n and (a × · · · × n) – 2, which differ from one another by 2 and are not divisible by any of the primes in the list of consecutive primes: 3, 5, 7, 11, 13, 17,19, . . ., n, will be twin primes larger than n and n – 2, our assumed largest existing twin prime pair, which is a contradiction of this assumption, thus implying or proving the infinitude of the twin primes. In this manner, i.e., by resorting to Algorithm 1 in Part 7, by continually adding more and more consecutive primes to the list of consecutive primes: 3, 5, 7, 11, 13, 17, 19, . . ., n, i.e., continually extending the value of n, and utilizing the formula: (a × · · · × n) – 2, as well as the formula: (a × · · · × n) – 4, to perform the computations a la Algorithm 1 in Part 7, many larger and larger twin primes can be found, all the way to infinity, in parallel with the infinitude of the list of consecutive primes: 3, 5, 7, 11, 13, 17, 19, . . . of which the twin primes are a part together with other prime pairs, wherein the twin primes are not likely to be finite (as is evident from Appendix 2) and can be expected to be infinite. (Algorithm 2 in Part 7 may also be utilized for this purpose but it is evidently a longer and less efficient method.) A largest existing twin prime pair is indeed an impossibility. The twin primes are infinite.

Appendix 1

Note: The (only) even prime 2 is omitted from the list of consecutive primes: 3, 5, 7, 11, 13, 17, 19, . . ., n – 2, n stated in the paper, wherein n and n – 2 are assumed to be the largest existing twin prime pair.

The list of newly created primes, and twin primes for n = 5, 7, 11, 13, 17, 19, . . . (n = 19 being the maximum limit achievable with a hand-held calculator) is as follows:

1. For n = 5, we get the following new primes/new twin primes:

$$(3\times5)-2=13\ (\alpha)$$
$$(3\times5)-4=11\ (\beta)$$

2. For n = 7, we get the following new primes/new twin primes:

$$(3\times5\times7)-2=103\ (\alpha)$$
$$(3\times5\times7)-4=101\ (\beta)$$

3. For n = 11, we get the following new primes/new twin primes:

$$(3\times5\times7\times11)-2=1,153\ (\alpha)$$
$$(3\times5\times7\times11)-4=1,151\ (\beta)$$

4. For n = 13, we get the following new prime and composite number with its prime factors:

$(3\times5\times7\times11\times13)-2=15,013\ (\alpha)$ – Prime Number
$(3\times5\times7\times11\times13)-4=15,011\ (\beta)$ – Composite Number ($=17\times883$, with 17 pairing with 19 to form a twin primes pair and 883 pairing with 881 to form another twin primes pair)

5. For n = 17, we get the following new primes/new twin primes:

$$(3\times5\times7\times11\times13\times17)-2=255,253\ (\alpha)$$
$$(3\times5\times7\times11\times13\times17)-4=255,251\ (\beta)$$

6. For n = 19, we get the following new prime and composite number with its prime factors:

$(3\times5\times7\times11\times13\times17\times19)-2=4,849,843\ (\alpha)$ – Prime Number
$(3\times5\times7\times11\times13\times17\times19)-4=4,849,841\ (\beta)$ – Composite Number ($=43\times112,787$, with 43 pairing with 41 to form a twin primes pair while 112, 787 is a stand – alone prime)

Results of α and β Above

1) α above generates 6 new primes (13; 103; 1,153; 15,013; 255,253; 4,849,843), nil composite numbers.

2) β above generates 4 new primes (11; 101; 1,151; 255,251), 2 composite numbers (15,011 = 17 × 883; 4,849,841 = 43 × 112,787).
3) α and β above together produce 4 pairs of new twin primes (13 and 11; 103 and 101; 1,153 and 1,151; 255,253 and 255,251).
4) The prime factors of α and β above form 3 pairs of new twin primes with prime partners which differ from them by 2 (19 and 17; 43 and 41; 883 and 881).
5) All the new twin primes in (3) and (4) above are larger than n and n – 2, the assumed largest existing twin prime pair, which is indirect proof of the infinitude of the twin primes.

Why It Is Impossible for Any n and n – 2 to Be the Largest Existing Twin Primes Pair

$\alpha = (3 \times 5 \times 7 \times 11 \times 13 \times 17 \times 19 \times \cdots \times n) - 2$, and $\beta = (3 \times 5 \times 7 \times 11 \times 13 \times 17 \times 19 \times \cdots \times n) - 4$ will never be divisible by any of the consecutive prime numbers in the list: 3, 5, 7, 11, 13, 17, 19, . . ., n, whether they are prime or composite (non-prime and divisible by prime numbers or prime factors). This means that none of the consecutive prime numbers in the list: 3, 5, 7, 11, 13, 17, 19, . . ., n can ever be factors of α and β, and α and β must be new primes/twin primes larger than all the consecutive prime numbers in the list: 3, 5, 7, 11, 13, 17, 19, . . ., n, or if they were composite (non-prime and divisible by prime numbers or prime factors), their prime factors (and "twin prime" partners which differ from them by 2) must be larger than all the consecutive prime numbers in the list: 3, 5, 7, 11, 13, 17, 19, . . ., n.

This all implies that no n and n – 2 (if n – 2 were also a prime number) in any list of consecutive prime numbers can ever possibly be the largest existing twin prime pair, since all the new primes/twin primes produced or generated by α and β will always be larger than n and n – 2. That is, a largest existing twin prime pair is an impossibility, which implies the infinitude of the list of the primes/twin primes.

In other words, by the above, which explains why all the new primes/twin primes, which α and β by the nature of their composition are capable of producing or generating, will always be larger than n and n – 2, no n and n – 2 in any list of consecutive prime numbers: 3, 5, 7, 11, 13, 17, 19, . . ., n can ever possibly be the largest existing twin prime pair, i.e., a largest existing twin prime pair is indeed an impossibility, thus implying the infinitude of the list of the twin primes.

This is a very important inference.

Regardless of how long the list of the twin prime pairs is, it is possible to find some new twin prime pairs which will always be larger than n and n – 2, our assumed largest existing twin prime pair – the largest twin prime pair in our assumed finite list of the twin prime pairs, with α and β, which is indirect proof of the infinity of the twin primes. In fact, by the same principle, many twin odd integers found between n and α, which differ from one another by 2 and are not divisible by any of the primes in the list of consecutive primes: 3, 5, 7, 11, 13, 17, 19, ..., n, will be twin prime pairs larger than n and n – 2, our assumed largest existing twin prime pair, which is a contradiction of this assumption, hence implying or proving the infinitude of the twin primes. (Refer to Algorithm 1, as well as Algorithm 2, in Part 7.)

Appendix 2

Anecdotal Evidence of the Infinity of the Twin Primes

TOP TWIN PRIMES IN 2000, 2001, 2007, and 2009

In the year 2000, $4648619711505 \times 2^{60000} \pm 1$ (18,075 digits) had been the top twin prime pair which had been discovered. In the year 2001, it only ranked eighth in the list of top 20 twin prime pairs, with $318032361 \cdot 2^{107001} \pm 1$ (32,220 digits) topping the list. In the year 2007, in the list of top 20 twin prime pairs, $318032361 \cdot 2^{107001} \pm 1$ (32,220 digits) ranked eighth, while $4648619711505 \times 2^{60000} \pm 1$ (18,075 digits) was nowhere to be seen; $2003663613 * 2^{195000} - 1$ and $2003663613 * 2^{195000} + 1$ (58,711 digits), which was discovered on January 15, 2007, by Eric Vautier (from France) of the Twin Prime Search (TPS) project in collaboration with PrimeGrid (BOINC platform), was at the top of the list. As at August 2009, $65516468355 \cdot 2^{333333} - 1$ and $65516468355 \cdot 2^{333333} + 1$ (100,355 digits) is at the top of the list of top 20 twin prime pairs, while $318032361 \cdot 2^{107001} \pm 1$ (32,220 digits) ranks 11th`, and $2003663613 * 2^{195000} - 1$ and $2003663613 * 2^{195000} + 1$ (58,711 digits) ranks second in this list.

We can expect larger twin primes than these extremely large twin primes, much larger ones, infinitely larger ones, to be discovered in due course.

List of prime pairs for the first 2,500 consecutive primes, 2 to 22,307, ranked according to their frequencies of appearance

S. no.	Ranking	Prime pairs	No. of pairs	Percentage
(1)	1	Prime pair separated by 6 integers	482	19.29%
(2)	2	Prime pair separated by 4 integers	378	15.13%
(3)	3	Prime pair separated by 2 integers (t. p.)	376	15.05%
(4)	4	Prime pair separated by 12 integers	267	10.68%
(5)	5	Prime pair separated by 10 integers	255	10.20%
(6)	6	Prime pair separated by 8 integers	229	9.16%
(7)	7	Prime pair separated by 14 integers	138	5.52%
(8)	8	Prime pair separated by 18 integers	111	4.44%
(9)	9	Prime pair separated by 16 integers	80	3.20%
(10)	10	Prime pair separated by 20 integers	47	1.88%
(11)	11	Prime pair separated by 22 integers	46	1.84%
(12)	12	Prime pair separated by 30 integers	24	0.96%
(13)	13	Prime pair separated by 28 integers	19	0.76%
(14)	14	Prime pair separated by 24 integers	16	0.64%
(15)	15	Prime pair separated by 26 integers	10	0.40%
(16)	16	Prime pair separated by 34 integers	9	0.36%
(17)	17	Prime pair separated by 36 integers	5	0.20%
(18)	18	Prime pair separated by 32 integers	2	0.08%

(continued)

S. no.	Ranking	Prime pairs	No. of pairs	Percentage
(19)	18	Prime pair separated by 40 integers	2	0.08%
(20)	19	Prime pair separated by 42 integers	1	0.04%
(21)	19	Prime pair separated by 52 integers	1	0.04%
Total no. of prime pairs in list: 2,498				

It is evident in the above list that the primes pairs separated by 6 integers, 4 integers, and 2 integers (twin primes), among the 21 classifications of primes pairs separated by from 2 integers to 52 integers (prime pairs separated by 38 integers, 44 integers, 46 integers, 48 integers, and 50 integers are not among them, but, they are expected to appear further down in the infinite list of the primes), are the most dominant, important. There is a long list of other prime pairs, besides those shown in the above list, which also play a part as the building blocks of the infinite list of the integers.

The list of the integers is infinite. The list of the primes is also infinite. The infinite primes are the building blocks of the infinite integers – the infinite odd integers are all either primes or composites of primes, and the infinite even integers, except for 2 which is a prime, are all also composites of primes. Therefore, all the prime pairs separated by the integers of various magnitudes, as described above, can never all be finite. If there is any possibility at all for any of these prime pairs to be finite, there is only the possibility that a number of these prime pairs are finite (but never all of them). However, will it have to be the prime pairs separated by 2 integers or twin primes (which are the subject of our investigation here), which are the only prime pair, or one among a number of prime pairs, which are finite? Why question only the infinity of the prime pairs separated by 2 integers, the twin primes? Are not the infinities of the prime pairs separated by 8 integers and more, whose frequencies of appearance are lower, as compared to those of the prime pairs which are separated by 6, 4 and 2 integers respectively, in the above list of prime pairs, more questionable? Why single out only the twin primes? (There are at least 18 other prime pairs, separated by from 8 integers to 52 integers, whose respective infinities should be more suspect, as is evident from the above list of prime pairs, if any infinities should be doubted. Evidently, the prime pairs separated by 2 integers (twin primes) are not that likely to be finite.)

The above represents anecdotal evidence that the twin primes are infinite, which is a ratification of the actual proof given earlier.

Part 2

Keywords: random, probability, predicted, found, induction

Theorem. There is an infinite number of twin primes.

Numerical Evidence

Many believe that the twin primes are infinite. In fact, twin prime pairs could easily be found among the integers. There is evidently no region of the natural number system so remote that it lies beyond the largest twin prime pair. It is even possible to forecast the approximate number of twin prime pairs found in any region of the natural number system.

The occurrence of twin prime pairs is evidently unpredictable or random. This means that the chance of 2 numbers x and $x + 2$ being prime (twin primes) is somewhat similar to the chance of getting heads on 2 successive tosses of a coin. If 2 successive tosses of a coin are independent, the chance of success of obtaining heads for the 2 successive tosses of the coin is the product of the chances of success of obtaining a head for each toss of the coin. As each coin has probability ½ of coming up heads with a toss, 2 coins would have probability ½ x ½ = ¼ of coming up a pair of heads with a toss.

The prime number theorem, which had been proven, states that if n is a large number, and we select a number x at random between 0 and n, the chance that x is prime would be approximately $1/\log n$, the larger n is, the better would be the approximation given by $1/\log n$ to the proportion of primes in the numbers up to n. Like 2 coins coming up heads, the chance that both x and $x + 2$ are prime (twin primes) would be approximately $1/(\log n)^2$. That is, there would be approximately $n/(\log n)^2$ twin prime pairs between 0 and n. As n goes to infinity, this fraction approaches infinity. This represents a quantitative version of the twin primes conjecture.

As $x + 2$ being prime depends on the fact that x is already prime, we should modify the estimate $n/(\log n)^2$ to $(1.32032...)n/(\log n)^2$.

The following is a comparison between the twin primes predicted by the above formula and the twin primes found, where the agreement is evidently very good:

Interval	Twin primes	
	Predicted	Found
100,000,000 – 100,150,000	584	601
1,000,000,000 – 1,000,150,000	461	466
10,000,000,000 – 10,000,150,000	374	389
100,000,000,000 – 100,000,150,000	309	276
1,000,000,000,000 – 1,000,000,150,000	259	276
10,000,000,000,000 – 10,000,000,150,000	221	208
100,000,000,000,000 – 100,000,000,150,000	191	186
1,000,000,000,000,000 – 1,000,000,000,150,000	166	161

All this represents numerical evidence that the twin primes are infinite as we could find more twin prime pairs whenever we look for them.

Proof.

Lemma. According to the principle of induction in set theory, if a set of natural numbers contains 1, and if it contains n + 1 whenever it contains a number n, then it must contain every natural number, e.g., induction proves that every natural number is a product of primes.

By the above lemma, there is an infinitude of twin primes in the infinite set of the integers; twin primes are found in all the above subsets of integers, from 100,000,000 – 100,150,000 to 1,000,000,000,000,000 – 1,000,000,000,150,000, and many very much larger twin primes have also been found in the infinite set of the integers, e.g., $318032361 \cdot 2^{107001} \pm 1$ (32,220 digits), $2003663613{*}2^{195000}–1$ and $2003663613{*}2^{195000}+1$ (58,711 digits), and $65516468355 \cdot 2^{333333}–1$ and $65516468355 \cdot 2^{333333}+1$ (100,355 digits), to name a few (this long list of twin primes, especially the large twin primes, are obtainable only with the help of modern computer technology) – the principle of induction implies that there must be an infinite number of twin primes, ranging from the twin prime pair, 3 and 5 (smallest twin prime pair), 5 and 7, 11 and 13, 17 and 19, . . ., $65516468355 \cdot 2^{333333}$ – 1 and $65516468355 \cdot 2^{333333}+1$ (100,355 digits – largest twin prime pair discovered as at August 2009), ... upward to infinity, in the infinite set of the integers – in other words, the twin primes conjecture must be true.

Part 3

Keywords: building blocks, symmetry, infinity

Theorem. The list of the twin prime pairs is infinite.

Proof.

Lemma. According to the precepts of fractal geometry and group theory, symmetry is a very important, intrinsic part of nature. There is symmetry all around us and within us. There is evident symmetry in human bodies, the structures of viruses and bacteria, polymers and ceramic materials, the permutations of numbers, the universe and many others, even the movements of prices in financial markets, the growths of populations, the sound of music, the flow of blood through our circulatory system, the behavior of people en masse, and so on. In other words, regularity, pattern, order, uniformity, or symmetry is evident everywhere.

The reasoning here makes use of a very important idea in fractal geometry and group theory, namely, symmetry.

A prime number is an integer which is divisible only by 1 and itself, e.g., 2, 3, 7, 19, etc. Twin prime pairs are 2 primes that differ from one another by 2, e.g., 5 and 7, 11 and 13, 17 and 19, and 29 and 31, etc. A composite number or non-prime is a product of primes or prime factors, e.g., the composite numbers 15 is the product of 2 primes, 3 and 5 (15 = 3 × 5), and 231 is the product of 3 primes, 3, 7, and 11 (231 = 3 × 7 × 11), etc. The integers or whole numbers are either primes or composites and are infinite.

The primes, which Euclid had proven to be infinite, are the atoms or building blocks of the infinite integers or whole numbers, which comprise of the infinite list of the odd numbers that are all either primes or products of primes (i.e., composites), and the infinite list of the even numbers that are all products of primes (i.e., composites, with the exception of 2 which is a prime, e.g., 6 = 2 × 3, 8 = 2 × 2 × 2, and 10 = 2 × 5, etc.). The infinite list of the integers or whole numbers may be classified as an infinite group, with various symmetries, subgroups and infinite elements, hidden within it. These various symmetries, subgroups and infinite elements, within this infinite group may be classified as follows:

(1) *Subgroup A*: Infinite consecutive primes such as 2, 3, 5, 7, 11, 13, 17, 19, 23, 29, and 31 to infinity, separated by 2 integers (twin primes), 4 integers, 6 integers, 8 integers, 10 integers, etc., which, incidentally, except for 2, are all odd numbers; this splitting up of the subgroup into infinite elements is shown below:
   (i) *Element A1*: Infinite list of all the prime pairs separated by 2 integers (twin primes) . . .. (Example: 17 and 19)
   (ii) *Element A2*: Infinite list of all the prime pairs separated by 4 integers/1 odd composite – single composite . . .. (Example: 79 and 83 separated by 81)
   (iii) *Element A3*: Infinite list of all the prime pairs separated by 6 integers/2 consecutive odd composites – twin composites . . .. (Example: 47 and 53 separated by 49 and 51)
   (iv) *Element A4*: Infinite list of all the prime pairs separated by 8 integers/3 consecutive odd composites – "triple" composites . . .. (Example: 359 and 367 separated by 361, 363 and 365)
   (v) *Element A5*: Infinite list of all the prime pairs separated by 10 integers/4 consecutive odd composites – "four-ple" composites . . .. (Example: 709 and 719 separated by 711, 713, 715, and 717)

(2) *Subgroup B*: Infinite consecutive odd composites such as 9, 15, 21, 25, 27, 33, 35, 39, 45, and 49 to infinity, of "various sizes" sandwiched between 2 primes; this splitting up of the subgroup into infinite elements is shown below:
   (i) *Element B1*: Infinite list of all "1 odd composite sandwiched between 2 primes single composite" (Example: 9 sandwiched between the primes 7 and 11)
   (ii) *Element B2*: Infinite list of all "2 consecutive odd composites sandwiched between 2 primes – twin composites" (Example: 253 and 255 sandwiched between the primes 251 and 257)
   (iii) *Element B3*: Infinite list of all "3 consecutive odd composites sandwiched between 2 primes – "triple" composites" (Example: 685, 687, and 689 sandwiched between the primes 683 and 691)

(iv) *Element B4*: Infinite list of all "4 consecutive odd composites sandwiched between 2 primes – "four-ple" composites" (Example: 2,769, 2,771, 2,773, and 2,775 sandwiched between the primes 2,767 and 2,777)

(v) *Element B5*: Infinite list of all "5 consecutive odd composites sandwiched between 2 primes – "five-ple" composites" (Example: 19,291, 19,293, 19,295, 19,297, and 19,299 sandwiched between the primes 19,289 and 19,301)

(3) *Subgroup C*: Infinite consecutive odd composites separated by 4 integers and 6 integers respectively; this splitting up of the subgroup into the 2 infinite elements is shown below:

1) *Element C1*: Infinite list of all "2 consecutive odd composites separated by 4 integers/1 prime" (Example: 209 and 213 separated by the prime 211)

2) *Element C2*: Infinite list of all "2 consecutive odd composites separated by 6 integers/2 primes" (Example: 279 and 285 separated by the twin primes 281 and 283)

(4) *Subgroup D*: Infinite single primes and twin primes separating 2 consecutive odd composites; this splitting up of the subgroup into the 2 infinite elements is shown below:

(i) *Element D1*: Infinite list of all the single primes separating 2 consecutive odd composites (Example: 23 separating the 2 consecutive odd composites 21 and 25)

(ii) *Element D2*: Infinite list of all the twin primes separating 2 consecutive odd composites (Example: 11 and 13 separating the 2 consecutive odd composites 9 and 15)

(5) *Subgroup E*: Infinite consecutive even composites such as 4, 6, 8, 10, 12, 14, 16, 18, 20, 22, etc. to infinity, all separated by only 2 integers; this subgroup may be classified as a single infinite element. There is always 1 even number between a twin prime pair, which is separated by 2 integers, a prime and a composite which are separated by 2 integers, and 2 composites which are separated by 2 integers. That is, the even numbers are always found in Subgroup A, Subgroup B, Subgroup C and Subgroup D above always evenly spaced out in consecutive order by 2 integers.

There is an evident symmetry in the above-mentioned infinite group, which would be broken if any of the elements within were to be finite. There are close interlinks between all the various infinite elements in all the five subgroups above, e.g., the infinity of the list of all the prime pairs each separated by 6 integers (Element A3/Subgroup A) implies the infinity of the list of all the "2 consecutive odd composites sandwiched between 2 primes – twin composites" (Element B2/Subgroup B) and vice versa, the infinity of the list of all the prime pairs each separated by 2 integers (twin primes) (Element A1/Subgroup A) implies the infinity of the list of all the "2 consecutive odd composites separated by 6 integers/2 primes" (Element C2/Subgroup C) and vice versa, the infinity of the list of all the infinite elements (A1, A2, A3, etc. to infinity) in Subgroup A above, which represents the infinity of the list of the primes which Euclid had in fact proven, implies the infinity of the list of all the "2 consecutive odd composites separated by 4 integers/1 prime" (Element C1/Subgroup C) and vice versa, the infinity of the list of all the "2 consecutive odd composites separated by 6 integers/2 primes" (Element C2/Subgroup C) implies the infinity of the list of all the twin primes separating 2 consecutive odd composites (Element D2/Subgroup D) and vice versa, the infinity of all the lists of all the infinite elements (A1, A2, A3, . . ., B1, B2, B3, . . ., C1 and C2, D1 and D2) in Subgroup A, Subgroup B, Subgroup C and Subgroup D above implies the infinity of the list of the consecutive even composites, i.e., 4, 6, 8, 10, 12, 14, 16, 18, 20, 22, etc. to infinity (Subgroup E), which we know to be true in any case, and vice versa.

Subgroup A and Subgroup B above are practically "mirror" images of one another – they represent the viewing of the primes and the composites from 2 variant angles – the infinitude, or, finiteness of either implies the infinitude, or, finiteness of the other; the same applies to both Subgroup C and Subgroup D above. It is similar to the following way of viewing a glass which is partially filled: this glass could be described as "half full" or "half empty" if it is half filled, "three-quarter full" or "one-quarter empty" if it is three-quarter filled, or "one-quarter full" or "three-quarter empty" if it is one-quarter filled, and so on.

It is evident that the infinitude, or, finiteness of any one of the above-mentioned elements would imply the infinitude, or, finiteness of the other element that is interlinked with it and vice versa. All these infinite elements are evidently entangled together and complementary, being all the infinite building blocks of the infinite integers or whole numbers. The infinity of the list of the integers or whole numbers, the primes included, in fact implies that all these various elements within it are infinite, and vice versa, since all these various elements are closely interlinked and could not do without each other. Therefore, the breaking of the evident intrinsic symmetry of this whole infinite group, i.e., the infinite list of the integers or whole numbers, due to the finiteness of any of the elements within it, could not be possible.

We pose a very important question: Besides questioning whether the infinite list of all the prime pairs separated by 2 integers (twin primes) is really infinite, should we not also be questioning whether the following are really infinite?

(a) Infinite lists of all the prime pairs separated respectively by 4 integers, 6 integers, 8 integers, 10 integers and sequentially larger integers (as in Subgroup A above).
(b) Infinite lists of all the respective consecutive odd composites of "various sizes" sandwiched between 2 primes (as in Subgroup B above).
(c) The 2 infinite lists with respectively "2 consecutive odd composites separated by 4 integers/1 prime" and "2 consecutive odd composites separated by 6 integers/2 primes" (as in Subgroup C above).
(d) The 2 infinite lists of respective single primes and twin primes separating 2 consecutive odd composites (as in Subgroup D above).
(e) Infinite list of the consecutive even composites all separated by only 2 integers (as in Subgroup E above).

Could there possibly be any symmetry-breaking in the above-mentioned infinite group whence one or more of the elements within it would be finite? In particular, could there be a possibility for the symmetry of this infinite group to be broken due to the finiteness of Element A1 (i.e., the finiteness of the twin primes) within it? Since the above-mentioned group, i.e., the list of the integers or whole numbers, is infinite, it is indeed not possible for all of these elements to be finite. And, there is no evident reason to account for why any of these elements, especially Element A1, i.e., the list of primes separated by 2 integers, or, twin primes, should be finite. In fact, all these infinite elements are like the slabs of various sizes in a building. They are all necessary for the construction of the infinite building known as the "infinite list of the integers or whole numbers" and should thus all be infinite, wherein the symmetry of the infinite group, i.e., the infinite list of the integers or whole numbers, would be preserved. The following explanation should make it clear that all these elements in the above-mentioned infinite group have to be infinite.

We here examine a hypothetical case wherein one of the elements within the infinite group is finite. Let us assume that Element A1, the list of primes separated by 2 integers, or, twin primes, in Subgroup A above, which is our target object, is finite. Let us look at the following list of consecutive integers or whole numbers where the bolded numbers are primes and the bolded, italicized numbers are twin primes:

201, 202, 203, 204, 205, 206, 207, 208, 209, 210, **211**, 212, 213, 214, 215, 216, 217, 218, 219, 220, 221, 222, **223**, 224, 225, 226, ***227***, 228, ***229***, 230, 231, 232, **233**, 234, 235, 236, 237, 238, ***239***, 240, ***241***, 242, 243, 244, 245, 246, 247, 248, 249, 250, **251**, 252, 253, 254, 255, 256, **257**, 258, 259, 260, 261, 262, **263**, 264, 265, 266, 267, 268, ***269***, 270, ***271***, 272, 273, 274, 275, 276, **277**, 278, 279, 280, ***281***, 282, ***283***, 284, 285, 286, 287, 288, 289, 290, 291, 292, **293**, 294, 295, 296, 297, 298, 299, 300, 301

Now, what would happen if the list of the twin primes (Element A1) were finite? Let us, e.g., assume that 269 and 271 are the largest existing twin prime pair – there are no twin primes larger than they. And let us look at the above list of consecutive integers or whole numbers now:

201, 202, 203, 204, 205, 206, 207, 208, 209, 210, **211**, 212, 213, 214, 215, 216, 217, 218, 219, 220, 221, 222, **223**, 224, 225, 226, ***227***, 228, ***229***, 230, 231, 232, **233**, 234, 235, 236, 237, 238, ***239***, 240, ***241***, 242, 243, 244, 245, 246, 247, 248, 249, 250, **251**, 252, 253, 254, 255, 256, **257**, 258, 259, 260, 261, 262, **263**, 264, 265, 266, 267, 268, ***269***, 270, ***271***, 272, 273, 274, 275, 276, **277**, 278, 279, 280, 282, 284, 285, 286, 287, 288, 289, 290, 291, 292, **293**, 294, 295, 296, 297, 298, 299, 300, 301

Since 269 and 271 are the largest existing twin primes, the twin prime pair, 281 and 283, would not exist anymore, resulting in 2 gaps (between the integers 280 and 284) in the above list of consecutive integers or whole numbers. Thus, the earlier complete list of consecutive integers or whole numbers has become incomplete and discontinuous as a result – 2 building blocks, 281 and 283, which are twin primes (primes separated by 2 integers), are missing from this list now – the list has now become asymmetrical, i.e., symmetry-breaking has occurred. As the list of integers or whole numbers is always complete, continuous, infinite and symmetrical, such a list with 2 missing consecutive odd numbers which are both primes (these missing consecutive odd numbers are twin primes) is an absurdity – it could never exist. Therefore, a largest existing twin prime pair, such as 269 and 271 mentioned above, is an absurdity – a largest existing twin prime pair could never exist, and it is evident here that, of necessity, in order that the list of the integers or whole numbers would be complete, continuous, infinite and symmetrical, the list of the twin primes (primes separated by 2 integers), as well as the lists of the prime pairs separated by 4 integers, 6 integers, 8 integers, 10 integers and so on in arithmetically ascending order, have to be infinite. That is, all the elements in Subgroup A above have to be infinite.

By a similar reasoning, the same has to also apply to all the elements in Subgroup B, Subgroup C, Subgroup D and Subgroup E above.

Therefore, by the above lemma, all the elements in Subgroup A, Subgroup B, Subgroup C, Subgroup D and Subgroup E above are infinite.

It is clear that the twin primes are infinite.

Part 4

Keywords: various prime pairs, building blocks, fraction of infinity

Theorem. The twin primes are infinite.

Proof.

Lemma. A fraction of infinity is also infinite (see appendix).
The list of the primes, which are the building blocks of the integers, had been proven by Euclid to be infinite. The question now is whether the list of the twin primes is also infinite.

It would be appropriate to conduct an examination of the prime pairs separated by 2 integers (commonly known as twin primes) vis-à-vis the other prime pairs separated by more than 2 integers, e.g., 4 integers, 6 integers, 8 integers and more. For this purpose, we select a reasonably large list of consecutive primes, which may be regarded as a basic unit of the infinite list of the primes; we examine a compilation of such data obtained from, say, the list of the first 2,500 consecutive primes – 2 to 22,307 – which is as follows:

List of prime pairs for the first 2,500 consecutive primes, 2 to 22,307, ranked according to their frequencies of appearance

S. no.	Ranking	Prime pairs	No. of pairs	Percentage
(1)	1	Prime pair separated by 6 integers	482	19.29%
(2)	2	Prime pair separated by 4 integers	378	15.13%
(3)	3	Prime pair separated by 2 integers (t. p.)	376	15.05%
(4)	4	Prime pair separated by 12 integers	267	10.68%
(5)	5	Prime pair separated by 10 integers	255	10.20%
(6)	6	Prime pair separated by 8 integers	229	9.16%
(7)	7	Prime pair separated by 14 integers	138	5.52%
(8)	8	Prime pair separated by 18 integers	111	4.44%
(9)	9	Prime pair separated by 16 integers	80	3.20%
(10)	10	Prime pair separated by 20 integers	47	1.88%
(11)	11	Prime pair separated by 22 integers	46	1.84%
(12)	12	Prime pair separated by 30 integers	24	0.96%
(13)	13	Prime pair separated by 28 integers	19	0.76%
(14)	14	Prime pair separated by 24 integers	16	0.64%
(15)	15	Prime pair separated by 26 integers	10	0.40%
(16)	16	Prime pair separated by 34 integers	9	0.36%
(17)	17	Prime pair separated by 36 integers	5	0.20%
(18)	18	Prime pair separated by 32 integers	2	0.08%
(19)	18	Prime pair separated by 40 integers	2	0.08%
(20)	19	Prime pair separated by 42 integers	1	0.04%
(21)	19	Prime pair separated by 52 integers	1	0.04%
Total no. of prime pairs in list: 2,498				

It is evident in the above list that the prime pairs separated by 6 integers, 4 integers and 2 integers (twin primes), among the 21 classifications of prime pairs separated by from 2 integers to 52 integers (prime pairs separated by 38 integers, 44 integers, 46 integers, 48 integers, and 50 integers are not among them, but, they are expected to appear further down in the infinite list of the primes), are the most dominant, important. There is a long list of other prime pairs, besides those shown in the above list, which also play a part as the building blocks of the infinite list of the integers. We shall prove the infinity of all these various building blocks below.

The list of the integers is infinite. The list of the primes is also infinite. The infinite primes are the building blocks of the infinite integers – the infinite odd integers are all either primes or composites of primes, and the infinite even integers, except for 2 which is a prime, are all also composites of primes. Therefore, all the prime pairs separated by the integers of various magnitudes, as described above, could never all be finite. If there is any possibility at all for any of these prime pairs to be finite, there is only the possibility that a number of these prime pairs are finite (but never all of them). However, would it have to be the prime pairs separated by 2 integers or twin primes (which are the subject of our investigation here), which are the only prime pair, or, one among a number of prime pairs, which are finite? Why question only the infinity of the prime pairs separated by 2 integers, the twin primes? Are not the infinities of the prime pairs separated by 8 integers and more, whose frequencies of appearance are lower, as compared to those of the prime pairs which are separated by 6, 4, and 2 integers, respectively, in the above list of prime pairs, more questionable? Why single out only the twin primes? (There are at least 18 other prime pairs, separated by from 8 integers to 52 integers, whose respective infinities should be more suspect, as is evident from the above list of prime pairs, if any infinities should be doubted. It is evident that the prime pairs separated by 2 integers (twin primes) are not that likely to be finite.)

All of the above lists of prime pairs which are respectively separated by from 2 integers to 52 integers are each, respectively, a fraction of the infinite list of the primes: the list of prime pairs separated by 2 integers (twin primes) is a fraction of the infinite list of the primes, the list of prime pairs separated by 4 integers is also a fraction of the infinite list of the primes, the list of prime pairs separated by 6 integers is a fraction of the infinite list of the primes too, and so on all the way down to the list of prime pairs separated by 52 integers.

Therefore, by the above lemma, all these various lists of prime pairs separated by integers of various magnitudes are each also infinite.

Appendix – Euclid's Proof of the Infinity of the Pythagorean Triples: A Fraction of Infinity Is Also Infinite

A Pythagorean triple is a set of 3 integers wherein 1 number squared added to another number squared equals the third number squared, e.g., 3^2 (9) + 4^2 (16) = 5^2 (25) below.

Euclid's proof of the infinity of the Pythagorean triples begins with the statement that the difference between 2 successive square numbers is always an odd number, as is evident below:

Column 1	Column 2		Column 3
1^2	1		
		}	3 (difference between 4 and 1)
2^2	4		
		}	5 (difference between 9 and 4)
3^2	9		
		}	7 (difference between 16 and 9)
4^2	16		
		}	9 (difference between 25 and 16)

(continued)

Column 1	Column 2		Column 3
5^2	25		
		}	11 (difference between 36 and 25)
6^2	36		
		}	13 (difference between 49 and 36)
7^2	49		
		}	15 (difference between 64 and 49)
8^2	64		
		}	17 (difference between 81 and 64)
9^2	81		
		}	19 (difference between 100 and 81)
10^2	100		

Every one of the infinity of odd numbers (in Column 3 above) could be added to a particular square number (in Column 2 above) to give another square number, e.g., 3 in Column 3 above could be added to 1 in Column 2 above to give the square number 4, 5 in Column 3 above could be added to 4 in Column 2 above to give the square number 9, 7 in Column 3 above could be added to 9 in Column 2 above to give the square number 16, and so on. A fraction of the infinite odd numbers in Column 3 above are themselves square numbers, e.g., the odd number 9 in Column 3 above is a square number, and is the only square number in the list of odd numbers shown there, representing there a fraction of the infinite odd numbers which are square numbers. However, a fraction of infinity is also infinite.

Therefore, there are also an infinity of odd square numbers (in Column 3 above) which could each be added to another square number (in Column 2 above) to give another square number, i.e., there is an infinitude of Pythagorean triples.

Part 5

Keywords: matched; building blocks; recurrent, identical mathematical structures

Theorem. There is an infinitude of twin prime pairs.

Proof 1.

Lemma. By Georg Cantor's definition of infinity something that is undeniably smaller than an infinite quantity could also be infinite. Cantor, who had worked alongside Hilbert, defined infinity as the size of the never-ending list of counting numbers (1, 2, 3, 4, 5, 6, 7, 8, . . .). Anything that is comparable in size is equally infinite.

By this definition, the number of odd counting numbers, that would intuitively appear to be smaller, is also infinite. It could be shown that the quantity of counting numbers and the quantity of odd numbers are comparable as we could pair off each counting number with a corresponding odd number, as follows:

$$\begin{array}{ccccccccc} 1 & 2 & 3 & 4 & 5 & 6 & 7 & 8 & 9\ldots \\ \downarrow & \downarrow & \downarrow & \downarrow & \downarrow & \downarrow & \downarrow & \downarrow & \downarrow\ldots \\ 1 & 3 & 5 & 7 & 9 & 11 & 13 & 15 & 17\ldots \end{array}$$

If every member of the counting numbers list could be matched with a member of the odd numbers list then the two lists must be the same size. As the list of counting numbers is infinite the list of odd numbers must also be infinite.

The primes (with the exception of the number "2" which is prime by definition only) and twin primes are also odd numbers and belong to the list of the odd numbers. As the list of the odd numbers is infinite the list of the primes and twin primes must also be infinite. (In fact, every odd number (in the infinite list of odd numbers) is either a prime number or a composite of prime numbers (i.e., it consists of prime factors), the primes being the "atoms" or building blocks of the odd numbers (the primes could also be regarded as the building blocks of all the whole numbers or integers, including the even numbers – vide the Goldbach conjecture, which, it appears, virtually all mathematicians believe to be true, all the even numbers with the exception of 2 are each the sum of two prime numbers – all the even numbers with the exception of 2 are also each the product of prime numbers (composite).) The twin primes could be found scattered all over the infinite list of odd numbers. All this means that just by the expediency of enumerating the infinite odd numbers, we could "generate" an infinite number of primes and twin primes. With regard to the twin primes, it is as if the innumerable "twin primes" fruits are waiting to be plucked from the infinitely big "odd numbers" tree.

Proof 2.

Lemma. The development of the self-similarity concept has brought fame to Mitchell Feigenbaum, who has worked in the Los Alamos Laboratory in the early 1970s. The "distinguished presence" of the twin primes among the prime numbers, as described in the paper, is analogous to the "self-similar mathematical pattern or structure" (which is the shape of the fig tree itself) of the various parts of the fig tree, i.e., its trunk to bough section, bough to branch section, branch to twig section and twig to twiglet section, in Feigenbaum's famous fig tree example, and such self-similar mathematical pattern or structure, or fractal characteristic, could also be found in other aspects of nature, e.g., waves, turbulence or chaos, the structures of viruses and bacteria, polymers and ceramic materials, the universe and many others, even the movements of prices in financial markets, the growths of populations, the sound of music, the flow of blood through our circulatory system, the behavior of people en masse, etc., which have all spawned a relatively new and important branch of mathematics with wide practical applications known as fractal geometry, which has been pioneered by Benoit Mandelbrot. As a matter of fact, self-similarity or fractal characteristic could be regarded as the fundamental mathematical aspect found in practically everything in nature including the numbers such as the prime numbers and twin primes which are the subjects of our investigation here, and this new branch of mathematics, fractal geometry, besides having a great practical impact on us also gives us a deeper vision of the universe in which we live and our place in it.

By Dirichlet's theorem, the arithmetic progressions (arithmetic sequences) of terms a + im and (a + 2) + im, i = 0, 1, 2, 3, 4, . . ., contain infinitely many primes if a and a + 2 have no factors or divisors in common with m other than +1 and −1 (i.e., the integers, a and m, are relatively prime or coprime, as well as the integers, a + 2 and m). We now, e.g., let a = 5 and m = 12 (we could always assign values to a and m such that both a and a + 2 have no factors or divisors in common with m other than + 1 and −1). Then we have the following sequences (a + im, i = 0, 1, 2, 3, 4, . . ., and (a + 2) + im, i = 0, 1, 2, 3, 4, . . .):

5, *17*, *29*, *41*, *53*, 65, 77, *89*, *101*, *113*, 125, *137*, *149*, 161, *173*, . . .

7, *19*, *31*, *43*, 55, *67*, *79*, 91, *103*, 115, *127*, *139*, *151*, *163*, 175, . . .

The integers in the above sequences which are italicized are prime numbers. The first sequence contains infinitely many primes. The second sequence contains infinitely many primes too. The corresponding integers in both the sequences, e.g., 5 and 7, 17 and 19, 29 and 31, and so on, differ from one another by 2, which is a characteristic of the twin primes, and a number of twin primes is evident in both the sequences.

Since the twin primes (5 and 7, 17 and 19, 29 and 31, 41 and 43, 101 and 103, 137 and 139, and 149 and 151) are evident in both the sequences, by the infinitude of the primes in both the sequences as postulated by Dirichlet's Theorem, and the above lemma, which postulates that the elements of an infinite subset of an infinite set contain all the recursive significant properties of that set unless the process that selects the elements of the subset directly excludes a property, infinitely many of them must appear thereafter in both the sequences; so, there must be infinitely many twin primes, the twin primes representing a characteristic or feature of the prime numbers, the recurrent, identical mathematical structures postulated by the well-established self-similarity concept.

Part 6

Keywords: induction; contradiction; recurrent, identical mathematical structures

Theorem. There is an infinitude of twin primes.

Proof 1.

Lemma. The development of the self-similarity concept has brought fame to Mitchell Feigenbaum, who has worked in the Los Alamos Laboratory in the early 1970s. The "distinguished presence" of the twin primes among the prime numbers, as described in the paper, is analogous to the "self-similar mathematical pattern or structure" (which is the shape of the fig tree itself) of the various parts of the fig tree, i.e., its trunk to bough section, bough to branch section, branch to twig section and twig to twiglet section, in Feigenbaum's famous fig tree example, and such self-similar mathematical pattern or structure, or, fractal characteristic, could also be found in other aspects of nature, e.g., waves, turbulence or chaos, the structures of viruses and bacteria, polymers and ceramic materials, the universe and many others, even the movements of prices in financial markets, the growths of populations, the sound of music, the flow of blood through our circulatory system, the behavior of people en masse, etc., which have all spawned a relatively new and important branch of mathematics with wide practical applications known as fractal geometry, which has been pioneered by Benoit Mandelbrot. As a matter of fact, self-similarity or fractal characteristic could be regarded as the fundamental mathematical aspect found in practically everything in nature including the numbers such as the prime numbers and twin primes which are the subjects of our investigation here, and this new branch of mathematics, fractal geometry, besides having a great practical impact on us also gives us a deeper vision of the universe in which we live and our place in it.

We could generate an infinite list of primes, Q, by utilizing Euclid's indirect (*reductio ad absurdum*) method of proving the infinity of the primes. In his proof, he first assumed that there is a finite list of prime numbers and then showed that there must exist an infinite number of additions to this list. The proof is as follows:

Assume that n is the largest existing prime number in the following list of primes:

$$x = 2,\ 3,\ 5,\ 7,\ 11,\ 13,\ 17,\ 19,\ \ldots,\ n$$

There could be a prime, q, which is larger than n, computed as follows:

$$q = (2 \times 3 \times 5 \times 7 \times 11 \times 13 \times 17 \times 19 \times \cdots \times n) + 1$$

This number, q, could not be divided exactly by any of our primes (from 2 to n) and must be either a prime number itself which is larger than n or must be divisible by prime factors each of which is larger than n, both of which cases contradicting our original assumption that n is the largest existing prime, thus implying the infinity of the primes. In this manner we have added to our original list of primes, which is x. We could now repeat the process by including our new prime, q, or its prime factors if q is not prime in our list and generate a new number that is either a prime number that is larger than q or a composite number with prime factors that are each larger than q, i.e., we could generate some other new prime or primes that are not on our list of known primes. No matter how long our list of prime numbers is it is possible to find some new prime or primes. Thus, the list of primes is never-ending and infinite. In other words, Q is an infinite list of prime numbers.

We could in fact modify Euclid's indirect method slightly, but use the same reasoning, to generate another infinite list of prime numbers, P. (This method would have had served Euclid equally well in proving the infinity of the primes.) This proof takes the following form:

Assume that n is the largest existing prime in the following list of prime numbers:

$$x = 2, 3, 5, 7, 11, 13, 17, 19, \ldots, n$$

There could be a prime number, p, which is larger than n, computed as follows:

$$p = (2 \times 3 \times 5 \times 7 \times 11 \times 13 \times 17 \times 19 \times \cdots \times n) - 1$$

(p is either a prime that is larger than n, or it has prime factors each of which is larger than n; therefore, n could not be the largest existing prime, and P is an infinite list of prime numbers.)

It is evident that Q – P = 2. In other words, $q_1 - p_1 = 2$, $q_2 - p_2 = 2$, $q_3 - p_3 = 2$, $q_4 - p_4 = 2$, and so on. This is the characteristic of twin primes.

We would here match P and Q for n = 2 to 19, which is as follows:

n = 2 to 19	*P*	*Q*
1] (2 × 3) ± 1	5 = Prime	7 = Prime (direct twin primes)
2] (2 × 3 × 5) ± 1	29 = Prime	31 = Prime (direct twin primes)
3] (2 × 3 × 5 × 7) ± 1	209 = 11 x 19 (indirect twin primes: 11 and 13; 17 and19)	211 = Prime
4] (2 × 3 × 5 × 7 × 11) ± 1	2,309 = Prime	2,311 = Prime (direct twin primes)
5] (2 × 3 × 5 × 7 × 11 × 13) ± 1	30,029 = Prime (indirect twin primes: 59 and 61)	30,031 = 59 × 509
6] (2 × 3 × 5 × 7 × 11 × 13 × 17) ± 1	510,509 = 61 × 8,369 (indirect twin primes: 59 and 61)	510,511 = 19 × 97 × 277 (indirect twin primes: 17 and 19)
7] (2 × 3 × 5 × 7 × 11 × 13 × 17 × 19) ± 1	9,699,689 = 53 × 197 × 929 (indirect twin primes: 197 and 199)	9,699, 691 = 347 × 27,953 (indirect twin primes: 347 and 349)

It is evident that P and Q, for $n = 2$ to 19, yield three pairs of direct twin primes and five pairs of different indirect twin primes which are the result of pairing the prime factors of the p's and the q's with the prime numbers which differ from each of them respectively by 2. Moreover, all these direct and indirect twin primes, a total of 8 pairs of them, are found between n and Q and are of course larger than n, the assumed largest existing prime number; they are scattered among the 25 newly found prime numbers (which are all larger than n). (Note: The above computations are performed with an ordinary hand-held calculator which is capable of yielding results up to $n = 19$ only. For results for $n > 19$, a more powerful calculator or computer would be needed.)

Beside these direct and indirect twin primes between n and Q, more twin primes larger than n, and relatively plenty of them too, could be generated between n and Q by resorting to the Sieve of Eratosthenes – we shall call these twin primes the extraneous twin primes. For example, in (3) above, where $n = 7$ and $Q = 211$, 41 and 43, 71 and 73, 101, and 103, are extraneous twin primes which are all larger than n ($n = 7$) and $n - 2 = 5$, 5, and 7 here being twin primes, and in (5) above, where $n = 13$ and $Q = 30{,}031$, 107, and 109, 137 and 139, 227 and 229, 311, and 313, etc., are extraneous twin primes which are all larger than n ($n = 13$) and $n - 2 = 11$, 11, and 13 being twin primes. Hence, if we assume that any of these $n - 2$'s and n's are the largest existing twin primes in the spirit of Euclid we would encounter contradictions from the extraneous twin primes (as well as the direct twin primes and/or the indirect twin primes) which are all larger (these being indirect proofs of the infinitude of the twin primes, which would be described below).

By induction:

P contains an infinite listing of direct primes, the p's, and prime factors of the p's wherever the p's are non-prime, as proven above.

Q also contains an infinite listing of direct primes, the q's, and prime factors of the q's wherever the q's are non-prime, as proven above.

$Q - P = 2$, which is a characteristic of the twin primes.

Both P and Q contain a number of direct twin primes (the result of the pairing of the direct primes in P and Q) and indirect twin primes (the result of the pairing of the prime factors of the p's and the q's in both P and Q with the prime numbers which differ from each of them respectively by 2) for $n = 2$ to 19, between n and Q, as is shown in the above example – in this case, a total of eight pairs of them scattered among the twenty-five newly found prime numbers, and everyone of these, twin primes and newly found prime numbers, is larger than n, the assumed largest existing prime, a contradiction. Beside these direct and indirect twin primes, more twin primes, the evidently more abundant extraneous twin primes between n and Q, are generated by the Sieve of Eratosthenes – these extraneous twin primes are all larger than n as well. All these are in line with the concept behind our proofs by contradiction, the indirect proofs – in these indirect proofs, contradiction implies an infinite number of primes, not excluding the twin primes, which are two primes that differ from one another by 2. (The indirect proof of the twin primes' infinitude is described below.) From the above-mentioned example of twin primes, and many, many other examples of twin primes beside this, it is evident that twin primes represent a characteristic or feature of the prime numbers. This is in accordance with the above lemma, upon which the method of renormalization in perturbation theory is based, developed by Mitchell Feigenbaum in the 1970s, which postulates that there is a tendency of identical mathematical structures to recur on many levels. Within a given structure, there would be smaller copies of the same structure, their sizes being determined by the scaling factor. Feigenbaum found that at the utmost tips of the fig tree, there is some mathematical structure which remains the same when its size is changed (enlarged) by a scaling factor of 4.669, which is found to be a constant like pi (3.142); this structure is the shape of the fig tree itself; in other words, little whorls could be found within big whorls. Renormalization has been a well-established technique in chaos theory/fractal geometry and is a mathematical trick which functions rather like a microscope, zooming in on the self-similar structure, removing any approximations, and filtering out everything else. All this shows the universality of some features of chaos. That is, some kind of order or pattern could be found in or is inherent in disorder or chaos. Likewise, the "distinguished" appearances of the twin primes could be found among the chaotic (disorderly, random, pattern-less) array of infinite prime numbers, the twin primes being comparable to the little whorls found within the big whorls – they represent the recurrent, identical mathematical structures postulated by the above lemma.

Since by Euclid's proof the list of the primes is infinite, and, since the twin primes represent a characteristic or feature of the prime numbers as is described above, it could thus be concluded that there is an infinitude of twin primes. This is the result of the "reflection" principle. (For example, the characteristic of a mountain or infinite volume of sand is reflected in the characteristic of some grains of sand found there so that studying the characteristic of some grains of sand found there is sufficient for deducing the characteristic of the mountain or infinite volume of sand. The twin primes found in P and Q above could be taken to represent the "grains of sand" found in the "infinite volume of sand" which is represented by the infinite list of the primes. The "grains of sand" (which are in effect a subset) and the "infinite volume of sand" (which are in effect a set), here, share this same characteristic – the presence of twin primes within them.)

Thus, in general, some of the direct primes in P and Q would pair to form direct twin primes and some of the prime factors of the p's and the q's in both P and Q would pair with other prime numbers (which differ from each of them respectively by 2) to form indirect twin primes, not to mention the presence of the relatively abundant extraneous twin primes generated by the Sieve of Eratosthenes. All these results are representative of the aforementioned characteristic or "pattern" of the prime numbers, and they are in accordance with the above lemma – they represent the recurrent, identical mathematical structures postulated by the above lemma.

By the above lemma and the "reflection" principle, the infinity of the primes in P (direct primes and prime factors) and the infinity of the primes in Q (direct primes and prime factors) imply that there is an infinitude of twin primes (direct twin primes, indirect twin primes and extraneous twin primes) generated by both P and Q together, and the Sieve of Eratosthenes.

Hence, it is clear that the twin primes are infinite.

Proof 2.

The following is indirect proof (proof by contradiction) of the infinity of the twin primes. As P and Q contain an infinitude of twin primes (direct and indirect), including the extraneous twin primes between n and Q generated by the Sieve of Eratosthenes, which are all larger than n, the assumed largest existing prime number, as proven above, this implies that if this n, and n – 2, are both prime numbers (twin primes), they could not be the largest existing pair of twin primes. If we assume that these n and n – 2 are the largest existing twin primes, we would be contradicted by the above-mentioned proof. This is the indirect proof of the infinity of the twin primes. In the above-mentioned example, for n = 2 to 19, our respective assumed largest existing pairs of twin primes (n – 2 and n), 3 and 5, 5 and 7, 11 and 13, and 17 and 19, are superseded (and "contradicted") by 29 and 31 (direct twin primes); 11 and 13, and, 17 and 19 (indirect twin primes); 59 and 61 (indirect twin primes); and 197 and 199, and, 347 and 349 (indirect twin primes), respectively; not to mention the respective larger extraneous twin primes generated by the Sieve of Eratosthenes. In other words, it could be expected that all the direct twin primes and the indirect twin primes so produced (or generated), and the extraneous twin primes generated by the Sieve of Eratosthenes, as described above, would be larger than n (the assumed largest existing prime number), or n – 2 and n (the assumed largest existing pair of twin primes) – this is incontrovertible indirect proof of the infinitude of the primes and twin primes. With the two formulae, $(2 \times 3 \times 5 \times 7 \times 11 \times 13 \times 17 \times 19 \times \cdots \times n) - 1$ and $(2 \times 3 \times 5 \times 7 \times 11 \times 13 \times 17 \times 19 \times \cdots \times n) + 1$ (and the help of the Sieve of Eratosthenes), we could indeed generate two infinite lists of prime numbers, with an infinite number of twin primes (direct, indirect and extraneous) scattered here and there all over the place.

We state the following theorem:

If we let n represent any successive prime number in a list of consecutive prime numbers which commences from the prime number 2 and assume it to be the largest existing prime number, then by the following statements:

$$p = (2\times3\times5\times7\times11\times13\times17\times19\times \cdots n) - 1$$

$$q = (2\times3\times5\times7\times11\times13\times17\times19\times \cdots n) + 1$$

since $q - p = 2$, if both p and q are prime numbers, in which case they are twin primes, they would be larger than n, and if n – 2 is also a prime number, in which case these n and n – 2 are twin primes, and if we then assume that these n and n – 2 are the largest existing twin prime pair, we would find that these twin primes, p and q, would be larger than these n and n – 2, the assumed largest existing twin prime pair, hence contradicting the assumption that these n and n – 2 are the largest existing twin prime pair and proving the infinitude of the twin primes according to the principle of "reductio ad absurdum"; if, on the other hand, either of these p and q or both these p and q are not prime (i.e., they are composite numbers), then the prime factors of either of these p and q, or both these p and q would be larger than these n and n – 2, and if any of these prime factors pair with other prime numbers that differ from them by 2 then we would have a twin prime pair or pairs consisting of these prime factor or factors of p and/or q and another prime number or numbers that differ from these prime factor or factors by 2 which would be larger than these n and n – 2, the assumed largest existing twin prime pair, thereby contradicting this assumption and proving the infinity of the twin primes through the principle of "reductio ad absurdum"; this rule would always apply whenever p and q are both primes, i.e., they are twin primes, or whenever p and/or q are non-prime, i.e., they are composite numbers, and their prime factor or factors pair with another prime number or numbers that differ from them by 2 to form a twin prime pair or pairs, when, at the same time, n and n – 2 are also prime numbers, i.e., twin primes. In other words: If p and q are both primes, i.e., twin primes, or if a prime factor or factors of p and/or q pair with another prime number or numbers that differ from them by 2 to form a twin prime pair or pairs (when p and/or q are composite, i.e., non-prime), then n and n – 2 could not be the largest existing twin prime pair if it happened that n and n – 2 are both primes, i.e., twin primes.

Proof 3.

The following describes two other ways of proving the infinity of the primes:

(1) Whatever number one picks, there is a prime larger than it.
(2) Pick a number – N.
(3) Multiply all the positive integers starting with 1 and ending with N.
(4) I.e., form the factorial of N (written N!).
(5) What one gets is divisible by every number up to N.
(6) When one adds 1 to N!, the result:
   (a) can't be a multiple of 2 (because it leaves 1 over, when you divide by 2).
   (b) can't be a multiple of 3 (because it leaves 1 over, when you divide by 3).
   (c) can't be a multiple of 4 (because it leaves 1 over, when you divide by 4).

   .

   .

   .

   (n) can't be a multiple of N (because it leaves 1 over, when you divide by N).
(7) In other words, N! + 1, if it is divisible at all (other than by 1 and itself) only is divisible by numbers greater than N.
(8) So, either N! + 1 is itself prime, or its prime divisors are greater than N.

(9) In either case, there must exist a prime bigger than N.
(10) The process holds no matter what number N is.
(11) Whatever N is, there is a prime greater than N.
(12) Thus, the infinitude of the primes. (Generalization)

However, if we modify (6) above as follows the conclusion would still be the same:

(6) When one minuses 1 from N!, the result:
   (a) can't be a multiple of 2 (because it leaves 1 over, when you divide by 2).
   (b) can't be a multiple of 3 (because it leaves 2 over, when you divide by 3).
   (c) can't be a multiple of 4 (because it leaves 3 over, when you divide by 4).
   .
   .
   .
   (n) can't be a multiple of N (because it leaves N – 1 over, when you divide by N).
(7) In other words, N! – 1, if it is divisible at all (other than by 1 and itself) only is divisible by numbers greater than N.
(8) So, either N! – 1 is itself prime, or its prime divisors are greater than N.
(9) In either case, there must exist a prime bigger than N.
(10) The process holds no matter what number N is.
(11) Whatever N is, there is a prime greater than N.
(12) Thus, the infinitude of the primes. (Generalization)

Therefore: When N! – 1 and N! + 1 are both primes, or when any of their prime divisors are partnered by primes which differ from them by 2 the result would be new twin prime pairs. By the infinitude of the primes generated by both N! + 1 and N! – 1 (which differ from one another by 2, a characteristic of the twin primes), as described above, and the well-established self-similarity concept stated above, there would be an infinitude of such twin prime pairs, the twin primes being a characteristic or feature of the prime numbers, the recurrent, identical mathematical structures postulated by the self-similarity concept.

Proof 4.

All odd numbers are either prime numbers or composite numbers (i.e., they contain prime factors), the prime numbers being the "atoms" or building blocks of the odd numbers. Based on this premise, we carry on with the following steps:

(1) It is possible to generate new prime numbers. All we have to do is to first create a composite number by multiplying the consecutive odd numbers together, starting from the lowest odd number, 3 (which is a prime number), in consecutive, ascending order, and ending at any odd number (N), for example (a hand-held calculator yields results up to N = 17 only):

(i) $3 \times 5 = 15$ (N = 5)
(ii) $3 \times 5 \times 7 = 105$ (N = 7)
(iii) $3 \times 5 \times 7 \times 9 = 945$ (N = 9)
(iv) $3 \times 5 \times 7 \times 9 \times 11 = 10{,}395$ (N = 11)
(v) $3 \times 5 \times 7 \times 9 \times 11 \times 13 = 135{,}135$ (N = 13)
(vi) $3 \times 5 \times 7 \times 9 \times 11 \times 13 \times 15 = 2{,}027{,}025$ (N = 15)
(vii) $3 \times 5 \times 7 \times 9 \times 11 \times 13 \times 15 \times 17 = 34{,}459{,}425$ (N = 17)

.
.
.

(n) $3 \times 5 \times 7 \times 9 \times 11 \times 13 \times 15 \times 17 \times \cdots \times N$

(2) The results in (1) above would be composite numbers which are divisible by any of their odd factors (primes and multiples of primes), for example, 3 or 5 in (i), and 3, 5, 7, 9, 11, 13, 15 or 17 in (vii).
(3) Next, we attempt to make each of these results non-composite, i.e., turn each of these results into a prime number, by taking away 2 from each of them to get result X (which could turn out to be a prime number).
(4) Then, we take away 2 from X, and we could get another prime number, Y (which would of course be a twin prime if it is a prime number and if X is also a prime number).
(5) For example:
(a) In (1)(i) above, we would get the following twin primes:

$$(3 \times 5) - 2 = 13 \quad (X)$$
$$(3 \times 5) - 2 - 2 = 11 \; (Y)$$

(b) In (1)(ii) above, we would get the following twin primes:

$$(3 \times 5 \times 7) - 2 = 103 \quad (X)$$
$$(3 \times 5 \times 7) - 2 - 2 = 101 \; (Y)$$

(c) In (1)(iii) above, we would get the following composite number and prime:

$$(3 \times 5 \times 7 \times 9) - 2 = 943 \quad (X) - \text{Composite Number}$$
$$(3 \times 5 \times 7 \times 9) - 2 - 2 = 941 \; (Y) - \text{Prime Number}$$

(d) In (1)(iv) above, we would get the following composite number and prime:

$$(3 \times 5 \times 7 \times 9 \times 11) - 2 = 10{,}393 \quad (X) - \text{Composite Number}$$
$$(3 \times 5 \times 7 \times 9 \times 11) - 2 - 2 = 10{,}391 \; (Y) - \text{Prime Number}$$

(e) In (1)(v) above, we would get the following composite number and prime:

$$(3 \times 5 \times 7 \times 9 \times 11 \times 13) - 2 = 135{,}133 \quad (X) - \text{Composite Number}$$
$$(3 \times 5 \times 7 \times 9 \times 11 \times 13 \times 15) - 2 = 2{,}027{,}023 \; (Y) - \text{Prime Number}$$

(f) In (1)(vi) above, we would get the following twin primes:

$$(3\times5\times7\times9\times11\times13\times15)-2=2,027,023 \quad (X)$$
$$(3\times5\times7\times9\times11\times13\times15)-2-2=2,027,021 \quad (Y)$$

(g) In (1)(vii) above, we would get the following composite number and prime:

$$(3\times5\times7\times9\times11\times13\times15\times17)-2=34,459,423 \quad (X) – \text{Prime Number}$$
$$(3\times5\times7\times9\times11\times13\times15\times17)-2-2=34,459,421 \quad (Y) – \text{Composite Number}$$

.

.

.

(6) X and Y would be larger than N and would never be divisible by any of the consecutive odd numbers, 3 to N. They are either prime numbers (which may pair to become twin primes) or composite numbers. The fact that X and Y are not divisible by any of the consecutive odd numbers, 3 to N, means that if they are each composite numbers their prime factors would be larger than N, and if each of these prime factors pair with another prime to form a twin prime pair, this twin pair would of course also be larger than N.

(7) Prime numbers and twin primes are likely to appear in X and Y for the following reasons:

(a) The lists of numbers to be multiplied together to create composite numbers in (1) above do not include even numbers and all the numbers are consecutive odd numbers (whereas Euclid's list of prime numbers in his indirect proof also includes the prime 2 which is an even number), leaving relatively little room or opportunity for composite numbers with prime factors larger than N to form after executing steps (3) and (4) above. The inclusion of even number(s) would cause an increase in the numerical value of the composite number consisting of odd factors, this composite number being only divisible by odd numbers, for example:

In the following list of numbers which includes even numbers: 2 × 3 × 4 × 5 × 6 × 7 × 8 × 9 × 10 × 11, the composite of odd numbers (3 × 5 × 7 × 9 × 11 = 10,395), 10,395, would be increased in numerical value by (2 × 4 × 6 × 8 × 10 = 3,840) 3,840 times to become (10,395 × 3,840 = 39,916,800) 39,916,800, due to the inclusion of the even numbers, 2, 4, 6, 8 and 10. (If we, for example, attempt to convert the result of multiplying the composite of odd numbers by the even numbers, in this case 39,916,800, into twin primes by subtracting 1 from it (39,916,800 – 1 = 39,916,799 which is not a prime but a composite number) and then subtracting 2 from the last result (39,916,799 – 2 = 39,916,797 which is also a composite number), we are more likely to end up with composite numbers rather than primes or twin primes.)

(b) X and Y are each smaller than the respective composites of the consecutive odd numbers, 3 to N (shown in (1) above), by 2 and 4 respectively, evidently leaving even less room or opportunity for composite numbers with prime factors larger than N to form after executing steps (3) and (4).

(c) (5) above evidently shows that X and Y are likely to contain primes and twin primes.

(8) Since X and Y if they are primes or if they pair to become twin primes, the prime factors of X and/or Y if X and/or Y are non-prime (composite) and any twin pairs formed by the prime factors of X and Y with other primes would be larger than N, this means that N would never be the largest existing prime number if it is prime and N – 2 and N would never be the largest existing twin prime pair if they are both primes. This is indirect proof of the infinity of the primes and twin primes. In the indirect proof, the proof by contradiction, we assume that N, and N – 2 and N are the largest existing prime and twin prime pair, respectively. Then we find a larger prime and twin prime pair to contradict this assumption and prove the infinity of the primes and twin primes.

(9) In the above-mentioned manner, we could go on creating new primes and twin primes ad infinitum.

(10) Thus, by (8) above and the well-established self-similarity concept stated above, we conclude the infinity of the primes and twin primes.

Appendix

The Sieve of Eratosthenes

The Sieve of Eratosthenes is an algorithm for making tables of primes. The process of determining all the primes not greater than a number N is carried out by writing down all the numbers from 2 to N, removing those after 2 which are multiples of 2, those after 3 which are multiples of 3, and continuing until all multiples of primes not greater than the square root of N, except the primes themselves, have been removed. Only prime numbers would remain.

The following is an example of the generation of primes (and twin primes) by the Sieve of Eratosthenes:

The following numbers from 2 to 50 are sieved by removing those which are multiples of 2, 3, 5 and 7 (square root of 50):

1, 2, 3, ***4,*** 5, ***6,*** 7, ***8, 9, 10,*** 11, ***12,*** 13, ***14, 15, 16***
17, ***18,*** 19, ***20, 21, 22,*** 23, ***24, 25, 26, 27, 28,*** 29, ***30,***
31, ***32, 33, 34, 35, 36,*** 37, ***38, 39, 40,*** 41, ***42,*** 43, ***44,***
45, 46, 47, ***48, 49, 50***

The bold, italicized numbers have been sieved and removed.
The following prime numbers remain after sieving:
2, 3, 5, 7, 11, 13, 17, 19, 23, 29, 31, 37, 41, 43, and 47

The following are twin primes:
3 and 5, 5 and 7, 11 and 13, 17 and 19, 29 and 31, and 41 and 43

Part 7

Keywords: prime factors, indivisible, point at infinity, new twin primes

Theorem. The twin primes are infinite.

Numerical Evidence

The following algorithms would be able to generate or sieve all the twin primes in any range of odd numbers which are all larger than those in the list of known consecutive primes/twin primes; these two important algorithms would provide plenty of numerical evidence that the twin primes are infinite:

Algorithm 1

We would provide an example with Items (1) to (3) from the following list of products of consecutive primes/twin primes, which should be sufficient for our purpose here:

1) $3 \times 5 = 15$
2) $3 \times 5 \times 7 = 105$
3) $3 \times 5 \times 7 \times 11 = 1{,}155$
4) $3 \times 5 \times 7 \times 11 \times 13 = 15{,}\ 015$
5) $3 \times 5 \times 7 \times 11 \times 13 \times 17 = 255{,}255$
6) $3 \times 5 \times 7 \times 11 \times 13 \times 17 \times 19 = 4{,}849{,}845$

The example is as follows:

1) For $3 \times 5 = 15$, we would find all the consecutive pairs of odd numbers between 5 and 15 which differ from one another by 2 and are not divisible by any of the consecutive primes/twin primes 3 and 5 in the list of consecutive primes/twin primes 3×5 whose product is 15.

   There is only 1 pair of odd numbers between 5 and 15 which differ from one another by 2 and are not divisible by the consecutive primes/twin primes 3 and 5 in the list of consecutive primes/twin primes 3×5 – they are the twin primes 11 and 13.
2) Similarly, for $3 \times 5 \times 7 = 105$, we would find all the consecutive pairs of odd numbers between 7 and 105 which differ from one another by 2 and are not divisible by any of the consecutive primes/twin primes 3, 5 and 7 in the list of consecutive primes/twin primes $3 \times 5 \times 7$ whose product is 105.

The consecutive pairs of odd numbers between 7 and 105 which differ from one another by 2 and are not divisible by the consecutive primes/twin primes 3, 5 and 7 are the following consecutive twin primes:

(a) 11 and 13
(b) 17 and 19
(c) 29 and 31
(d) 41 and 43
(e) 59 and 61
(f) 71 and 73
(g) 101 and 103

3) Similarly, in this final case, for 3 × 5 × 7 × 11 = 1,155, we would find all the consecutive pairs of odd numbers between 11 and 1,155 which differ from one another by 2 and are not divisible by any of the consecutive primes/twin primes 3, 5, 7 and 11 in the list of consecutive primes/twin primes 3 × 5 × 7 × 11 whose product is 1,155. Many of the consecutive pairs of odd numbers between 11 and 1,155 which differ from one another by 2 and are not divisible by the consecutive primes/twin primes 3, 5, 7 and 11 are twin primes (while the rest are primes larger than 3, 5, 7 and 11 and/or composite numbers whose prime factors are each larger than 3, 5, 7, and 11), some of which are as follows:
   (a) 17 and 19
   (b) 29 and 31
   (c) 41 and 43
   (d) 59 and 61
   (e) 71 and 73
   (f) 101 and 103
   (g) 107 and 109
   (h) 137 and 139
   (i) 149 and 151
   (j) 179 and 181
   (k) and so on to 1,151 and 1,153

In this way, we would also be able to achieve the following:

1) For 3 × 5 × 7 × 11 × 13 = 15,015, find all the consecutive twin primes between 13 and 15,015.
2) For 3 × 5 × 7 × 11 × 13 × 17 = 255,255, find all the consecutive twin primes between 17 and 255,255.
3) For 3 × 5 × 7 × 11 × 13 × 17 × 19 = 4,849,845, find all the consecutive twin primes between 19 and 4,849,845.

Algorithm 2

We would, similar to Algorithm 1 above, also provide an example with Items (1) to (3) from the following list of products of consecutive primes/twin primes, which should be sufficient for our purpose here:

1) $3 \times 5 = 15$
2) $3 \times 5 \times 7 = 105$
3) $3 \times 5 \times 7 \times 11 = 1{,}155$
4) $3 \times 5 \times 7 \times 11 \times 13 = 15, 015$
5) $3 \times 5 \times 7 \times 11 \times 13 \times 17 = 255{,}255$
6) $3 \times 5 \times 7 \times 11 \times 13 \times 17 \times 19 = 4{,}849{,}845$

.
.
.

The example is as follows:

1) For $3 \times 5 = 15$, we would first find all the consecutive pairs of even numbers between 5 and 15 which differ from one another by 2 and are not divisible by any of the consecutive primes/twin primes 3 and 5 in the list of consecutive primes/twin primes 3×5. Then we deduct each of these consecutive pairs of even numbers which are not divisible by any of the consecutive primes/twin primes 3 and 5 from the product of these consecutive primes/twin primes 3×5 which is 15. The results would each be 1 pair of twin primes, 1 prime and 1 composite of primes, or 2 composites of primes. In this way, we would be able to find all the consecutive twin primes between 5 and 15.
   a. There is only 1 pair of even numbers between 5 and 15 which differ from one another by 2 and are not divisible by any of the consecutive primes/twin primes 3 and 5 in the list of consecutive primes/twin primes 3×5 – they are the pair 2 and 4.
   b. The following is the result after we deduct this pair of even numbers 2 and 4 which are not divisible by any of the consecutive primes/twin primes 3 and 5 from the product of these consecutive primes/twin primes 3×5 which is 15:
      (a) $15 - 2$ and $15 - 4$: 13 and 11 (twin primes)

2) Similarly, for $3 \times 5 \times 7 = 105$, we would first find all the consecutive pairs of even numbers between 7 and 105 which differ from one another by 2 and are not divisible by any of the consecutive primes/twin primes 3, 5, and 7 in the list of consecutive primes/twin primes $3 \times 5 \times 7$, which are as follows:
   (a) 2 and 4
   (b) 32 and 34
   (c) 44 and 46
   (d) 62 and 64
   (e) 74 and 76

(f) 86 and 88
(g) 92 and 94

Then we deduct each of these consecutive pairs of even numbers which are not divisible by any of the consecutive primes/twin primes 3, 5 and 7 from the product of these consecutive primes/twin primes 3 × 5 × 7 which is 105. The results would each be 1 pair of twin primes, 1 prime and 1 composite of primes, or 2 composites of primes. In this way, we would be able to find all the consecutive twin primes between 7 and 105, which are as follows:

(a) 105 – 2 and 105 – 4: 103 and 101 (twin primes)
(b) 105 – 32 and 105 – 34: 73 and 71 (twin primes)
(c) 105 – 44 and 105 – 46: 61 and 59 (twin primes)
(d) 105 – 62 and 105 – 64: 43 and 41 (twin primes)
(e) 105 – 74 and 105 – 76: 31 and 29 (twin primes)
(f) 105 – 86 and 105 – 88: 19 and 17 (twin primes)
(g) 105 – 92 and 105 – 94: 13 and 11 (twin primes)

3) Similarly, in this final case, for 3 × 5 × 7 × 11 = 1,155, we would first find all the consecutive pairs of even numbers between 11 and 1,155 which differ from one another by 2 and are not divisible by any of the consecutive primes/twin primes 3, 5, 7 and 11 in the list of consecutive primes/twin primes 3 × 5 × 7 × 11, some of which are as follows:

(a) 2 and 4
(b) 32 and 34
(c) 62 and 64
(d) 74 and 76
(e) 92 and 94
(f) 116 and 118
(g) 122 and 124
(h) 134 and 136
(i) and so on to 1,136 and 1,138

Next we deduct each of these consecutive pairs of even numbers which are not divisible by any of the consecutive primes/twin primes 3, 5, 7, and 11 from the product of these consecutive primes/twin primes 3 × 5 × 7 × 11 which is 1,155. The results would each be 1 pair of twin primes, 1 prime and 1 composite of primes, or 2 composites of primes. In this way, we would be able to find all the consecutive twin primes between 11 and 1,155, some of which are as follows:

(a) 1,155 – 2 and 1,155 – 4: 1,153 and 1,151 (twin primes)
(b) 1,155 – 32 and 1,155 – 34: 1,123 (prime) and 1,121 (composite of primes which are

each larger than 3, 5, 7, and 11 = 19 × 59)

(c) 1,155 – 62 and 1,155 – 64: 1,093 and 1,091 (twin primes)

(d) 1,155 – 74 and 1,155 – 76: 1,081 and 1,079 (composite of primes which are each larger than 3, 5, 7, and 11 = 23 × 47) (composite of primes which are each larger than 3, 5, 7, and 11 = 13 × 83)

(e) 1,155 – 92 and 1,155 – 94: 1,063 and 1,061 (twin primes)

(f) 1,155 – 116 and 1,155 – 118: 1,039 (prime) and 1,037 (composite of primes which are each larger than 3, 5, 7 and 11 = 17 × 61)

(g) 1,155 – 122 and 1,155 – 124: 1,033 and 1,031 (twin primes)

(h) 1,155 – 134 and 1,155 – 136: 1,021 and 1,019 (twin primes)

(i) and so on to 1,155 – 1,136 and 1,155 – 1,138: 19 and 17 (twin primes)

In like manner, we would also be able to achieve the following:

1) For 3 × 5 × 7 × 11 × 13 = 15,015, find all the consecutive twin primes between 13 and 15,015.
2) For 3 × 5 × 7 × 11 × 13 × 17 = 255,255, find all the consecutive twin primes between 17 and 255,255.
3) For 3 × 5 × 7 × 11 × 13 × 17 × 19 = 4,849,845, find all the consecutive twin primes between 19 and 4,849,845.

By utilizing any of the above algorithms, we would be able to find many twin primes which are all larger than those in any chosen list of consecutive primes/twin primes, i.e., we would be able to generate many larger and larger twin primes with these algorithms.

Proof.

Lemma. According to the principle of induction in set theory, if a set of natural numbers contains 1, and if it contains n + 1 whenever it contains a number n, then it must contain every natural number, e.g., induction proves that every natural number is a product of primes.

We could have a "feel" of the infinity of the twin primes when we look at the lists of the top twin primes discovered to-date, which constitutes an argument for the twin primes' infinity; incidentally, this was more or less the kind of proof that had been used to settle the Four-Color problem in 1976 by Wolfgang Haken and Kenneth Appel, which relied on the computing powers of powerful computers – 1,200 hours of computer time, in fact. Both Haken and Appel had studied the work of Heinrich Heesch who had claimed that the infinity of infinitely variable maps could be constructed from some finite number of finite maps and that by studying these "building block" maps (these basic maps being the equivalent of the electron, proton and neutron, the fundamental objects from which all else could be constructed) it could be possible to get a hold on the general problem. Using the same reasoning Haken and Appel then reduced the four-color problem to 1,482 "building block" configurations, whereby proving that these 1,482 maps were four-colorable would imply that all maps would be four-colorable. Likewise, the twin primes' infinitude could be regarded as proven if a pair or more of twin primes were discovered which exceed a certain designated finite limit, a limit which is so expansive that it is practically indescribable, unimaginable and unnamable, such that it could be regarded as representing infinity itself – the pair of twin primes which represents this limit would be in effect a pair of googolplexes and virtually the point or focus at infinity. (Here, the designated finite limit could be treated as approximately equal to infinity.) According to the approximation theorem (twentieth. century), any two numbers are approximately equal – this holds whenever the range of approximation is greater than the modulus (positive value) of the difference of the numbers. In geometry, infinity is regarded as a "location", a finite entity, for instance, parallel lines could be said to intersect at a point at infinity, and parallel planes at a line at infinity; the asymptote to a curve could be regarded as intersecting the curve at infinity. The idea of infinity as a location had been introduced by Johann Kepler, who pointed out that a parabola could be regarded as an ellipse or a hyperbola with one focus at infinity. The idea had been developed by Girard Desargues in his formulation of projective geometry, which assumed the existence of an ideal point at infinity. This equivalency link between infinity and finiteness also appears in infinite series, which are series with an unlimited number of terms. Infinite series could be either divergent or convergent. Convergence is an important feature of a series, i.e., series that are convergent play a major role in mathematics. (It is thus necessary to be able to test a series for convergence by such tests as the Cauchy convergence test, the Cauchy integral test, Abel's test and Dirichlet's test.) An infinite series which converges to a finite sum is known as a convergent series.

We could thus prove the infinitude of the twin primes by using any of the above algorithms, preferably the evidently more efficient Algorithm 1, with the help of powerful computers, to find a pair or more of googolplex twin primes beyond the designated point or focus at infinity which is also a pair of googolplex twin primes, as is described above. (As at August 2009, $65516468355 \cdot 2^{333333}$–1 and $65516468355 \cdot 2^{333333} + 1$ (100,355 digits) is at the top of the list of top 20 twin prime pairs, while 2003663613*2-^{195000}–1 and 2003663613*2^{195000} + 1 (58,711 digits), which was the top twin prime pair in 2007, ranks second in this list. By utilizing any of the above algorithms, preferably Algorithm 1, and the aid of powerful computers it would be possible to find larger and larger twin primes than these.) This would be a computer-assisted proof similar to the above-described proof of the Four-Color problem in 1976 by Wolfgang Haken and Kenneth Appel. By the above lemma, the discovery of these larger googolplex twin primes beyond the above-said point or focus at infinity would imply that the twin primes are infinite. This would be a constructive, substantiated proof of the infinity of the twin primes. It would evidently be difficult to accept a proof of the twin primes conjecture without having to confirm or check the validity of the logic by computing a sufficiently long list of twin primes, even to the extent of looking out for counter-examples.

Hence, the great importance of the above algorithms.

Also, we could have an indirect proof, a proof by contradiction, for no matter how large the twin primes we discover with any of the 2 algorithms are, we could find more twin primes with any of the 2 algorithms which would always be larger than the last pair of twin primes we discovered – all these newly found twin primes would be evidently larger than all the primes and twin primes in the list of products of consecutive primes/twin primes utilized to generate these new twin primes. (Refer to Part 1.)

All this would constitute a well substantiated constructive proof of the infinity of the twin primes.

Conclusion

An array of methods has been adopted in this paper in solving the twin primes problem.

The inductive method, which is a well-established proof, is one of the methods utilized. The following lends support to this inductive proof of the infinity of the twin primes: (a) The characteristic of a mountain or infinite volume of sand is reflected in the characteristic of some grains of sand found there so that studying the characteristic of some grains of sand found there is enough for deducing the characteristic of the mountain or infinite volume of sand, to ascertain the quality of a batch of products it is only necessary to inspect some carefully selected samples from that batch of products and not everyone of the products and to carry out a population census, i.e., find out the characteristics of a population, it is only necessary to carry out a survey on some carefully selected respondents and not the whole population; in like manner, by the same principle, we just need to study a carefully selected list of integers and their associated primes and twin primes and deduce by induction whether the twin primes would always turn up, appear infinitely, in the list which is itself infinite – this act is rather like extrapolation. (For example, there are 376 pairs of twin primes (752 primes) found within the 2,500 consecutive primes from 2 to 22,307 – this means that 30.08%, which is sizeable, of the 2,500, not a small quantity, consecutive primes are twin primes; 3, 5, and 7 are the only "triple" primes found. There is no regularity in pattern in the appearance of the twin primes, except that the intervals between consecutive twin primes vary greatly by from 4 integers to 370 integers – the intervals between the consecutive twin primes increase and decrease, and then increase and decrease again, by turns, giving rise to a graph that is characterized by many peaks, i.e., the curve is rough and nonlinear, making its description (hence, forecast of the twin primes) by differential equations practically impossible. By the principle of induction in this case we could deduce that the twin primes would be infinite.) (b) Therefore, if x is a subset of y and if x is a list of prime numbers while y is another list of prime numbers, the characteristic presence of twin primes in x suggests the characteristic presence of twin primes in y, so that if y is an infinite list of prime numbers, whence the prime numbers in it run to infinity, so do the twin primes in it. In fact, induction plays an important part in a number of the proofs.

The other argument used to prove the twin primes' infinity is the indirect (*reductio ad absurdum*) method, which had been used by Euclid and other mathematicians after him. Logically, one or two examples of "contradiction" should be sufficient proof of infinity, for it does not make sense to have a need for an infinite number of cases of "contradiction", as our proof would then have to be infinitely and impossibly long, an absurdity. This method of proof is "proof by implication" as a result of "contradiction" – which is a "short-cut" and smart way in proving infinity, instead of "proving infinity by counting to infinity", which is ludicrous and impossible. Hence, one or two cases of "contradiction" should be sufficient for implying that there would be an infinitude of twin primes, which of course also tacitly implies that there would be an infinitude of the number of cases of such "contradiction". (Euclid evidently had this logical point in mind when he formulated the indirect (*reductio ad absurdum*) proof of the infinity of the primes.) This method of proof had been cleverly used by a number of mathematicians, not the least by the great German mathematician, David Hilbert. For example, Hilbert had used an indirect method (the "reductio ad absurdum" proof) to prove Gordan's theorem without having to show an actual "construction", a proof which had been accepted by his peers.

Another proof involves the use of Cantor's "one-to-one correspondence" method of proving infinity, which is well accepted by number theorists.

Another argument involves Dirichlet's theorem and the well-established self-similarity concept.

There is also the involvement of concepts from set theory, group theory, geometry, and so on.

The paper describes at least nine ways of finding or generating twin primes. Importantly, it presents two algorithms for generating or sieving all the twin primes in any range of odd numbers – by utilizing any of these two algorithms, we would be able to find many twin primes which are all larger than those in any chosen list of consecutive primes, i.e., we would be able to generate many larger and larger twin primes. This is indeed significant. There is evidently some deep meaning in the ease with which the twin primes turn up, as is manifest in the nine ways of obtaining them. It is thus evident that the twin primes are an inherent characteristic of the infinite prime numbers (as well as odd numbers), a characteristic which could be regarded as "self-similar" or "fractal". A twin prime pair is in effect any pair of odd numbers which differ from one another by 2 and are indivisible by any number except itself, the negative of itself, +1 and −1 (i.e., the pair of odd numbers are prime numbers). Any consecutive odd numbers or odd numbers that differ from one another by 2 are therefore potential prime numbers, as well as potential twin primes, and the likelihood of them being prime is infinite (vide Euclid's proof and Dirichlet's Theorem), i.e., the primes would always be found among them and would be there all the way to infinity (the primes being evidently the "atoms" or building blocks of all the whole numbers or integers, i.e., all the odd numbers and even numbers – every odd number or integer is either a prime number or composite of prime numbers (i.e., the integer has prime factors), and every even number is the sum of two prime numbers (vide the Goldbach conjecture which, it appears, practically all mathematicians believe to be true), as well as the product of prime numbers (composite)); hence, the likelihood of them being twin primes is infinite as well (the twin primes being an inherent property of the infinite prime numbers – as well as odd numbers – the recurrent, identical mathematical structures postulated by the well-established self-similarity concept).

So far, there has not been any indication or confirmation that the number of twin primes is finite and the so-called largest existing pair of twin primes has not been found and confirmed (which of course would be impossible to find and confirm if the twin primes were infinite). On the other hand, practically everyone could intuit that the number of twin primes is infinite. Furthermore, there are much proofs or indications that there is an infinitude of twin primes. At the end now of a long chain of reasoning, we conclude that the list of the twin primes is infinite.

It is evident here that the twin primes conjecture could be approached in a number of different ways. All this could be food for further thought about the conjecture.

References

[1] D. Burton, 1980, Elementary Number Theory, Allyn & Bacon

[2] R. Courant and H. Robbins, revised by I. Stewart, 1996, What Is Mathematics? An Elementary Approach to Ideas and Methods, Oxford University Press

[3] M. J. Feigenbaum, 1980, Universal Behavior in Nonlinear Systems, Los Alamos Science, No. 1

[4] G. H. Hardy and E. M. Wright, 1979, An Introduction to Theory of Numbers, Oxford, England: Clarendon Press

[5] D. H. Lehmer, 1914, List of Prime Numbers from 1 to 10,006,721, Publication No. 165, Carnegie Institution of Washington, Washington, D.C.

[6] M. E. Lines, 1986, A Number for Your Thoughts, Adam Hilger

[7] B. B. Mandelbrot, 1977, The Fractal Geometry of Nature, W. H. Freeman

[8] R. McWeeny, 2002, Symmetry: An Introduction to Group Theory and Its Applications, Dover

[9] J. M. T. Thompson and H. B. Stewart, 1986, Nonlinear Dynamics and Chaos, New York: John Wiley

[10] World Wide Web

1 (b) The Existence of the Twin Primes

Abstract

The author had published a paper on the solutions for the twin primes conjecture in an international mathematics journal in 2003. This paper approaches the twin primes problem through the analysis of the intrinsic nature of the prime numbers.

MSC: 11-XX (Number Theory)

Keywords: characteristic of the primes, separation of the primes, intrinsically infinite

Theorem. The twin primes are infinite.

Proof.
We note a very important intrinsic characteristic of the primes. Like all the houses in a neighborhood or location which are separated from each other by the number of houses between them, the primes are also separated from each other by the number of integers separating them. The closest will of course be the prime neighbors separated by 2 integers (i.e., twin primes), followed next in proximity by the prime neighbors separated by 4 integers, then by the prime neighbors separated by 6 integers, the prime neighbors separated by 8 integers, the prime neighbors separated by 10 integers, the prime neighbors separated by 12 integers, and so on, by larger and larger intervals, as is shown in the appendix. The twin primes are actually comparable to two closest neighbors living just next door to one another. There will always be two closest next-door neighbors, neighbors living two doors away, neighbors living three doors away, neighbors living four doors away, neighbors living five doors away, neighbors living six doors away, and so on, by greater and greater intervals, in any neighborhood or residential area; there will always be different intervals separating all the houses in a neighborhood or location. Similarly, in the infinite list of the primes, there will always be different intervals separating all the primes, ranging from the smallest interval of 2 integers (in the case of the twin primes), 4 integers, 6 integers, 8 integers, 10 integers, 12 integers, and more and more integers, which is an intrinsic characteristic of the primes. In other words, there will always be intervals of various magnitudes or sizes (i.e., intervals of various numbers of integers) between, separating, all the primes in the infinite list of the primes, and each of these intervals of various magnitudes or sizes can be expected to be infinite as the list of the primes is infinite. The twin primes, which we are examining here, are not likely to be finite (as is evident from the appendix), and should be infinite; in fact, to say that the twin primes are finite is like saying that next-door neighbors who are closest, in a neighborhood or residential area, are rare and limited, which is absurd.

Hence, the conclusion that the twin primes are infinite.

Appendix

Anecdotal Evidence of the Infinity of the Twin Primes

Top Twin Primes in 2000, 2001, 2007, and 2009

In the year 2000, $4648619711505 \times 2^{60000} \pm 1$ (18,075 digits) had been the top twin prime pair which had been discovered. In the year 2001, it only ranked eighth in the list of top 20 twin prime pairs, with $318032361 \cdot 2^{107001} \pm 1$ (32,220 digits) topping the list. In the year 2007, in the list of top 20 twin prime pairs, $318032361 \cdot 2^{107001} \pm 1$ (32,220 digits) ranked eighth, while $4648619711505 \times 2^{60000} \pm 1$ (18,075 digits) was nowhere to be seen; $2003663613 * 2^{195000} - 1$ and $2003663613 * 2^{195000} + 1$ (58,711 digits), which was discovered on January 15, 2007, by Eric Vautier (from France) of the Twin Prime Search (TPS) project in collaboration with PrimeGrid (BOINC platform), was at the top of the list. As at August 2009, $65516468355 \cdot 2^{333333} - 1$ and $65516468355 \cdot 2^{333333} + 1$ (100,355 digits) is at the top of the list of top 20 twin prime pairs, while $318032361 \cdot 2^{107001} \pm 1$ (32,220 digits) ranks 11th·, and $2003663613 * 2^{195000} - 1$ and $2003663613 * 2^{195000} + 1$ (58,711 digits) ranks second in this list.

We can expect larger twin primes than these extremely large twin primes, much larger ones, infinitely larger ones, to be discovered in due course.

List of primes pairs for the first 2,500 consecutive primes, 2 to 22,307, ranked according to their frequencies of appearance

S. no.	Ranking	Prime pairs	No. of pairs	Percentage
(1)	1	Prime pair separated by 6 integers	482	19.29%
(2)	2	Prime pair separated by 4 integers	378	15.13%
(3)	3	Prime pair separated by 2 integers (t. p.)	376	15.05%
(4)	4	Prime pair separated by 12 integers	267	10.68%
(5)	5	Prime pair separated by 10 integers	255	10.20%
(6)	6	Prime pair separated by 8 integers	229	9.16%
(7)	7	Prime pair separated by 14 integers	138	5.52%
(8)	8	Prime pair separated by 18 integers	111	4.44%
(9)	9	Prime pair separated by 16 integers	80	3.20%
(10)	10	Prime pair separated by 20 integers	47	1.88%
(11)	11	Prime pair separated by 22 integers	46	1.84%
(12)	12	Prime pair separated by 30 integers	24	0.96%
(13)	13	Prime pair separated by 28 integers	19	0.76%
(14)	14	Prime pair separated by 24 integers	16	0.64%
(15)	15	Prime pair separated by 26 integers	10	0.40%
(16)	16	Prime pair separated by 34 integers	9	0.36%
(17)	17	Prime pair separated by 36 integers	5	0.20%
(18)	18	Prime pair separated by 32 integers	2	0.08%
(19)	18	Prime pair separated by 40 integers	2	0.08%
(20)	19	Prime pair separated by 42 integers	1	0.04%
(21)	19	Prime pair separated by 52 integers	1	0.04%

Total no. of prime pairs in list: 2,498

It is evident in the above list that the prime pairs separated by 6 integers, 4 integers and 2 integers (twin primes), among the 21 classifications of prime pairs separated by from 2 integers to 52 integers (prime pairs separated by 38 integers, 44 integers, 46 integers, 48 integers and 50 integers are not among them, but, they are expected to appear further down in the infinite list of the primes), are the most dominant, important. There is a long list of other prime pairs, besides those shown in the above list, which also play a part as the building blocks of the infinite list of the integers.

The list of the integers is infinite. The list of the primes is also infinite. The infinite primes are the building blocks of the infinite integers – the infinite odd integers are all either primes or composites of primes, and the infinite even integers, except for 2 which is a prime, are all also composites of primes. Therefore, all the prime pairs separated by the integers of various magnitudes, as described above, can never all be finite. If there is any possibility at all for any of these prime pairs to be finite, there is only the possibility that a number of these prime pairs are finite (but never all of them). However, will it have to be the prime pairs separated by two integers or twin primes (which are the subject of our investigation here), which are the only prime pair, or one among a number of prime pairs, which are finite? Why question only the infinity of the prime pairs separated by two integers, the twin primes? Are not the infinities of the prime pairs separated by eight integers and more, whose frequencies of appearance are lower, as compared to those of the prime pairs which are separated by six, four, and two integers, respectively, in the above list of prime pairs, more questionable? Why single out only the twin primes? (There are at least 18 other prime pairs, separated by from 8 integers to 52 integers, whose respective infinities should be more suspect, as is evident from the above list of prime pairs, if any infinities should be doubted. Evidently, the prime pairs separated by two integers (twin primes) are not that likely to be finite.)

References

[1] D. Burton, 1980, Elementary Number Theory, Allyn & Bacon

[2] R. Courant and H. Robbins, Revised By I. Stewart, 1996, What Is Mathematics? An Elementary Approach to Ideas and Methods, Oxford University Press

[3] G. H. Hardy and E. M. Wright, 1979, An Introduction to Theory of Numbers, Oxford, England: Clarendon Press

[4] D. H. Lehmer, 1914, List of Prime Numbers from 1 to 10,006,721, Publication No. 165, Carnegie Institution of Washington, Washington, D.C.

[5] M. E. Lines, 1986, A Number for Your Thoughts, Adam Hilger

[6] World Wide Web

1 (c) The Infinity of the Twin Primes

Abstract

The author had published a paper on the solutions for the twin primes conjecture in an international mathematics journal in 2003. This paper approaches the twin primes problem through the analysis of the composite numbers.

MSC: 11-XX (Number Theory)
Keywords: composite numbers, unique factorization, prime factors

Theorem. The twin primes are infinite.

Proof.
The integers or whole numbers comprise of both the even and odd integers, and both the prime numbers and composite numbers, which are all infinite. (The primes had been proven to be infinite by Euclid long ago, and the composites of primes are also infinite, which is implied of course by the infinitude of the primes.) The even integers after 2 are composites of even, or, both even and odd, primes (there is only one even prime, i.e., 2, and it is always present in the composites which make up the even integers), e.g., $4 = 2 \times 2$, $8 = 2 \times 2 \times 2$, $12 = 2 \times 2 \times 3$, $30 = 2 \times 3 \times 5$, $32 = 2 \times 2 \times 2 \times 2 \times 2$, $36 = 2 \times 2 \times 3 \times 3$, while the odd integers are comprised of both the primes, e.g., 3, 5, 7, 11, 13, 17, and 19, and composites of odd primes, e.g., $9 = 3 \times 3$, $15 = 3 \times 5$, $21 = 3 \times 7$, $27 = 3 \times 3 \times 3$, $35 = 5 \times 7$, $57 = 3 \times 19$.

All the primes in the infinite list of primes are separated by integers ranging from 2 (in the case of the twin primes), 4, 6, 8, 10, 12, and upwards, as is shown in the appendix. The question is whether the twin primes, i.e., the prime pairs separated by two integers, are infinite. However, as is evident from the appendix, the list of the twin primes is not likely to be finite and can be expected to be infinite. We will proceed to prove this.

The Prime Number Theorem, which describes the distribution of the primes, is familiar. The primes have been known as the basic units, atoms or building blocks of the integers, both even and odd ones. We will show the indispensable role of the twin primes, i.e., the prime pairs separated by 2 integers, as well as the prime pairs separated by 4 integers, 6 integers, 8 integers, 10 integers, 12 integers, and more, and the other primes, in the formation or construction of the composite numbers. We will first of all do so by analyzing the construction or composition of the composite numbers, both even and odd ones – we will do so by deconstructing a number of composite numbers. A composite number theorem which describes the distribution of the composite numbers, as well as their building blocks, the prime numbers, is a possibility. We note that each composite number, whether even or odd, is the unique product of a particular, unique set of prime numbers only, and no product of a different set of primes can ever possibly result in the same composite number. We demonstrate this fact through some examples, which are as follows:

Even Composite Numbers

(i) 10 = 2 × 5 (only)
(ii) 34 = 2 × 17 (only)
(iii) 106 = 2 × 53 (only)
(iv) 258 = 2 × 3 × 43 (only)
(v) and so on to infinity

Odd Composite Numbers

(i) 21 = 3 × 7 (only)
(ii) 99 = 3 × 3 × 11 (only)
(iii) 325 = 5 × 5 × 13 (only)
(iv) 451 = 11 × 41 (only)
(v) and so on to infinity

It is clear from these examples that no products of other primes can ever possibly produce the above-mentioned composites besides the products of the primes shown above, which are unique.

Lemma.

This is in accordance with the fundamental theorem of arithmetic or unique factorization theorem, which states that there is only one possible combination of primes which will multiply together to produce any particular number, e.g., the only combination of primes which will produce the number 2,079 is: 3 × 3 × 3 × 7 × 11.

In the same manner, the following numbers are also uniquely factorized:
63 = 3 × 3 × 7 (only)
153 = 3 × 3 × 17 (only)
1,021,020 = 2 × 2 × 3 × 5 × 7 × 11 × 13 × 17 (only)

In other words, every positive whole number which is not prime can be broken up into prime factors, and this can happen in only one way.

By this lemma, we can produce an infinitude of unique (only ones possible) composite numbers by “playing around” with or manipulating any list of prime numbers. For example, with the following list of primes:

2, 3, 5, 7, 11, 13, 17, 19

we obtain the following unique (only ones possible) composites:

39 = 3 × 13 (only)
78 = 2 × 3 × 13 (only)
182 = 2 × 7 × 13 (only)

2,261 = 7 × 17 × 19 (only)
3,230 = 2 × 5 × 17 × 19 (only)
46,189 = 11 × 13 × 17 × 19 (only)
62,985 = 3 × 5 × 13 × 17 × 19 (only)
746,130 = 2 × 3 × 5 × 7 × 11 × 17 × 19 (only), and so on.

Similarly, with the following twin primes (in bold), and other primes, we obtain the following unique (only ones possible) composites:

---

15 = **3** × **5** (only)
35 = **5** x **7** (only)
70 = 2 × **5** × **7** (only)
143 = **11** × **13** (only)
286 = 2 × **11** × **13** (only)
323 = **17** × **19** (only)
104,329 = **17** × **19** × **17** × **19** (only)
344,285 = 3 × 11 × **17** × **19** × **17** × **19** (only)
46,189 = **11** × **13** × **17** × **19** (only) (13 and 17 are a prime pair separated by 4 integers)
3,417,986 = 2 × **11** × **13** × **17** × **19** × 37 (only) (13 and 17 are a prime pair separated by 4 integers)
10,403 = **101** × **103** (only)
11,663 = **107** × **109** (only)
22,499 = **149** × **151** (only)
121,103 = **347** × **349** (only)
435 = **3** × **5** × 29 (only)
34,320 = 2 × 2 × 2 × 2 × **3** × **5** × **11** × **13** (only)
62,418 = 2 × 3 × **101** × **103** (only)
12,621,939 = 3 × 11 × 17 × **149** × **151** (only)
1,616,615 = **5** × **7** × **11** × **13** × **17** × **19** (only) (7 and 11 and 13 and 17 are each a prime pair separated by 4 integers)
74,364,290 = 2 × **5** × **7** × **11** × **13** × **17** × **19** × 23 (only) (7 and 11 and 13 and 17 are each a prime pair separated each a prime pair separated by 4 integers)
419,868 = 2 × 2 × 3 × 3 × **107** × **109** (only)
1,331,584 = 2 × 2 × 2 × 2 × 2 × 2 × 2 ×
101 × **103** (only)
1,001,623 = 7 × **17** × **19** × 443 (only)
and so on to infinity

The following is the product of primes for composite numbers in abbreviated form:

$$c = \prod^{p}$$

where p is the prime.

In the formation or construction of these unique (only ones possible) composites described above, there will be an infinitude of combinations of the twin prime pairs, which we are here examining, with the other primes in the infinite list of the primes. As is shown above, a unique (only one possible) composite can be the product of a twin prime pair (e.g., 3 × 5). It can be the product of a twin prime pair with another prime or primes (e.g., 3 × 5 × 19 × 79). It can be the product of a twin prime pair with itself (e.g., 3 × 5 × 3 × 5). It can be the product of a twin prime pair with itself and another prime or primes (e.g., 3 × 5 × 3 × 5 × 23 × 89). It can be the product of a twin prime pair with another twin prime pair or other twin prime pairs (e.g., 11 × 13 × 227 × 229 × 461 × 463). It can be the product of a twin prime pair with another twin prime pair or other twin prime pairs and another prime or primes (e.g., 11 × 13 × 227 × 229 × 461 × 463 × 701 × 1,459 × 2,447). It is evident that due to the infinitude of the primes, the amount of such combinations will be infinite.

The reasoning to be brought up hereafter is subtle. There should be an infinitude of products of prime pairs separated by 2 integers (i.e., twin primes, e.g., 29 and 31 and 41 and 43), prime pairs separated by 4 integers (e.g., 7 and 11 and 13 and 17), prime pairs separated by 6 integers (e.g., 1,117 and 1,123 and 1,861 and 1,867), prime pairs separated by 8 integers (e.g., 2,459 and 2,467 and 4,289 and 4,297), prime pairs separated by 10 integers (e.g., 6,691 and 6,701 and 10,321 and 10,331), prime pairs separated by 12 integers (e.g., 9,649 and 9,661 and 11,399 and 11,411), and so on, and other primes from the infinite list of the primes, involved in the formation or construction of the composite numbers, since all these prime pairs separated by intervals of various magnitudes or sizes (from 2 integers in the case of the twin primes, 4 integers, 6 integers, 8 integers, 10 integers, 12 integers, and so on), and the other primes from the infinite list of the primes are evidently the building blocks of different configurations which are necessary for the construction of the infinite unique (only ones possible) composites through the unique (only one possible) product of prime pairs separated by intervals of various magnitudes or sizes and other primes as well, all these primes being the unique (only ones possible) factors of the unique (only ones possible) composites, in line with the above lemma. The formation or construction of such unique (only ones possible) composites, which are infinite, can be described, e.g., as follows:

$$c = p \times p_1 \times p_2 \times p_3 \times p_4 \times \ldots \quad (p_1 \,\&\, p_2,\ p_3 \,\&\, p_4 \text{ being twin primes pairs})$$

It can thus be stated that the infinite list of the composite numbers comprises of the respective infinite lists of products of prime pairs separated by 2 integers (twin primes), prime pairs separated by 4 integers, prime pairs separated by 6 integers, prime pairs separated by 8 integers, prime pairs separated by 10 integers, prime pairs separated by 12 integers, and so on, and other primes.

All the respective lists of products relating to the prime pairs separated by intervals of various magnitudes or sizes and other primes which produce composites, which have been described above, are unique, the only ones of their kind possible, obtainable only through unique factorization, which cannot be replaced or substituted through multiplying other prime numbers or factors (to produce the same composite number), the prime pairs separated by intervals of various magnitudes or sizes and the other primes being the building blocks of different configurations which are necessary for the construction of the infinite unique (only ones possible) composites through the unique (only one possible) product of prime pairs separated by intervals of various magnitudes or sizes and other primes as well, which implies that an infinitude of them, since they are the building blocks of different configurations, needs to exist in order that the list of the composites, of which they are the building blocks of different configurations, is infinite, *more so* since they are unique (only ones possible) and cannot be replaced or substituted through multiplying other prime numbers or factors (to produce the same composite number). Since unique factorization means that the products of all these prime pairs separated by intervals of various magnitudes or sizes and other primes as well will only produce unique (only ones possible) composites, in order for there to be larger and larger unique (only ones possible) composites to be produced we will need to get products of larger and larger prime pairs (i.e., prime pairs separated by 2 integers (twin primes), prime pairs separated by 4 integers, prime pairs separated by 6 integers, prime pairs separated by 8 integers, prime pairs separated by 10 integers, prime pairs separated by 12 integers, and so on) and larger and larger primes. To cater for the production of these larger and larger unique (only ones possible) composites to infinity, the supply of the prime pairs separated by 2 integers (twin primes), prime pairs separated by 4 integers, prime pairs separated by 6 integers, prime pairs separated by 8 integers, prime pairs separated by 10 integers, prime pairs separated by 12 integers, and so on, all the way upward, and the other primes, has also to be infinite, e.g., certain larger and larger unique (only ones possible) composites must necessarily have larger prime pairs separated by 2 integers (twin primes), larger prime pairs separated by 4 integers, larger prime pairs separated by 6 integers, larger prime pairs separated by 8 integers, larger prime pairs separated by 10 integers, larger prime pairs separated by 12 integers, and so on, and other larger primes, as their *unique prime factors*, or *only possible prime factors*, which implies that all these prime pairs separated by intervals of various magnitudes or sizes and the other primes need to be infinite in order to cater for the production of the infinite list of the larger and larger unique (only ones possible) composites. The infinite list of integers or whole numbers comprises of the infinite list of the prime numbers and the infinite list of the composite numbers, as is described above, which means that all these building blocks of different configurations are also the building blocks of the infinite integers or whole numbers.

What forms will these building blocks of different configurations, these prime pairs separated by intervals of various magnitudes or sizes and the other primes, these *unique prime factors*, take, in the formation or construction of the unique (only ones possible) composites? Any prime pair in the group can be the product of one another. The prime pair can also be the product of itself with another prime or primes. It can also be the product of itself with itself. It can be the product of itself with itself and another prime or primes too. It can be the product of itself with another prime pair or other prime pairs as well. It can also be the product of itself with another prime pair or other prime pairs and another prime or primes. It is clear that due to the infinitude of the primes the amount of such combinations will be infinite. In this manner, an infinite quantity of larger and larger unique (only ones possible) composites can be produced, evidently owing to the fact that the primes involved are *unique prime factors* and larger primes. The product of a larger prime pair will always produce a larger composite number than the product of a smaller prime pair, e.g., the product of the larger twin prime pair 17 × 19 produces the larger unique (only one possible) composite number 323, while the product of the smaller twin prime pair 11 × 13 produces the smaller unique (only one possible) composite number 143. This fact implies that since the composites are infinite, the prime pairs separated by intervals of various magnitudes or sizes and the other primes, which are the composites' *unique prime factors*, will also be infinite, the infinite list of these prime pairs separated by intervals of various magnitudes and sizes and the other primes providing larger and larger prime factors for producing the larger and larger unique (only ones possible) composites. If we liken the infinite list of the integers or whole numbers to an infinitely large jigsaw puzzle, there will be an infinitude of "gaps of certain sizes", or composites of certain values, that can only be filled up by the products of certain unique (only ones possible) primes from among the prime pairs separated by 2 integers (twin primes), prime pairs separated by 4 integers, prime pairs separated by 6 integers, prime pairs separated by 8 integers, prime pairs separated by 10 integers, prime pairs separated by 12 integers, and so on, and the other primes – these unique (only ones possible) primes are the *unique prime factors* of the unique (only ones possible) composites so constructed by the products of these unique (only ones possible) primes, as per the above lemma. This necessitates the infinitude of these unique (only ones possible) primes, i.e., the prime pairs separated by 2 integers (twin primes), prime pairs separated by 4 integers, prime pairs separated by 6 integers, prime pairs separated by 8 integers, prime pairs separated by 10 integers, prime pairs separated by 12 integers, and so on, and the other primes, for the purpose of filling up the "gaps of certain sizes", or composites of certain values, which are infinite.

We bring up the following example of 10 larger and larger unique (only ones possible) composites which are the results of the products of 10 consecutive larger and larger twin prime pairs to show the "unique factorization" effect:

(i) 15 = 3 × 5 (only)
(ii) 35 = 5 × 7 (only)
(iii) 143 = 11 × 13 (only)
(iv) 323 = 17 × 19 (only)

(v) 899 = 29 × 31 (only)
(vi) 1,763 = 41 × 43 (only)
(vii) 3,599 = 59 × 61 (only)
(viii) 5,183 = 71 × 73 (only)
(ix) 10,403 = 101 × 103 (only)
(x) 11,663 = 107 × 109 (only)

.
.
.

All the primes, or building blocks of different configurations, being *unique prime factors* of the unique (only ones possible) composites, are a necessity, a must-have, essential, compulsory, and indispensable, for the construction of the latter, since there can never be any replacements or substitutes for them. The larger the unique (only ones possible) composites are, the larger their prime factors (i.e., these mentioned *unique prime factors*) therefore have to be. Since the unique (only ones possible) composites increase to infinity, it implies that these *unique prime factors* (which include the twin primes that are the subject of our examination), which are a necessity and are not substitutable, also have to increase to infinity. In the case of the twin primes, as in the case of the other primes in the infinite list of the primes, larger and larger twin primes, as well as larger and larger prime pairs separated by larger intervals than the twin primes' and other larger and larger primes, will be needed here and there for the production of larger and larger unique (only ones possible) composites. Thus, the twin primes, being *unique prime factors* of the infinite unique (only ones possible) composites, together with the other prime pairs separated by larger intervals and the other primes, are necessarily infinite.

In summing up, we state that since the twin primes, together with the other prime pairs separated by larger intervals and the other primes, are not replaceable or substitutable by other primes, and are a necessity, a must-have, essential, compulsory, and indispensable, for the construction of the unique (only ones possible) composites, being the latter's atoms or building blocks of different configurations, the latter's *unique prime factors* as per the above lemma, and since these unique (only ones possible) composites are infinite, it is implied that the twin primes, the other prime pairs separated by larger intervals and the other primes, being the latter's atoms or building blocks of different configurations, also have to be infinite, larger and larger twin primes, prime pairs separated by larger intervals, and other primes, being essential, compulsory and necessary for the formation or construction of the larger and larger unique (only ones possible) composites and not replaceable or substitutable, becoming in effect the *unique prime factors*, or *only possible prime factors*, of the latter, *more so* that they have to be around all the time, to be infinite, since they are essential, compulsory, necessary and cannot be replaced or substituted by other primes (to produce the same composite number) besides being the atoms or building blocks of different configurations of the unique (only ones possible) composites.

Appendix

Anecdotal Evidence of the Infinity of the Twin Primes

Top Twin Primes in 2000, 2001, 2007, and 2009

In the year 2000, $4648619711505 \times 2^{60000} \pm 1$ (18,075 digits) had been the top twin prime pair which had been discovered. In the year 2001, it only ranked eighth in the list of top 20 twin prime pairs, with $318032361 \times 2^{107001} \pm 1$ (32,220 digits) topping the list. In the year 2007, in the list of top 20 twin prime pairs, $318032361 \cdot 2^{107001} \pm 1$ (32,220 digits) ranked eighth, while $4648619711505 \times 2^{60000} \pm 1$ (18,075 digits) was nowhere to be seen; $2003663613 * 2^{195000} - 1$ and $2003663613 * 2^{195000} + 1$ (58,711 digits), which was discovered on January 15, 2007, by Eric Vautier (from France) of the Twin Prime Search (TPS) project in collaboration with PrimeGrid (BOINC platform), was at the top of the list. As at August 2009, $65516468355 \cdot 2^{333333} - 1$ and $65516468355 \cdot 2^{333333} + 1$ (100,355 digits) is at the top of the list of top 20 twin prime pairs, while $318032361 \cdot 2^{107001} \pm 1$ (32,220 digits) ranks 11th, and $2003663613 * 2^{195000} - 1$ and $2003663613 * 2^{195000} + 1$ (58,711 digits) ranks second in this list.

We can expect larger twin primes than these extremely large twin primes, much larger ones, infinitely larger ones, to be discovered in due course.

List of prime pairs for the first 2,500 consecutive primes, 2 to 22,307, ranked according to their frequencies of appearance

S. no.	Ranking	Prime pairs	No. of pairs	Percentage
(1)	1	Prime pair separated by 6 integers	482	19.29%
(2)	2	Prime pair separated by 4 integers	378	15.13%
(3)	3	Prime pair separated by 2 integers (t. p.)	376	15.05%
(4)	4	Prime pair separated by 12 integers	267	10.68%
(5)	5	Prime pair separated by 10 integers	255	10.20%
(6)	6	Prime pair separated by 8 integers	229	9.16%
(7)	7	Prime pair separated by 14 integers	138	5.52%
(8)	8	Prime pair separated by 18 integers	111	4.44%
(9)	9	Prime pair separated by 16 integers	80	3.20%
(10)	10	Prime pair separated by 20 integers	47	1.88%
(11)	11	Prime pair separated by 22 integers	46	1.84%
(12)	12	Prime pair separated by 30 integers	24	0.96%
(13)	13	Prime pair separated by 28 integers	19	0.76%
(14)	14	Prime pair separated by 24 integers	16	0.64%
(15)	15	Prime pair separated by 26 integers	10	0.40%
(16)	16	Prime pair separated by 34 integers	9	0.36%
(17)	17	Prime pair separated by 36 integers	5	0.20%
(18)	18	Prime pair separated by 32 integers	2	0.08%
(19)	18	Prime pair separated by 40 integers	2	0.08%
(20)	19	Prime pair separated by 42 integers	1	0.04%
(21)	19	Prime pair separated by 52 integers	1	0.04%

Total no. of prime pairs in list: 2,498

It is evident in the above list that the prime pairs separated by 6 integers, 4 integers and 2 integers (twin primes), among the 21 classifications of prime pairs separated by from 2 integers to 52 integers (prime pairs separated by 38 integers, 44 integers, 46 integers, 48 integers, and 50 integers are not among them, but, they are expected to appear further down in the infinite list of the primes), are the most dominant, important. There is a long list of other prime pairs, besides those shown in the above list, which also play a part as the building blocks of the infinite list of the integers.

The list of the integers is infinite. The list of the primes is also infinite. The infinite primes are the building blocks of the infinite integers – the infinite odd integers are all either primes or composites of primes, and the infinite even integers, except for 2 which is a prime, are all also composites of primes. Therefore, all the prime pairs separated by the integers of various magnitudes, as described above, can never all be finite. If there is any possibility at all for any of these prime pairs to be finite, there is only the possibility that a number of these prime pairs are finite (but never all of them). However, will it have to be the prime pairs separated by 2 integers or twin primes (which are the subject of our investigation here), which are the only prime pair, or one among a number of prime pairs, which are finite? Why question only the infinity of the prime pairs separated by 2 integers, the twin primes? Are not the infinities of the prime pairs separated by 8 integers and more, whose frequencies of appearance are lower, as compared to those of the prime pairs which are separated by 6, 4, and 2 integers, respectively, in the above list of prime pairs, more questionable? Why single out only the twin primes? (There are at least 18 other prime pairs, separated by from 8 integers to 52 integers, whose respective infinities should be more suspect, as is evident from the above list of prime pairs, if any infinities should be doubted. Evidently, the prime pairs separated by 2 integers (twin primes) are not that likely to be finite.)

References

[1] D. Burton, 1980, Elementary Number Theory, Allyn & Bacon
[2] R. Courant and H. Robbins, Revised By I. Stewart, 1996, What Is Mathematics? An Elementary Approach to Ideas and Methods, Oxford University Press
[3] G. H. Hardy and E. M. Wright, 1979, An Introduction to Theory of Numbers, Oxford, England: Clarendon Press
[4] D. H. Lehmer, 1914, List of Prime Numbers from 1 to 10,006,721, Publication No. 165, Carnegie Institution of Washington, Washington, D.C.
[5] M. E. Lines, 1986, A Number for Your Thoughts, Adam Hilger
[6] World Wide Web

2 (a) A Rumination of the Goldbach Conjecture (This Paper Had Been Published in an International Mathematics Journal in 2012)

Abstract

The Goldbach conjecture is a challenging prime number riddle. This paper offers some thoughts on the conjecture, provides some numerical evidence and describes some approaches.

MSC: 11-XX (Number Theory)

Keywords: regularity of pattern; recurrent, identical mathematical structures; building blocks; "reflection" principle

Introduction

The problem of whether there is an infinitude of cases of even numbers which are each the sum of 2 primes is an inherently difficult one to solve, as infinity (normally symbolized by: ∞) is a difficult concept and is against common sense. It is impossible to count, calculate or live to infinity, perhaps with the exception of God. Infinity is a nebulous idea and appears to be only an abstraction devoid of any actual practical meaning. How do we quantify infinity? How big is infinity? The difficulty of the problem of infinity has been compounded by Georg Cantor who proved that there are actually different sizes to infinity, an idea so bizarre to many mathematicians that he was attacked for his ideas during much of his career. The attack was so serious that he suffered mental illness and severe depression. However, after his death his ideas became widely accepted as the only consistent, accurate and powerful definition of infinity. Hilbert had honored him by saying, "No one shall drive us from the paradise Cantor has created for us." Nevertheless, in this paper offering solutions for infinity, in this case the infinity of the even numbers which are each the sum of 2 primes, incontrovertible evidence that the peculiar characteristics of the prime numbers themselves contribute to the infinite "generation" of such even numbers would be put forward.

Solution

Christian Goldbach, tutor to the teenage Czar Peter II, had examined dozens of even numbers and noticed that he could split all of them into the sum of 2 primes. Thus, his conjecture that every even number after the number 2 is the sum of 2 prime numbers, for example:
$4 = 2 + 2$

6 = 3 + 3
8 = 3 + 5
10 = 3 + 7 and 5 + 5
12 = 5 + 7
14 = 3 + 11 and 7 + 7
16 = 3 + 13 and 5 + 11
18 = 5 + 13 and 7 + 11
20 = 3 + 17 and 7 + 13
50 = 19 + 31
100 = 53 + 47
21,000 = 17 + 20,983
.
.
.

Computer searches completed in 2000 had verified that all even numbers up to 400 trillion (4×10^{14}), which is not a small list, are sums of two primes, while in 2008, a distributed computer search ran by Tomas Oliveira e Silva, a researcher at the University of Aveiro, Portugal, had further verified the Goldbach conjecture up to 12×10^{17}. But is the conjecture valid?

Theorem. Every even number after 2 is the sum of 2 primes.

Proof.
By Euclid's proof, there is an infinitude of primes; that is, the list of primes: 2, 3, 5, 7, 11, 13, 17, 19, 23, 29, 31, 37, . . ., continues to infinity.

Goldbach's conjecture states that every even number after the number 2 is the sum of 2 primes. How do we prove this?

First of all, we ask a "reversed" question here (as opposed to Goldbach's conjecture). We ask whether all the prime numbers in the infinite list of prime numbers would combine with each other to form a regular, continuous (*without breaks or gaps*) and infinite list of even numbers. This would lead to our proof.

Let us now take a subset of primes from the infinite set of prime numbers, say, all the primes found in the set of integers ranging from 1 to 50, that is, the following subset of prime numbers:

2, 3, 5, 7, 11, 13, 17, 19, 23, 29, 31, 37, 41, 43, and 47.

Then, we conduct a close examination of how this subset of prime numbers "behaves", that is, how the prime numbers combine (one-to-one) with each other to form even numbers, and observe whether there is any "regularity of pattern" in the way they do so. Here, we look at how the prime numbers from 2 to 47 combine with each other to form even numbers.

We could observe the primes from 2 to 47 "generating" a regular, continuous (*without breaks or gaps*) list of even numbers from 4 to 94. This is a regular, continuous list of even numbers, the even numbers becoming evidently progressively more repetitious. For example, there are five discernable combinations of primes/partitions for the even number 48, which is as follows:

a) 19 + 29 = 48
b) 17 + 31 = 48
c) 11 + 37 = 48
d) 5 + 43 = 48
e) 7 + 41 = 48

and there are five discernable combinations of primes/partitions for the even number 54, which is as follows:

a) 17 + 37 = 54
b) 13 + 41 = 54
c) 11 + 43 = 54
d) 7 + 47 = 54
e) 23 + 31 = 54

And many others.

There appears to be a "regularity of pattern" in the way the even numbers "pop up".

From this, we could thus conclude the following characteristic or "pattern" of the prime numbers: They would combine (one-to-one) with each other to form a regular, continuous (*without breaks or gaps*) list of even numbers, with "overwhelming repetitions" all over the place.

Next, we select another subset of primes from the infinite set of prime numbers. We would here take the prime numbers from the set of integers 51 to 100, which is just "next to" the set of integers 1 to 50 from which we have taken our first subset of prime numbers, to be our second subset of primes. This second subset of prime numbers is as follows:

53, 59, 61, 67, 71, 73, 79, 83, 89, and 97.

As usual, we conduct a close examination of how this second subset of prime numbers "behaves", that is, how they combine (one-to-one) with each other to form even numbers, and observe whether there is "regularity of pattern" in the way they do so.

Here, as earlier, we could observe the primes from 53 to 97 "generating" a regular, continuous (*without breaks or gaps*) list of even numbers ranging from 56 to 194.

This regular, continuous list of even numbers is also evidently progressively more repetitious. For example, there are eight discernable combinations of primes/partitions for the even number 90, which is as follows:

a) 53 + 37 = 90
b) 59 + 31 = 90
c) 61 + 29 = 90
d) 67 + 23 = 90
e) 71 + 19 = 90
f) 73 + 17 = 90
g) 79 + 11 = 90
h) 83 + 7 = 90

and there are eight discernable combinations of primes/partitions for the even number 120, which is as follows:

a) 53 + 67 = 120
b) 59 + 61 = 120
c) 67 + 53 = 120
d) 73 + 47 = 120
e) 79 + 41 = 120
f) 83 + 37 = 120
g) 89 + 31 = 120
h) 97 + 23 = 120

And many others.

There appears to be a "regularity of pattern" in the way the even numbers "pop up" here – as a matter of fact, this "regularity of pattern" resembles that found in the earlier listing.

The list of even numbers "generated" by this second subset of prime numbers, that is, the regular, continuous list of even numbers ranging from 56 to 194, even overlaps (by a wide margin) the list of even numbers "generated" by the first subset of primes (2 to 47), that is, the regular, continuous list of even numbers ranging from 4 to 94.

From this listing also, from the "characteristics" found in all these listings, where the "regularity of pattern" of the appearance of the even numbers is evident, we could deduce the following characteristic of the prime numbers: The prime numbers would combine (one-to-one) with each other to form a regular, continuous (*without breaks or gaps*) list of even numbers, with "overwhelming repetitions" all over the place. This could be further confirmed by studying the even numbers "generated", e.g., by the three subsets of primes by combining with other prime numbers, including the prime numbers before them, for the three consecutive sets of integers, 101 to

150, 151 to 200, and 201 to 250, with “overwhelming repetitions” all over the place (see Item (1) in the data below, where there are also much further examples).

Lemma. The well-established self-similarity concept, which was developed by Mitchell Feigenbaum in the 1970s and which brought him fame, upon which the method of renormalization in perturbation theory is based, postulates that there is a tendency of identical mathematical structures to recur on many levels. Within a given structure, there would be smaller copies of the same structure, their sizes being determined by the scaling factor. Feigenbaum found that at the utmost tips of the fig tree, there is some mathematical structure which remains the same when its size is changed (enlarged) by a scaling factor of 4.669, which is found to be a constant like pi (3.142); this structure is the shape of the fig tree itself; in other words, little whorls could be found within big whorls. Renormalization has been a well-established technique in chaos theory/fractal geometry and is a mathematical trick which functions rather like a microscope, zooming in on the self-similar structure, removing any approximations, and filtering out everything else. All this shows the universality of some features of chaos. That is, some kind of order or pattern could be found in or is inherent in disorder or chaos. In other words, the elements of an infinite subset of an infinite set contain all the recursive significant properties of that set unless the process which selects the elements of the subset directly excludes a property.

To make it simpler, we re-phrase this concept as follows: The characteristic of a mountain or infinite volume of sand is reflected in the characteristic of some grains of sand found there so that studying the characteristic of some grains of sand found there is sufficient for deducing the characteristic of the mountain or infinite volume of sand. Likewise, if x is a subset of y and if x is a list of prime numbers while y is another list of prime numbers, the characteristic presence of the even numbers “generated” by all the primes in x suggests (or reflects) the characteristic presence of even numbers “generated” by all the primes in y, so that if y is an infinite list of prime numbers, whence the prime numbers in it run to infinity, so do the even numbers “generated” by all the primes in it. What is described here is actually the “reflection” principle.

Therefore, by the above-mentioned principle, all the above-mentioned selected subsets of primes in the infinite set of primes would each reflect (present an image of, or display something which has similarity with) the characteristic of this infinite set of primes; that is, all the infinite primes in the infinite set of primes, including its infinite subsets such as the selected subsets mentioned above, would combine (one-to-one) with each other to form a regular, continuous (*without breaks or gaps*) and infinite (implied by the infinitude of the primes (vide Euclid’s proof) and the even numbers themselves) list of even numbers. It is evident that the higher up the infinite list of primes we go, the more “overwhelming” or dense the (one-to-one) combinations of primes (in the formation of even numbers) would become, the number of permutations of the combinations of primes tending toward infinity (with the infinity of the prime numbers), as a study with further selected subsets of primes would reveal. A study of the even numbers "generated" by all these subsets of primes would show that the higher up the subsets of primes we go the more "overwhelmingly" the even numbers "generated" would repeat themselves and overlap. This is (very) significant. Though the infinitude of the prime numbers would ensure that there would always be new even numbers being "generated", there is also the "fear" that there might be gaps, breaks or lack of continuity in the even numbers thus "generated". But, it is evident that these more and more profuse repetitions and overlaps of the even numbers thus "generated" by the primes the higher up the infinite list of prime numbers we go "ensure" that such gaps or breaks would not appear between the even numbers "generated" – they "ensure" that the even numbers thus "generated" by the primes in the infinite list of primes would be regular, continuous, *without breaks or gaps*, and in consecutive running order. Also (very) significant is the great number of new even numbers that each of the primes in these subsets of primes helps to "generate". This "profuse generation" of "regular batches" of even numbers by the prime numbers represents a characteristic or feature of the prime numbers, a universal "pattern" or feature of the "chaotic" infinite prime numbers, or recurrent, identical mathematical structures, which is all in accordance with the above lemma. (This "pattern of behavior" of the prime numbers, as described in the paper, is analogous to the "self-similar mathematical pattern or structure" (which is the shape of the fig tree itself) of the various parts of the fig tree, that is, its trunk to bough section, bough to branch section, branch to twig section and twig to twiglet section, in Feigenbaum's famous fig tree example, and such self-similar mathematical pattern or structure, or fractal characteristic, could also be found in other aspects of nature, for example, waves, turbulence or chaos, the structures of viruses and bacteria, polymers and ceramic materials, the universe and many others, even the movements of prices in financial markets, the growths of populations, the sound of music, the flow of blood through our circulatory system, the behavior of people en masse, and so on, which have all spawned a relatively new and important branch of mathematics with wide practical applications known as fractal geometry, which has been pioneered by Benoit Mandelbrot. As a matter of fact, self-similarity or fractal characteristic could be regarded as the fundamental mathematical aspect found in practically everything in nature including the numbers such as the prime numbers and the even numbers which are the subjects of our investigation here, and this new branch of mathematics, fractal geometry, besides having a great practical impact on us also gives us a deeper vision of the universe in which we live and our place in it.) In other words, by the above lemma the infinity of the prime numbers implies the infinity of the "profuse generation" of "regular batches" of even numbers by the prime numbers, that is, the validity of the Goldbach conjecture.

Here, we take a close look at the following data:

(1) *No. of old/repeated (also appeared earlier) even numbers/overlaps "generated" (by the additions/ combinations of two primes), for integers 1 to 1,250 (see Appendix 1 for example of computation method)*

(a) Set of Integers, 1 to 50, with 14 primes within it = not applicable
(aa) Percentage increase in repetition = not applicable
(b) Set of integers, 51 to 100, with 10 primes within it = **20** repeated even nos.
(bb) Percentage increase in repetition = not applicable
(c) Set of integers, 101 to 150, with 10 primes within it = 46 repeated even nos.
(cc) Percentage increase in repetition = (46 – 20) ÷ 20 × 100% = **130%**
(d) Set of integers, 151 to 200, with 11 primes within it = 73 repeated even nos.
(dd) Percentage Increase In Repetition = (73 – 46) ÷ 46 × 100% = 58.7%
(e) Set of integers, 201 to 250, with 7 primes within it = 93 repeated even nos.
(ee) Percentage Increase In Repetition = (93 – 73) ÷ 73 × 100% = 27.4%
(f) Set of integers, 251 to 300, with 9 primes within it = 115 repeated even nos.
(ff) Percentage increase in repetition = (115 – 93) ÷ 93 × 100% = 23.66%
(g) Set of integers, 301 to 350, with 8 primes within it = 139 repeated even nos.
(gg) Percentage increase in repetition = (139 – 115) ÷ 115 × 100% = 20.87%
(h) Set of integers, 351 to 400, with 8 primes within it = 172 repeated even nos.
(hh) Percentage increase in repetition = (172 – 139) ÷ 139 × 100% = 23.74%
(i) Set of integers, 401 to 450, with 9 primes within it = 196 repeated even nos.
(ii) Percentage increase in repetition = (196 – 172) ÷ 172 × 100% = 13.95%
(j) Set of integers, 451 to 500, with 8 primes within it = 220 repeated even nos.
(jj) Percentage increase in repetition = (220 – 196) ÷ 196 × 100% = 12.24%
(k) Set of integers, 501 to 550, with 6 primes within it = 247 repeated even nos.
(kk) Percentage increase in repetition = (247 – 220) ÷ 220 × 100% = 12.27%
(l) Set of integers, 551 to 600, with 8 primes within it = 268 repeated even nos.
(ll) Percentage increase in repetition = (268 – 247) ÷ 247 × 100% = 8.5%
(m) Set of integers, 601 to 650, with 9 primes within it = 298 repeated even nos.
(mm) Percentage increase in repetition = (298 – 268) ÷ 268 × 100% = 11.19%
(n) Set of integers, 651 to 700, with 7 primes within it = 320 repeated even nos.
(nn) Percentage increase in repetition = (320 – 298) ÷ 298 × 100% = 7.38%
(o) Set of integers, 701 to 750, with 7 primes within it = 340 repeated even nos.
(oo) Percentage increase in repetition = (340 – 320) ÷ 320 × 100% = 6.25%
(p) Set of integers, 751 to 800, with 7 primes within it = 367 repeated even nos.
(pp) Percentage increase in repetition = (367 – 340) ÷ 340 × 100% = 7.94%
(q) Set of integers, 801 to 850, with 7 primes within it = 392 repeated even nos.
(qq) Percentage increase in repetition = (392 – 367) ÷ 367 × 100% = 6.81%
(r) Set of integers, 851 to 900, with 8 primes within it = 412 repeated even nos.
(rr) Percentage increase in repetition = (412 – 392) ÷ 392 × 100% = 5.1%
(s) Set of integers, 901 to 950, with 7 primes within it = 433 repeated even nos.
(ss) Percentage increase in repetition = (433 – 412) ÷ 412 × 100% = 5.1%
(t) Set of integers, 951 to 1,000, with 7 primes within it = 470 repeated even nos.
(tt) Percentage increase in repetition = (470 – 433) ÷ 433 × 100% = 8.55%
(u) Set of integers, 1,001 to 1,050, with 8 primes within it = 492 repeated even nos.
(uu) Percentage increase in repetition = (492 – 470) ÷ 470 × 100% = 4.68%
(v) Set of integers, 1,051 to 1,100, with 8 primes within it = 523 repeated even nos.
(vv) Percentage increase in repetition = (523 – 492) ÷ 492 × 100% = 6.3%
(w) Set of integers, 1,101 to 1,150, with 5 primes within it = 545 repeated even nos.
(ww) Percentage increase in repetition = (545 – 523) ÷ 523 × 100% = 4.21%
(x) Set of integers, 1,151 to 1,200, with 7 primes within it = 553 repeated even nos.

(continued)

(xx)	Percentage increase in repetition = (553 – 545) ÷ 545 × 100% = **1.47%**
(y)	Set of integers, 1,201 to 1,250, with 8 primes within it = **592** repeated even nos.
(yy)	Percentage increase in repetition = (592 – 553) ÷ 553 × 100% = **7.05%**

It could be seen that on the whole the No. Of Old/Repeated (Also Appeared Earlier) Even Numbers/Overlaps "Generated" (By The Additions/Combinations Of Two Primes) increases progressively from 20 in (b) to 592 in (y), while it could be seen that the Percentage Increase In Repetition on the whole thins out from 130% in (cc) to 7.05% in (yy), with the lowest percentage increase of 1.47% found in (xx). This statistical trend or feature is not surprising and represents (very) significant evidence that lends support to the validity of the Goldbach conjecture – the infinitude of both the primes and the even numbers implies that the above overlaps increase progressively to infinity.

(2)	*Density of new even numbers "generated" (see Appendix 1 for example of computation method)*	
	(a)	Set of integers, 51 to 100, with 10 primes within it = 5 new even nos. per prime no. (No. of new even nos. "generated" = 50. no. of primes = 10.)
	(b)	Set of integers, 101 to 150, with 10 primes within it = 5.2 new even nos. per prime no. (No. of new even nos. "generated" = 52. no. of primes = 10.)
	(c)	Set of integers, 151 to 200, with 11 primes within it = **4.55 new even nos. per prime no.** (No. of new even nos. "generated" = 50. no. of primes = 11.)
	(d)	Set of integers, 201 to 250, with 7 primes within it = 6 new even nos. per prime no. (No. of new even nos. "generated" = 42. no. of primes = 7.)
	(e)	Set of integers, 251 to 300, with 9 primes within it = 5.78 new even nos. prime no. (No. of new even nos. "generated" = 52. no. of primes = 9.)
	(f)	Set of integers, 301 to 350, with 8 primes within it = 7 new even nos. per prime no. (No. of new even nos. "generated" = 56. no. of primes = 8.)
	(g)	Set of integers, 351 to 400, with 8 primes within it = 6 new even nos. per prime no. (No. of new even nos. "generated" = 48. no. of primes = 8.)
	(h)	Set of integers, 401 to 450, with 9 primes within it = 5.78 new even nos. per prime no. (No. of new even nos. "generated" = 52. no. of primes = 9.)
	(i)	Set of integers, 451 to 500, with 8 primes within it = 6.25 new even Nos. per prime no. (No. of new even nos. "generated" = 50. no. of primes = 8.)
	(j)	Set of integers, 501 to 550, with 6 primes within it = 8 new even nos. per prime no. (No. of new even nos. "generated" = 48. no. of primes = 6.)
	(k)	Set of integers, 551 to 600, with 8 primes within it = 6.5 new even nos. per prime no. (No. of new even nos. "generated" = 52. no. of primes = 8.)
	(l)	Set of integers, 601 to 650, with 9 primes within it = 5.33 new even nos. per prime no. (No. of new even nos. "generated" = 48. no. of primes = 9.)
	(m)	Set of integers, 651 to 700, with 7 primes within it = 6.29 new even nos. per prime no. (No. of new even nos. "generated" = 44. no. of primes = 7.)
	(n)	Of integers, 701 to 750, with 7 primes within it = 7.43 new even nos. per prime no. (No. of new even nos. "generated" = 52. no. of primes = 7.)
	(o)	Set of integers, 751 to 800, with 7 primes within it = 7.71 new even nos.per prime no.

(continued)

	(No. of new even nos. “generated” = 54. no. of primes = 7.)
(p)	Set of integers, 801 to 850, with 7 primes within it = 6 new even nos. per prime no. (No. of new even nos. “generated” = 42. no. of primes = 7.)
(q)	Set of integers, 851 to 900, with 8 primes within it = 6 new even nos. per prime no. (No. of new even nos. “generated” = 48. no. of primes = 8.)
(r)	Set of integers, 901 to 950, with 7 primes within it = 8.57 new even nos. per prime no. (No. of new even nos. “generated” = 60. no. of primes = 7.)
(s)	Set of integers, 951 to 1,000, with 7 primes within it = 7.14 new even nos. per prime no. (No. of new even nos. “generated” = 50. no. of primes = 7.)
(t)	Set of integers, 1,001 to 1,050, with 8 primes within it = 6.5 new even nos. per prime no. (No. of new even nos. “generated” = 52. no. of primes = 8.)
(u)	Set of integers, 1,051 to 1,100, with 8 primes within it = 6 new even nos per prime no. (No. of new even nos. “generated” = 48. no. of primes = 8.)
(v)	Set of integers, 1,101 to 1,150, with 5 primes within it = 6.4 new even nos per prime no. (No. of new even nos. “generated” = 32. no. of primes = 5.)
(w)	Set of integers, 1,151 to 1,200, with 7 primes within it = **9.14 new even nos. per prime no.** (No. of new even nos. “generated” = 64. no. of primes = 7.)
(x)	Set of integers, 1,201 to 1,250, with 8 primes within it = 7 new even nos. per prime no. (No. of new even nos. “generated” = 56. no. of primes = 8.)

Average density for the above 24 items ((a) to (x)) = 155.54 ÷ 24 = 6.48 new even nos. per prime no.

Maximum density = 9.14 new even nos. per prime no. (No. of new even nos. “generated” = 64. No. of primes = 7.)

Minimum density = 4.55 new even nos. per prime no. (No. of new even nos. “generated” = 50. No. of primes = 11.)

Such a “profuse generation” of “regular batches” of even numbers by the prime numbers is (very) significant and represents a characteristic or feature of the prime numbers, a universal “pattern” or feature of the “chaotic” infinite prime numbers (or recurrent, identical mathematical structures), which is excellently in accordance with the above lemma. This lends further support to the validity of the Goldbach conjecture, which, as stated above, is implied by both the infinitude of the primes and the even numbers.

There is indeed further incontrovertible proof which is obtainable by analyzing a number of even numbers; e.g., we could split a group of 240 even consecutive numbers, from 4 to 482, into eight equal batches (30 even numbers per batch) and analyze the batches, which buttresses the above evidence that the infinite quantity of primes would “generate” a regular, continuous (*without breaks or gaps*) and infinite list of even numbers. The density of distribution or prime additions/combinations per even number evidently become greater and greater the higher up the infinite list of the even numbers we go – this increase in density evidently represents a definite pattern in the "behavior" of the prime numbers. This pattern is (very) significant and is discernible in the following example:

(1) *First batch of 30 even numbers (4 to 62)* (*See Appendix 2 for example of computation method*)
   a) Maximum no. of prime additions/combinations per even number = 5
   b) Minimum no. of prime additions/combinations per even number = 1
   c) Density of distribution = average prime additions/combinations per even number = **2.77** prime additions/combinations per even number

(2) *Second batch of 30 even numbers (64 to 122)*
   a) Maximum no. of prime additions/combinations per even number = 14
   b) Minimum no. of prime additions/combinations per even number = 2
   c) Density of distribution = average prime additions/combinations per even number = **6.1** prime additions/combinations per even number
   d) Percentage increase in density of distribution = (6.1 – 2.77) ÷ 2.77 × 100% = 120.22%

(3) *Third batch of 30 even numbers (124 to 182)*
   a) Maximum no. of prime additions/combinations per even number = 16
   b) Minimum no. of prime additions/combinations per even number = 4
   c) Density of distribution = average prime additions/combinations per even number = **9.07** prime additions/combinations per even number
   d) Percentage increase in density of distribution = (9.07 – 6.1) ÷ 6.1 × 100% = 48.69%

(4) *Fourth batch of 30 even numbers (184 to 242)*
   a) Maximum no. of prime additions/combinations per even number = 22
   b) Minimum no. of prime additions/combinations per even number = 5
   c) Density of distribution = average prime additions/combinations per even number = **10.53** prime additions/combinations per even number
   d) Percentage increase in density of distribution = (10.53 – 9.07) ÷ 9.07 × 100% = 16.1%

(5) *Fifth batch of 30 even numbers (244 to 302)*
   a) Maximum no. of prime additions/combinations per even number = 21
   b) Minimum no. of prime additions/combinations per even number = 7
   c) Density of distribution = average prime additions/combinations per even number = **12.37** prime additions/combinations per even number
   d) Percentage increase in Density of distribution = (12.37 – 10.53) ÷ 10.53 × 100% = 17.47%

(6) *Sixth batch of 30 even numbers (304 to 362)*
   a) Maximum no. of prime additions/combinations per even number = 27
   b) Minimum no. of prime additions/combinations per even number = 7
   c) Density of distribution = average prime additions/combinations per even
   d) Number = **13.77** prime additions/combinations per even number
   e) Percentage increase in density of distribution = (13.77 – 12.37) ÷ 12.37 × 100% = 11.32%

(7) *Seventh batch of 30 even numbers (364 to 422)*
   a) Maximum no. of prime additions/combinations per even number = 30
   b) Minimum no. of prime additions/combinations per even number = 7
   c) Density of distribution = average prime additions/combinations per even number = **15.23** prime additions/combinations per even number
   d) Percentage increase in density of distribution = (15.23 − 13.77) ÷ 13.77 × 100% = 10.6%

(8) *Eighth batch of 30 even numbers (424 to 482)*
   a) Maximum no. of prime additions/combinations per even number = 30
   b) Minimum no. of prime additions/combinations per even number = 9
   c) Density of distribution = average prime additions/combinations per even number = **16.93** prime additions/combinations per even number
   d) Percentage increase in density of distribution = (16.93 − 15.23) ÷ 15.23 × 100% = 11.16%

The density of distribution is expected to increase to infinity, though the percentage increase in density of distribution is expected to thin out toward infinity – it could be seen above to increase from 2.77 prime additions/combinations per even number for batch of even numbers, 4 to 62, all the way up to 16.93 prime additions/combinations per even number for batch of even numbers, 424 to 482. This is nevertheless (very) significant evidence that lends support to the validity of the Goldbach conjecture. Also, the maximum no. of prime additions/combinations per even number and the minimum no. of prime additions/combinations per even number could be seen to range from 5 and 1, respectively, for batch of even numbers, 4 to 62, to 30 and 9, respectively, for batch of even numbers, 424 to 482. This trend of "upward increase" of the (maximum and minimum) numbers of prime additions/combinations for each even number implies that at some points toward infinity the numbers of prime additions/combinations for each even number could be thousands, millions, billions, trillions, and more, if only we have the computing power to compute/check such prime additions/combinations. This is (very) significant too and is also evidence that lends support to the validity of the Goldbach conjecture. By the infinitude of the primes and even numbers and the above lemma, these "patterns", as described here, would be there all the way to infinity, which would be in accordance with the Goldbach conjecture.

The one-to-one additions/combinations of the primes in the formation of even numbers do evidently become more and more "overwhelming" or profuse the higher up the infinite list of even numbers/prime numbers we go, thereby assuring an infinite, regular and consecutive supply of even numbers, as is evident from the example just above. This, together with the above-described evidently more and more profuse repetitions and overlaps of the even numbers "generated" by the primes the higher up the infinite list of prime numbers we go (refer to no. of old/repeated (also appeared earlier) even numbers/overlaps "generated" (by the additions/combinations of two primes), for integers 1 to 1,250 above), goes on to show that the Goldbach conjecture becomes, evidently, even stronger and stronger the higher up the infinite list of prime numbers/even numbers we go. Here, we have in fact approached the problem from two different, but somewhat related, angles – by a statistical analysis of the "behavior" of the primes in the formation of even numbers, and a statistical analysis of the even numbers "generated" as a result. The statistical data thus obtained are indeed found to greatly support the Goldbach conjecture, evidently the more so the higher up the infinite list of prime numbers/even numbers we go, and by the infinitude of the primes and even numbers and the above lemma there would be an infinitude of such statistical data thus obtained. Hence, by virtue of these imposing statistical trends, plus the statistical trend that a prolific number of new even numbers are always being "generated" (refer to density of new even numbers "generated" above), as well as the infinitude of the prime numbers and the even numbers, together with the above lemma, we affirm the validity of the Goldbach conjecture.

Reversing the "reversed way", we hereby affirm that every even number after the number 2 in the infinite list of even numbers is a combination or sum of 2 primes. In fact, the prime numbers are the building blocks or "atoms" of all the even numbers – and more – the prime numbers are the building blocks of all the integers or whole numbers: every even number (with the exception of 2) is the sum of 2 prime numbers, and every odd number is either a prime number, or a composite of prime numbers (that is, the odd number has prime factors). It is truly the peculiar characteristics of the prime numbers themselves (as described above, whose distribution could in fact be predicted by the prime number theorem which had been proven, implying some pattern or fractal nature in the prime numbers as per the above lemma), which could be regarded as a self-similar or fractal feature as such, that are responsible for the Goldbach conjecture being true. By induction the Goldbach conjecture has been proven true – the above constitutes proof of the Goldbach conjecture (which ought to be known as the Goldbach theorem instead).

This proof could be extended here. It has been mentioned above that the Goldbach conjecture had been tested and found to be correct for every even number up to 12×10^{17} by computer searches completed in 2008. Thus, by the above lemma, and the infinitude of the primes and even numbers, this long list of consecutive even numbers up to 12×10^{17} reflects (indicates or implies) the fact that all the infinite even numbers above 12×10^{17} would each be the sum of 2 primes. (This is in accordance with the "reflection" principle stated above, wherein it is mentioned that the characteristic of a mountain or infinite volume of sand is reflected in the characteristic of some grains of sand found there so that studying the characteristic of some grains of sand found there is enough for deducing the characteristic of the mountain or infinite volume of sand.)

Conclusion

We declare that the Goldbach conjecture is true – every even number after the number 2 is indeed the sum of 2 primes.

This paper should be able to provide some food for further thought on the conjecture.

Appendix 1

(20) *Set of integers, 1,201 to 1,250, with 8 primes within it*

(a) Primes: 1,201; 1,213; 1,217; 1,223; 1,229; 1,231; 1,237 and 1,249
(b) No. of primes: 8
(c) No. of even numbers "generated" (including repetitions) by the 8 primes = 648 (1,204 [1,201 + 3] to 2,498 [1,249 + 1,249])
(d) No. of new even numbers "generated" = 56 (2,388 to 2,498)
(e) No. of old/repeated (also appeared in (19) above, with some also having appeared in (18), (17), (16), (15), (14), (13), (12), (11), (10), (9), and (8) above) even numbers "generated" (i.e., repetitions/overlaps) = 592 (1,204 To 2,386)
(f) Density of new even numbers "generated" = (d) ÷ 8 primes = 56 ÷ 8 = 7 new even numbers per prime number

Appendix 2

(8) *Eighth batch of 30 even numbers (424 to 482)*

(a) 424: no. of above-mentioned prime additions/combinations = 12
(b) 426: no. of above-mentioned prime additions/combinations = 21
(c) 428: no. of above-mentioned prime additions/combinations = **9**
(d) 430: no. of above-mentioned prime additions/combinations = 14
(e) 432: no. of above-mentioned prime additions/combinations = 19
(f) 434: no. of above-mentioned prime additions/combinations = 14
(g) 436: no. of above-mentioned prime additions/combinations = 11
(h) 438: no. of above-mentioned prime additions/combinations = 22
(i) 440: no. of above-mentioned prime additions/combinations = 15
(j) 442: no. of above-mentioned prime additions/combinations = 13
(k) 444: no. of above-mentioned prime additions/combinations = 22
(l) 446: no. of above-mentioned prime additions/combinations = 12
(m) 448: no. of above-mentioned prime additions/combinations = 13
(n) 450: no. of above-mentioned prime additions/combinations = 29
(o) 452: no. of above-mentioned prime additions/combinations = 14
(p) 454: no. of above-mentioned prime additions/combinations = 12
(q) 456: no. of above-mentioned prime additions/combinations = 26

(continued)

(r) 458: no. of above-mentioned prime additions/combinations = **9**
(s) 460: no. of above-mentioned prime additions/combinations = 17
(t) 462: no. of above-mentioned prime additions/combinations = **30**
(u) 464: no. of above-mentioned prime additions/combinations = 13
(v) 466: no. of above-mentioned prime additions/combinations = 14
(w) 468: no. of above-mentioned prime additions/combinations = 26
(x) 470: no. of above-mentioned prime additions/combinations = 16
(y) 472: no. of above-mentioned prime additions/combinations = 14
(z) 474: no. of above-mentioned prime additions/combinations = 24
(aa) 476: no. of above-mentioned prime additions/combinations = 14
(bb) 478: no. of above-mentioned prime additions/combinations = 12
(cc) 480: no. of above-mentioned prime additions/combinations = **30**
(dd) 482: no. of above-mentioned prime additions/combinations = 11
(i) Maximum no. of prime additions/combinations = 30
(ii) Minimum no. of prime additions/combinations = 9
(iii) Total no. of prime additions/combinations for (a) to (dd) = 508
(iv) Total no. of even numbers = 30
(v) Density of distribution = average prime additions/combinations per even number = (iii) ÷ (iv) = 508 ÷ 30 = 16.93 Prime additions/combinations per even number

References

[1] D. Burton, 1980, Elementary Number Theory, Allyn & Bacon

[2] R. Courant and H. Robbins, Revised by I. Stewart, 1996, What Is Mathematics? An Elementary Approach to Ideas and Methods, Oxford University Press

[3] M. J. Feigenbaum, 1980, Universal Behavior in Nonlinear Systems, Los Alamos Science, No. 1

[4] G. H. Hardy and E. M. Wright, 1979, An Introduction to Theory of Numbers, Oxford, England: Clarendon Press

[5] D. H. Lehmer, 1914, List of Prime Numbers from 1 to 10,006,721, Publication No. 165, Carnegie Institution of Washington, Washington, D.C.

[6] M. E. Lines, 1986, A Number for Your Thoughts, Adam Hilger

[7] B. B. Mandelbrot, 1977, The Fractal Geometry of Nature, W. H. Freeman

[8] R. McWeeny, 2002, Symmetry: An Introduction to Group Theory and Its Applications, Dover

[9] J. M. T. Thompson and H. B. Stewart, 1986, Nonlinear Dynamics and Chaos, New York: John Wiley

[10] World Wide Web

2 (b) On the Goldbach Conjecture (This Paper Had Been Published in an International Mathematics Journal in 2012)

Abstract

The author had published a paper on the solutions for the twin primes conjecture in an international mathematics journal in 2003. This paper, which comprises three parts that are each self-contained, presents various approaches to the Goldbach conjecture, which is a related problem.

MSC: 11-XX (Number Theory)

Keywords: partitions; induction; contradiction; elimination; symmetry; different approaches

Introduction

The expected mode of solving the Goldbach conjecture appears to be the utilization of advanced calculus or analysis, e.g., by the summation, or integration, of the reciprocals involving directly or indirectly the primes to see whether they converge or diverge, in order to get a "feel" of the pattern of the distribution of the primes. But, such a method of solving the problem has evidently not succeeded so far. Some other approach or approaches could be more appropriate. This paper brings up a number of such approaches.

Part 1

Theorem. Every even number after 2 is the sum of 2 primes.

Proof 1.

Lemma. By Euclid's proof the primes are infinite.
The prime number theorem, which had been proven, states that the limit of the quotient of the 2 functions $\pi(n)$ and $n/\log n$ as n approaches infinity is 1, which is expressed by the formula:

$$\lim_{n\to\infty} \pi(n)/(n/\log n) = 1$$

where $\pi(n)$ is approximately equal to $(n/\log n)$.

The function $\pi(n)$ represents the number of primes less than or equal to the number n. This function measures the distribution of the prime numbers. With it, we compute the ratio $n/\pi(n)$ which says what fraction of the numbers up to a given point are primes. (It is actually the reciprocal of this fraction.) The following is the result of a computation:

n	$\pi(n)$	$n/\pi(n)$
10	4 (a)	2.5
100	25 (b)	4.0
1,000	168 (c)	6.0
10,000	1,229 (d)	8.1
100,000	9,592 (e)	10.4
1,000,000	78,498 (f)	12.7
10,000,000	664,579 (g)	15.0
100,000,000	5,761,455 (h)	17.4
1,000,000,000	50,847,534 (i)	19.7
10,000,000,000	455,052,512 (j)	22.0

It is noticeable that as one moves from 1 power of 10 to the next, the ratio $n/\pi(n)$ increases by about 2.3, e.g., 22.0 – 19.7 = 2.3. As $\log_e 10 = 2.30258\ldots$, we may thus regard $\pi(n)$ as approximately equal to $n/\log n$.

We have the following partitions with the primes described in the "$\pi(n)$" column above:

1) With (a) above, we have the following "prime + prime = even number" combinations:

a) prime a + prime a: 4 × 4 "prime + prime" combinations
b) prime a + prime b: 4 × 25 "prime + prime" combinations
c) prime a + prime c: 4 × 168 "prime + prime" combinations
d) prime a + prime d: 4 × 1,229 "prime + prime" combinations
e) prime a + prime e: 4 × 9,592 "prime + prime" combinations
f) prime a + prime f: 4 × 78,498 "prime + prime" combinations
g) prime a + prime g: 4 × 664,579 "prime + prime" combinations
h) prime a + prime h: 4 × 5,761,455 "prime + prime" combinations
i) prime a + prime i: 4 × 50,847,534 "prime + prime" combinations
j) prime a + prime j: 4 × 455,052,512 "prime + prime" combinations

For example, for (j) above, a prime described in (a) in the "$\pi(n)$" column above plus a prime described in (j) in the "$\pi(n)$" column above give an even number, and there are 4 x 455,052,512 such "prime + prime = even number" combinations.

2) With (b) above, we have the following "prime + prime = even number" combinations:

a) prime b + prime a: 25 × 4 "prime + prime" combinations
b) prime b + prime b: 25 × 25 "prime + prime" combinations
c) prime b + prime c: 25 × 168 "prime + prime" combinations
d) prime b + prime d: 25 × 1,229 "prime + prime" combinations
e) prime b + prime e: 25 × 9,592 "prime + prime" combinations

f) prime b + prime f: 25 × 78,498 “prime + prime” combinations
g) prime b + prime g: 25 × 664,579 “prime + prime” combinations
h) prime b + prime h: 25 × 5,761,455 “prime + prime” combinations
i) prime b + prime i: 25 × 50,847,534 “prime + prime” combinations
j) prime b + prime j: 25 × 455,052,512 “prime + prime” combinations

3) With (c) above, we have the following “prime + prime = even number” combinations:
a) prime c + prime a: 168 × 4 “prime + prime” combinations
b) prime c + prime b: 168 × 25 “prime + prime” combinations
c) prime c + prime c: 168 × 168 “prime + prime” combinations
d) prime c + prime d: 168 × 1,229 “prime + prime” combinations
e) prime c + prime e: 168 × 9,592 “prime + prime” combinations
f) prime c + prime f: 168 × 78,498 “prime + prime” combinations
g) prime c + prime g: 168 × 664,579 “prime + prime” combinations
h) prime c + prime h: 168 × 5,761,455 “prime + prime” combinations
i) prime c + prime i: 168 × 50,847,534 “prime + prime” combinations
j) prime c + prime j: 168 × 455,052,512 “prime + prime” combinations

4) With (d) above, we have the following “prime + prime = even number” combinations:
a) prime d + prime a: 1,229 × 4 “prime + prime” combinations
b) prime d + prime b: 1,229 × 25 “prime + prime” combinations
c) prime d + prime c: 1,229 × 168 “prime + prime” combinations
d) prime d + prime d: 1,229 × 1,229 “prime + prime” combinations
e) prime d + prime e: 1,229 × 9,592 “prime + prime” combinations
f) prime d + prime f: 1,229 × 78,498 “prime + prime” combinations
g) prime d + prime g: 1,229 × 664,579 “prime + prime” combinations
h) prime d + prime h: 1,229 × 5,761,455 “prime + prime” combinations
i) prime d + prime i: 1,229 × 50,847,534 “prime + prime” combinations
j) prime d + prime j: 1,229 × 455,052,512 “prime + prime” combinations

5) With (e) above, we have the following “prime + prime = even number” combinations:
a) prime e + prime a: 9,592 × 4 “prime + prime” combinations
b) prime e + prime b: 9,592 × 25 “prime + prime” combinations
c) prime e + prime c: 9,592 × 168 “prime + prime” combinations
d) prime e + prime d: 9,592 × 1,229 “prime + prime” combinations
e) prime e + prime e: 9,592 × 9,592 “prime + prime” combinations
f) prime e + prime f: 9,592 × 78,498 “prime + prime” combinations
g) prime e + prime g: 9,592 × 664,579 “prime + prime” combinations
h) prime e + prime h: 9,592 × 5,761,455 “prime + prime” combinations
i) prime e + prime i: 9,592 × 50,847,534 “prime + prime” combinations
j) prime e + prime j: 9,592 × 455,052,512 “prime + prime” combinations

6) With (f) above, we have the following “prime + prime = even number” combinations:
a) prime f + prime a: 78,498 × 4 “prime + prime” combinations
b) prime f + prime b: 78,498 × 25 “prime + prime” combinations
c) prime f + prime c: 78,498 × 168 “prime + prime” combinations
d) prime f + prime d: 78,498 × 1,229 “prime + prime” combinations
e) prime f + prime e: 78,498 × 9,592 “prime + prime” combinations
f) prime f + prime f: 78,498 × 78,498 “prime + prime” combinations
g) prime f + prime g: 78,498 × 664,579 “prime + prime” combinations
h) prime f + prime h: 78,498 × 5,761,455 “prime + prime” combinations
i) prime f + prime i: 78,498 × 50,847,534 “prime + prime” combinations
j) prime f + prime j: 78,498 × 455,052,512 “prime + prime” combinations

7) With (g) above, we have the following “prime + prime = even number” combinations:
a) prime g + prime a: 664,579 × 4 “prime + prime” combinations
b) prime g + prime b: 664,579 × 25 “prime + prime” combinations
c) prime g + prime c: 664,579 × 168 “prime + prime” combinations
d) prime g + prime d: 664,579 × 1,229 “prime + prime” combinations
e) prime g + prime e: 664,579 × 9,592 “prime + prime” combinations
f) prime g + prime f: 664,579 × 78,498 “prime + prime” combinations
g) prime g + prime g: 664,579 × 664,579 “prime + prime” combinations
h) prime g + prime h: 664,579 × 5,761,455 “prime + prime” combinations
i) prime g + prime i: 664,579 × 50,847,534 “prime + prime” combinations
j) prime g + prime j: 664,579 × 455,052,512 “prime + prime” combinations

8) With (h) above, we have the following “prime + prime = even number” combinations:
a) prime h + prime a: 5,761,455 × 4 “prime + prime” combinations
b) prime h + prime b: 5,761,455 × 25 “prime + prime” combinations
c) prime h + prime c: 5,761,455 × 168 “prime + prime” combinations
d) prime h + prime d: 5,761,455 × 1,229 “prime + prime” combinations
e) prime h + prime e: 5,761,455 × 9,592 “prime + prime” combinations
f) prime h + prime f: 5,761,455 × 78,498 “prime + prime” combinations
g) prime h + prime g: 5,761,455 × 664,579 “prime + prime” combinations
h) prime h + prime h: 5,761,455 × 5,761,455 “prime + prime” combinations
i) prime h + prime i: 5,761,455 × 50,847,534 “prime + prime” combinations
j) prime h + prime j: 5,761,455 × 455,052,512 “prime + prime” combinations

9) With (i) above, we have the following “prime + prime = even number” combinations:
a) prime i + prime a: 50,847,534 × 4 “prime + prime” combinations
b) prime i + prime b: 50,847,534 × 25 “prime + prime” combinations
c) prime i + prime c: 50,847,534 × 168 “prime + prime” combinations
d) prime i + prime d: 50,847,534 × 1,229 “prime + prime” combinations
e) prime i + prime e: 50,847,534 × 9,592 “prime + prime” combinations

f) prime i + prime f: 50,847,534 × 78,498 "prime + prime" combinations
g) prime i + prime g: 50,847,534 × 664,579 "prime + prime" combinations
h) prime i + prime h: 50,847,534 × 5,761,455 "prime + prime" combinations
i) prime i + prime i: 50,847,534 × 50,847,534 "prime + prime" combinations
j) prime i + prime j: 50,847,534 × 455,052,512 "prime + prime" combinations

10) With (j) above, we have the following "prime + prime = even number" combinations:
a) prime j + prime a: 455,052,512 × 4 "prime + prime" combinations
b) prime j + prime b: 455,052,512 × 25 "prime + prime" combinations
c) prime j + prime c: 455,052,512 × 168 "prime + prime" combinations
d) prime j + prime d: 455,052,512 × 1,229 "prime + prime" combinations
e) prime j + prime e: 455,052,512 × 9,592 "prime + prime" combinations
f) prime j + prime f: 455,052,512 × 78,498 "prime + prime" combinations
g) prime j + prime g: 455,052,512 × 664,579 "prime + prime" combinations
h) prime j + prime h: 455,052,512 × 5,761,455 "prime + prime" combinations
i) prime j + prime i: 455,052,512 × 50,847,534 "prime + prime" combinations
j) prime j + prime j: 455,052,512 × 455,052,512 "prime + prime" combinations

The above partitions/"prime + prime = even number" combinations are evidently progressively more "overwhelming", dense, and repetitive. That is, the Goldbach conjecture becomes evidently progressively stronger and stronger toward infinity. It is not surprising that computer searches completed in 2000 had verified that all even numbers up to 400 trillion (4×10^{14}), which is not a small list, are sums of 2 primes, while in 2008, a distributed computer search ran by Tomas Oliveira e Silva, a researcher at the University of Aveiro, Portugal, had further verified the Goldbach conjecture up to 12×10^{17}, which is a long, impressive list.

The infinitude of the primes, as per the above lemma, together with the infinitude of the even numbers, however imply that the above partitions/"prime + prime = even number" combinations would become increasingly more "overwhelming", dense, and repetitive toward infinity (the Goldbach conjecture becoming evidently stronger and stronger the higher up the infinite list of prime numbers/even numbers we go), hence "ensuring" the continuity (without any breaks or gaps) of the even numbers generated, and would be so all the way to infinity, thus proving that every even number after 2 is the sum of 2 primes.

Proof 2.

Lemma. According to the principle of complete induction in set theory, if a set of natural numbers contains 1 and for each n, it contains $n + 1$ whenever it contains all numbers less than $n + 1$, then it must contain every natural number, e.g., complete induction proves that every natural number is a product of primes.

By the above lemma, every even number after 2 in the infinite set of the integers is the sum of 2 primes; as per the distributed computer search completed in 2008 at the University of Aveiro, Portugal, stated above, 12×10^{17} in the infinite set of the integers is the largest even number found to be the sum of 2 primes while all the consecutive even numbers before it, from 4 to $(12 \times 10^{17}) - 2$, which is not a small list of numbers (it is in fact a long, impressive list, obtainable only with the help of modern computer technology), are also found to be sums of 2 primes – the principle of complete induction implies that all even numbers after 12×10^{17} in the infinite set of the integers must also be sums of 2 primes, i.e., it implies that every even number after 2 in the infinite set of the integers must be the sum of 2 primes – in other words, the Goldbach conjecture must be true.

Proof 3.
Lemma. By Euclid's proof the primes are infinite.

We make use of the proof by "reductio ad absurdum" here. For this indirect proof, we assume that the Goldbach conjecture is false. (Before we proceed further, we should again note that a long, impressive list of consecutive even numbers, from 4 to 12 x 10^{17}, had already been verified to be sums of 2 primes, and these partitions/"prime + prime = even number" combinations would become increasingly more "overwhelming", dense, and repetitive toward infinity (the Goldbach conjecture becoming evidently stronger and stronger the higher up the infinite list of prime numbers/even numbers we go), as is described above. The moot question now is, of course, whether after 12×10^{17} there would be an even number in the infinite list of even numbers which is the last, or largest, even number which is the sum of 2 primes – this largest even number, if it exists (thereby proving the falsehood of the Goldbach conjecture), must (of necessity) be the sum of 2 primes which are each the largest existing prime. Before we continue, this point should be clearly held in mind.) This assumption implies that there is a limit to the even numbers which are sums of 2 primes and that there is a largest even number (e) which is, and must necessarily be, the sum of 2 primes that are each the largest existing prime ($e = x + x$, this largest even number, e, representing the ultimate limit of the even numbers which are sums of 2 primes, the 2 primes which add up to give e being of necessity each the largest existing prime (x)). This is of course a contradiction of the above lemma, which would imply that the lemma is false. But the lemma cannot be false – it is in fact a theorem (which had been proven by Euclid); there cannot be a largest existing prime (x) – the primes are infinite. This means that our assumption that the Goldbach conjecture is false is untenable and that the Goldbach conjecture must be true, i.e., every even number after 2 must be the sum of 2 primes. As a matter of fact, the above lemma implies that there would be an infinite number of double primes which sum up to an even number.

By both induction and contradiction the Goldbach conjecture is hence proved.

Part 2

Theorem. Every even number after 2 is the sum of 2 primes.

Proof 1. Every even number after 2 is the sum of 2 odd numbers. Every odd number is either a prime which is odd or a composite – product of primes which are odd; notably, every prime with the exception of 2 is an odd number. Every even number after 2 is also a composite, but, a composite with at least 1 even prime factor, namely, 2, while the rest of its prime factors are odd, i.e., it is an even composite.

Therefore, every even number after 2 is the sum of 2 primes which are odd and/or the sum of 1 prime which is odd and 1 odd composite whose prime factors are odd and/or the sum of 2 odd composites whose prime factors are odd, besides being an even composite with at least 1 even prime factor, namely, 2, while the rest of its prime factors are odd.

Lemma. By Euclid's proof, the primes are infinite; this implies that there would be an infinitude of sums of 2 primes as per the Goldbach conjecture. The even numbers, which are sums of 2 primes as per the conjecture, are also infinite. Thus, there are an infinite number of even numbers which are sums of 2 primes, both the even numbers and sums of 2 primes being infinite.

Corollary. The odd numbers, which are either prime, every prime with the exception of 2 being an odd number, or composite (have prime factors which are odd), are infinite; this implies that there would be an infinite number of sums of 2 odd numbers, each of which is equal to an even number. Hence, as there is an infinitude of even numbers which are sums of 2 primes, as per the above lemma, and as all primes with the exception of 2 are odd numbers, there are an infinite number of even numbers which are sums of 2 odd numbers that are prime, all the even numbers, sums of 2 odd numbers and primes being infinite; i.e., every even number after 2 is also the sum of 2 odd numbers that are prime.

We thereby see the close interlink or relationship between the primes, even numbers and odd numbers, which are all infinite, which is significant.

The following are thus evident:

a) Every sum of 2 primes which are odd numbers is equal to an even number, as is below in consecutive order:

$2 + 2 = 1 + 3 = \mathbf{4}$
$3 + 3 = 1 + 5 = \mathbf{6}$
$3 + 5 = 1 + 7 = \mathbf{8}$
$5 + 5 = 3 + 7 = \mathbf{10}$
$5 + 7 = 1 + 11 = \mathbf{12}$
$7 + 7 = 3 + 11 = 1 + 13 = \mathbf{14}$
$3 + 13 = 5 + 11 = \mathbf{16}$
$7 + 11 = 5 + 13 = 1 + 17 = \mathbf{18}$

7 + 13 = 3 + 17 = 1 + 19 = **20**

11 + 11 = 3 + 19 = 5 + 17 = 11 + 11 = **22**

11 + 13 = 5 + 19 = 7 + 17 = 1 + 23 = **24**

13 + 13 = 3 + 23 = 7 + 19 = **26**

11 + 17 = 5 + 23 = **28**

13 + 17 = 11 + 19 = 7 + 23 = 1 + 29 = **30**

3 + 29 = 13 + 19 = 1 + 31 = **32**

17 + 17 = 3 + 31 = 5 + 29 = 11 + 23 = 17 + 17 = **34**

17 + 19 = 5 + 31 = 7 + 29 = 13 + 23 = **36**

19 + 19 = 7 + 31 = 1 + 37 = **38**

3 + 37 = 11 + 29 = 17 + 23 = **40**

19 + 23 = 5 + 37 = 11 + 31 = 13 + 29 = 1 + 41 = **42**

3 + 41 = 7 + 37 = 13 + 31 = 1 + 43 = **44**

23 + 23 = 3 + 43 = 5 + 41 = 17 + 29 = **46**

5 + 43 = 7 + 41 = 11 + 37 = 17 + 31 = 19 + 29 = 1 + 47 = **48**

3 + 47 = 7 + 43 = 13 + 37 = 19 + 31 = **50**

23 + 29 = 5 + 47 = 11 + 41 = **52**

7 + 47 = 11 + 43 = 13 + 41 = 17 + 37 = 23 + 31 = 1 + 53 = **54**

3 + 53 = 13 + 43 = 19 + 37 = **56**

29 + 29 = 5 + 53 = 11 + 47 = 17 + 41 = 29 + 29 = **58**

29 + 31 = 7 + 53 = 13 + 47 = 17 + 43 = 19 + 41 = 23 + 37 = 1 + 59 = **60**

31 + 31 = 3 + 59 = 19 + 43 = 1 + 61 = **62**

3 + 61 = 5 + 59 = 11 + 53 = 17 + 47 = 23 + 41 = **64**

5 + 61 = 7 + 59 = 13 + 53 = 19 + 47 = 23 + 43 = 29 + 37 = **66**

7 + 61 = 31 + 37 = 1 + 67 = **68**

3 + 67 = 11 + 59 = 17 + 53 = 23 + 47 = 29 + 41 = **70**

5 + 67 = 11 + 61 = 13 + 59 = 19 + 53 = 29 + 43 = 31 + 41 = 1 + 71 = 72

37 + 37 = 3 + 71 = 7 + 67 = 13 + 61 = 31 + 43 = 37 + 37 = 1 + 73 = **74**

3 + 73 = 5 + 71 = 17 + 59 = 23 + 53 = 29 + 47 = 76

37 + 41 = 5 + 73 = 7 + 71 = 11 + 67 = 31 + 47 = 37 + 41 = 78

7 + 73 = 13 + 67 = 19 + 61 = 37 + 43 = 1 + 79 = **80**

41 + 41 = 3 + 79 = 11 + 71 = 23 + 59 = 29 + 53 = **82**

41 + 43 = 5 + 79 = 11 + 73 = 13 + 71 = 17 + 67 = 23 + 61 = 31 + 53 = 37 + 47 = 1 + 83 = **84**

43 + 43 = 3 + 83 = 7 + 79 = 13 + 73 = 19 + 67 = 43 + 43 = **86**

5 + 83 = 17 + 71 = 29 + 59 = 41 + 47 = **88**

7 + 83 = 11 + 79 = 17 + 73 = 19 + 71 = 23 + 67 = 29 + 61 = 31 + 59 = 37 + 53 = 43 + 47 = 1 + 89 = **90**

3 + 89 = 13 + 79 = 19 + 73 = 31 + 61 = 1 + 91 = **92**

47 + 47 = 5 + 89 = 11 + 83 = 23 + 71 = 41 + 53 = 47 + 47 = **94**

5 + 91 = 7 + 89 = 13 + 83 = 17 + 79 = 23 + 73 = 29 + 67 = 37 + 59 = 43 + 53 = **96**

7 + 91 = 19 + 79 = 31 + 67 = 37 + 61 = 1 + 97 = **98**

47 + 53 = 3 + 97 = 11 + 89 = 17 + 83 = 29 + 71 = 41 + 59 = 47 + 53 = **100**

5 + 97 = 11 + 91 = 13 + 89 = 19 + 83 = 23 + 79 = 29 + 73 = 31 + 71 = 41 + 61 = 43 + 59 = 1 + 101 = **102**

b) Every sum of 1 prime which is an odd number and 1 odd composite which is the product of primes which are odd, is equal to the sum of 2 primes which are odd numbers, which are all each equal to an even number, as is below in consecutive order:

3 + **9** = 5 + 7 = 1 + 11 = **12**
5 + **9** = 3 + 11 = 7 + 7 = 1 + 13 = **14**
7 + **9** = 3 + 13 = 5 + 11 = **16**
3 + **15** = 7 + 11 = 5 + 13 = 1 + 17 = **18**
11 + **9** = 3 + 17 = 7 + 13 = 1 + 19 = **20**
13 + **9** = 3 + 19 = 5 + 17 = 11 + 11 = **22**
3 + **21** = 11 + 13 = 5 + 19 = 7 + 17 = 1 + 23 = **24**
17 + **9** = 3 + 23 = 7 + 19 = 13 + 13 = **26**
19 + **9** = 5 + 23 = 11 + 17 = **28**
5 + **25** = 13 + 17 = 11 + 19 = 7 + 23 = 1 + 29 = **30**
23 + **9** = 3 + 29 = 13 + 19 = 1 + 31 = **32**
7 + **27** = 17 + 17 = 3 + 31 = 5 + 29 = 11 + 23 = 17 + 17 = **34**
3 + **33** = 17 + 19 = 5 + 31 = 7 + 29 = 13 + 23 = **36**
29 + **9** = 7 + 31 = 19 + 19 = 1 + 37 = **38**
31 + **9** = 3 + 37 = 11 + 29 = 17 + 23 = **40**
3 + **39** = 19 + 23 = 5 + 37 = 11 + 31 = 13 + 29 = 1 + 41 = **42**
5 + **39** = 3 + 41 = 7 + 37 = 13 + 31 = 1 + 43 = **44**
37 + **9** = 3 + 43 = 5 + 41 = 17 + 29 = 23 + 23 = **46**
3 + **45** = 5 + 43 = 7 + 41 = 11 + 37 = 17 + 31 = 19 + 29 = 1 + 47 = **48**
41 + **9** = 3 + 47 = 7 + 43 = 13 + 37 = 19 + 31 = **50**
43 + **9** = 5 + 47 = 11 + 41 = 23 + 29 = **52**
5 + **49** = 7 + 47 = 11 + 43 = 13 + 41 = 17 + 37 = 23 + 31 = 1 + 53 = **54**
47 + **9** = 3 + 53 = 13 + 43 = 19 + 37 = **56**
3 + **55** = 29 + 29 = 5 + 53 = 11 + 47 = 17 + 41 = 29 + 29 = **58**
5 + **55** = 29 + 31 = 7 + 53 = 13 + 47 = 17 + 43 = 19 + 41 = 23 + 37 = 1 + 59 = **60**
53 + **9** = 3 + 59 = 19 + 43 = 31 + 31 = 1 + 61 = **62**
7 + **57** = 3 + 61 = 5 + 59 = 11 + 53 = 17 + 47 = 23 + 41 = **64**
11 + **55** = 5 + 61 = 7 + 59 = 13 + 53 = 19 + 47 = 23 + 43 = 29 + 37 = **66**
59 + **9** = 7 + 61 = 31 + 37 = 1 + 67 = **68**
61 + **9** = 3 + 67 = 11 + 59 = 17 + 53 = 23 + 47 = 29 + 41 = **70**
3 + **69** = 5 + 67 = 11 + 61 = 13 + 59 = 19 + 53 = 29 + 43 = 31 + 41 = 1 + 71 = **72**
5 + **69** = 37 + 37 = 3 + 71 = 7 + 67 = 13 + 61 = 31 + 43 = 37 + 37 = 1 + 73 = **74**
67 + **9** = 3 + 73 = 5 + 71 = 17 + 59 = 23 + 53 = 29 + 47 = **76**
3 + **75** = 37 + 41 = 5 + 73 = 7 + 71 = 11 + 67 = 31 + 47 = 37 + 41 = **78**
71 + **9** = 7 + 73 = 13 + 67 = 19 + 61 = 37 + 43 = 1 + 79 = **80**
73 + **9** = 3 + 79 = 11 + 71 = 23 + 59 = 29 + 53 = 41 + 41 = **82**

3 + **81** = 41 + 43 = 5 + 79 = 11 + 73 = 13 + 71 = 17 + 67 = 23 + 61 = 31 + 53 = 37 + 47 = 1 + 83 = **84**
5 + **81** = 43 + 43 = 3 + 83 = 7 + 79 = 13 + 73 = 19 + 67 = 43 + 43 = **86**
79 + **9** = 5 + 83 = 17 + 71 = 29 + 59 = 41 + 47 = **88**
3 + **87** = 7 + 83 = 11 + 79 = 17 + 73 = 19 + 71 = 23 + 67 = 29 + 61 = 31 + 59 = 37 + 53 = 43 + 47 = 1 + 89 = **90**
83 + **9** = 3 + 89 = 13 + 79 = 19 + 73 = 31 + 61 = 1 + 91 = **92**
7 + **87** = 47 + 47 = 5 + 89 = 11 + 83 = 23 + 71 = 41 + 53 = 47 + 47 = **94**
3 + **93** = 5 + 91 = 7 + 89 = 13 + 83 = 17 + 79 = 23 + 73 = 29 + 67 = 37 + 59 = 43 + 53 = **96**
89 + **9** = 7 + 91 = 19 + 79 = 31 + 67 = 37 + 61 = 1 + 97 = **98**
91 + **9** = 3 + 97 = 11 + 89 = 17 + 83 = 29 + 71 = 41 + 59 = 47 + 53 = **100**
3 + **99** = 5 + 97 = 11 + 91 = 13 + 89 = 19 + 83 = 23 + 79 = 29 + 73 = 31 + 71 = 41 + 61 = 43 + 59 = 1 + 101 = **102**

c) Every sum of 2 odd composites which are products of primes which are odd, is equal to the sum of 2 primes which are odd numbers, which are all each equal to an even number, as is below in consecutive order:

9 + **9** = 5 + 13 = 7 + 11 = 1 + 17 = **18**
9 + **15** = 5 + 19 = 7 + 17 = 11 + 13 = 1 + 23 = **24**
15 + **15** = 7 + 23 = 11 + 19 = 13 + 17 = 1 + 29 = **30**
9 + **25** = 7 + 27 = 17 + 17 = 3 + 31 = 5 + 29 = 11 + 23 = 17 + 17 = **34**
15 + **21** = 5 + 31 = 7 + 29 = 13 + 23 = 17 + 19 = **36**
15 + **25** = 3 + 37 = 11 + 29 = 17 + 23 = **40**
21 + **21** = 5 + 37 = 11 + 31 = 13 + 29 = 19 + 23 = 1 + 41 = **42**
9 + **35** = 3 + 41 = 7 + 37 = 13 + 31 = 1 + 43 = **44**
21 + **25** = 3 + 43 = 5 + 41 = 17 + 29 = 23 + 23 = **46**
9 + **39** = 5 + 43 = 7 + 41 = 11 + 37 = 17 + 31 = 19 + 29 = 1 + 47 = **48**
25 + **25** = 3 + 47 = 7 + 43 = 13 + 37 = 19 + 31 = **50**
25 + **27** = 5 + 47 = 11 + 41 = 23 + 29 = **52**
27 + **27** = 7 + 47 = 11 + 43 = 13 + 41 = 17 + 37 = 23 + 31 = 1 + 53 = **54**
21 + **35** = 3 + 53 = 13 + 43 = 19 + 37 = **56**
9 + **49** = 29 + 29 = 5 + 53 = 11 + 47 = 17 + 41 = 29 + 29 = **58**
27 + **33** = 7 + 53 = 13 + 47 = 17 + 43 = 19 + 41 = 23 + 37 = 29 + 31 = 1 + 59 = **60**
27 + **35** = 31 + 31 = 3 + 59 = 19 + 43 = 1 + 61 = **62**
9 + **55** = 3 + 61 = 5 + 59 = 11 + 53 = 17 + 47 = 23 + 41 = **64**
33 + **33** = 5 + 61 = 7 + 59 = 13 + 53 = 19 + 47 = 23 + 43 = 29 + 37 = **66**
33 + **35** = 7 + 61 = 31 + 37 = 1 + 67 = **68**
35 + **35** = 3 + 67 = 11 + 59 = 17 + 53 = 23 + 47 = 29 + 41 = **70**
9 + **63** = 5 + 67 = 11 + 61 = 13 + 59 = 19 + 53 = 29 + 43 = 31 + 41 = 1 + 71 = **72**
35 + **39** = 3 + 71 = 7 + 67 = 13 + 61 = 31 + 43 = 37 + 37 = 1 + 73 = **74**
21 + **55** = 3 + 73 = 5 + 71 = 17 + 59 = 23 + 53 = 29 + 47 = **76**

39 + **39** = 5 + 73 = 7 + 71 = 11 + 67 = 31 + 47 = 37 + 41 = **78**
15 + **65** = 7 + 73 = 13 + 67 = 19 + 61 = 37 + 43 = 1 + 79 = **80**
25 + **57** = 41 + 41 = 3 + 79 = 11 + 71 = 23 + 59 = 29 + 53 = **82**
39 + **45** = 5 + 79 = 11 + 73 = 13 + 71 = 17 + 67 = 23 + 61 = 31 + 53 = 37 + 47 = 41 + 43 = 1 + 83 = **84**
9 + **77** = 43 + 43 = 3 + 83 = 7 + 79 = 13 + 73 = 19 + 67 = 43 + 43 = **86**
25 + **63** = 5 + 83 = 17 + 71 = 29 + 59 = 41 + 47 = **88**
45 + **45** = 7 + 83 = 11 + 79 = 17 + 73 = 19 + 71 = 23 + 67 = 29 + 61 = 31 + 59 = 37 + 53 = 43 + 47 = 1 + 89 = **90**
15 + **77** = 3 + 89 = 13 + 79 = 19 + 73 = 31 + 61 = 1 + 91 = **92**
45 + **49** = 5 + 89 = 11 + 83 = 23 + 71 = 41 + 53 = 47 + 47 = **94**
9 + **87** = 5 + 91 = 7 + 89 = 13 + 83 = 17 + 79 = 23 + 73 = 29 + 67 = 37 + 59 = 43 + 53 = **96**
49 + **49** = 7 + 91 = 19 + 79 = 31 + 67 = 37 + 61 = 1 + 97 = **98**
49 + **51** = 3 + 97 = 11 + 89 = 17 + 83 = 29 + 71 = 41 + 59 = 47 + 53 = **100**
51 + **51** = 5 + 97 = 11 + 91 = 13 + 89 = 19 + 83 = 23 + 79 = 29 + 73 = 31 + 71 = 41 + 61 = 43 + 59 = 1 + 101 = **102**

d) From (a), (b), and (c) above, we have the even numbers from 4 to 102 . . . composed as follows:

1) **4** = 2 + 2 = 1 + 3 (sum of 2 primes only)
2) **6** = 3 + 3 = 1 + 5 (sum of 2 primes only)
3) **8** = 3 + 5 = 1 + 7 (sum of 2 primes only)
4) **10** = 5 + 5 = 3 + 7 (sum of 2 primes only)
5) **12** = 5 + 7 = 1 + 11 = **3** + **9** (sum of 1 prime and 1 odd composite)
6) **14** = 3 + 11 = 7 + 7 = 1 + 13 = **5** + **9** (sum of 1 prime and 1 odd composite)
7) **16** = 3 + 13 = 5 + 11 = **7** + **9** (sum of 1 prime and 1 odd composite)
8) **18** = 5 + 13 = 7 + 11 = 1 + 17 = **3** + **15** (sum of 1 prime and 1 odd composite) = **9** + **9** (sum of 2 odd composites)
9) **20** = 3 + 17 = 7 + 13 = 1 + 19 = **11** + **9** (sum of 1 prime and 1 odd composite)
10) **22** = 3 + 19 = 5 + 17 = 11 + 11 = **13** + **9** (sum of 1 prime and 1 odd composite)
11) **24** = 5 + 19 = 7 + 17 = 11 + 13 = 1 + 23 = **3** + **21** (sum of 1 prime and 1 odd composite) = **9** + **15** (sum of 2 odd composites)
12) **26** = 3 + 23 = 7 + 19 = 13 + 13 = **17** + **9** (sum of 1 prime and 1 odd composite)
13) **28** = 5 + 23 = 11 + 17 = **19** + **9** (sum of 1 prime and 1 odd composite)
14) **30** = 7 + 23 = 11 + 19 = 13 + 17 = 1 + 29 = **5** + **25** (sum of 1 prime and 1 odd composite) = **15** + **15** (sum of 2 odd composites)
15) **32** = 3 + 29 = 13 + 19 = 1 + 31 = **23** + **9** (sum of 1 prime and 1 odd composite)
16) **34** = 17 + 17 = 3 + 31 = 5 + 29 = 11 + 23 = 17 + 17 = 7 + **27** (sum of 1 prime and 1 odd composite) = **9** + **25** (sum of 2 odd composites)
17) **36** = 5 + 31 = 7 + 29 = 13 + 23 = 17 + 19 = **3** + **33** (sum of 1 prime and 1 odd composite) = **15** + **21** (sum of 2 odd composites)
18) **38** = 7 + 31 = 19 + 19 = 1 + 37 = **29** + **9** (sum of 1 prime and 1 odd composite)

19) **40** = 3 + 37 = 11 + 29 = 17 + 23 = **31** + **9** (sum of 1 prime and 1 odd composite) = **15** + **25** (sum of 2 odd composites)
20) **42** = 5 + 37 = 11 + 31 = 13 + 29 = 19 + 23 = 1 + 41 = **3** + **39** (sum of 1 prime and 1 odd composite) = **21** + **21** (sum of 2 odd composites)
21) **44** = 3 + 41 = 7 + 37 = 13 + 31 = 1 + 43 = **5** + **39** (sum of 1 prime and 1 odd composite) = **9** + **35** (sum of 2 odd composites)
22) **46** = 3 + 43 = 5 + 41 = 17 + 29 = 23 + 23 = **37** + **9** (sum of 1 prime and 1 odd composite) = **21** + **25** (sum of 2 odd composites)
23) **48** = 5 + 43 = 7 + 41 = 11 + 37 = 17 + 31 = 19 + 29 = 1 + 47 = **3** + **45** (sum of 1 prime and 1 odd composite) = **9** + **39** (sum of 2 odd composites)
24) **50** = 3 + 47 = 7 + 43 = 13 + 37 = 19 + 31 = **41** + **9** (sum of 1 prime and 1 odd composite) = **25** + **25** (sum of 2 odd composites)
25) **52** = 5 + 47 = 11 + 41 = 23 + 29 = **43** + **9** (sum of 1 prime and 1 odd composite) = **25** + **27** (sum of 2 odd composites)
26) **54** = 7 + 47 = 11 + 43 = 13 + 41 = 17 + 37 = 23 + 31 = 1 + 53 = **5** + **49** (sum of 1 prime and 1 odd composite) = **27** + **27** (sum of 2 odd composites)
27) **56** = 3 + 53 = 13 + 43 = 19 + 37 = **47** + **9** (sum of 1 prime and 1 odd composite) = **21** + **35** (sum of 2 odd composites)
28) **58** = 29 + 29 = 5 + 53 = 11 + 47 = 17 + 41 = 29 + 29 = **3** + **55** (sum of 1 prime and 1 odd composite) = **9** + **49** (sum of 2 odd composites)
29) **60** = 7 + 53 = 13 + 47 = 17 + 43 = 19 + 41 = 23 + 37 = 29 + 31 = 1 + 59 = **5** + **55** (sum of 1 prime and 1 odd composite) = **27** + **33** (sum of 2 odd composites)
30) **62** = 3 + 59 = 19 + 43 = 31 + 31 = 1 + 61 = **53** + **9** (sum of 1 prime and 1 odd composite) = **27** + **35** (sum of 2 odd composites)
31) **64** = 3 + 61 = 5 + 59 = 11 + 53 = 17 + 47 = 23 + 41 = **7** + **57** (sum of 1 prime and 1 odd composite) = **9** + **55** (sum of 2 odd composites)
32) **66** = 5 + 61 = 7 + 59 = 13 + 53 = 19 + 47 = 23 + 43 = 29 + 37 = **11** + 55 (sum of 1 prime and 1 odd composite) = **33** + **33** (sum of 2 odd composites)
33) **68** = 7 + 61 = 31 + 37 = 1 + 67 = **59** + **9** (sum of 1 prime and 1 odd composite) = **33** + **35** (sum of 2 odd composites)
34) **70** = 3 + 67 = 11 + 59 = 17 + 53 = 23 + 47 = 29 + 41 = **61** + **9** (sum of 1 prime and 1 odd composite) = **35** + **35** (sum of 2 odd composites)
35) **72** = 5 + 67 = 11 + 61 = 13 + 59 = 19 + 53 = 29 + 43 = 31 + 41 = 1 + 71 = **3** + **69** (sum of 1 prime and 1 odd composite) = **9** + **63** (sum of 2 odd composites)
36) **74** = 3 + 71 = 7 + 67 = 13 + 61 = 31 + 43 = 37 + 37 = 1 + 73 = **5** + **69** (sum of 1 prime and 1 odd composite) = **35** + **39** (sum of 2 odd composites)
37) **76** = 3 + 73 = 5 + 71 = 17 + 59 = 23 + 53 = 29 + 47 = **67** + **9** (sum of 1 prime and 1 odd composite) = **21** + **55** (sum of 2 odd composites)
38) **78** = 5 + 73 = 7 + 71 = 11 + 67 = 31 + 47 = 37 + 41 = **3** + **75** (sum of 1 prime and 1 odd composite) = **39** + **39** (sum of 2 odd composites)
39) **80** = 7 + 73 = 13 + 67 = 19 + 61 = 37 + 43 = 1 + 79 = **71** + **9** (sum of 1 prime and 1 odd composite) = **15** + **65** (sum of 2 odd composites)

40) **82** = 3 + 79 = 11 + 71 = 23 + 59 = 29 + 53 = 41 + 41 = **73** + **9** (sum of 1 prime and 1 odd composite) = **25** + **57** (sum of 2 odd composites)
41) **84** = 5 + 79 = 11 + 73 = 13 + 71 = 17 + 67 = 23 + 61 = 31 + 53 = 37 + 47 = 41 + 43 = 1 + 83 = **3** + **81** (sum of 1 prime and 1 odd composite) = **39** + **45** (sum of 2 odd composites)
42) **86** = 43 + 43 = 3 + 83 = 7 + 79 = 13 + 73 = 19 + 67 = 43 + 43 = **5** + **81** (sum of 1 prime and 1 odd composite) = **9** + 77 (sum of 2 odd composites)
43) **88** = 5 + 83 = 17 + 71 = 29 + 59 = 41 + 47 = **79** + **9** (sum of 1 prime and 1 odd composite) = **25** + **63** (sum of 2 odd composites)
44) **90** = 7 + 83 = 11 + 79 = 17 + 73 = 19 + 71 = 23 + 67 = 29 + 61 = 31 + 59 = 37 + 53 = 43 + 47 = 1 + 89 = **3** + **87** (sum of 1 prime and 1 odd composite) = **45** + **45** (sum of 2 odd composites)
45) **92** = 3 + 89 = 13 + 79 = 19 + 73 = 31 + 61 = 1 + 91 = **83** + **9** (sum of 1 prime and 1 odd composite) = **15** + 77 (sum of 2 odd composites)
46) **94** = 5 + 89 = 11 + 83 = 23 + 71 = 41 + 53 = 47 + 47 = 7 + **87** (sum of 1 prime and 1 odd composite) = **45** + **49** (sum of 2 odd composites)
47) **96** = 5 + 91 = 7 + 89 = 13 + 83 = 17 + 79 = 23 + 73 = 29 + 67 = 37 + 59 = 43 + 53 = **3** + **93** (sum of 1 prime and 1 odd composite) = **9** + **87** (sum of 2 odd composites)
48) **98** = 7 + 91 = 19 + 79 = 31 + 67 = 37 + 61 = 1 + 97 = **89** + **9** (sum of 1 prime and 1 odd composite) = **49** + **49** (sum of 2 odd composites)
49) **100** = 3 + 97 = 11 + 89 = 17 + 83 = 29 + 71 = 41 + 59 = 47 + 53 = **91** + **9** (sum of 1 prime and 1 odd composite) = **49** + **51** (sum of 2 odd composites)
50) **102** = 5 + 97 = 11 + 91 = 13 + 89 = 19 + 83 = 23 + 79 = 29 + 73 = 31 + 71 = 41 + 61 = 43 + 59 = 1 + 101 = **3** + **99** (sum of 1 prime and 1 odd composite) = **51** + **51** (sum of 2 odd composites)

(The above is only a partial or incomplete listing of sums of 1 prime and 1 odd composite, and sums of 2 odd composites, each of which is equal to the sum of 2 primes as well as an even number. For example, in the list of compositions for the even numbers 4 to 102 . . . above, in Item (48), we could also have other "combinations" such as: **98** = 7 + 91 = 19 + 79 = 31 + 67 = 37 + 61 = 1 + 97 = ***25*** + ***73*** *(sum of 1 prime and 1 odd composite)* = ***21*** + *77 (sum of 2 odd composites),* and so on, in Item (49), we could also have other "combinations" such as: **100** = 3 + 97 = 11 + 89 = 17 + 83 = 29 + 71 = 41 + 59 = 47 + 53 = ***31*** + ***69*** *(sum of 1 prime and 1 odd composite)* = ***45*** + ***55*** *(sum of 2 odd composites),* and so on, and in Item (50), we could also have other "combinations" such as: **102** = 5 + 97 = 11 + 91 = 13 + 89 = 19 + 83 = 23 + 79 = 29 + 73 = 31 + 71 = 41 + 61 = 43 + 59 = 1 + 101 = ***17*** + ***85*** *(sum of 1 prime and 1 odd composite)* = ***21*** + ***81*** *(sum of 2 odd composites),* and so on That is, there are more "combinations" than those shown in the above listing.)

In (d) above, in the list of compositions for the 50 consecutive even numbers 4 to 102 . . ., the even numbers 4, 6, 8, and 10 are only formed through the summing of 2 primes and not at all through the summing of 1 prime and 1 odd composite, or the summing of 2 odd composites, which are impossibilities here. These sums of 2 primes are present (always present) throughout the whole list of compositions, from 4 right through to 102, while this is not the case for the sums of 1 prime and 1 odd composite, and the sums of 2 odd composites.

We reason here by the process of elimination, through analyzing the information in (d) above which pertains to the compositions of the 50 consecutive even numbers 4 to 102 . . . taken from the infinite list of even numbers. We stated at the beginning the following about the even numbers after 2:

Firstly, every even number after 2 is:

A) The sum of 2 odd numbers.

(Every odd number is either a prime which is odd or a composite – product of primes which are odd.

Notably, every prime with the exception of 2 is an odd number.)

Secondly, every even number after 2 is also (the below-mentioned is the logical consequence of (A) above):

1) The sum of 2 primes which are odd.
2) And/or the sum of 1 prime which is odd and 1 odd composite whose prime factors are odd.
3) And/or the sum of 2 odd composites whose prime factors are odd.

Evidently, at least 1 of (1), (2), and (3) above has to be the "atom" or building block of the even numbers. In (d) above, we observe the following:

i) All the 50 consecutive even numbers 4 to 102 . . . in (d) above taken from the infinite list of even numbers are sums of 2 primes.
ii) It is impossible for each of the even numbers 4, 6, 8, and 10 in (d) above to be the sum of 1 prime which is odd and 1 odd composite whose prime factors are odd.
iii) It is impossible for each of the even numbers 4, 6, 8, 10, 12, 14, 16, 20, 22, 26, 28, 32, and 38 in (d) above to be the sum of 2 odd composites whose prime factors are odd.

It is evident from (i), (ii), and (iii) above that neither (2) nor (3) can be the "atom" or building block of the even numbers since they are "incomplete". As (1) – the sum of 2 primes which are odd – is "complete", i.e., always present in the 50 consecutive even numbers 4 to 102 . . . in (d) above, unlike (2) and (3), it evidently is the "atom" or building block of the even numbers. That is, every even number after 2 is evidently the sum of 2 primes which are odd. In fact, a distributed computer search completed in 2008 at the University of Aveiro, Portugal, had verified this for all even numbers up to 12×10^{17}, which is not a small list. Definitely, due respectively to (ii) and (iii) above, we cannot say that every even number after 2 is the sum of 1 prime which is odd and 1 odd composite whose prime factors are odd, or every even number after 2 is the sum of 2 odd composites whose prime factors are odd.

By the above lemma and corollary, the infinitudes of the primes, even numbers, and odd numbers indeed imply that there are an infinite number of sums of two primes which are odd numbers, which are each equal to an even number. As the sums of two primes which are odd numbers are evidently the "atoms" or building blocks of the even numbers, it also implies that they are infinite, since the even numbers are infinite.

Hypothetically, if on the other hand just one of the three items stated above, primes, even numbers and odd numbers, were finite, the above-said sums of two primes which are odd numbers, each of which is equal to an even number, would be finite. The primes, even numbers, and odd numbers are evidently intricately linked, with the primes playing the part of building blocks of both the even and odd numbers through various "combinations" as is described below. However, as the primes, even numbers and odd numbers are intricately linked, the finiteness (or infinity) of any one of them implies the finiteness (or infinity) of the other two, and vice versa. These three items are evidently "close comrades-in-arm" working together to give special meaning to the integers. As these three are all infinite, it indeed implies that there is an infinitude of even numbers which are infinitely the sums of two primes that are odd and infinite.

Proof 2.

Lemma. According to the precepts of fractal geometry and group theory, symmetry is a very important, intrinsic part of nature. There is symmetry all around us and within us. There is evident symmetry in human bodies, the structures of viruses and bacteria, polymers and ceramic materials, the permutations of numbers, the universe and many others, even the movements of prices in financial markets, the growths of populations, the sound of music, the flow of blood through our circulatory system, the behavior of people en masse, and so on In other words, regularity, pattern, order, uniformity or symmetry is evident everywhere.

The above-mentioned most basic, always present sums of two primes, each of which is equal to an even number, which are evidently the "atoms" or building blocks of the even numbers, are characterized by the feature of symmetry (in 2008, a distributed computer search ran by Tomas Oliveira e Silva, a researcher at the University of Aveiro, Portugal, had verified that all even numbers up to 12×10^{17}, which is no small list of numbers, are sums of two primes, a regularity, uniformity, order, pattern, symmetry). Thus, by the above lemma, every even number after 2 is naturally or inherently the sum of 2 primes, i.e., there is an infinitude of sums of 2 primes which are each equal to an even number.

Hence, the confirmation of the following generalization pertaining to the integers, whereby it is indeed evident that the primes play a very important role:

Let a prime = p, and, a composite = c = p × p . . ..

a) Every even number after 2 = p + p = *c = *p × p . . . and/V = c + p = (p × p . . .) + p and/V = c + c = (p × p . . .) + (p × p . . .) (in *c = *p × p . . . here, which is an even composite, 1 or more of the p's are 2, the only even prime, e.g., 6 = 2 × 3, 8 = 2 × 2 × 2, 10 = 2 × 5, 18 = 2 × 3 × 3, 20 = 2 × 2 × 5, 24 = 2 × 2 × 2 × 3)

b) Every odd number = p V c = p × p . . . (in c = p × p . . . here, which is an odd composite, like the c = p × p . . .'s in (a) above, all the p's are odd, e.g., 9 = 3 × 3, 15 = 3 × 5, 21 = 3 × 7, 25 = 5 × 5, 63 = 3 × 3 × 7, 99 = 3 × 3 × 11, . . .)

It is easy to see that the Goldbach conjecture is valid, i.e., every even number after 2 is the sum of 2 primes.

Part 3

Theorem. Every even number after 2 is the sum of 2 primes.

Proof. The prime numbers are evidently the atoms or building blocks of the integers. The integers are either primes (not divisible by other integers except 1) or composites (divisible by other integers, e.g., the prime numbers), and even (the sums of 2 primes as conjectured by Goldbach) or odd (primes, or, composites whereby they are divisible by prime factors). Therefore, to determine whether the conjecture that every even number (except the number 2) is the sum of 2 primes is true, it would be appropriate to analyze the evident atoms or building blocks of the even numbers, viz., the prime numbers. For the solution to this conjecture we note that the primes (vide Euclid's proof) and the even numbers are infinite, which implies that this conjecture should be true.

We here analyze the "behavior" of the first 2,400 consecutive prime numbers (divided into 12 batches of consecutive primes, each subsequent batch with an increment of 200 primes), leaving out 2 (because it is an even prime) and commencing with 3, which is the 2nd. consecutive prime, the latter to be the first prime in our list of 2,400 consecutive primes (3 to 21,391), as follows:

(1) *200 Consecutive Primes from 3 to 1,229*
- (a) Even numbers (obtained by summing of 2 primes) = 6 to 2,458
- (b) No. of even numbers = 1,227
- (c) No. of primes = 200
- (d) Average no. of even numbers "generated" by each of these 200 consecutive primes = 1,227 ÷ 200 = **6.14**
- (e) No. of summings of 2 primes/permutations (3 + 3, 3 + 5, 3 + 7, 3 + 11, etc.) for these 200 primes = 200 × 200 = 40,000
- (f) Average no. of summings of 2 primes/permutations for each of the 1,227 even numbers = 40,000 ÷ 1,227 = **32.60**

(2) *400 Consecutive Primes from 3 to 2,749*
- (a) Even numbers (obtained by summing of 2 primes) = 6 to 5,498
- (b) No. of even numbers = 2,747
- (c) No. of primes = 400

(d) Average no. of even numbers "generated" by each of these 400 consecutive primes = 2,747 ÷ 400 = **6.87**

(e) No. of summings of 2 primes/permutations (3 + 3, 3 + 5, 3 + 7, 3 + 11, etc.) for these 400 primes = 400 × 400 = 160,000

(f) Average no. of summings of 2 primes/permutations for each of the 2,747 even numbers = 160,000 ÷ 2,747 = **58.25**

(3) *600 Consecutive Primes from 3 to 4,421*

(a) Even numbers (obtained by summing of 2 primes) = 6 to 8,842

(b) No. of even numbers = 4,419

(c) No. of primes = 600

(d) Average no. of even numbers "generated" by each of these 600 consecutive primes = 4,419 ÷ 600 = **7.37**

(e) No. of summings of 2 primes/permutations (3 + 3, 3 + 5, 3 + 7, 3 + 11, etc.) for these 600 primes = 600 × 600 = 360,000

(f) Average no. of summings of 2 primes/permutations for each of the 4,419 even numbers = 360,000 ÷ 4,419 = **81.47**

(4) *800 Consecutive Primes from 3 to 6,143*

(a) Even numbers (obtained by summing of 2 primes) = 6 to 12,286

(b) No. of even numbers = 6,141

(c) No. of primes = 800

(d) Average no. of even numbers "generated" by each of these 800 consecutive primes = 6,141 ÷ 800 = **7.68**

(e) No. of summings of 2 primes/permutations (3 + 3, 3 + 5, 3 + 7, 3 + 11, etc.) for these 800 primes = 800 × 800 = 640,000

(f) Average no. of summings of 2 primes/permutations for each of the 6,141 even numbers = 640,000 ÷ 6,141 = **104.22**

(5) *1,000 Consecutive Primes from 3 to 7,927*

(a) Even numbers (obtained by summing of 2 primes) = 6 to 15,854

(b) No. of even numbers = 7,925

(c) No. of primes = 1,000

(d) Average no. of even numbers "generated" by each of these 1,000 consecutive primes = 7,925 ÷ 1,000 = **7.93**

(e) No. of summings of 2 primes/permutations (3 + 3, 3 + 5, 3 + 7, 3 + 11, etc.) for these 1,000 primes = 1,000 × 1,000 = 1,000,000

(f) Average no. of summings of 2 primes/permutations for each of the 7,925 even numbers = 1,000,000 ÷ 7,925 = **126.18**

(6) *1,200 Consecutive Primes from 3 to 9,739*

(a) Even numbers (obtained by summing of 2 primes) = 6 to 19,478

- (b) No. of even numbers = 9,737
- (c) No. of primes = 1,200
- (d) Average no. of even numbers "generated" by each of these 1,200 consecutive primes = 9,737 ÷ 1,200 = **8.11**
- (e) No. of summings of 2 primes/permutations (3 + 3, 3 + 5, 3 + 7, 3 + 11,etc.) for these 1,200 primes = 1,200 × 1,200 = 1,440,000
- (f) Average no. of summings of 2 primes/permutations for each of the 9,737 even numbers = 1,440,000 ÷ 9,737 = **147.89**

(7) *1,400 Consecutive Primes from 3 to 11,677*

(a) Even numbers (obtained by summing of 2 primes) = 6 to 23,354

- (b) No. of even numbers = 11,675
- (c) No. of primes = 1,400
- (d) Average no. of even numbers "generated" by each of these 1,400 consecutive primes = 11,675 ÷ 1,400 = **8.34**
- (e) No. of summings of 2 primes/permutations (3 + 3, 3 + 5, 3 + 7, 3 + 11, etc.) for these 1,400 primes = 1,400 × 1,400 = 1,960,000
- (f) Average no. of summings of 2 primes/permutations for each of the 11,675 even numbers = 1,960,000 ÷ 11,675 = **167.88**

(8) *1,600 Consecutive Primes from 3 to 13,513*

(a) Even numbers (obtained by summing of 2 primes) = 6 to 27,026

- (b) No. of even numbers = 13,511
- (c) No. of primes = 1,600
- (d) Average no. of even numbers "generated" by each of these 1,600 consecutive primes = 13,511 ÷ 1,600 = **8.44**
- (e) No. of summings of 2 primes/permutations (3 + 3, 3 + 5, 3 + 7, 3 + 11, etc.) for these 1,600 primes = 1,600 × 1,600 = 2,560,000
- (f) Average no. of summings of 2 primes/permutations for each of the 13,511 even numbers = 2,560,000 ÷ 13,511 = **189.48**

(9) *1,800 Consecutive Primes from 3 to 15,413*

(a) Even numbers (obtained by summing of 2 primes) = 6 to 30,826

- (b) No. of even numbers = 15,411
- (c) No. of primes = 1,800
- (d) Average no. of even numbers "generated" by each of these 1,800 consecutive primes = 15,411 ÷ 1,800 = **8.56**

(e) No. of summings of 2 primes/permutations (3 + 3, 3 + 5, 3 + 7, 3 + 11, etc.) for these 1,800 primes = 1,800 × 1,800 = 3,240,000

(f) Average no. of summings of 2 primes/permutations for each of the 15,411 even numbers = 3,240,000 ÷ 15,411 = **210.24**

(10) *2,000 Consecutive Primes from 3 to 17,393*

(a) Even numbers (obtained by summing of 2 primes) = 6 to 34,786

(b) No. of even numbers = 17,391

(c) No. of primes = 2,000

(d) Average no. of even numbers "generated" by each of these 2,000 consecutive primes = 17,391 ÷ 2,000 = **8.70**

(e) No. of summings of 2 primes/permutations (3 + 3, 3 + 5, 3 + 7, 3 + 11, etc.) for these 2,000 primes = 2,000 × 2,000 = 4,000,000

(f) Average no. of summings of 2 primes/permutations for each of the 17,391 even numbers = 4,000,000 ÷ 17,391 = **230.00**

(11) *2,200 Consecutive Primes from 3 to 19,427*

(a) Even numbers (obtained by summing of 2 primes) = 6 to 38,854

(b) No. of even numbers = 19,425

(c) No. of primes = 2,200

(d) Average no. of even numbers "generated" by each of these 2,200 consecutive primes = 19,425 ÷ 2,200 = **8.83**

(e) No. of summings of 2 primes/permutations (3 + 3, 3 + 5, 3 + 7, 3 + 11, etc.) for these 2,200 primes = 2,200 × 2,200 = 4,840,000

(f) Average no. of summings of 2 primes/permutations for each of the 19,425 even numbers = 4,840,000 ÷ 19,425 = **249.16**

(12) *2,400 Consecutive Primes from 3 to 21,391*

(a) Even numbers (obtained by summing of 2 primes) = 6 to 42,782

(b) No. of even numbers = 21,389

(c) No. of primes = 2,400

(d) Average no. of even numbers "generated" by each of these 2,400 consecutive primes = 21,389 ÷ 2,400 = **8.91**

(e) No. of summings of 2 primes/permutations (3 + 3, 3 + 5, 3 + 7, 3 + 11, etc.) for these 2,400 primes = 2,400 × 2,400 = 5,760,000

(f) Average no. of summings of 2 primes/permutations for each of the 21,389 even numbers = 5,760,000 ÷ 21,389 = **269.30**

There would evidently be more and more profuse repetitions and overlaps of the even numbers "generated" by the primes the higher up the infinite list of prime numbers we go, which is significant.

We compare all the (d)s and (f)s in (1) to (12) above, which is as follows:
(d) Average no. of even numbers "generated" by each of the consecutive primes in (1) to (12) above, as follows according to the listings (1) to (12):
(1) **6.14**
(2) **6.87**
(3) **7.37**
(4) **7.68**
(5) **7.93**
(6) **8.11**
(7) **8.34**
(8) **8.44**
(9) **8.56**
(10) **8.70**
(11) **8.83**
(12) **8.91**

(f) Average no. of summings of 2 primes/permutations for each of the even numbers in (1) to (12) above, as follows according to the listings (1) to (12):
(1) **32.60**
(2) **58.25**
(3) **81.47**
(4) **104.22**
(5) **126.18**
(6) **147.89**
(7) **167.88**
(8) **189.48**
(9) **210.24**
(10) **230.00**
(11) **249.16**
(12) **269.30**

The following is evident from the above information:
(A): (d) Average no. of even numbers "generated" by each of the consecutive primes in the above 12 listings increases continually all the way from the list: (1) 200 consecutive primes from 3 to 1,229 to the list: (12) 2,400 consecutive primes from 3 to 21,391, from **6.14** even numbers per prime number in List (1) to **8.91** even numbers per prime number in List (12).
(B): (f) Average no. of summings of 2 primes/permutations for each of the even numbers in the above 12 listings increases continually all the way from the list: (1)

200 consecutive primes from 3 to 1,229 to the list: (12) 2,400 consecutive primes from 3 to 21,391, from **32.60** number of summings of 2 primes/permutations per even number in List (1) to **269.30** number of summings of 2 primes/permutations per even number in List (12).

Proof 1.
Lemma. According to the principle of complete induction in set theory, if a set of natural numbers contains 1 and for each n, it contains $n+1$ whenever it contains all numbers less than $n+1$, then it must contain every natural number, e.g., complete induction proves that every natural number is a product of primes.

By induction, we now deduce the following:

The larger the list of consecutive primes becomes, the greater would be the average number of even numbers "generated" by each of the primes in the list of consecutive primes (inferred from (A) above).

The larger the list of consecutive primes becomes, the greater would be the average number of summings of 2 primes/permutations for each of the even numbers in the infinite list of even numbers (inferred from (B) above).

Furthermore, the Goldbach conjecture had been tested and found to be correct for every even number up to 12×10^{17}, which is not a small list, by a distributed computer search carried out at the University of Aveiro, Portugal, in 2008.

As the primes and the even numbers are infinite, by the above lemma and all the above deductions and information, it could be inferred that the increases stated in (A) and (B) above, with the even numbers each being the sum of 2 primes, continue to infinity, i.e., the Goldbach conjecture becomes stronger and stronger the higher up the infinite list of prime numbers/even numbers we go – all the way to infinity.

The validity of the Goldbach conjecture is thereby confirmed – every even number after 2 is the sum of 2 primes.

Proof 2.
Next, we resort to the proof by contradiction. The above deduction would be reversed if, e.g., the following takes place (which is the reversal of the above-mentioned information):

(A): (d) Average no. of even numbers "generated" by each of the consecutive primes in the above 12 listings decreases continually all the way from the list: (1) 200 consecutive primes from 3 to 1,229 to the list: (12) 2,400 consecutive primes from 3 to 21,391, from **8.91** even numbers per prime number in List (1) to **6.14** even numbers per prime number in List (12).

(B): (f) Average no. of summings of 2 primes/permutations for each of the even numbers in the above 12 listings decreases continually all the way from the list: (1) 200 consecutive primes from 3 to 1,229 to the list: (12) 2,400 consecutive primes from 3 to 21,391, from **269.30** number of summings of 2 primes/permutations per even number in List (1) to **32.60** number of summings of 2 primes/permutations per even number in List (12).

If this reversed state happens, the implication is that there would reach a point when there are no more batches of 2 prime numbers summing together to form even numbers, in which case the Goldbach conjecture would be false. Evidently this would happen when the prime numbers are finite. As the prime numbers are infinite (as Euclid had proved long ago) this would never happen.

Since the above information indicate otherwise, and the prime numbers are infinite, we accept the above induction/deduction and infer that the Goldbach conjecture could not be false, i.e., the Goldbach conjecture is true, and every even number (except 2) is indeed the sum of two prime numbers. This concludes the proof by contradiction.

Thus, by both induction and contradiction or *reductio ad absurdum* the validity of the Goldbach conjecture is proved.

Conclusion

A number of methods have been adopted in this paper in proving the Goldbach conjecture.

The inductive method, which is a well-established proof, is one of the methods utilized. The following lends support to this inductive proof of the Goldbach conjecture: (a) The characteristic of a mountain or infinite volume of sand is reflected in the characteristic of some grains of sand found there so that studying the characteristic of some grains of sand found there is enough for deducing the characteristic of the mountain or infinite volume of sand, to ascertain the quality of a batch of products it is only necessary to inspect some carefully selected samples from that batch of products and not everyone of the products and to carry out a population census, i.e., find out the characteristics of a population, it is only necessary to carry out a survey on some carefully selected respondents and not the whole population; in like manner, by the same principle, we just need to study a carefully selected list of even numbers, find out whether they are all sums of two primes and deduce by induction whether all even numbers after this list would also be sums of two primes – this act is rather like extrapolation. (For example, a distributed computer search completed in 2008 at the University of Aveiro, Portugal, had confirmed that every even number up to 12×10^{17}, which is no small list of numbers, is the sum of two primes. By the principle of induction in this case we could deduce that all the even numbers after 12×10^{17} would also be sums of two primes.) (b) Thus, in this way every even number after 2 could be reasonably proved to be the sum of two primes. In fact, induction plays an important part in a number of the proofs.

The other argument used to prove the conjecture is the indirect (*reductio ad absurdum*) method, which had been used by Euclid and other mathematicians after him. Logically, one or two examples of "contradiction" should be sufficient proof of infinity, for it does not make sense to have a need for an infinite number of cases of "contradiction", as our proof would then have to be infinitely and impossibly long, an absurdity. This method of proof is "proof by implication" as a result of "contradiction" – which is a "short-cut" and smart way in proving infinity, instead of "proving infinity by counting to infinity", which is ludicrous and impossible. Hence, one or two cases of "contradiction" should be sufficient for implying that there would be an infinitude of even numbers which are sums of two primes, which of course also tacitly implies that there would be an infinitude of the number of cases of such "contradiction". (Euclid evidently had this logical point in mind when he formulated the indirect (*reductio ad absurdum*) proof of the infinity of the primes.) This method of proof had been cleverly used by a number of mathematicians, not the least by the great German mathematician, David Hilbert. For example, Hilbert had used an indirect method (the "reductio ad absurdum" proof) to prove Gordan's theorem without having to show an actual "construction", a proof which had been accepted by his peers.

There is also the involvement of concepts from set theory, group theory, geometry, etc.

One important query here, which many might not have considered: What if the list of prime numbers is not infinite? Of course, if that is the case, the Goldbach conjecture would be false. It would then have been absurd for the Goldbach conjecture to have been conceived at all. However, the list of primes is infinite (vide Euclid's proof). This gives credence to the Goldbach conjecture.

A very important related point, in fact a most important point, must be highlighted here. If the Goldbach conjecture were indeed false, there must be an ultimate (largest) even number which is (and must necessarily be) the result of the summation of 2 primes that are each the largest existing prime. It must be noted that this is actually an impossibility, as there can never be a largest existing prime – by Euclid's proof, the primes are infinite (refer to Part 1, Proof 3 above). Hence, the Goldbach conjecture cannot be false, and by both *reductio ad absurdum* (contradiction), and induction (wherein all even numbers up to 12×10^{17}, not a small list, had been confirmed to be sums of two primes), has to be true.

Another important point is that the Goldbach conjecture becomes evidently stronger and stronger the higher up the infinite list of prime numbers/even numbers we go, as has been shown above. Thus, by implication, induction, extrapolation, it could be concluded that the Goldbach conjecture is valid – that every even number after 2 is the sum of two primes.

So far, there has been no indication or confirmation at all that the number of even numbers after the number 2 which are each the sum of two primes is finite and the largest existing even number which is the sum of two primes has not been found and confirmed. (This would of course be the case if the Goldbach conjecture is true.) Also, no counter-example (i.e., an even number which is never the sum of two primes) has been found so far. On the other hand, practically everyone could intuit that the list of even numbers after the number 2 which are each the sum of two primes is infinite. Besides, the evidence, as shown in this paper, is overwhelmingly in support of the infinity of this list.

We have no other more logical choice but to take the stand that every even number after the number 2 is the sum of two prime numbers.

In conclusion, we state that the Goldbach conjecture is true – every even number after the number 2 is indeed the sum of two primes.

It is evident here that the Goldbach conjecture could be approached in a number of different ways. All this could be food for further thought about the conjecture.

References

[1] D. Burton, 1980, Elementary Number Theory, Allyn & Bacon

[2] R. Courant and H. Robbins, Revised By I. Stewart, 1996, What Is Mathematics? An Elementary Approach to Ideas and Methods, Oxford University Press

[3] M. J. Feigenbaum, 1980, Universal Behavior in Nonlinear Systems, Los Alamos Science, No. 1

[4] G. H. Hardy and E. M. Wright, 1979, An Introduction to Theory of Numbers, Oxford, England: Clarendon Press

[5] D. H. Lehmer, 1914, List of Prime Numbers from 1 to 10,006,721, Publication No. 165, Carnegie Institution of Washington, Washington, D.C.

[6] M. E. Lines, 1986, A Number for Your Thoughts, Adam Hilger

[7] B. B. Mandelbrot, 1977, The Fractal Geometry of Nature, W. H. Freeman

[8] R. McWeeny, 2002, Symmetry: An Introduction to Group Theory and Its Applications, Dover

[9] J. M. T. Thompson and H. B. Stewart, 1986, Nonlinear Dynamics and Chaos, New York: John Wiley

[10] World Wide Web

3 (a) The Problem of Turbulence

Abstract

The Navier-Stokes differential equations describe the motion of fluids which are incompressible. The three-dimensional Navier-Stokes equations misbehave very badly although they are relatively simple-looking. The solutions could wind up being extremely unstable even with nice, smooth, reasonably harmless initial conditions. A mathematical understanding of the outrageous behavior of these equations would dramatically alter the field of fluid mechanics. This paper describes why the three-dimensional Navier-Stokes equations are not solvable, i.e., the equations cannot be used to model turbulence, which is a three-dimensional phenomenon.

The general equations of motion for a viscous fluid were obtained by Sir George Stokes in 1845. The following is the fundamental equation (in vectorial form) governing the flow of a viscous fluid:

$$\partial \mathrm{v}/\partial \mathrm{t} + (\mathrm{v}.^{\nabla})\mathrm{v} = -1/\mathrm{p}^{\nabla}\ \mathrm{Pe} - {}^{\nabla}\varphi + \eta\,/\mathrm{p}^{\nabla}2\mathrm{v}$$

where v is the velocity of the fluid (as a function of position), Pe the pressure, φ the gravitational potential, p the density and η the viscosity.

The scientist normally makes a forecast of the outcome of a flow and uses the Navier-Stokes equations to model this forecast. However, in the instance of turbulence, making this forecast will be fraught with difficulty, if it can be carried out at all. Putting it another way, if turbulence could be forecasted, predicted and described by the Navier-Stokes equations it could not be turbulence, for turbulence implies puzzlement, lack of order or pattern and lack of predictability.

The Navier-Stokes equations are nonlinear due to the acceleration terms such as $u\partial u/\partial x$. As a result, the solution to these equations may not be unique. For instance, the flow between two rotating cylinders can be solved using the Navier-Stokes equations to treat a relatively simple flow with circular streamlines; it can also be a flow with streamlines which are like a spring wound around the cylinders as a torus; there are also more complex flows which are solutions to the Navier-Stokes equations, all satisfying the identical boundary conditions.

For simple geometries, the Navier-Stokes equations can be solved with relative ease. However, the equations cannot be solved for a turbulent flow even for the simplest of examples. A turbulent flow is highly unsteady, nonlinear and three-dimensional and therefore requires that the three velocity components be specified at all points in a region of interest at some initial time, say $t = 0$. But, even for the simplest geometry, such information will be almost impossible to obtain. Therefore, the solutions for turbulent flows have to be left to the experimentalist and are not attempted by solving the Navier-Stokes equations.

References

[1] Chaos: Making a New Science, Viking Press, James Gleick, 1987
[2] Nonlinear Dynamics and Chaos, John Wiley, J. M. T. Thompson & H. B. Stewart, 1986
[3] An Introduction To Chaotic Dynamical Systems, Robert L. Devaney, Benjamin- Cummings, 1986

3 (b) The Mathematical Theory of Turbulence or Chaos (This Paper Had Been Published in an International Journal in 2010)

Abstract

The motion of fluids which are incompressible could be described by the Navier-Stokes differential equations. However, the three-dimensional Navier-Stokes equations for modeling turbulence misbehave very badly although they are relatively simple-looking. The solutions could wind up being extremely unstable even with nice, smooth, reasonably harmless initial conditions. A mathematical understanding of the outrageous behavior of these equations would greatly affect the field of fluid mechanics. In this paper a reasoned, practical approach toward resolving the issue is adopted and a practical, statistical kind of mathematical solution is proposed.

Keywords: approximation; simulation; statistical data; forecast

1 Navier-Stokes Equation

The general equations of motion for a viscous fluid had been obtained by Sir George Stokes in 1845. The fundamental equation (in vectorial form) governing the flow of a viscous fluid is as follows:

$$\partial v/\partial t + (v.^{\nabla})v = -1/p^{\nabla}Pe - {}^{\nabla}\varphi + \eta/p^{\nabla}2v$$

where v is the velocity of the fluid (as a function of position), Pe the pressure, φ the gravitational potential, p the density and η the viscosity.

A fluid in motion could be characterized by its velocity field (velocity as a function of position). However, because of the complex nature of the forces affecting fluids (in general, forces of both compression and viscosity) the result of applying basic principles such as Newton's second law is a set of nonlinear equations. Computational methods therefore play a large part in fluid dynamics. Newton's second law states that the rate of change of momentum p of a body equals the total force F acting upon it, as is described by the following equation:

$$F = \partial p/\partial t$$

If, as is normally the case, the mass of the body is constant, $F = \partial(mv)/\partial t$ reduces to $F = m\partial v/\partial t$ or $F = ma$, where a is the acceleration of the body. Note that the force and acceleration are vectors. The first law is the null case of the second law (if $F = 0$ then $a = 0$).

The Navier-Stokes equation is a miracle of brevity, relating a fluid's velocity, pressure, density, and viscosity. In two dimensions, fluid flow governed by this partial differential equation is deterministic and predictable. But this equation fails when the fluid becomes turbulent as turbulence represents three-dimensional flow of the fluid, for which the Navier-Stokes equation does very poorly. Whereas fluid flow under normal conditions tends to be laminar, in turbulence it becomes irregular and develops eddies, ripples and whorls. But yet there is some sort of order found within this disorder or turbulence which could be described as self-similar or fractal. What mathematical technique could be used to describe this state?

The Navier-Stokes equations are nonlinear and do not submit to any general method of solution. Each new problem has to be carefully formulated as to geometry and proper boundary conditions. Then some scheme of attack might be adopted with the hope of reaching a solution. In most cases all attempts to obtain an exact solution fail. Approximate solutions have to make do. In a few cases exact solutions could be obtained. The possibility that perhaps the flow of the fluid is unidirectional, i.e., v (x, y, t) = 0, is not an assumption. It is rather an intuitive guess which is pursued until we either find a solution or become convinced that it does not lead to a solution, in which case we mark it as an unsuccessful trial.

Substitution of viscosity in the Navier-Stokes equations with viscosity = 0 reduces them to a form called the Euler equations:

$$p\,Dq/Dt = pg - \nabla p \text{ (in vectorial form)}$$

The Euler equations had been formulated earlier than the Navier-Stokes equations and considered an approximation. The Euler equations are of the first order and cannot in general satisfy the boundary conditions. We could therefore conclude that the Euler equations do not form a good approximation near a rigid boundary. Far from a boundary and where viscosity = 0 is a fair estimate, they have an important role as approximations and are generally easier to solve than the full Navier-Stokes equations.

The Navier-Stokes equations do need for their solution initial conditions as well as boundary conditions. The following are proper boundary conditions for a velocity on a rigid boundary:

$$q_n = q_t = 0$$

where q_n is the normal component of the velocity relative to the solid boundary, and q_t is the tangential component. These conditions are also termed the no-penetration ($q_n = 0$) and no-slip ($q_t = 0$) viscous boundary conditions. When the region occupied by the fluid is not closed, i.e., the fluid is not completely confined, additional conditions are still required on some surfaces which completely enclose the domain of the solution. These might represent some real physical surfaces or they might be chosen quite arbitrarily, provided the velocity on them is known. The pressure, which is also a dependent variable, also requires boundary conditions. The Navier-Stokes equations are then satisfied and we now know the resulting pressure field. This flow can exist only if the obtained pressure is possible. An acceptable boundary condition might be: $p = p_\infty = \text{const}$ at $r \to \infty$, which then implies: $p = p_\infty - pQ^2/8\Pi^2.1/r^2$. We also note that in the solution for the pressure there is no trace of the viscosity. This pressure therefore also satisfies the Euler equations. (As viscosity in a fluid enables it to smooth out or overcome the ripples, eddies and whorls of turbulence, a viscous fluid is in effect not so much affected by turbulence than a non-viscous fluid. Thus, the Navier-Stokes equations, as they relate to viscous fluids, present a better solution for incompressible fluids which are viscous and subject to turbulence than the Euler equations for non-viscous fluids [1].)

2 Newton's Law of Viscosity

The following equation is known as Newton's law of viscosity:

$$T_{yx} = \mu \, du/dy$$

A fluid that obeys this law is called a Newtonian fluid. This equation states that in unidirectional flow the shear stress in a Newtonian fluid is directly proportional to the transverse velocity gradient, du/dy, which is also known as the rate of shear strain or the rate of shear deformation.

There is no obvious reason why real fluids should obey this law. In fact, there are more fluids that do not obey this equation than those that do. Fluids that do not obey this law are called non-Newtonian. It is fortunate that the three most abundant fluids, air, water and petroleum, obey Newton's law of viscosity rather closely. The typical non-Newtonian fluids are paints, polymer solutions and melts, blood and many liquid food products, such as jellies and soups [1].

3 Computing of Reynolds Numbers

The following is the formula for the Reynolds number:

$$Re = v \, L \, p/n$$

where p (kg/m^3) is the fluid's density, v (m/s) is a typical fluid velocity, L (m) is some characteristic length, e.g., diameter of a pipe through which the fluid flows, and n (kgm/s^2) is the coefficient of viscosity (Gooey substances have a higher n value than runny ones like petrol. Viscous force = ability to smooth out turbulent whorls, eddies and ripples.).

The Reynolds number is the ratio between two forces: the initial force and the viscous force (frictional force). When the Reynolds number is below a few hundred, the flow of the fluid is smooth. When the Reynolds number exceeds about 2000 to 3000, the flow is completely turbulent. Between these values, the flow is sometimes smooth, sometimes turbulent. The greater the viscosity of the fluid is, the greater is its capability in overcoming or smoothing out the whorls, eddies, and ripples of turbulence [1].

4 Fluid Flow

According to the law of continuity, water flowing from a wide pipe to a narrower pipe speeds up, while water flowing from a narrow pipe to a wider one slows down, and the slow-moving water in the wide pipe would always have a higher pressure than the fast-moving water in the narrower pipe. Turbulent fluid flow could cause a pipe to give way.

Some equations pertaining to fluid flow are as follows:

$$\text{Energy (E)} = \text{Mass (m)} \times \text{Speed } (v^2) \tag{i}$$

$$\text{Altitude (A)} + \text{Energy (E)} = \text{Constant} \tag{ii}$$

$$\text{Energy (E)} = \text{Density of Fluid (p)} \times \text{Speed } (v^2) \tag{iii}$$

$$\text{Pressure (P)} + \text{Energy (E)} = \text{Constant} \tag{iv}$$

$$\text{Pressure (P)} + \text{Density of Fluid (p)} \times \text{Speed } (v^2) = \text{Constant} \qquad \text{(v) [1]}$$

5 Model for Turbulence

The logistic equation has been a popular method used for the modeling of turbulence and chaos. The formula had been invented by the Belgian mathematician, Pierre Francois Verhulst, and is very simple. But because the process has to be repeated over and over again it ends up being extremely complicated. The equation works in the following way:

For instance, if X is the population now, then the population next year is given by:

$$\text{Xnext} = rX(1 - X)$$

where r is some constant which could be adjusted according to the population being modeled. It is simplest if values of X between 0 and 1 are taken, so that 1 is the maximum population and 0 represents extinction. We might, for instance, take an arbitrary value for r of 2.6, and begin thus.

Let $X = 0.2$. Then $1 - X = 0.8$, and $X(1 - X) = 0.2 \times 0.8 = 0.16$.

Then multiply this result by 2.6 and we would get 0.416.

Repeat the process. Start with $X = 0.416$ and we would get 0.6317. The population increases.

Start with 0.6317 and we would get 0.6049. The population falls.

Start with 0.6049 and we would get 0.6214. The population goes up again.

Repeating or iterating this process over and over again we would obtain the following population figures:

0.6117, 0.6176, 0.6141, 0.6162, 0.6150, 0.6156, 0.6152, 0.6155, 0.6153, 0.6154, 0.6153, 0.6154, 0.6154, 0.6154.

The population rises and falls but converges on a fixed number.

Scientists have however tried to model turbulence or chaos using this equation. But, does this model really describe turbulence or chaos [2–5]?

6 Analyzing Motions of Fluids with Fourier Analysis

The surface water in a wave moves in a circular path at an angular velocity $w = 2\pi/T$ where T is the period of rotation. Deeper water moves in ellipses of decreasing size and increasing eccentricity.

The superpositions of simple sinusoidal oscillations in fluids could produce more complicated patterns of oscillations. The inverse mathematical operation of Fourier analysis could reduce any complicated oscillation into a sum of its simple sinusoidal components, each with a different period and amplitude.

With waves, there is a phenomenon which oscillates both in time and space. It might seem that this would considerably complicate any mathematical attempt to describe the superposition of waves. We could practically analyze complicated wave shapes either by freezing them in time or by freezing them in space. In the time domain, we obtain the wave's frequency components, while in the space domain we get the corresponding spectrum of wavelengths. These two approaches could each stand on their own, one being transformable to the other, because the product $f\lambda$ is a constant (the wave speed), the longer wavelengths corresponding to the lower frequencies (i.e., the longer periods). The wavelength spectrum could be computed directly from the frequency spectrum by noting that for each harmonic component $\lambda = v/f = v/T$ – this holds for all of the wave's Fourier components (e.g., $\lambda_o = 3$ m, $f_0 = 20$ cycles/s and wave speed $v = 60$ m/s) [1].

7 Fourier Series and Circles

We could interpret a Fourier series geometrically as the projection of a system of superimposed circular motions. A circle rides upon the nest of spinning circles beneath it, and we project the motion of a point on its circumference onto a line. The result is a periodic but distinctly non-sinusoidal motion which could be described mathematically as follows:

$$y\,(t) = (1)\sin\,(2\pi t/T) + (1/3)\sin\,(3.2\pi t/T) + (1/5)\sin\,(5.2\pi t/T) + (1/7)\sin\,(7.2\pi t/T)$$

where T = fundamental period of oscillation = time for one rotation of the largest circle.

This is the Fourier series which has been truncated after the fourth term for greater simplicity. If more terms are added (i.e., more circles turning upon circles), the resulting graph would more nearly approximate a series of alternating horizontal line segments [1].

8 Probability Waves and Turbulence

We now examine several related important ideas in quantum theory. Schrodinger had found an equation that could be applied to any physical system in which the mathematical form of the energy is known, which is as follows:

$$\partial^2 \psi / \partial x^2 + 8\pi^2 m/h^2 \ (E - V)\psi = 0$$

where ∂^2 is the second derivative with respect to x, x is the position of the particle, ψ is the Schrodinger wave function, or the probability amplitude for an electron in the state n to scatter into the direction m, E is energy, and V is potential energy.

The Schrodinger equation is a deterministic time-symmetrical description of nature. In classical mechanics, when one says that a quantum system is in a particular "state", one means that the state is a point in phase space. It is here described by a wave function whose evolution over time is expressed by the following equation:

$$ih/2\pi \, \partial\psi \, (t)/\partial t = H_{op} \, \psi \, (t)$$

This equation identifies the time derivative of the Schrodinger wave function ψ with the action of the Hamiltonian operator on ψ. It is not derived but assumed at the start, and could thus be validated only by experiment. In quantum theory, it is the fundamental law of nature. Here, ψ is the probability amplitude for an electron – it is only an abstraction and has no physical reality. ψ is also, in a sense, the electron's own intensity wave. When it is squared and the absolute value is taken, it turns out to be a physical probability of the associated particle's presence.

Later, Born stated that the probability of the existence of a state is given by the square of the normalized amplitude of the individual wave function (i.e., ψ^2). This was another new concept, i.e., the probability that a certain quantum state exists. Born had said there were no more exact answers in atomic theory, but just probabilities. The wave Ψ determines the likelihood that the electron would be in a particular position, and has no physical reality unlike the electromagnetic field.

Dirac had posited that light could be treated as waves or particles. In fact, in quantum mechanics, particles are regarded as waves. The behavior of these particles could be predicted, as it were, and they are therefore known as probability waves or Dirac wave particles. A wave/particle duality is present. However, when the particle is not observed, it remains a wave (a probability wave), but upon being observed it becomes a particle.

The following is the formal solution of the Schrodinger equation:

$$\psi(t) = U(t)\psi(0)$$

where $U(t) = e^{-iHt}$, U (t) is the evolution operator which links the value of the wave function at time t to that at the initial time $t = 0$. Both future and past play the same part, since $U(t_1)\,U(t_2) = U(t_1 + t_2)$, whatever the sign of t_1 and t_2. This property defines a dynamical group.

This description of how quantum particles behave could not be strictly applied to the macro world of fluid flow. Despite this, the above "probability" principles pertaining to how quantum particles behave could be somewhat broadly adapted and used as a guide for the interpretation of turbulent fluid flow.

In another important theory, the Uncertainty Principle, which was propounded by Heisenberg, it is posited that the very act of observing a quantum particle affects its behavior. According to this theory, the position and the momentum of an elementary particle could not be known simultaneously. The reason for this is that if an electron could be held still long enough for its position to be determined, then its momentum could no longer be determined. A special point is that the product of two uncertainties (or spreads of possible values) is always at least a certain minimum number. From the de Broglie/Einstein relation, $\Delta p \sim h/\lambda$, Heisenberg obtained the imprecision in the momentum. Multiplying the two inaccuracies together, he showed that the product, $\Delta x\ \Delta p$, would always be greater than or equal to ($\geq$) a certain amount, which is as follows:

$$(\Delta x)\ (\Delta p) \geq (\lambda)\ (h/\lambda) \geq h, \text{ or, } \ldots \qquad \text{(i)}$$
$$\Delta x\ \Delta p \geq h \qquad \text{(ii)}$$

where Δp and h/λ represent the de Broglie relation, and Δx and λ are from the diffraction limit.

The frustrated researcher looking for certainty must always make a compromise, knowledge gained about time, for instance, is paid for in uncertainty about frequency and vice versa. Though one does not notice Heisenberg Uncertainty Principle in one's everyday experience with the gross macroscopic world, the wave/particle duality defeats the atomic experimentalist who looks for perfection.

Another important implication of uncertainty which is worthy of comment is its effect on causality – the relation of cause to effect. Cause produces an effect. In classical physics if one understands fully the nature of a particular cause, one could then predict the effect. Cause and effect and predictability were cornerstones of classical physics and now they were under question. If it is impossible to measure precisely both the position and velocity of an electron (or any other particle) at the same moment, then it is also impossible to predict exactly where that electron would be at any given time afterward. An experimenter could send off two electrons in the same direction, at the same speed, and they would not necessarily end up in the same location. In the language of physics, the same cause could bring about different effects. There are serious philosophical consequences in this idea.

All this in effect represents "chaotic" behavior in the quantum world. We have to remember that chaotic behavior is beyond prediction; the above-mentioned describes the behavior of quantum particles in a probabilistic, uncertain sort of way – it represents a branch of physics known as statistical mechanics. The corollary of this is the macro-world phenomenon of three-dimensional turbulent fluid flow which defies the solution of the Navier-Stokes equations. Perhaps, a probability function, Φ, should be incorporated in the Navier-Stokes equation, like the case for the quantum particles [6, 7, 14].

9 Nature of Turbulence

It has been found that the transition from steady state through several splittings or bifurcations to chaos is similar to phase transitions – transitions that take place when a substance changes from a gas to a liquid or a liquid to a solid – all these transitions are also similar in that scaling is involved. It is believed that the presence of strange attractors, which could be defined as an endless path in phase space where the future depends sensitively on the initial conditions, is responsible for the presence of turbulence or chaos. A strange attractor possesses the following characteristics:
(i) It is generated by a simple set of differential equations.
(ii) It attracts and therefore all nearby trajectories in phase space converge toward it.
(iii) It has a great or very sensitive dependence on initial conditions, i.e., tiny differences or errors in the initial conditions lead quickly to large differences in the trajectory.
(iv) It is fractal, i.e., there is self-similarity or some familiar pattern within it.

The problem with fluid flow in conditions of turbulence is that the path taken by the fluid is continuous but nowhere differentiable. Turbulence gives rise to whorls, eddies, and ripples in the fluid. However, there is a self-similar structure or pattern within the fluid – whorls would be found within whorls. This is in accordance with the well-established self-similarity concept which had been developed by Mitchell Feigenbaum in the 1970s and which brought him fame. According to this concept, there is a tendency of identical mathematical structures to recur on many levels, and within a given structure there would be smaller copies of the same structure, their sizes being determined by the scaling factor, which is 4.669 and found to be a constant like pi (3.142). (This means there is some kind of order or pattern found in turbulence, chaos, or disorder.) If we were to plot a curve to describe the fluid's movement under conditions of turbulence we could expect the curve to be rough and nonlinear (which means it is not possible to derive the differential equations for describing this curve, thus making predictions concerning the fluid's movement very difficult if not impossible). However, viscous fluids, for which the Navier-Stokes equations are formulated, are able to smooth out or overcome the ripples, eddies and whorls of turbulence, as mentioned above, the more viscous the fluids the more able they are in doing so. Hence, the more viscous the fluid the more successful the Navier-Stokes equations should be in describing the motion of the fluid.

If turbulence or chaos could be predicted by a mathematical equation or equations then it is not really turbulence or chaos. One should bear in mind that chaos, as the term implies, results in disorder, having no discernable pattern, confusion and puzzlement, which is the contrary of the state of being orderly, having an obvious pattern, being deterministic and being predictable. Nevertheless, whatever method we adopt for describing turbulence or chaos we still need to confirm its validity through experiments, just as the validity of the Navier-Stokes equations for two-dimensional fluid motions has been confirmed by physical experiments. Thus, we should first proceed with the physical experiments to get a better understanding of turbulence or chaos in incompressible fluids [8–10, 21].

10 Fluid Viscosity and Turbulence

The more viscous (sticky or gooey) the incompressible fluid is the less able it is to flow or move freely, the less runny it is, and, as stated above, the more able it is to overcome or smooth out the whorls, eddies and ripples of turbulence. A fluid might be so viscous and its flow so restricted that it appears almost like a solid. Such fluids which come to mind are, e.g., paints and polymer solutions. What does all this imply?

It is evident that there is a threshold or cut-off point in the viscosity of a fluid at and above which turbulence does not affect it much, which means that the effects of turbulence are only marked if a fluid has a viscosity that is less than this threshold or cut-off point. At or beyond this threshold, this cut-off point, one could expect the Navier-Stokes equations not to fare too badly.

How would the introduction of a large initial force affect a viscous fluid whose viscosity is at or above this threshold? If a large initial force is introduced, this fluid could be expected to remain “lumpy” instead of moving more freely or becoming more runny, i.e., turbulence could not be expected to appear.

The opposite is true below this threshold. Below this cut-off point or threshold, i.e., if the fluid is not too viscous, with the initial force large enough, the fluid’s Reynolds number stands a chance of exceeding 2000 to 3000, with turbulence setting in, and whorls, eddies and ripples forming in the fluid. The Navier-Stokes equations would encounter difficulties [1, 10].

11 Two-Dimensional Flows and Their Solutions

Solutions of the Navier-Stokes equations result in velocity vectors, q, and pressures, p, which satisfy both the momentum equations and the continuity equation. If one were given such a combination [q, p], one could check whether it constitutes a solution by substitution into the equations. How such a solution is found is something else. Any general step leading to this goal is helpful. For two-dimensional flows, it is possible to get rid of the continuity equation from the system of equations by using only functions which satisfy the continuity equation. This elimination is a formal step toward a solution. The functions which affect this elimination are the stream functions.

A flow could be defined as two-dimensional when its description in Cartesian coordinates shows no z-component of the velocity and no dependence on the z-coordinate. A flow like this could be described in the z = 0 plane, with the velocity vector and the streamlines lying in this plane. Moreover, the z = C planes, which are parallel to the z = 0 plane, display a flow pattern which is identical to that in the z = 0 plane. The z = 0 plane is called the representative plane [1].

12 Three-Dimensional Flows and Practical Solutions

Historically, the dynamics of turbulence or chaos is the corollary of Poincare's "three-body" problem concerning planetary motions which Poincare found to be very complex and unsolvable.

Geometrically, turbulence in fluids cannot be described by the use of the "Poincare section", i.e., there is no periodic solution for it. (In the use of the "Poincare section", for there to be a periodic solution, a circular curve must return to the section at its exact starting point. In the condition of turbulence the curve would not return to its exact starting point and there is therefore no periodic solution.)

Our solutions here are to be implemented as much as possible in the spirit of the Navier-Stokes equations that pertain to the "behavior" of an incompressible viscous fluid in three dimensions (x, y, z). We aim to overcome the obstacle encountered by the Navier-Stokes equations in the three-dimensional case.

At least two methods of experimentation could be carried out to get a good picture of turbulence in incompressible fluids. This "picture" is however only an "approximation", with a "factor of accuracy" – which comprises an upper limit and a lower limit. The first method involves bouncing laser beams off a reflective strip that is immersed in the turbulent viscous fluid, making use of the Doppler effect, to determine the velocity of the strip, and, thus, the velocity of the fluid. However, this method only provides data obtained from one point, axis, or dimension, though we could extrapolate the data for the other two dimensions to get a three-dimensional picture. The second method involves the simulation of fluid motions in conditions of turbulence with powerful computers. The fact that turbulence or chaos is unpredictable implies that the result of fluid motions in turbulent conditions is anybody's guess. The guesses could vary widely in scope but each of them has a "probability" of being correct. A statistical method of analyzing fluid motions in conditions of turbulence would be strongly recommended, which is in some way similar in principle to the above-mentioned statistical mechanics. Though statistical mechanics could be used as an example for describing the behaviors of incompressible fluids under conditions of turbulence (e.g., the way it is being used to predict the behaviors and positions of quantum particles – only in a probabilistic manner), applying its methods is bound to face practical difficulties. (How do we assign "probability ratings" to the behaviors of quantum particles? What are the bases for such "ratings"? Are the "ratings" based on experimental proofs or data, which should be the case?) A suggestion is to adopt a method of analyzing fluid flows under conditions of turbulence based on statistical data obtained through actual, rigorous experiments, the utilization of statistical and interpolation/extrapolation methods, and sound common sense or logic.

The second method, which the author strongly recommends, involves carrying out an experiment through computer simulation. Computer simulation is very powerful and is commonly used today. Instead of carrying out an experiment involving real fluids and real turbulence, which would be cumbersome, we would simulate this experiment with powerful computers, which would be more practicable and would cut down costs and save time considerably. (In computer simulations electrons are actually being utilized to represent the object or objects being simulated and these simulations could be carried out in three dimensions – in this case viscous, incompressible fluid in turbulence with Reynolds numbers above 3000 being simulated in three dimensions as is described below.) As the flow of the fluid is completely turbulent when the Reynolds number exceeds about 2000 to 3000, we could via our powerful computers simulate fluid flows under conditions of turbulence for Reynolds numbers of, e.g., 3050, 3300, 3550, 3800, 4050, 4300, 4550, 4800, and 5050, respectively, and thereby obtain the velocity results of fluid motion (by the process of iteration) for each of these Reynolds numbers.

We would use three powerful computers or work stations which work at very high speeds for these computer simulations. For each of the above-mentioned nine Reynolds numbers (3050, 3300, 3550, 3800, 4050, 4300, 4550, 4800, and 5050) the three computers or work stations would carry out the following. The first computer would perform the simulation of fluid movement under conditions of turbulence and the fluid's velocity at a point, P, in the fluid would be measured in the x direction or axis (front view or plane) by a probe at this point, P, and recorded. The second computer would perform the same simulation but at a different plane or view, say, the side view (y), and the fluid's velocity at point P in the fluid would be measured in the y direction or axis (side view or plane) by a probe at point P and recorded. The third computer would perform the same simulation at a yet different plane or view, now, the top view (z), and the fluid's velocity at point P in the fluid would be measured in the z direction or axis (top view or plane) by a probe at point P and recorded. Hence,

we now have the fluid's velocities (v = m/s) at each of the three planes or axes, x, y, z, for the same Reynolds number, which are as follows:
(i) The fluid's velocity in the x axis – v (x)
(ii) The fluid's velocity in the y axis – v (y)
(iii) The fluid's velocity in the z axis – v (z)

For each of the nine Reynolds numbers, the three computers or work stations would each carry out (iterate) the simulations, say, 1 million times (the more simulations the better) to produce, respectively, 1 million values for v (x), 1 million values for v (y), and 1 million values for v (z).

(The principles involved in measuring the velocities of the fluid in the x, y, z directions or axes are explained here. The probes mentioned above should be made of thin material and plate-like and should offer minimal resistance against the flow of the fluid, i.e., offer minimal interference with the flow of the fluid. For each axis or plane, the flat surface of the probe should face the direction of the axis. The velocity of the fluid at each axis, plane or dimension is a function of the fluid pressure on the flat surface of the probe, the higher the fluid pressure the higher the velocity of the fluid and vice versa. The probe in each axis or plane should be able to measure both positive (+) velocities and negative (–) velocities, i.e., velocities in the opposite directions.)

This process for all the above-mentioned 9 Reynolds numbers would be carried out, giving a total of 27 million simulations, and 9 million values for v (x), 9 million values for v (y), and 9 million values for v (z).

We next apply the statistical method of time-series analysis (which is a method of statistical analysis designed to eliminate seasonal variations in trends, used for forecasting and prediction, but implemented with some modification here). For each of the above-mentioned 9 Reynolds numbers, we compute (with the aid of the computer) the moving averages (m) for the 1 million v (x)'s, 1 million v (y)'s and 1 million v (z)'s, regardless of whether v (x), v (y), and v (z) are positive (+) or negative (–), by only adding (no subtracting) and dividing, as follows:
(i) $v_1(x) + v_2(x) \div 2 = m_1(x)$
(ii) $v_1(x) + v_2(x) + v_3(x) \div 3 = m_2(x)$
(iii) $v_1(x) + v_2(x) + v_3(x) + v_4(x) \div 4 = m_3(x)$

$$(^*q)\ v_1(x) + v_2(x) + v_3(x) + v_4(x) + \ldots + v_{q+1}(x) \div q + 1 = m_q(x)$$

$(^*: q = 999{,}999)$

whereby $m_1(x), m_2(x), m_3(x), \cdots, m_q(x)$ are the 999,999 moving averages (m) for v (x).

(The same applies to the 1 million v (y)'s and the 1 million v (z)'s.)

Thus, we have 999,999 (q) moving averages (m) each for each of the 1 million v (x)'s, 1 million v (y)'s and 1 million v (z)'s for each of the above-mentioned 9 Reynolds numbers (giving a total of 26,999,973 moving averages (m) for all the above-mentioned 9 Reynolds numbers).

For each of the above-mentioned 9 Reynolds numbers, substituting 999,999 with q, we get the qth moving averages (m_q) for v (x), v (y) and v (z), which are actually each respectively the 999,999th average velocity of the fluid in each of the three dimensions or axes, x, y, z, which are as follows:

(i) m_q (x)
(ii) m_q (y)
(iii) m_q (z)

Upon the completion of all the computer simulations and computations of the moving averages (m), we would produce a statistical table with the fluid velocities, m_q (x), m_q (y), and m_q (z), in the three dimensions or axes, x (front view), y (side view), and z (top view), for each of the nine Reynolds numbers: 3050, 3300, 3550, 3800, 4050, 4300, 4550, 4800, and 5050. For each of the three velocities, m_q (x), m_q (y), and m_q (z), for each of these nine Reynolds numbers, we would include an upper limit of accuracy, u (x), u (y), and u (z) (each of which is respectively the largest moving average (m) that has been obtained through carrying out the 1 million simulations for each of the three dimensions or axes, x, y, z), and a lower limit of accuracy, l (x), l (y), and l (z) (each of which is respectively the smallest moving average (m) which has been obtained through carrying out the 1 million simulations for each of the three dimensions or axes, x, y, z). For example, for each of these nine Reynolds numbers we would have the following velocities with their respective upper and lower limits of accuracy:

(i) x dimension or axis – m_q (x), u (x), l (x)
(ii) y dimension or axis – m_q (y), u (y), l (y)
(iii) z dimension or axis – m_q (z), u (z), l (z)

However, for the Reynolds numbers between these nine Reynolds numbers, i.e., the "intermediate" Reynolds numbers (such as 3200, 4400, and 4950), we would obtain the velocity results with their respective upper and lower limits of accuracy, i.e., the m_q's and their respective u's and l's, through interpolation (i.e., estimate the velocity results and their respective upper and lower limits of accuracy – some methods of interpolation are the Lagrange interpolation and the Gregory-Newton interpolation).

With the above-mentioned statistical data for the nine Reynolds numbers we would apply the rules of vector calculus to obtain for each of these nine Reynolds numbers the "resultant" velocity of the fluid in three dimensions or axes (x, y, z) and its upper and lower limits of accuracy. For example, we could obtain the "resultant" velocity and its upper and lower limits of accuracy for unidirectional fluid flow in the three dimensions or axes (x, y, z) as follows:

$$m_q\,(x,\,y,\,z),\,u\,(x,\,y,\,z),\,l\,(x,\,y,\,z)$$
$$=+m_q\,(x)+m_q\,(y)+m_q\,(z);+u\,(x)+u\,(y)+u\,(z);+l\,(x)+l\,(y)+l\,(z)$$

where m_q (x, y, z) is the "resultant" velocity of the three moving averages m_q (x), m_q (y) and m_q (z), whereby m_q (x) would be treated as positive (+) if more than 50% of the 1 million v (x)'s are positive (+) and treated as negative (–) if more than 50% of the 1 million v (x)'s are negative (–), m_q (y) would be treated as positive (+) if more than 50% of the 1 million v (y)'s are positive (+) and treated as negative (–) if more than 50% of the 1 million v (y)'s are negative (–), and m_q (z) would be treated as positive (+) if more than 50% of the 1 million v (z)'s are positive (+) and treated as negative (–) if more than 50% of the 1 million v (z)'s are negative (–).

Hence, when more than 50% of the 1 million v (x)'s, v (y)'s or v (z)'s are negative (–), i.e., move in the opposite direction, and their moving average (m_q) is hence treated as negative (–), we have to subtract m_q (x), m_q (y) or m_q (z), e.g., if m_q (z) is negative (–), we have to compute the "resultant" velocity m_q (x, y, – z) as follows:

$$m_q(x,\ y,\ -z)=+m_q(x)+m_q(y)-m_q(z),\ \text{i.e., subtract } m_q(z) \text{ from} +m_q(x)+m_q(y),$$

with its upper and lower limits of accuracy computed as follows:

(i) u (x, y, – z) = + u (x) + u (y) – u (z), where u (x) is the largest moving average (m (x)) among the 999,999 (q) moving averages (m) for the 1 million v (x)'s, u (y) is the largest moving average (m (y)) among the 999,999 (q) moving averages (m) for the 1 million v (y)'s, and u (z) is the largest moving average (m (z)) among the 999,999 (q) moving averages (m) for the 1 million v (z)'s.

(ii) l (x, y, – z) = + l (x) + l (y) – l (z), where l (x) is the smallest moving average (m (x)) among the 999,999 (q) moving averages (m) for the 1 million v (x)'s, l (y) is the smallest moving average (m (y)) among the 999,999 (q) moving averages (m) for the 1 million v (y)'s, and l (z) is the smallest moving average (m (z)) among the 999,999 (q) moving averages (m) for the 1 million v (z)'s.

"Velocity" diagrams (or vector diagrams) could be produced for all the possible "resultant" velocities indicated below, showing their directions of flow, which vary:

(i) For + m_q (x) + m_q (y) + m_q (z); + u (x) + u (y) + u (z); + l (x) + l (y) + l (z), we get the "resultant" velocity m_q (x, y, z), u (x, y, z), l (x, y, z).

(ii) For – m_q (x) – m_q (y) – m_q (z); – u (x) – u (y) – u (z); – l (x) – l (y) – l (z), we get the "resultant" velocity m_q (– x, – y, – z), u (– x, – y, – z), l (– x, – y, – z).

(iii) For – m_q (x) – m_q (y) + m_q (z); – u (x) – u (y) + u (z); – l (x) – l (y) + l (z), we get the "resultant" velocity m_q (– x, – y, z), u (– x, – y, z), l (– x, – y, z) (where – m_q (x) – m_q (y) > + m_q (z)).

(iv) For + m_q (x) – m_q (y) – m_q (z); + u (x) – u (y) – u (z); + l (x) – l (y) – l (z), we get the "resultant" velocity m_q (x, – y, – z), u (x, – y, – z), l (x, – y, – z) (where – m_q (y) – m_q (z) > + m_q (x)).

(v) For – m_q (x) + m_q (y) – m_q (z); – u (x) + u (y) – u (z); – l (x) + l (y) – l (z), we get the "resultant" velocity m_q (– x, y, – z), u (– x, y, – z), l (– x, y, – z) (where – m_q (x) – m_q (z) > + m_q (y)).

(vi) For – m_q (x) + m_q (y) + m_q (z); – u (x) + u (y) + u (z); – l (x) + l (y) + l (z), we get the "resultant" velocity m_q (– x, y, z), u (– x, y, z), l (– x, y, z) (where + m_q (y) + m_q (z) > – m_q (x)).

(vii) For + m_q (x) – m_q (y) + m_q (z); + u (x) – u (y) + u (z); + l (x) – l (y) + l (z), we get the "resultant" velocity m_q (x, – y, z), u (x, – y, z), l (x, – y, z) (where + m_q (x) + m_q (z) > – m_q (y)).

(viii) For + m_q (x) + m_q (y) – m_q (z); + u (x) + u (y) – u (z); + l (x) + l (y) – l (z), we get the "resultant" velocity m_q (x, y, – z), u (x, y, – z), l (x, y, – z) (where + m_q (x) + m_q (y) > – m_q (z)).

So far, in the above "velocity" scenarios, for those with either, one negative velocity (–m_q) and two positive velocities (+m_q), or one positive velocity (+m_q) and two negative velocities (–m_q), it is assumed that for the former the two positive velocities (+m_q) combined together are larger than the one negative velocity (–m_q), and for the latter, it is assumed that the two negative velocities (–m_q) combined together are larger than the one positive velocity (+m_q). Now, what happens when for the former, the two positive velocities (+m_q) combined together are equal to the one negative velocity (–m_q), and for the latter, the two negative velocities (–m_q) combined together are equal to the one positive velocity (+m_q)? In each of these cases, the "resultant" velocity (m_q (x, y, z), u (x, y, z), l (x, y, z)) would be equal to zero, i.e., there would be no "resultant" velocity.

What happens when the negative velocity, – m_q (x), – m_q (y), or – m_q (z), is larger than the other two positive velocities combined together, e.g., – m_q (x) > + m_q (y) + m_q (z), – m_q (y) > + m_q (x) + m_q (z), or – m_q (z) > + m_q (x) + m_q (y)? For such cases we have the following "resultant" velocities, whose directions of flow differ from each other:

(i) For – m_q (x) + m_q (y) + m_q (z); – u (x) + u (y) + u (z); – l (x) + l (y) + l (z), where – m_q (x) > + m_q (y) + m_q (z), we get the "resultant" velocity m_q (– x, y, z), u (– x, y, z), l (– x, y, z).

(ii) For + m_q (x) – m_q (y) + m_q (z); + u (x) – u (y) + u (z); + l (x) – l (y) + l (z), where – m_q (y) > + m_q (x) + m_q (z), we get the "resultant" velocity m_q (x, – y, z), u (x, – y, z), l (x, – y, z).

(iii) For + m_q (x) + m_q (y) – m_q (z); + u (x) + u (y) – u (z); + l (x) + l (y) – l (z), where – m_q (z) > + m_q (x) + m_q (y), we get the "resultant" velocity m_q (x, y, – z), u (x, y, – z), l (x, y, – z).

What happens then when the positive velocity, + m_q (x), + m_q (y), or + m_q (z) is larger than the other two negative velocities combined together, e.g., + m_q (x) > – m_q (y) – m_q (z), + m_q (y) > – m_q (x) – m_q (z), or, + m_q (z) > – m_q (x) – m_q (y)? For such cases we have the following "resultant" velocities, whose directions of flow differ from each other:

(i) For + m_q (x) – m_q (y) – m_q (z); + u (x) – u (y) – u (z); + l (x) – l (y) – l (z), where + m_q (x) > – m_q (y) – m_q (z), we get the "resultant" velocity m_q (x, – y, – z), u (x, – y, – z), l (x, – y, – z).

(ii) For – m_q (x) + m_q (y) – m_q (z); – u (x) + u (y) – u (z); – l (x) + l (y) – l (z), where + m_q (y) > – m_q (x) – m_q (z), we get the "resultant" velocity m_q (– x, y, – z), u (– x, y, – z), l (– x, y, – z).

(iii) For – m_q (x) – m_q (y) + m_q (z); – u (x) – u (y) + u (z); – l (x) – l (y) + l (z), where + m_q (z) > – m_q (x) – m_q (y), we get the "resultant" velocity m_q (– x, – y, z), u (– x, – y, z), l (– x, – y, z).

For the three velocities in the x, y, z dimensions or axes, i.e., m_q (x), m_q (y) and m_q (z), there are 14 possible "resultant" velocities, excluding the null "resultant" velocities, of which there are six possible cases (which are not expected to be likely to occur in turbulence), as shown above. Each of these "resultant" velocities would have an upper limit of accuracy (u (x, y, z)) and a lower limit of accuracy (l (x, y, z)). For fluid velocities (m_q (x), m_q (y), or m_q (z)) in the opposite direction their values are negative. We could expect turbulence to be characterized by any of these 14 "resultant" velocities.

Nevertheless, there is a possibility that one or more of the velocities, m_q (x), m_q (y) and m_q (z), might be equal to zero, though the chances of this occurring in turbulence might be remote:

(i) If, e.g., m_q (z) = 0, then m_q (x, y, z = 0), u (x, y, z = 0), l (x, y, z = 0) would be a "resultant" velocity in two dimensions or axes only, with the fluid moving in the two dimensions or axes, x and y, only.

(ii) For m_q (x) = 0, we get the "resultant" velocity m_q (x = 0, y, z), u (x = 0, y, z), l (x = 0, y, z).

(iii) For m_q (y) = 0, we get the "resultant" velocity m_q (x, y = 0, z), u (x, y = 0, z), l (x, y = 0, z).

As in the three-dimensional case above, there is the possibility in the two-dimensional case that one or both of the two velocities are negative or positive velocities. If, in the case whereby one of the two velocities is negative while the other is positive, the negative velocity is equal to the positive velocity (of which there are six possible cases), then the "resultant" velocity is null (zero). The following are the possible "resultant" velocities:

(i) For m_q (z) = 0, we get the "resultant" velocity m_q (x, – y, z = 0), u (x, – y, z = 0), l (x, – y, z = 0), when m_q (x) is positive while m_q (y) is negative, whereby – m_q (y) > + m_q (x).

(ii) For m_q (z) = 0, we get the "resultant" velocity m_q (x, – y, z = 0), u (x, – y, z = 0), l (x, – y, z = 0), when m_q (x) is positive while m_q (y) is negative, whereby + m_q (x) > – m_q (y).

(iii) For m_q (z) = 0, we get the "resultant" velocity m_q (– x, y, z = 0), u (– x, y, z = 0), l (– x, y, z = 0), when m_q (y) is positive while m_q (x) is negative, whereby – m_q (x) > + m_q (y).

(iv) For m_q (z) = 0, we get the "resultant" velocity m_q (– x, y, z = 0), u (–x, y, z = 0), l (– x, y, z = 0), when m_q (y) is positive while m_q (x) is negative, whereby + m_q (y) > – m_q (x).

(v) For m_q (z) = 0, we get the "resultant" velocity m_q (– x, – y, z = 0), u (– x, – y, z = 0), l (– x, – y, z = 0), when m_q (x) and m_q (y) are both negative.

(vi) For m_q (z) = 0, we get the "resultant" velocity m_q (x, y, z = 0), u (x, y, z = 0), l (x, y, z = 0), when m_q (x) and m_q (y) are both positive.

(vii) For m_q (x) = 0, we get the "resultant" velocity m_q (x = 0, – y, z), u (x = 0, – y, z), l (x = 0, – y, z), when m_q (z) is positive while m_q (y) is negative, whereby – m_q (y) > + m_q (z).

(viii) For m_q (x) = 0, we get the "resultant" velocity m_q (x = 0, – y, z), u (x = 0, – y, z), l (x = 0, – y, z), when m_q (z) is positive while m_q (y) is negative, whereby + m_q (z) > – m_q (y).

(ix) For m_q (x) = 0, we get the "resultant" velocity m_q (x = 0, y, – z), u (x = 0, y, – z) l (x = 0, y, – z), when m_q (y) is positive while m_q (z) is negative, whereby – m_q (z) > + m_q (y).

(x) For m_q (x) = 0, we get the "resultant" velocity m_q (x = 0, y, – z), u (x = 0, y, – z), l (x = 0, y, – z), when m_q (y) is positive while m_q (z) is negative, whereby + m_q (y) > – m_q (z).

(xi) For m_q (x) = 0, we get the "resultant" velocity m_q (x = 0, – y, – z), u (x = 0, – y, – z), l (x = 0, – y, – z), when m_q (y) and m_q (z) are both negative.

(xii) For m_q (x) = 0, we get the "resultant" velocity m_q (x = 0, y, z), u (x = 0, y, z), l (x = 0, y, z), when m_q (y) and m_q (z) are both positive.

(xiii) For m_q (y) = 0, we get the "resultant" velocity m_q (– x, y = 0, z), u (– x, y = 0, z), l (– x, y = 0, z), when m_q (z) is positive while m_q (x) is negative, whereby – m_q (x) > + m_q (z).

(xiv) For m_q (y) = 0, we get the "resultant" velocity m_q (– x, y = 0, z), u (– x, y = 0, z), l (– x, y = 0, z), when m_q (z) is positive while m_q (x) is negative, whereby + m_q (z) > – m_q (x).

(xv) For m_q (y) = 0, we get the "resultant" velocity m_q (x, y = 0, – z), u (x, y = 0, – z), l (x, y = 0, – z), when m_q (x) is positive while m_q (z) is negative, whereby – m_q (z) > + m_q (x).

(xvi) For m_q (y) = 0, we get the "resultant" velocity m_q (x, y = 0, – z), u (x, y = 0, – z), l (x, y = 0, – z), when m_q (x) is positive while m_q (z) is negative, whereby + m_q (x) > – m_q (z).

(xvii) For m_q (y) = 0, we get the "resultant" velocity m_q (–- x, y = 0, – z), u (–- x, y = 0, – z), l (–-x, y = 0, – z), when m_q (x) and m_q (z) are both negative.

(xviii) For m_q (y) = 0, we get the "resultant" velocity m_q (x, y = 0, z), u (x, y = 0, z), l (x, y = 0, z), when m_q (x) and m_q (z) are both positive.

(xix) For m_q (x) = 0 and m_q (y) = 0, we get the "resultant" velocity m_q (x = 0, y = 0, z), u (x = 0, y = 0, z), l (x = 0, y = 0, z) in one dimension or axis, which is positive.

(xx) For m_q (x) = 0 and m_q (y) = 0, we get the "resultant" velocity m_q (x = 0, y = 0, – z), u (x = 0, y = 0, – z), l (x = 0, y = 0, – z) in one dimension or axis, which is negative.

(xxi) For m_q (x) = 0 and m_q (z) = 0, we get the "resultant" velocity m_q (x = 0, y, z = 0), u (x = 0, y, z = 0), l (x = 0, y, z = 0) in one dimension or axis, which is positive.

(xxii) For m_q (x) = 0 and m_q (z) = 0, we get the "resultant" velocity m_q (x = 0, – y, z = 0), u (x = 0, – y, z = 0), l (x = 0, – y, z = 0) in one dimension or axis, which is negative.

(xxiii) For m_q (y) = 0 and m_q (z) = 0, we get the "resultant" velocity m_q (x, y = 0, z = 0), u (x, y = 0, z = 0), l (x, y = 0, z = 0) in one dimension or axis, which is positive.

(xxiv) For m_q (y) = 0 and m_q (z) = 0, we get the "resultant" velocity m_q (- x, y = 0, z = 0), u (– x, y = 0, z = 0), l (–x, y = 0, z = 0) in one dimension or axis, which is negative.

After all the simulations and computations of moving averages (m) for the 27 million fluid velocities (v) in the x, y, z dimensions or axes, we firstly obtain a table, which would act as a rough statistical guide for fluid flow under conditions of turbulence, with the following statistical data, for each of the nine Reynolds numbers (3050, 3300, 3550, 3800, 4050, 4300, 4550, 4800, and 5050):

(1) (i) The fluid's average velocity in the x dimension or axis – m_q (x)
 (ii) This fluid average velocity's upper limit of accuracy – u (x) (i.e., largest moving average (m) for v (x))
 (iii) This fluid average velocity's lower limit of accuracy – l (x) (i.e., smallest moving average (m) for v (x))

(2) (i) The fluid's average velocity in the y dimension or axis – m_q (y)
 (ii) This fluid average velocity's upper limit of accuracy – u (y) (i.e., largest moving average (m) for v (y))
 (iii) This fluid average velocity's lower limit of accuracy – l (y) (i.e., smallest moving average (m) for v (y))

(3) (i) The fluid's average velocity in the z dimension or axis – m_q (z)
 (ii) This fluid average velocity's upper limit of accuracy – u (z) (i.e., largest moving average (m) for v (z))
 (iii) This fluid average velocity's lower limit of accuracy – l (z) (i.e., smallest moving average (m) for v (z))

(4) (i) The fluid's "resultant" velocity in the x, y, z dimensions or axes – m_q (x, y, z) (i.e., +/– m_q (x) +/– m_q (y) +/– m_q (z))

(ii) This fluid "resultant" velocity's upper limit of accuracy – u (x, y, z) (i.e., +/– m (x) +/– m (y) +/– m (z))
(iii) This fluid "resultant" velocity's lower limit of accuracy – l (x, y, z) (i.e., +/– m (x) +/– m (y) +/– m (z))

The "velocity" diagrams depicting the various "resultant" velocities and the directions of flow should also be included, a total of 27 diagrams, three "velocity" diagrams for each of the nine Reynolds numbers – one for m_q (x, y, z), one for u (x, y, z) and one for l (x, y, z). This table may be titled "Statistical Data of an Incompressible Fluid's Velocities for the Nine Reynolds Numbers: 3050, 3300, 3550, 3800, 4050, 4300, 4550, 4800, and 5050".

For the "intermediate" Reynolds numbers, e.g., 3200, 4400, and 4950, we would interpolate the fluid's respective velocities and their upper and lower limits of accuracy with the above-mentioned statistical table (statistical data of an incompressible fluid's velocities for the nine Reynolds numbers: 3050, 3300, 3550, 3800, 4050, 4300, 4550, 4800, and 5050) and the complementary table (velocity results of the 27 million simulations in the X, Y, Z dimensions) described below as guides, i.e., estimate them – some methods of interpolation that could be adopted are the Lagrange interpolation and the Gregory-Newton interpolation. For velocities for Reynolds numbers outside the range of the above nine Reynolds numbers (3050–5050), e.g., for Reynolds numbers 2500 and 5200, we would extrapolate them, including their upper and lower limits of accuracy, using the same tables as guides.

With the above techniques (and with proper simulations) we thus have a statistical table of fluid flow velocity results (Statistical Data of an Incompressible Fluid's Velocities for the Nine Reynolds Numbers: 3050, 3300, 3550, 3800, 4050, 4300, 4550, 4800, and 5050) under conditions of turbulence (with their respective ranges of possibilities or probabilities, i.e., their respective upper limits and lower limits of accuracy) for the various Reynolds numbers which are more or less rigorously based on actual experiments (simulations in this case), a table which is somewhat like, e.g., the statistical tables for t-distribution and chi-squared distribution. In the above-mentioned experiments (or simulations), we could expect more accurate results with more simulations being carried out (more data being collected), the more the simulations carried out the more accurate the results are likely to be. This represents a practical, logical way of roughly approximating the velocities of fluid motions under conditions of turbulence (in three dimensions) at the various Reynolds numbers. Given a fluid velocity (m_q (x, y, z), u (x, y, z), l (x, y, z)) from the above-mentioned statistical table, we could in principle tell (or predict) for a particular Reynolds number the position (position after distance traveled, d_2, relative to the position at $d_1 = 0$) of an object, e.g., an aluminum strip, which could be taken to represent a section of the fluid, carried along by the fluid in motion, at a point of time, $t_2 > 0$, with the initial time being $t_1 = 0$, or the distance ($\{d_2 > 0\} - \{d_1 = 0\}$) this object traveled at a point of time, $t_2 > 0$, with the initial time being $t_1 = 0$, by computation with the equation: $d_2 - d_1 = (m_q$ (x, y, z), u (x, y, z), l (x, y, z)) x (t_2 – t_1). However, in this instance, wherein turbulence rules, this prediction of position or distance is not expected to be accurate (but should be regarded only as a rough approximation) and should be subject to the statistical rule of "probability" – in this case the upper limit ({u (x, y, z)} x {t_2 – t_1}) and the lower limit ({l (x, y, z)} x {t_2 – t_1}) of the accuracy of the result obtained, as indicated by the above-mentioned statistical table. The "velocity" diagrams in the statistical table would act as a guide with regard to the positioning of the object and the direction it would travel. Here the movement of the object, which is the aluminum strip, in effect represents the movement of the fluid which carries it along. However, the data reflected in this first statistical table (statistical data of an incompressible fluid's velocities for the nine Reynolds numbers: 3050, 3300, 3550, 3800, 4050, 4300, 4550, 4800, and 5050) should be regarded only as a rough approximation. With the further assistance of the complementary table (velocity results of the 27 million simulations in the X, Y, Z dimensions) described below this approximation could be refined.

With the data from the above-mentioned statistical table (statistical data of an incompressible fluid's velocities for the nine Reynolds numbers: 3050, 3300, 3550, 3800, 4050, 4300, 4550, 4800, and 5050) it is possible to plot a curve for the fluid flow velocity results (m_q (x, y, z)) for the above-mentioned nine Reynolds numbers. If this curve is smooth and linear, which is unlikely, from the gradient of the slope of this curve the differential equations for this curve could be derived, whereby forecasts or predictions would be possible. However, if this curve is rough and nonlinear, which is likely to be the case, then these differential equations would not be obtainable, and we would have to rely on the above-mentioned table of fluid flow velocity results (statistical data of an incompressible fluid's velocities for the nine Reynolds numbers: 3050, 3300, 3550, 3800, 4050, 4300, 4550, 4800, and 5050), and the complementary table (velocity results of the 27 million simulations in the X, Y, Z dimensions) described below, as statistical guides for our approximation (including interpolation and extrapolation) of velocity results for fluid flows at the various Reynolds numbers. Both these statistical tables with their "velocity" diagrams should complement one another and should be a great help for this approximation process, the more copious the data available there the more effective the approximation should be. It is thus possible to approximate not only fluid velocities but directions of fluid motions as well (with the aid of the "velocity" diagrams). We could continue to plot graphs or curves with these statistical data and attempt to interpolate or extrapolate with them, looking out for trends or patterns, and arrive at some forecasts or predictions, which is expected to be a difficult task. By doing this we could at least get a "feel" or intuitive understanding of the whole situation, which should make the job of forecasting or predicting the outcome easier, an evidently challenging undertaking.

The above-mentioned statistical table, statistical data of an incompressible fluid's velocities for the nine Reynolds numbers: 3050, 3300, 3550, 3800, 4050, 4300, 4550, 4800, and 5050, presents the average velocities for the nine Reynolds numbers. These average velocities each represents the velocity "trend" for each of the nine Reynolds numbers. However, carrying out the approximation for the respective velocities for the respective Reynolds numbers, wherein complete turbulence is involved, might not be that easy. It should be remembered that the Navier-Stokes equations fare very badly when turbulence sets in, when a viscous, incompressible fluid behaves in an unpredictable, irregular, chaotic, and "nonlinear" manner. To complement the first statistical table (statistical data of an incompressible fluid's velocities for the nine Reynolds numbers: 3050, 3300, 3550, 3800, 4050, 4300, 4550, 4800, and 5050), it would be appropriate to have a proper tabulation of all the 27 million fluid velocities (v) in the x, y, z dimensions or axes obtained through the simulations in another table (or booklet), which may be titled "Velocity results of the 27 million simulations in the X, Y, Z dimensions", which shows whole ranges of velocities (minimum velocity, maximum velocity, and all the velocities between the minimum and the maximum, a total of 27 million velocities (v's), 3 million velocities (v's) for each of the nine Reynolds numbers), with their order of listing in this table (or booklet) exactly the same as the order they appeared in during the simulations, with indications whether they are positive (+) or negative (–) velocities, and the 999,999 (q) moving averages (m) each for each of the 1 million v (x)'s, 1 million v (y)'s, and 1 million v (z)'s for each of the nine Reynolds numbers (giving a total of 26,999,973 moving averages (m) for all the nine Reynolds numbers) included. The "velocity" diagrams described above, which would also be helpful as guides, should be incorporated, e.g., as an appendix, in this table (or booklet). With this additional table and its "velocity" diagrams acting as a further guide, the task of approximating (including interpolating and extrapolating) the velocities for the various Reynolds numbers (for turbulence) would be made easier. With these two tables there are now more or less solid and realistic data to carry out the approximation (including interpolation and extrapolation) – for both fluid velocities, and directions of fluid motions. The approximations could also include probabilities of occurrence (an example of which is presented below). However, sound common sense, good intuition and a sharp eye for patterns and details are important for the approximation (including interpolation and extrapolation). Since turbulence is a much complex phenomenon, the approximation (including interpolation and extrapolation) should be carried out with great care and patience. There are two choices here now. We could supplant the approximation method of the Navier-Stokes equations with the above-mentioned statistical method. Alternatively, we could use this statistical method as a complement to the Navier-Stokes equations, whereby we might have the "best of both worlds" [2–5, 11–23].

13 Conclusion

There is a possibility, however remote, that turbulence or chaos when viewed en masse, on a very large scale, would appear to be smooth, present some sort of pattern or appear to have some order. According to the precepts of fractal geometry, a relatively new branch of mathematics pioneered by Benoit Mandelbrot, phenomena which appear random, when viewed en masse, display some orderliness and pattern, which could be termed “fractal”. Thus, a new mathematical technique for describing the flow of an incompressible viscous fluid in three dimensions, i.e., in turbulence, a probably statistical one involving large samples of data, like the method described just above, is a logical step.

As for the case of computer simulation, it is becoming more and more popular. Simulation has been used in the physical sciences, as well as the engineering sciences, e.g., in aeronautical design, electronic circuit design and mechanical design. The author himself has considerable experience with computer simulation. Simulation has generally proven to be cost-saving, time-saving, and effective. In engineering applications, e.g., simulation has made it unnecessary to produce or manufacture the prototypes for any new product designs for the purpose of feasibility studies, which could be costly affairs. The feasibility studies are now carried out directly through the simulation exercises, whereby it is possible to quickly find out whether the designs would work or not. The only serious obstacle to simulation appears to be the cost of the simulation software itself, which could come up to many thousands of dollars, so that a cost/benefit analysis for using the software is necessary. The benefits and cost-savings for using the software should outweigh the cost of the software itself in order for its use to be justifiable. The other problem might be a technical one; some simulation software are complex and not user-friendly, requiring a long learning-curve, and these are usually the more powerful software with more built-in features. Nevertheless, once these software have been mastered their use would bring about relatively greater and more effective results. Some might feel that such powerful, and hence more expensive, software are an overkill and prefer to go for something at the lower end which is also cheaper. Such software usually have powerful features such as allowing one to have “walk-through” views and “inside-out” views of an object which would be physically impossible otherwise. Such are the great powers of simulation software now. The simulation of turbulence with powerful software and computers would certainly prove to be very useful.

As nonlinear equations such as the Navier-Stokes equations have to rely on approximation and exact solutions are highly unlikely, especially for turbulent fluid movement, a statistical guide based on actual data collected such as the guides or tables described just above is definitely a boon. A statistical guide that is based on reliable data collected and which has been put to the test and fine-tuned would be in a better position to lead us to more accurate approximations than the Navier-Stokes equations.

From the data in the above-mentioned statistical tables it is now possible, e.g., to make approximations or estimates of various fluid velocities (as well as the directions of various fluid motions) with various probabilities of occurrence for the various Reynolds numbers, such as the following:

(1) x m/s (accompanied by the appropriate "velocity" diagram): a per cent probability of occurrence
(2) y m/s (accompanied by the appropriate "velocity" diagram): b per cent probability of occurrence
(3) z m/s (accompanied by the appropriate "velocity" diagram): c per cent probability of occurrence

The Navier-Stokes equations do not have any allowance for such probabilistic approximations. These partial differential equations have been found to be solvable for the two-dimensional case, i.e., for each of the equations a function (or solution) could be found that, when substituted for the dependent variable in the equation, leads to an identity. But for three dimensions finding such functions (or solutions) has been a problem. In other words, though the formulas could be found to describe a two-dimensional fluid motion, such formulas are not available or obtainable for the three-dimensional case, e.g., the case for turbulent fluid motion. Evidently, the geometry of a three-dimensional fluid motion is rather complex, while finding the formulas to describe this complex three-dimensional movement of fluid is a difficult task. To understand this difficulty, consider the motion of a speck of particle carried along in a flowing fluid. The speck could be thrust first in one direction, then another, and another, and so on, sometimes moving in a fairly straight line, other times spiraling around as the current takes it along. This movement of the speck is also the movement of the fluid which carries it along. It is a three-dimensional movement that appears chaotic or turbulent – it is rather complicated and, therefore, not amenable to a description by some formulas. The Navier-Stokes equation makes use of differentiation to obtain the rate of change of some changing quantity, and in order to do this, the value or position or path of that quantity has to be given by an appropriate formula. Differentiation then acts upon this formula to produce another formula which gives the rate of change. Since the formulas for the three-dimensional case, e.g., turbulent fluid motion, are unobtainable the only recourse is approximation. Is there any hope of discovering these formulas in the future? The author is much pessimistic. Let us look again at the case of turbulent fluid motion, an essentially three-dimensional phenomenon. As stated earlier, a curve describing a fluid's movement under conditions of turbulence could be expected to be rough and nonlinear thus making it not possible to derive the differential equations (or formulas) for describing this curve, making predictions relating to the fluid's movement very difficult if not impossible. Moreover, turbulence or chaos implies disorder, irregularity, lack of discernable pattern, confusion, and puzzlement. This implies that turbulence could never be described by formulas or differential equations, and to be able to obtain the formulas or differential equations for turbulence and thus be able to make predictions relating to turbulence would mean that the so-called turbulence is not really turbulence at all, a contradiction. (Do the editors of our dictionaries then have to revise the meanings for turbulence and chaos? According to the *Encyclopedic World Dictionary* published by Paul

Hamlyn, "turbulence" could be defined as "the haphazard secondary motion due to eddies within a moving fluid", or "irregular motion of the atmosphere, as that indicated by gusts and lulls in the wind", "turbulent flow" is "fluid flow in which the motion at any point varies rapidly in magnitude and direction", and "chaos" could be defined as "utter confusion or disorder, wholly without organization or order", or "the infinity of space or formless matter supposed to have preceded the existence of the ordered universe".) It is thus absurd for us to expect the Navier-Stokes equations to have the solution for a three-dimensional phenomenon such as turbulent or chaotic fluid motion. It is hence appropriate to have another mathematical technique to deal with this difficult situation instead, viz., the statistical method, taking into consideration the example of statistical mechanics in quantum theory. As the outcomes of turbulence or chaos could never be predicted with certainty and the predictions might not even be highly accurate, it is logical and practical to consider the possible outcomes of this phenomenon in a statistical or probabilistic way. Much attempts have already been carried out to understand turbulence or chaos, which is now a hot subject. The author thinks that even if we understand the causes and mechanics or physics of turbulence or chaos it would be naïve to believe it is possible to obtain the formulas for describing such a phenomenon, for, to say that we know how an object which causes confusion and puzzlement (a chaotic object) would behave is to say that we are not confused and puzzled by this object, a self-contradiction. As described above, the logistic equation has been a popular model for chaos, but, here again, it would be naïve, and self-contradicting, to believe that the logistic equation is a sufficient formula for making predictions relating to chaos. Some others might state that only when we could not predict the outcomes of a phenomenon could we regard that phenomenon as chaotic but once we could get those predictions the phenomenon is no more chaotic, which would imply that chaos is transitional and subjective. These are possibly the ones who believe that there is a solution for the Navier-Stokes equations in the three-dimensional case. Simply put, if chaos is predictable it is not chaos. Only when it is really unpredictable could it be chaos. Mathematically and objectively speaking too, it has never been found possible to derive the differential equations to describe a nonlinear phenomenon such as turbulence, as there is no regular pattern found in turbulence (which is fluid flow in which the motion at any point varies rapidly in magnitude and direction), and mathematics, which, in a sense, is the study and analysis of patterns, simply found nothing possible to study or analyze in turbulence since it does not display any discernable, set pattern or regularity, except for the presence of eddies, ripples and whorls, which in the terms of fractal geometry could be described as a fractal characteristic. The only plausible solution appears to be a statistical one, wherein there is some hope of discovering some meaningful patterns or orderly features when large samples of data are analyzed. (According to the precepts of fractal geometry, phenomena which appear random when viewed en masse display some orderliness and pattern which could be regarded as a fractal characteristic.) For example, the prime numbers are very random and haphazard entities, yet, when viewed en masse they display a regularity in the way they thin out, whereby it is affirmed that the number of primes not exceeding a given natural number n is approximately n/ln n, in the sense that the ratio of the number of such primes to n/ln n eventually approaches 1 as n becomes larger and larger, ln n being the natural logarithm (to the base e) of n (vide the prime number theorem proved in 1896 by Hadamard and Vallee-Poussin).

With statistical methods, such as that described above, it is now possible to evaluate turbulence on a probabilistic basis, which is a more practical and realistic way of looking at turbulence, whose outcomes are uncertain, irregular, haphazard and very difficult if not impossible to predict. As the Navier-Stokes equations fare very poorly in the three-dimensional case, i.e., in the case of turbulence, a new mathematical technique for making approximations for the three-dimensional case, such as the statistical one described above, should be given a chance to take over, to supplant the Navier-Stokes equations, or, at least, to complement them.

After all, the results of the differential equations, such as the Navier-Stokes equations, would still have to be confirmed by actual physical data, or physical experiment. The more direct, faster or more efficient method of understanding and making forecasts pertaining to the various levels of turbulence or chaos is evidently to execute a well-planned computer simulation exercise (or, at least, a well-planned physical experiment) and apply the proper statistical technique in the interpretation of the data which are obtained through the computer simulation exercise. Such a procedure has been described in detail above.

We should be realistic and refrain from expecting certainly true or almost certainly true forecasts from our "mathematization" of turbulence or chaos. It is more reasonable or realistic to expect forecasts with only some degree of probability of being true, and where the forecasts do indeed turn out to be totally true on occasions they should be regarded as rare phenomena not unlike the cases of some fortunate persons hitting the lotteries. In fact, if the day ever arrives when an equation is available for accurately forecasting the outcome of turbulence or chaos, turbulence or chaos would be obsolete. With the lack of a discernable pattern or patterns in turbulence or chaos, such an equation evidently would never be found. In other words, turbulence or chaos, due to its excessively nonlinear, excessively random or pattern-less, nature, would be practically impossible to describe geometrically through any mathematical equation, e.g., the Navier-Stokes differential equation. A sort of fixed, geometrical shape or pattern has to be discernable to be describable by a mathematical equation. However, in the case of turbulence or chaos, the object in turbulence evidently takes on "different geometrical shapes all the time". Differential equations such as the Navier-Stokes equation are thus bound to fail in helping us to understand turbulence. On the other hand, statistical methods evidently have more chance of success; no matter how slim the chance of success with statistical methods might be, it is evidently still better than that of other mathematical methods. It should be remembered that in the case of statistical methods, the greater the sample size utilized is the greater the chance of success would be. They are therefore the best hope for success for understanding turbulence [2–5, 11–23].

Appendix

The Self-Similarity Concept and Fractal Geometry

The formulation of the self-similarity concept has brought fame to Mitchell Feigenbaum, who had worked in the Los Alamos Laboratory in the early 1970s. This concept, upon which the method of renormalization in perturbation theory is based, postulates that there is a tendency of identical mathematical structures to recur on many levels. Within a given structure, there would be smaller copies of the same structure, their sizes being determined by the scaling factor. Feigenbaum found that at the utmost tips of the fig tree, there is some mathematical structure which remains the same when its size is changed (enlarged) by a scaling factor of 4.669, which is found to be a constant like pi (3.142). This structure is the shape of the fig tree itself. In other words, little whorls could be found within big whorls. Renormalization has been a well-established technique in chaos theory/fractal geometry and is a mathematical trick which functions rather like a microscope, zooming in on the self-similar structure, removing any approximations, and filtering out everything else. All this shows the universality of some features of chaos. That is, some kind of order or pattern could be found in or is inherent in disorder, turbulence, or chaos.

In Feigenbaum's famous fig tree example, for instance, there is a self-similar mathematical pattern or structure (which is the shape of the fig tree itself) in the various parts of the fig tree, i.e., its trunk to bough section, bough to branch section, branch to twig section and twig to twiglet section. Such self-similar mathematical pattern or structure, or fractal characteristic, could also be found in other aspects of nature, e.g., waves, turbulence or chaos, the structures of viruses and bacteria, polymers and ceramic materials, the universe and many others, even the movements of prices in financial markets, the growths of populations, the sound of music, the flow of blood through our circulatory system, the behavior of people en masse, etc., which have all spawned a relatively new and important branch of mathematics with wide practical applications known as fractal geometry, which has been pioneered by Benoit Mandelbrot. As a matter of fact, self-similarity or fractal characteristic could be regarded as the fundamental mathematical aspect found in practically everything in nature, and this new branch of mathematics, fractal geometry, besides having a great practical impact on us, also gives us a more profound vision of the universe in which we live and our place in it [21–23].

References

[1] A. Beiser, Concepts of Modern Physics, Fifth Edition, McGraw-Hill, 1995

[2] J. Gleick, Chaos: Making a New Science, Viking Press, 1987

[3] J. M. T. Thompson & H. B. Stewart, Nonlinear Dynamics and Chaos, John Wiley, 1986

[4] R. L. Devaney, An Introduction to Chaotic Dynamical Systems, Benjamin-Cummings, 1986

[5] J. Moser, Stable and Random Motions in Dynamical Systems, Princeton University Press, 1974

[6] A. Modinos, Quantum Theory of Matter, John Wiley & Sons, 1996

[7] P. A. M. Dirac, The Principles Of Quantum Mechanics, Oxford University Press, 1958

[8] I. Goldhirsh, Rapid Granular Flows, Annual Review of Fluid Mechanics, Volume 35, 2003, 267–293

[9] R. May, Simple Mathematical Models with Very Complicated Dynamics, Nature, Volume 261, 10 June 1976, 459–467

[10] H. L. Swinney & J. P. Gollub, The Transition to Turbulence, Physics Today, Vol. 31, No. 8, August 1978

[11] M. Courbage, Intrinsic Irreversibility of Kolmogorov Dynamical Systems, Physica, 1983

[12] M. Courbage & I. Progogine, Intrinsic Randomness and Intrinsic Irreversibility in Classical Dynamical Systems, Proceedings of the National Academy of Sciences, 80 (8), April 1983, 2412–2416

[13] N. E. Huang, Z. Shen, S. R. Long, A View Of Nonlinear Water Waves: The Hilbert Spectrum, Annual Review of Fluid Mechanics, Volume 31, 1999, 417–457

[14] B. Misra, I. Prigogine, M. Courbage, From Deterministic Dynamics to Probabilistic Descriptions, Physica, Volume 98A, 1979

[15] T. Petrosky & I. Progogine, Poincare Resonances and the Limits of Trajectory Dynamics, Proceedings of the National Academy of Sciences, 90 (20), October 1993, 9393–9397

[16] T. Petrosky & I. Progogine, Poincare Resonances and the Extension of Classical Dynamics, Chaos, Solitons and Fractals 5, 1995

[17] P. Moin and K. Mahesh, Direct Numerical Simulation: A Tool In Turbulence Research, Annual Review of Fluid Mechanics, Volume 30, 1998, 539–578

[18] C. Meneveau, J. Katz, Scale-Invariance and Turbulence Models for Large-Eddy Simulation, Annual Review of Fluid Mechanics, Volume 32, 2000, 1–32

[19] K. R. Sreenivasan, R.A. Antonia, The Phenomolgy of Small-Scale Turbulence, Annual Review of Fluid Mechanics, Volume 29, 1997, 435–472

[20] Z. Warhaft, Passive Scalars in Turbulent Flows, Annual Review of Fluid Mechanics, Volume 32, 2000, 203–240

[21] M. J. Feigenbaum, Universal Behavior in Nonlinear Systems, Los Alamos Science, Volume 1, No. 1, Summer 1980, 4–27

[22] B. B. Mandelbrot, The Fractal Geometry of Nature, W. H. Freeman, 1977

[23] H. Lauwerier, Fractals – Images of Chaos, Penguin, 1991

Appendix 11: Interesting Quotations on Logic and Thought – Something to Ponder About

Logic is a systematic method of coming to the wrong conclusion with confidence.

Anonymous

You can only find truth with logic if you have already found truth without it.

G. K. Chesterton

Logic must look after itself.

Ludwig Wittgenstein

The principles of logic and metaphysics are true simply because we never allow them to be anything else.

A. J. Ayer

Thinking is seeing.

Honore de Balzac

Everything worth thinking has already been thought, our concern must only be to try to think it through again.

Goethe

Nonsense is the logic of the insane.

Anonymous

Because of logic, people argue, quarrel and fight. Therefore, logic is no good.

Anonymous

When parties argue or debate, each is trying to show the other that he has more logic.

Anonymous

Nothing is more dangerous than an idea, when you only have one idea.

Alain

One of the greatest pains to human nature is the pain of a new idea.

Walter Bagehot

One can live in the shadow of an idea without grasping it.

Elizabeth Bowen

Many ideas grow better when transplanted into another mind than in the one where they sprang up.

Oliver Wendell Holmes Jr.

One can resist the invasion of an army; but one cannot resist the invasion of ideas.

Victor Hugo

https://doi.org/10.1515/9783112214367-072

"Dying for an idea", again, sounds well enough, but why not let the idea die instead of you.

Wyndham Lewis

Society goes on and on and on. It is the same with ideas.

Ramsay Macdonald

An idea isn't responsible for the people who believe in it.

Don Marquis

The human mind treats a new idea the way the body treats a strange protein – it rejects it.

Sir Peter Medawar

This creature Man, who in his own selfish affairs is a coward to the backbone, will fight for an idea like a hero.

George Bernard Shaw

It is not normally our ideas which make us optimists or pessimists, but it is our optimism or our pessimism, which is perhaps of a physiological or pathological origin . . . . which makes our ideas.

Miguel de Unamunce

His mind is open; yes, it is so open that nothing is retained; ideas simply pass through him.

F. H. Bradley

The mind is always fooled by the heart.

La Rochefoucauld

Little minds are interested in the extraordinary; great minds in the commonplace.

Elbert Hubbard

Brain: An apparatus with which we think that we think.

Ambrose Bierce

Learning without thought is labour lost, thought without learning is perilous.

Confucius

The most fluent talkers or most plausible reasoners are not always the justest thinkers.

William Hazlitt

A great many people think they are thinking when they are merely rearranging their prejudices.

William James

There is nothing either good or bad, but thinking makes it so.

William Shakespeare

People use thought only to justify their injustices, and they use words only to disguise their thoughts.

Voltaire

In order to draw a limit to thinking, we should have to be able to think both sides of this limit.

Ludwig Wittgenstein

The logic or reasoning behind a phenomenon or happening is a matter of interpretation or explanation, which varies from person to person, and, this is a problem as conflicts thus tend to abound.

Anonymous

Logic while obvious to some could fly past others.

Anonymous

In mathematics the reasoning or logic could be so abstract or divorced from reality as to seem artificial, contrived, and even ridiculous or absurd – at times it ends up making the apparently simple look complex and forbidding – ideas which could be easily understood could suddenly end up being obscure and incomprehensible after unfamiliar symbolism is used – in this manner nonsense could actually be made out to look grand and impressive, and many people could thus be unknowingly duped – it is as though the mathematician has placed a "smoke-screen" between his ideas and others rendering his ideas not easily "discernable" – perhaps, it is better to call the subject "madthematics" as it is so often maddeningly difficult and fear-inspiring – unlike the teacher whose task is apparently to make a topic or subject easy, accessible, and comprehensible, the mathematician could be regarded as an expert in making a simple or easy topic complex, obscure, difficult, and impressive, an example of bright intellect at work, just similar to the lawyer and his dense, incomprehensible "legalese" or the engineer and his high-flown jargon.

Anonymous

When several fools agree that something makes sense, it is logic.

Anonymous

Logic is a play of words and ideas.

Anonymous

On the logic of arithmetic/mathematics: We now have a ten-based number system, giving us even and odd numbers or integers, as well as prime numbers and composite numbers. What happens if our number system is not 10-based, but, e.g., 5-based,

7-based, 9-based, 11-based, 127-based, or even 1,003-based, etc.? For example, the consecutive numbers or integers for a seven-based number system might be as follows:

1, 2, 3, 4, 5, 6, 7, 11, 12, 13, 14, 15, 16, 17, 21, 22, 23, 24, 25, 26, 27, 31, 32, 33, 34, 35, 36, 37, . . .

The above consecutive progression of seven-based numbers or integers gives us the equivalent of our 1 to 28. Which in the above consecutive progression of seven-based numbers or integers could now be considered the even numbers, which the odd numbers, which the prime numbers and which the composite numbers? What does this tell us about mathematics and its logic? Is mathematics real or just an invention of the human mind?

Anonymous

It is better to have a lucky mind than a logical mind.

Anonymous

It is better to be an unhappy Aristotle than a happy fool.

Anonymous

Empirical fact is more reliable, more important than logic, which should not be overrated.

Author

Logic would be at its most fruitful stage if the people around were indeed logical, i.e., if logic were to be of any practical use in society it should first be understood, and appreciated, by others.

Author

Logic should bring expected or forecasted outcomes or results. Otherwise, it is not truly logic but simply creative imagination.

Author

Logic to the ordinary human being is something which makes sense. It is intuitive. What makes sense to a person is based on his unique experiences of past events, past experienced causes and effects. Therefore, logic is subjective. Thus, the logic of an alien from another universe, if it could be regarded as logic, due to his different experiences (perhaps experiences with the different physical laws, and, hence, the different causes and effects, in their alien universe), is likely to be different from human logic.

Author

Mathematics and, thus, mathematical logic may not exist in another universe. Mathematics has many axioms or assumptions and rules, which are thought to be self-evident, obvious or undoubtedly logical or true, and which are the result of the mathematician's creativity and imagination. All this is thus arbitrary. Therefore, if there were mathematics in an alien civilization in another universe it is likely to be different from human mathematics – at least, the axioms and rules of alien mathematics are likely to be different. How about alien mathematics which is successfully used to compute the outcomes of alien behaviors? This would be very important mathematics from the human point of view – this behavioral mathematics, that is, would help solve our great societal problems, if only such mathematics were attainable in this world.

Author

Logic accompanied by objectivity of thought is the most important, while logic without objectivity of thought is not only undesirable but dangerous – society would be better off if its members live with less prejudices, pet peeves, and axes to grind than more logic, i.e., more objectivity than more logic, though more objectivity accompanied by more logic would be great.

Author

Would you prefer to be a fool whose "intelligence" is appreciated by your fellowmen or a brilliant genius whom your fellowmen regard as a fool?

Author

Would logic be of any real importance to a person who is the sole human survivor and living thing in the world while the rest of the human race and all other living things such as animals are extinct?

Author

The result of logical reasoning should be certifiable by physical or experimental proof for the sake of acceptability, though this is not always possible. It might be possible for a mind, perhaps one belonging to an alien or extra-terrestrial, to know the truth or fact without the need for logical reasoning, perhaps by a leap of intuition (to a certain degree, the human mind appears to have this intuitive ability to know without reasoning, an ability which could probably be much further developed), this truth or fact itself might or might not be certifiable by physical or experimental proof, though an eventual turn of events might certify its validity as the truth or fact. Such a mind is analogous to that of an effectual fortune-teller or soothsayer. It is in effect a mind which would be much more powerful than a logical mind, which might have to strain or even damage its brain cells, and thus takes some time, to arrive at the truth. Therefore, logic, which apparently depends on intuition, needs not be the most important quality of a mind after all. There could possibly be an all-knowing or all-seeing mind, perhaps a very highly intuitive mind, perhaps belonging to an extra-terrestrial, which has no need for logic or logical reasoning at all. Such a mind would of course know the truth or fact in an instant as it needs not waste time on logic or logical reasoning, and arrive at the definite truth or fact even, unlike the person who uses logic or the logician, who might not arrive at the exact truth at all but perhaps only the probable or partial truth, or, worse, he might not be able to arrive at any truth at all, not even the probable or partial truth. Would it not be much, much better to know the truth or fact without the need for reasoning, the need for logic, whereby brain-workers, such as scientists and mathematicians, would probably become redundant? Who could say for certain that this is impossible, even in an extra-terrestrial world? We should thus not make such a big deal of logic after all. Perhaps, we should pay more attention to our intuitive ability to know truths or facts without any recourse to logic or reasoning and greatly develop this ability.

Author

Apparently, many aspects of life or nature are inconclusive, controversial, and undecidable by logical deduction due to the lack of facts or true, confirmed data, which might be unobtainable, unavailable or unconfirmable, and conclusions could only be arrived at by personal, subjective evaluation or personal judgment. Thus, in many instances, judgment alone would have a more decisive or important role to play than logical reasoning, though it would be fine if judgment could be backed by logical reasoning, at least a modicum of it. And, pure judgment is apparently as important as pure logic, and both should be utilized equally for the best possible outcome or result.

Author

It is ironical that that the smartest or most intelligent person in the world has to be evaluated and acknowledged as the smartest or most intelligent by others who are his intellectual inferiors. Should the evaluation not be the other way round, the smartest or most intelligent person having the moral right on account of his more superior intellect to evaluate or judge the intellects of his intellectual inferiors, i.e., his fellowmen? Are the not so intelligent really capable of understanding and evaluating the intellect of the most intelligent? Of course, the smartest or most intelligent is a relative thing – he is to be compared to all his fellowmen, who are less intelligent than he. Who really should be the one making this comparison, this evaluation? Could a weak-minded person, even a person of average intellect, e.g., be entrusted with this task? The irony of all this deserves scrutiny.

Author

The problems of logic or logical reasoning could be the result of the wrong facts or even untruths being used and/or misunderstanding or misinterpreting the meanings of words or symbols used or facts. Therefore, precise facts and precise terms (words or symbols with precise meanings) are important for logical reasoning and argument. In truth, if the facts and meanings of words or symbols used in the presentation of logical reasoning are clear, precise and unequivocal, the logical conclusion should be clear-cut and indisputable and there should not be a need for argument or disputation by the parties concerned; at the very least, argument and disputation should be minimized. It is grievous when the parties concerned each misunderstands or misconstrues what the other party actually has in mind as something else, resulting in argument, dispute, wrong decision, or wrong action, which is apparently a rather common occurrence.

Author

Objectivity is the most important, even more important than logic. With objectivity, one really knows what one thinks one knows. Without objectivity, with subjectivity or prejudice, one does not know that what one thinks one knows is falsehood or fake knowledge. Reasoning, even correct reasoning, carried out without objectivity but based on prejudices or falsehoods would result in wrong conclusions, and would be futile. However, correct reasoning carried out with an objective mind and is thereby based on truths would end up with correct conclusions. Thus, objectivity should precede logic.

Author

The most abstruse, most difficult, most incomprehensible logic is the logic of the spiritual or supernatural. Apparently, no scientific mind has as yet tried to understand, analyze this special aspect of nature, which should prove to be most challenging and perhaps impossible. The spiritual or supernatural is apparently a very great riddle waiting for a logical explanation. It would be a very great challenge for the world's smartest person to tackle this riddle.

One of the World's Smartest Persons

Mathematics demands proofs which are very difficult or rather impossible to achieve; proofs which would be easily accepted by the scientist would not be accepted by the mathematician. Take, for example, the Goldbach conjecture which states that every even number after the number 2 is the sum of 2 primes. The Goldbach conjecture has been verified to be true for all even numbers up to 400 trillion (4×10^{14}) by computer searches completed in 2000. The scientist would accept this as proof that the Goldbach conjecture is true, but not the mathematician. Ask yourself then, "What is the purpose of a proof?" A proof is an object which induces the feeling of certainty that a theory or conjecture is true. However, in the above-mentioned case, the rather overwhelming evidence obtained by computer searches completed in 2000 could provide any scientist this feeling of certainty that the Goldbach conjecture is true, but, the mathematician wants something that is more certain, that is, this evidence while being able to provide the scientist the feeling of certainty that the Goldbach conjecture is true would not be able to provide the same feeling of certainty in the mathematician, whose criterion for a proof is more stringent. On the other hand, a highly intuitive person, perhaps, an extra-terrestrial, could easily sense that the Goldbach conjecture is true, that is, the feeling of certainty that the Goldbach conjecture is true comes easily to him. There are many proofs or theorems in mathematics, the result of intricate methods of reasoning. The moot point is what makes the mathematician so certain that these proofs or theorems based on their respective somewhat labored or intricate reasoning are true. Is there any solid reason for believing that the labored or intricate reasoning is correct, which confirms that the proofs or theorems are correct? Are there any proofs that the proofs or theorems are correct? Are there any proofs that prove the proofs which prove that the proofs or theorems are correct, and so on and on, ad infinitum. Is it not just intuition at work, even for the mathematician, but at a "higher level" though? Reasoning with its many intricate steps as in abstract reasoning, for example, in mathematics, which requires a number of axioms, lemmas and theorems in order to arrive at the truth, implies a mind which is lacking in power. A really powerful, intuitive mind should be able to arrive at the truth, which might be confirmed by experiment, without the need for any form of reasoning, which is actually a waste of time and energy. Since human beings are evidently incapable of such feats, we could only hope to find such capabilities in some extra-terrestrials, or higher beings. Logic might thus be unnecessary and be redundant. Human beings might be simply too stupid, as they often have to crack their brains with the use of logic. It is clear that a brain for which the truth of a matter is very clear and obvious at first encounter is a much more superior, intelligent one than a brain which requires a long, complicated chain of reasoning to arrive at the same truth, for example, the human brain. Could human beings ever be so smart instead of spending so much time with logical deduction, such as in mathematics, which is actually unproductive? Such a superior, inhuman brain would in all likelihood have great intuitive, even telepathic, mind-reading, powers.

Anonymous

Axioms or self-evident statements – obvious statements without any need for formal proofs – are normally the starting-points for mathematical reasoning. To arrive at a logical, valid conclusion, the mathematical reasoning or proof might be convoluted and not so obvious. The problem here is that what is self-evident or obvious to one person might not be so to another person. For example, the axioms or starting-points of a mathematical reasoning though obvious to the mathematician performing the reasoning might not be so to another person. Moreover, to a much more intelligent mind than this mathematician who performs the reasoning, e.g., the mind of an extremely intelligent extra-terrestrial, the truth or validity of the mathematical statement which is not obvious to this mathematician performing the reasoning in order to confirm the validity of the mathematical statement could be very obvious or self-evident to the extremely intelligent extra-terrestrial. In other words, a not self-evident or obvious mathematical statement to the mathematician could be an axiom or self-evident statement to the extremely intelligent extra-terrestrial, which implies that to the extremely intelligent extra-terrestrial the mathematical reasoning performed by the mathematician in order to confirm the validity or truth of the mathematical statement is unnecessary and is redundant. In other words, this extremely intelligent extra-terrestrial knows without the need for reasoning which would be redundant; his mind would be very fast and sharp; practically all things whose truth or validity requires hard reasoning on the part of human beings to confirm, e.g., mathematics and science, would be self-evident, or axioms, to this extremely intelligent extra-terrestrial – he knows the mathematics and science without resorting to reasoning.

Author

Mathematical logic, such as proofs and theorems, as in the higher, more abstract aspects of mathematics, e.g., in topology, differential geometry or complex analysis, is evidently more subjective and less precise than the logic of the operations of arithmetic, a lower level of mathematics, namely, the operations of addition, deduction, multiplication, and division. In fact, all the logic of mathematics depends on how the mathematician interprets concepts and ideas. Though practically no one would dispute the logic of arithmetic, e.g., no one would dispute that 1 + 1 = 2, 2 – 1 = 1, 2 x 2 = 4, or 4 ÷ 2 = 2, even the interpretation of these "precise, indisputable" arithmetical operations is, strictly speaking, not precise and is therefore disputable. For example, as described in the earlier part of the book, to be really precise we should be specific in our arithmetical operations what the numbers or quantities in the arithmetical operations refer to or describe. For instance, 1 apple + 1 orange cannot = 2 apples, or 2 oranges. For example, only 1 apple of the same size (size x) + 1 apple of the same size (size x) can = 2 apples of the same size (size x). Though this precise logic would be obvious to practically everyone, it would not be a surprise that someone could dispute this logic, e.g., this person could claim that it is really impossible to have apples of exactly the same size – he could claim that there are bound to be some differences in size no matter how minute, and he is correct in this claim, which implies that our above arithmetical logic does not hold water. Logic therefore depends much on interpretation and acceptance of that interpretation, which implies subjectivity. This is certainly true of the higher, more abstract aspects of mathematics. For instance, the square root of minus one evidently does not make logical sense, but mathematicians have had interpreted it in such a way (by using a coordinate system) that it makes sense, and it even yields practical results when used in applied, practical mathematics (surprisingly), which explains why it has been accepted by mathematicians without question.

Author

It is apparent that human logic is the logic of self-interest or egoism. When people argue, they would often resort to a logic which would protect or even further their self-interests. It is apparent that intellectual honesty is often lacking in people. Despite being proud of their intelligence and their ability to reason, people often reason like fools with wrong, biased logic. The logic would often be there to confirm what they like to be confirmed, to confirm what would be of benefit or advantage to them. People often use different logic cleverly to argue opposing points, e.g., attorneys arguing cases in courts, negotiators, cheaters, liars, people who use clever, convincing, though wrong, logic to deceive others or even themselves, etc. It is apparent that logic is a nebulous, slippery object, very much influenced by self-interest or ego. In dealing with our fellow-beings we should not expect true, precise, objective logic to work wonders for us. Sad to say, wrong but convincing logic, especially logic that protects or enhances the interests of the other party, has better chances of getting the other party to buy your ideas. This is a sorry state for real logic. This state of affairs should be reversed with the help and cooperation of all right-thinking people. We should all be intellectually honest or objective. At least, we should all attempt to be so. However, it is apparent that there are many intellectually dishonest or prejudiced people who believe or think that they are intellectually honest and unbiased; these people would be difficult to deal with. This is a serious problem. Only when people are intellectually honest and unbiased when dealing with each other would there be real harmony and peaceful co-existence. For those who are really concerned about the well-being of our society, they should set an example by being intellectually honest, unbiased, open-minded, flexible in outlook, and able to give and take, and they should attempt to influence or lead their fellow-beings to be like-wise. We should all realize that logic itself is imperfect, nebulous, slippery, and problematic. Thus, we should not be so dogmatic or uncompromising about our logic, our ideas, our views. Do not be obstinate. Exercise flexibility. Be prepared to give and take. Be prepared to cooperate and encourage cooperation. Welcome and encourage teamwork. Think win-win. One would surely not want to be the only one who is happy while everyone else is suffering.

Author

Religious man has complete faith and trust in God whom he cannot and does not hear from and see. Why can't he have complete faith and trust in an evidently honest and good fellow-man whom he can hear from and see? That is, he has faith in something which may not or does not exist and little or no faith in something good which definitely exists. This is contradictory, illogical, and difficult to fathom.

Anonymous

Some people seem to enjoy good luck while some do not. What exactly is luck? What causes good luck or bad luck? Is luck just coincidence, or is it due to the influence of some higher power, a god perhaps? What is the logic of luck?

Anonymous

When using logic to convince a party to accept your point of view, make sure you use logic on a party who is capable of understanding and appreciating logic. Otherwise, you may as well talk logic to the concrete wall.

Author

Bibliography

[1] Principia Mathematica, Volumes 1, 2 & 3, B. Russell and A. N. Whitehead, Cambridge University Press, 1960

[2] On Formally Undecidable Propositions of Principia Mathematica And Related Systems, K. Godel, Dover Publications, 1992

[3] Introduction To Mathematical Philosophy, B. Russell, Allen and Unwin, 1919

[4] From Frege To Godel: A Sourcebook In Mathematical Logic, 1879–1931, J. van Heijenoort, Harvard University Press, 1967

[5] Set Theory, Logic And Their Limitations, M. Machover, Cambridge, University Press, 1996

[6] Paradoxes, M. Sainsbury, Cambridge University Press, 2009

[7] Logic And Knowledge, B. Russell, Allen and Unwin, 1968

[8] The Foundations Of Arithmetic, G. Frege, Blackwell, 1980

[9] Computation: Finite And Infinite Machines, Marvin Minsky, Prentice-Hall, 1967

[10] The Philosophy Of Mathematics, Edited By W. D. Hart, Oxford University Press, 1996

[11] Infinity, Jose Benardete, Clarendon Press, 1964

[12] Collected Works, L. E. J. Brouwer, North-Holland, 1975

[13] Elements Of Intuitionism, Michael Dummett, Clarendon Press, 2000

[14] Godel, Escher, Bach: An Eternal Golden Braid, Douglas Hofstadter, Basic Books, 1999

[15] Set Theory And The Continuum Hypothesis, Paul J. Cohen, W. A. Benjamin, 1966

[16] Randomness And Mathematical Proof, Gregory Chaitin, Scientific American, 47–52 (May, 1975)

[17] From Mathematics To Philosophy, Hao Wang, Humanities Press, 1974

[18] Georg Cantor, His Mathematics And Philosophy Of The Infinite, Joseph W. Dauben, Harvard University Press, 1990

[19] Mathematical Logic, Joseph R. Shoenfield, Addison-Wesley, 1967

https://doi.org/10.1515/9783112214367-073

Index

https://doi.org/10.1515/9783112214367-074

Printed and bound by CPI Group (UK) Ltd, Croydon, CR0 4YY
16/09/2026
14963056-0003